Mathematical Modelling of Heat and Mass Transfer Processes

Mathematics and Its Applications

Volume 348

Mathematical Modelling of Heat and Mass Transfer Processes

by

V. G. Danilov
V. P. Maslov
and
K. A. Volosov

Moscow Institute of Electronics and Mathematics,
Moscow, Russia

SPRINGER SCIENCE+BUSINESS MEDIA, B.V.

A C.I.P. Catalogue record for this book is available from the Library of Congress.

ISBN 978-94-010-4183-6 ISBN 978-94-011-0409-8 (eBook)
DOI 10.1007/978-94-011-0409-8

This is a completely revised and updated translation
of the original Russian work of the same title,
Moscow, Nauka © 1987. Translation by M. A. Shishkova.

Printed on acid-free paper

CONTENTS

PREFACE

In the present book the reader will find a review of methods for constructing a certain class of asymptotic solutions, which we call *self-stabilizing solutions.* This class includes solitons, kinks, traveling waves, etc. It can be said that either the solutions from this class or their derivatives are localized in the neighborhood of a certain curve or surface.

For the present edition, the book published in Moscow by the Nauka publishing house in 1987, was almost completely revised, essentially up-dated, and shows our present understanding of the problems considered.

The new results, obtained by the authors after the Russian edition was published, are referred to in footnotes. As before, the book can be divided into two parts: the methods for constructing asymptotic solutions (Chapters I–V) and the application of these methods to some concrete problems (Chapters VI–VII). In Appendix a method for justification some asymptotic solutions is discussed briefly.

The final formulas for the asymptotic solutions are given in the form of theorems. These theorems are unusual in form, since they present the results of calculations.

The authors hope that the book will be useful to specialists both in differential equations and in the mathematical modeling of physical and chemical processes.

The authors express their gratitude to Professor M. Hazewinkel for his attention to this work and his support.

The authors are especially grateful to M. A. Shishkova for editing, translating, and typesetting the continuously varying manuscript.

FROM THE PREFACE TO THE RUSSIAN EDITION

Dissipative structures introduced by I. Prigogine are the central objects of a new, rapidly developing field of science called *synergetics*. Actually, solutions of some standard problems, localized in space or, in a sense, "almost" localized, are considered. Such solutions propagate with time, and their structure varies slowly. They describe different phenomena such as flame spread, mould growth, crystal growth, motion of a drop along the inclined plane, a grove evolution, etc. Therefore, from the viewpoint of natural science, such solutions are no less interesting than solitons. The present monograph deals with such solutions, which we call *synergets*.

In this book the evolution of dissipative structures is studied by using constructive methods, i.e., by constructing asymptotic solutions of semilinear parabolic equations. The problem reduces to solving certain ordinary differential equations. As a rule, this is simpler than the investigation of the initial partial differential equation.

An important advantage of asymptotic methods is the possibility to examine equations with variable coefficients of rather general form by analytical methods and thus to investigate the stability of dissipative structures with respect to the properties of inhomogeneous media.

The authors wish to thank C. P. Kurdyumov, V. P. Myasnikov, O. S. Ryzhov, and V. V. Pukhnachev for useful contacts, comments, and advice. They also express their gratitude to S. Yu. Dobrokhotov, V. A. Tsupin, G. A. Omel'yanov, P. N. Zhevandrov, and the scientific editor of this book V. E. Nazaikinskii, with whom they have discussed the topics of this book at different times.

INTRODUCTION

The classical models of mathematical physics are based on the concept of a continuous (qualitatively) homogeneous medium. However, even the simplest observations show that objects and phenomena around us often have the form of localized structures, which appear, move, and interact. Localized structures (i.e., groups or families of objects) are naturally formed among the objects of the surrounding world. They are everywhere, from star systems to biological populations.

Though different in nature, macroscopic structures have much in common. The most general common property is the existence of a boundary that isolates the structure. There are two types of such structures, with sharp boundary and with fuzzy boundary.

A group of plants, for example, an oak grove gives an example of a structure with sharp boundary (see Figure 1). Here one can see three different regions: the space around the grove, the grove itself, where the density of trees is approximately stable, and a narrow (in contrast to the grove) boundary region (the border of the grove), where new trees grow intensively. There are no trees beyond the border of the grove.

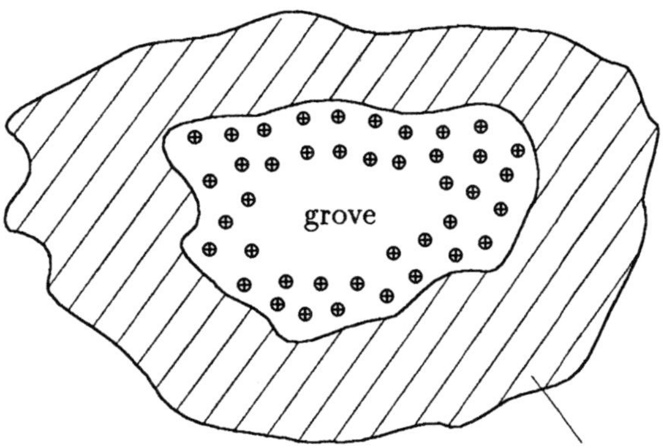

FIGURE 1

The process of combustion gives another example.

In chemistry the process of combustion is considered as parallel chemical reactions proceeding at different rates. In the case of two interacting gases, these reactions can be conditionally described as follows [33, 34]

$$A + B_2 \to AB + B, \qquad B + A_2 \to AB + A.$$

The slower reacting substance is usually called the *leading center.* The rate of reaction, in which this substance is involved, determines the rate of combustion. Ya. B. Zeldovich considered the case when the product of decay (dissociation), which forms the leading center, is in excess in the products of reaction. In this case, as Ya. B. Zeldovich showed, it is necessary to take into account the diffusion of leading centers into the flame region, which leads to a self-consistent motion of the flame front.

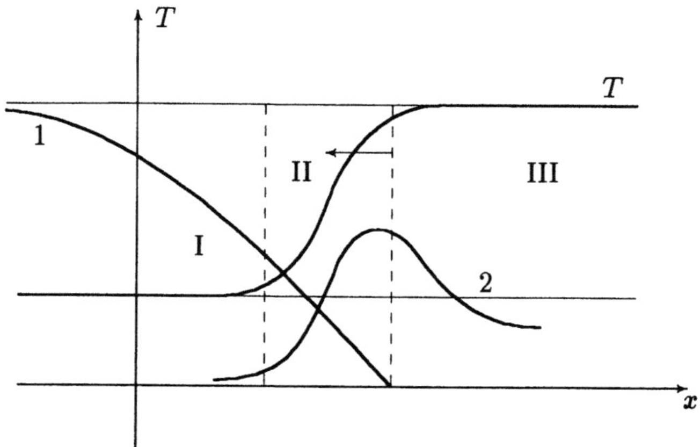

FIGURE 2. I — reagents, II — flame region, III — reaction products

Figure 2 shows the distributions of temperature T and reagent concentrations (curve 1 shows the concentration of substance A, curve 2 shows the concentration of substance B, i.e., of the leading center). Since heat is released in chemical reactions of combustion, in the region of reaction products the temperature is maximal. In the region of reagents the temperature is lower and the probability that a chemical reaction starts at this temperature is small.

There is a natural question: what is in common between different mathematical models describing the evolution of these structures? The answer is: the evolution of both structures is described by quasilinear parabolic equations, which mathematically express the balance relations characterizing these structures.

The equation describing the evolution of a grove has the form*:

$$\frac{\partial U}{\partial \tau} - D \operatorname{div} U \operatorname{grad} U - U(A - BU) = 0.$$

Here U is the density of plants, $A = A(\tau)$, $B = B(\tau)$ are given time-dependences describing the distribution of resources responsible for vital activity of the population and for laws of birth and death, D is the transport coefficient.

In dimensionless coordinates, this equation takes the form

$$\varepsilon \frac{\partial u}{\partial t} - \varepsilon^2 \operatorname{div} u \operatorname{grad} u - u(a - bu) = 0.$$

*The data characterizing this model were obtained by S. M. Semenov, who told us about them.

Here $\varepsilon = \tau_0 D/L^2$, where τ_0 is the characteristic time of grove life, $\tau_0 \sim 1000$ [years], L is the characteristic size of the grove, $L \sim 500$ [km], $D \sim 50 \div 200$ [km^2/year]. For these magnitudes of characteristic quantities, we have $\varepsilon \sim 0.2$.

The combustion process is described by the well-known Zeldovich equations [33, 34, 36]. One of the most famous model equations in dimensionless coordinates has the form

$$\varepsilon \frac{\partial T}{\partial t} - \varepsilon^2 \operatorname{div} \operatorname{grad} T - \gamma^2 T^2 (1 - T) = 0.$$

Here T is the temperature, $\gamma^2 = \gamma^2(x,t)$ is a given function, characterizing the medium nonhomogeneity, ε is a small parameter arising when we pass to dimensionless coordinates.

An obvious distinction between these equations is that the coefficient at the second-order derivative in the Zeldovich equation is constant, while in the equation describing the grove evolution this coefficient depends on an unknown function u and vanishes for $u = 0$.

From mathematical viewpoint, the equation, describing the grove evolution, has a localized solution (identically equal to zero outside a certain domain, see Figure 1) for any value of the parameter ε, while the solution of the Zeldovich equation is exponentially small as $\varepsilon \to 0$ outside the combustion region, i.e., this solution is localized only as $\varepsilon \to 0$.

The mathematical description of the evolution of a large class of structures is related to a special class of solutions to quasilinear parabolic equations.

From mathematical viewpoint, the class of problems considered is characterized by the localization and a finite velocity of perturbation propagation, i.e., the support of the solution is a closed subset of the domain, where the problem is posed. This support varies in time so that its boundary moves in space with a certain finite velocity.

Generally speaking, the solution has a weak discontinuity on the boundary of its support. Therefore we have a problem about a weak discontinuity propagation and a problem of constructing asymptotics with respect to a small parameter. In contrast to linear hyperbolic equations (any beforehand given singularity can propagate) and linear parabolic equations (any singularity is instantly smoothed out), in the case of quasilinear degenerate parabolic equations, we have a finite! number of types of singularities that can propagate. These singularities (in general position) have the form $[n]^\alpha$, where n is the distance along the normal to the weak discontinuity front (the boundary of the support), and the exponent $\alpha > 0$ is defined by the properties of nonlinear terms in the equation. The singularities are classified in Chapter III.

Note that this book is not concerned with processes related to transitions between different structures. In particular, we do not consider the Cauchy problem, we study the class of solutions that describe self-similarly evolving structures.

From the time of Euler, it is well-known that, in the general scheme for construction of asymptotic solutions with respect to a small parameter, the next approximations are expressed in terms of the derivatives of the leading term. Therefore, if the leading term is not sufficiently smooth, the next terms of the expansion possess singularities and the asymptotic expansion cannot be used. Thus, if solutions of quasilinear parabolic equations have a weak discontinuity, it is necessary to take into account the possible influence of the preceding terms on the following ones.

Now our aim is to present the ideas used for the construction of asymptotic solutions of quasilinear parabolic equations with small parameter at the derivatives.

Here we restrict ourselves to the case of uniformly bounded as $\varepsilon \to 0$ solutions. As an example consider the equation

$$\varepsilon\frac{\partial u}{\partial t} - \varepsilon^2\frac{\partial^2 u^k}{\partial x^2} - \gamma^2(x,t)(1-u^{k-1})u = 0, \qquad x \in \mathbf{R}. \tag{1}$$

Here $0 < \varepsilon < 1$ is a small parameter, $k > 1$, $x \in \mathbf{R}$, $t \in \mathbf{R}_+$, $\gamma = \gamma(x,t) \neq 0$, $\gamma \in C^\infty(\mathbf{R} \otimes \mathbf{R}_+)$. We denote $K(u) = ku^{k-1}$, $F(u) = u(1-u^{k-1})$.

The physical considerations naturally imply the following boundary conditions for equation (1):

$$u(x,t)\Big|_{x\to\infty} = 1, \qquad u\Big|_{x\to-\infty} = 0, \qquad K(u)\frac{\partial u}{\partial x}\Big|_{u\to+0} = 0. \tag{2}$$

For $\gamma = 1$, one of the exact nonnegative solutions of problem (1), (2) has the form

$$u(x,t,\varepsilon) = W(\tau)\big|_{\tau=S(x,t)/\varepsilon}, \tag{3}$$

$$W(\tau) = \begin{cases} (1-\exp\{-\tau/a\})^\alpha, & \tau \geqslant 0, \\ 0, & \tau < 0, \end{cases} \qquad \alpha = \frac{1}{k-1}, \qquad a = \alpha k,$$

where $S(x,t) = x + t$ is a linear function.

This solution is called a *simple wave* (see Figure 3.1, curve 3). As $\tau \to 0$, we have $W(\tau) = O(\tau^\alpha)$, and as $\tau \to \infty$, the function (3) tends to one exponentially (not like a power).

If $\gamma(x,t) \neq \text{const}$, i.e., if the medium properties vary in time and/or from point to point, then there is no exact solution in the form of a simple wave. However, one can construct an asymptotic solution in the form of a "distorted" wave. In this case the function S is linear only in x, and the function u has a more complicated form, namely,

$$u(x,t,\tau,\varepsilon) = W(x,t,\tau,\varepsilon)\big|_{\tau=S(x,t,\varepsilon)/\varepsilon},$$

where

$$W(x,t,\tau,\varepsilon) = \big[W_0(x,t,\tau) + \varepsilon W_1(x,t,\tau)\big] + O(\varepsilon^2), \qquad \nabla_x S\big|_{S=0} \neq 0 \tag{4}$$

(the latter relation is a nonlinear generalization of the condition that the coefficient at x does not vanish).

In (4) the function $W(x,t,\tau,\varepsilon)$ is continuously differentiable in x, t, ε for all $\tau > 0$, $\varepsilon > 0$ and satisfies the conditions

$$\frac{\partial^{\delta_1+\delta_2} W_i(x,t,\tau,\varepsilon)}{\partial t^{\delta_1}, \partial x^{\delta_2}}\bigg|_{\tau=S(x,t,\varepsilon)/\varepsilon} = O(1), \qquad \text{as} \quad \varepsilon \to 0, \quad \delta_1, \delta_2 \in \mathbb{Z}_+ \tag{5}$$

(i.e., the derivatives of the function W_i in slow variables x, t are bounded). Moreover, as $\tau \to \infty$, the derivatives of the function W_i with respect to τ decrease faster than any negative power of τ and uniformly in x, t.

Theorem 1. *The asymptotic solution of the problem (1), (2) has the form (4). The function $W_0(x,t,\tau)$ has the form*

$$W_0(x,t,\tau) \stackrel{\text{def}}{=} W_0(\tau) = \begin{cases} (1-\exp\{-\tau/\tau_0\})^\alpha, & \tau > 0, \quad \alpha = 1/(k-1), \\ 0, & \tau < 0, \quad \tau_0 = \alpha/k, \end{cases}$$

and satisfies the boundary-value problem

$$\begin{gathered} \frac{dW_0}{d\tau} - \frac{d^2W_0^k}{d\tau^2} - W_0(1-W_0^{k-1}) = 0, \\ W_0\big|_{\tau=0} = 0, \qquad W_0\big|_{\tau\to\infty} \to 1. \end{gathered} \tag{6}$$

The function S has the form

$$S(x,t,\varepsilon) = \beta^0(t)(x+\varphi^0(t)) + \varepsilon\beta^0\varphi^1(t),$$

where the functions β^0, φ^0 satisfy the system of equations

$$\beta^0(t) = \gamma(-\varphi^0(t),t), \qquad \beta^0\frac{d\varphi^0}{dt} = \gamma^2(-\varphi^0(t),t). \tag{7}$$

The functions β_1 and W_1 are defined in Lemma 1 below.

Lemma 1. *The function $W_1(x,t,\tau)$ has the form*

$$W_1 = \frac{1}{\gamma^2(-\varphi^0,t)}\frac{\partial W_0}{\partial\tau}\int_0^\tau\left(\int_0^{\tau'} f(t,\xi)\frac{\partial W_0}{\partial\xi}(\xi)[V(\xi)K(W_0(\xi))]^{-1}\,d\xi\right) \times V(\tau')\left(\frac{\partial W_0}{\partial\tau'}(\tau')\right)^{-2}d\tau', \tag{8}$$

where

$$V = \frac{1}{K^2(W_0(\tau))}\exp\left\{\int_0^\tau\frac{d\tau'}{K(W_0(\tau'))}\right\} = \left(1-\exp\left\{-\frac{\tau}{\tau_0}\right\}\right)^{\alpha-2}\exp\left\{\frac{\tau\alpha}{\tau_0}\right\}k^{-2},$$

$$f = -\frac{dW_0}{d\tau}\left(\frac{d}{dt}(\beta^0\varphi^1) + \frac{d\beta^0}{dt}\left(\frac{\tau}{\beta^0}-\varphi^1\right)\right) + \frac{\partial\gamma^2}{\partial x}\bigg|_{x=-\varphi^0}\left(\frac{\tau}{\beta^0}-\varphi^1\right)F(W_0),$$

$$F(W_0) = W_0(1-W_0^{k-1}).$$

The estimates hold

$$W_1 = O(\tau^\alpha), \qquad \tau\to 0,$$

$$W_1 = O\left(\tau^2\exp\left\{-\frac{\tau}{\alpha k}\right\}\right), \qquad \tau\to\infty.$$

The function $\varphi^1(t)$ satisfies the equation

$$\beta^0 I_0\frac{d\varphi^1}{dt} + \frac{\partial\gamma^2}{\partial x}(-\varphi^0,t)I_2\varphi^1 + (\beta^0)^{-1}\frac{d\beta^0}{dt}I_1 - (\beta^0)^{-1}\frac{\partial\gamma^2(-\varphi^0,t)}{\partial x}I_3 = 0, \tag{9}$$

where

$$I_0 = \int_0^\infty \frac{\left(\frac{dW_0}{d\tau}\right)^2}{V(\tau)K(W_0)}\,d\tau, \qquad I_1 = \int_0^\infty \frac{\tau\left(\frac{dW_0}{d\tau}\right)^2}{V(\tau)K(W_0)}\,d\tau,$$
$$I_2 = \int_0^\infty \frac{F(W_0)\frac{dW_0}{d\tau}}{V(\tau)K(W_0)}\,d\tau, \qquad I_3 = \int_0^\infty \frac{\tau F(W_0)\frac{dW_0}{d\tau}}{V(\tau)K(W_0)}\,d\tau.$$

Equation (9) *is equivalent to the following orthogonality condition*

$$\int_0^\infty \frac{1}{V(\tau)K(W_0(\tau))} f(t,\tau)\frac{\partial W_0(\tau)}{\partial \tau}\,d\tau = 0.$$

Before we prove Theorem 1, we recall that in the asymptotic theory of nonlinear equations there are two types of solutions. They differ by the form of the dependence on the fast variable τ. The solutions of the first kind are periodic in τ. Such solutions were studied in many works, see for example, [31, 98, 111, 112]. They are rapidly oscillating wave packets.

Solutions of another important type, namely, asymptotic stabilizing solutions, appeared when asymptotic solutions of solitary wave type (solitons and shock waves) were constructed, for example, in [55, 56, 58, 94, 99]. In this case, the function $W(\tau, x, t, \varepsilon)$ has the limit as $\tau \to \pm\infty$ and

$$|\tau|^N \frac{\partial^\nu}{\partial \tau^\nu} W(x,t,\tau,\varepsilon) \to 0 \qquad \forall \nu, \qquad N > 0. \tag{10}$$

In this book we consider solutions of nonlinear problems of the second kind. The method for constructing asymptotic nonsmooth stabilizing solutions, first presented in [51] for heat and mass transfer equations and equations of chemical kinetics has much in common with the Bogolyubov–Mitropolsky and Whitham methods. However, there are important distinctions. One of them concerns the fact that in problems about the construction of stabilizing asymptotic solutions we have not only a small operator of multiplication by the parameter ε, but also a small operator of multiplication by the function $S(x,t,\varepsilon)$ (in linear problems, similar considerations lead to the theory of complex germ [52, 53]).

Let us consider a simple example. Suppose $g(x,\varepsilon) = \varphi(x)\exp\{-x^2/\varepsilon\}$, where $x \in \mathbf{R}$, $\varphi(x)$ is a uniformly bounded function. Then

$$x^{2N} g(x,\varepsilon) = O(\varepsilon^N) \tag{11}$$

is uniform in x. Actually, the functions $g(x,\varepsilon)$ can be represented in the form

$$g(x,\varepsilon) = W(x,\tau)\big|_{\tau = x^2/\varepsilon}, \qquad W(x,\varepsilon) = \varphi(x)e^{-\tau}.$$

We have

$$|x^{2N}e^{-x^2/\varepsilon}\varphi(x)| \leqslant \varepsilon^N \max_x |\varphi(x)| \max_{\tau \geqslant 0}(\tau^N e^{-\tau}) = \varphi^N N^N e^{-N} \max_x |\varphi(x)|,$$

which yields (11).

Similarly, for stabilizing functions $W(x,t,S/\varepsilon,\varepsilon)$, we have

$$\left|S^N \frac{\partial^\nu W}{\partial \tau^\nu}\left(x,t,\frac{S}{\varepsilon},\varepsilon\right)\right| \leqslant \varepsilon^N \max\left|\tau^N \frac{\partial^\nu}{\partial \tau^\nu} W(x,t,\tau,\varepsilon)\right| = O(\varepsilon^N) \quad \forall \nu, \quad N > 0. \tag{12}$$

It should be pointed out that a modified Whitham method can be successfully used here since, in problems considered, the operator of multiplication by the function S is "small".

Note that, precisely as in the case of asymptotics considered in the theory of complex germ [52, 53], the asymptotic representation of the form (4) is not unique. For example, if

$$\widehat{S}(x,t,\varepsilon) = S(x,t,\varepsilon) + O(S^\nu(x,t,\varepsilon)), \qquad \nu > 1,$$

is uniform in x, t, ε, then, expanding the function $W_i(x,t,S/\varepsilon)$ by the Taylor formula and taking into account that the derivatives in τ decrease as $\tau \to \infty$, we get

$$\begin{aligned} W_i\left(x,t,\frac{\widehat{S}}{\varepsilon}\right) &= W_i\left(x,t,\frac{S(x,t,\varepsilon)}{\varepsilon} + \varepsilon^{-1} O(S^\nu(x,t,\varepsilon))\right) \\ &= W_i\left(x,t,\frac{S}{\varepsilon}\right) + \varepsilon^{\nu-1} \frac{\partial W_i}{\partial \tau}\left(x,t,\frac{S}{\varepsilon}\right) \frac{O(S^\nu)}{S^\nu} \left(\frac{S^\nu}{\varepsilon^\nu}\right) + \dots \\ &= W_i\left(x,t,\frac{S}{\varepsilon}\right) + \varepsilon^{\nu-1} \widehat{W}_i\left(x,t,\frac{S}{\varepsilon}\right) + \dots, \end{aligned}$$

where the function $\widehat{W}_i(x,t,\tau)$ has the same properties as the functions $W_i(x,t,\tau)$, in particular, has the same singularity in τ for $\tau = 0$.

Let $\varphi(t,\varepsilon)$ satisfy the equation

$$S(-\varphi,t,\varepsilon) = 0.$$

The conditions imposed on the function S imply that the function $\varphi(t,\varepsilon)$ smoothly depends on t and ε, in particular,

$$S(-\varphi(t,0),t,0) = 0,$$

i.e., the function $\varphi(t,0)$ determines (up to $O(\varepsilon)$) the position of the weak discontinuity of the solution.

Let us prove Theorem 1 and Lemma 1. We do not prove some estimates (this will be done in Chapter IV). Expanding the function $S(x,t,\varepsilon)$ according to the Taylor formula at the point $x = -\varphi(t,\varepsilon)$, we get

$$\begin{aligned} S(x,t,\varepsilon) = \beta(t,\varepsilon)(x+\varphi(t,\varepsilon)) + \beta_1(t,\varepsilon)(x+\varphi(t,\varepsilon))^2 + \dots \\ \dots + O((x+\varphi(t,\varepsilon))^N). \end{aligned} \tag{13}$$

From the preceding it is seen that, to construct an asymptotic solution up to mod $O(\varepsilon^2)$, it is sufficient to calculate only the first two summands in (13). Thus, without loss of generality, we set

$$S(x,t,\varepsilon) = \beta^0(t)(x+\varphi^0(t)) + \varepsilon\beta^0(t)\varphi^1(t).$$

Then we substitute the function (4) into equation (1) and set the sum of terms having the order ε^0 (uniformly in τ) equal to zero. Then, taking into account the above remark (see (12)), we obtain the equation

$$\beta^0\frac{d\varphi^0}{dt}\frac{\partial W_0}{\partial\tau} - (\beta^0)^2\frac{\partial}{\partial\tau}\left(K(W_0)\frac{\partial W_0}{\partial\tau}\right) - \gamma^2(-\varphi^0(t),t)F(W_0) = 0, \tag{14}$$

which will be called *standard*.

Here the following relations are taken into account:

$$\begin{aligned}\frac{\partial S}{\partial t} &= \beta^0\frac{d\varphi^0}{dt} + \frac{d\beta^0}{dt}(x+\varphi^0) + \varepsilon\frac{d}{dt}(\beta^0\varphi^1)\\ &= \beta^0\frac{d\varphi^0}{dt} + \varepsilon\frac{d}{dt}(\beta^0\varphi^1) + \varepsilon\frac{d\beta^0}{dt}\left(\frac{\tau}{\beta^0}-\varphi^1\right)\Big|_{\tau=S/\varepsilon},\\ \frac{\partial S}{\partial x} &= \beta^0, \qquad x+\varphi^0 = \varepsilon\left(\frac{\tau}{\beta^0}-\varphi^1\right)\Big|_{\tau=S/\varepsilon}.\end{aligned}$$

Now we set the sum of terms, having the order ε^1 uniformly in τ, equal to zero:

$$\begin{aligned}&\beta^0\frac{d\varphi^0}{dt}\frac{\partial W_1}{\partial\tau} - (\beta^0)^2\frac{\partial^2}{\partial\tau^2}(K(W_0)W_1) - \gamma^2(-\varphi^0(t),t)\frac{dF}{dW}(W_0)W_1\\ &= -\frac{\partial W_0}{\partial t} - \left[\frac{d}{dt}(\beta^0\varphi^1) + \frac{d\beta^0}{dt}\left(\frac{\tau}{\beta^0}-\varphi^1\right)\right]\frac{W_0}{d\tau} + \frac{\partial\gamma^2}{\partial x}\left(\frac{\tau}{\beta^0}-\varphi^1\right)F(W_0).\end{aligned} \tag{15}$$

This relation is based on the fact that, on the solutions considered, the function $F(W_0)$ tends to zero (as $\tau\to\infty$) faster than any negative power of τ. This fact will be explained in Chapter IV.

We change the variables in (14):

$$\xi = \frac{\gamma(-\varphi^0,t)}{\beta^0}\tau, \qquad \widetilde{W}_0(t,\xi) = W_0(t,\tau), \qquad \frac{d}{d\tau} = \frac{\gamma(-\varphi^0,t)}{\beta^0}\frac{\partial}{\partial\xi}.$$

In new variables the standard equation has the form

$$\begin{gathered} b(t)\frac{d\widetilde{W}_0}{d\xi} - \frac{d}{d\xi}\left(K(\widetilde{W}_0)\frac{d\widetilde{W}_0}{d\xi}\right) - F(\widetilde{W}_0) = 0,\\ b(t) = \frac{1}{\gamma(-\varphi^0,t)}\frac{d\varphi^0}{dt}.\end{gathered} \tag{16}$$

We shall prove that, since the derivatives of W_0 with respect to t must be bounded, $b\equiv\text{const}$.

Denote $G = \partial\widetilde{W}/\partial t$. The function G satisfies the linear equation

$$b\frac{dG}{d\xi} - \left[K(\widetilde{W}_0)\frac{d^2G}{d\xi^2} + 2\frac{dK(\widetilde{W}_0)}{d\widetilde{W}_0}\frac{d\widetilde{W}_0}{d\xi}\frac{dG}{d\xi}\right] - \left[\frac{d^2K(\widetilde{W}_0)}{d\widetilde{W}_0^2}\left(\frac{d\widetilde{W}_0}{d\xi}\right)^2 + \frac{dK(\widetilde{W}_0)}{d\widetilde{W}_0}\frac{d^2\widetilde{W}_0}{d\xi^2} + \frac{dF}{d\widetilde{W}_0}\right]G = -\frac{db}{dt}\frac{d\widetilde{W}_0}{d\xi}. \quad (17)$$

This variation equation with right-hand side corresponds to equation (16). The Wronskian of equation (17) has the form

$$V_1 = \frac{1}{K^2(\widetilde{W}_0)}\exp\left\{b\int\frac{d\xi'}{K(\widetilde{W}_0)}\right\}. \quad (18)$$

The solution of (17) has the form

$$G = \frac{d\widetilde{W}_0}{d\xi}\int\left(\frac{db}{dt}\int\frac{1}{V_1K(\widetilde{W}_0)}\left(\frac{d\widetilde{W}_0}{d\xi}\right)^2 d\xi\right)V_1\left(\frac{d\widetilde{W}_0}{d\xi}\right)^{-2} d\xi. \quad (19)$$

The following estimates hold (see Chapter III, IV):

$$\begin{aligned} \frac{dW_0}{d\xi} &= O(\xi^{\alpha-1}), \qquad \xi \to 0, \\ \frac{dW_0}{d\xi} &= O(e^{-\xi/k}), \qquad \xi \to \infty. \end{aligned} \quad (20)$$

Below we distinguish two cases.

Case 1. We assume that, for a certain value of t, the inequality holds

$$b(t)a_1 < 2a + 1, \qquad a_1 = \lim_{\tau\to 0}\left(\frac{K(W_0)}{\tau}\right)^{-1}. \quad (21)$$

Then the inner integral in (19) converges within the limits $(0, \infty)$ and the solution of equation (17) has the form

$$G = \frac{d\widetilde{W}_0}{d\xi}\int_0^{\xi} V_1(\xi')\left(\frac{d\widetilde{W}_0}{d\xi}\right)^{-2}\left(\frac{db}{dt}\int_0^{\xi'}\frac{1}{V_1K(\widetilde{W}_0)}\left(\frac{d\widetilde{W}_0}{d\xi}\right)^2 d\xi''\right) d\xi'. \quad (22)$$

The necessary condition for the solution $\widetilde{W}_1(\xi, t)$ to decrease as $\xi \to \infty$ is the relation

$$\frac{db}{dt}\int_0^{\infty}\frac{1}{V_1K(\widetilde{W}_0)}\left(\frac{d\widetilde{W}_0}{d\xi}\right)^2 d\xi = 0, \quad (23)$$

which coincides with the condition that the right-hand side of equation (17) and the solution of the adjoint homogeneous equation are orthogonal to each other.

Relation (23) holds only for $\partial b/\partial t = 0$ since, obviously, the integral in (23) does not vanish.

Case 2. As shown in Chapters III and IV, in this example the inequality

$$b(t)a_1 \geqslant 2\alpha + 1 \tag{24}$$

does not hold for any $k > 1$. Therefore

$$\big(\gamma(-\varphi^0, t)\big)^{-1}\frac{d\varphi^0}{dt} = b = \text{const}\,. \tag{25}$$

In our example the function $\widetilde{W}_0(\xi)$, satisfying equation (16) and the boundary conditions, exists only for $b = 1$ and is given by the formula:

$$\widetilde{W}_0(\xi) = \begin{cases} (1 - e^{-\xi/a})^{\alpha}, & \xi \geqslant 0, \\ 0, & \xi < 0, \quad a = \alpha k. \end{cases}$$

Recall that, by the construction proposed, the asymptotic solution has the form (4). Using the formulas obtained, we get

$$\begin{aligned} u(x,t,\varepsilon) &= \big[\widetilde{W}_0(\xi(\tau,t)) + \varepsilon\widetilde{W}_1(\xi(\tau,t),t)\big]\big|_{\tau=S/\varepsilon} \\ &= W_0\Big(\frac{\gamma}{\varepsilon}((x+\varphi^0)+\varepsilon\varphi^1)\Big) + \varepsilon W_1\Big(\frac{\gamma}{\varepsilon}((x+\varphi^0)+\varepsilon\varphi^1),t\Big). \end{aligned}$$

The latter relation implies that, though the function $\beta^0(t)$ is not yet defined when we analyze equations (14), (16) and, as is easy to understand, cannot be determined by the equations for the next corrections W_i, $i > 0$, this indeterminacy does not influence the final expression for the solution (the function β^0 is cancelled out). Therefore, from the very beginning, we can fix the choice of the function β^0. It turns out that the most convenient way is to add the equation

$$\beta^0(t) = \gamma(-\varphi^0, t) \tag{26}$$

to equation (25). Then, obviously, equations (25), (26) coincide with (7) and we have $\xi(\tau, t) = \tau$.

Let us consider the equation (15). By (7), it has the form

$$\gamma^2(-\varphi^0,t)\Big(\frac{\partial W_1}{\partial \tau} - \frac{d^2}{d\tau^2}(K(W_0)W_1) - \frac{dF}{dW}(W_0)W_1\Big) = f,$$

where f is the right-hand side of (15).

The solution of this equation is constructed similarly to that of (17).

In the case considered, i.e., for $a_1 < 2\alpha + 1$, the function $W_1 = W_1(\tau, t)$ has the form

$$W_1(\tau,t) = \big(\gamma^2(-\varphi^0,t)\big)^{-1}\frac{dW_0}{d\xi}\int_0^{\tau}\Big(\int_0^{\tau'}\frac{f\frac{dW_0}{d\tau}}{VK(W_0)}\,d\tau''\Big)\frac{V}{\big(\frac{dW_0}{d\tau}\big)^2}\,d\tau'.$$

The necessary and sufficient condition for the function W_1 to decrease as $\tau \to \infty$ is

$$\int_0^\infty \frac{f\frac{dW_0}{d\tau}}{VK(W_0)}\,d\tau = 0. \tag{27}$$

This integral converges (see Theorem 2.1, Chapter IV) and relation (27) allows to determine the function φ^1. The equation for this function is given in the statement of Lemma 1.

The theorem is proved.

Note that in this problem the maximal amplitude is independent of the velocity, therefore the algorithm for constructing asymptotic solutions is close (in its structure) to the algorithm for constructing asymptotics in linear problems (for example, to the WKB method). Namely, we substitute the solution of a given structure into the equation, set the coefficients at equal power of ε equal to each other and obtain a system of equations. In this system the first equation determines the "phase", the second equation determines the principal term of the expansion of "amplitude", etc. However, there is no principle of superposition of solutions (in the classical sense).

Following [20], we shall explain some results that generalize Theorem 1.

Consider the semilinear parabolic equation

$$Lu = \varepsilon\frac{\partial u}{\partial t} - \varepsilon^2\frac{\partial}{\partial x}\left(\lambda(x,t)K(u)\frac{\partial u}{\partial x}\right) - \gamma^2(x,t)F(u) = 0 \tag{28}$$

where $\lambda(x,t), \gamma(x,t) \geqslant \delta > 0 \in C^\infty$, $F(0) = F(1) = 0$, $K(u) \sim u^{k-1}$, $F(u) \sim u^q$, $u \to +0$, $k > 0$, $q > 0$, $k + q \geqslant 2$, $F'(1) \neq 0$, $K(u) > 0$ for $u > 0$ and either

(a) $k + q > 2$, $R(u) > 0$ for $u \in (0,1)$, $R(u) = K(u)F(u)$ or

(b) $k + q = 2$, $R(u) > 0$ for $u \in (0,1)$, $R'(0) > 0$, $R'(u) \leqslant R'(0)$ or

(c) $k + q > 2$, $q < 1$, $R(u) < 0$ for $u \in (0,1)$, $R'(u) \leqslant R'(1)$.

We shall consider localized solutions with values $0 \leqslant u \leqslant 1$ such that $u = 0$ for $x \leqslant x_f(t,\varepsilon)$, where $x_f(t,\varepsilon)$ is an unknown function depending on the initial conditions. The boundary conditions have the form

$$u\Big|_{x\to\infty} \to 1, \qquad u\Big|_{x=x_f} = 0, \qquad K(u)\frac{\partial u}{\partial x}\Big|_{x=x_f} = 0. \tag{29}$$

It turns out that in both cases (b) and (c) the Cauchy problem for equation (28) is ill-posed in the class of additive asymptotic solutions (see Theorem 2 below). In case (c) the ill-posedness can be eliminated if we consider the global solution matched with the solution of equation (28), linearized in a neighbourhood of $u = 1$.

In Chapter IV we show that the boundary-value problem for the ordinary differential equation

$$\begin{gathered} b\chi' - (K(\chi)\chi')' - F(\chi) = 0, \qquad b = \text{const}, \\ \chi(\tau)\big|_{\tau=0} = 0, \qquad \chi(\tau)\big|_{\tau\to\infty} \to 1, \qquad K(\chi)\chi'\big|_{\tau=0} = 0, \qquad \chi' \geqslant 0, \end{gathered} \tag{30}$$

is closely related to problem (28), (29). In case (a) the solution $\chi = \chi(\tau)$ exists for a certain value of b known as the Zeldovich constant, in case (b) the solution exists for $b \geqslant b_{\min} = 2\sqrt{|R'_\chi|_{\chi=0}|}$ and in case (c) for $b \leqslant -b_{\min} = -2\sqrt{R'_\chi(1)}$ (see Chapter I and [107]). In case (a) the asymptotic solution of problem (28), (29) is given by the formula:

$$u(x,t,\varepsilon) = \left[\chi(\tau) + \varepsilon\chi'(\tau)\int_0^\tau \frac{V}{(\chi')^2}\int_a^{\tau'} \frac{f(t,\tau'')\chi'}{VK(\chi)}\,d\tau''\,d\tau' + \dots\right]\Bigg|_{\tau=S/\varepsilon}, \tag{31}$$

under the condition

$$\int_0^\infty \frac{f(t,\tau)\chi'(\tau)}{VK(\chi)}\,d\tau = 0. \tag{32}$$

Here $a = 0$, V is the Wronskian of the first variation equation and $f(\tau,t)$ is the right-hand side of this equation (see Chapter IV),

$$S = \beta(x+\varphi^0) + \varepsilon\beta\varphi^1 + \varepsilon^2\beta\varphi^2 + \dots, \qquad \beta = \left(\frac{\gamma^2(-\varphi^0,t)}{\lambda(-\varphi^0,t)}\right)^{1/2},$$

$\varphi^0_t = b\gamma^2(-\varphi^0,t)/\beta$ and φ^1 is defined by (32) which has the meaning of the orthogonality condition. Relation (32) also implies the estimates for the second term χ_1 in (31) (similar estimates hold for all terms of the asymptotic expansion):

$$\begin{aligned} &\chi_1 = O(\tau^\alpha), && \tau \to 0, && \alpha = (k-1)^{-1},\\ &\chi_1 = O(\tau^{-N}), && \tau \to \infty, && \forall N > 0. \end{aligned} \tag{33}$$

In both cases (b) and (c) the following statement holds [20].

Theorem 2. *For any $N > 0$ there exist two functions $u^i_N(x,t\varepsilon)$ $(i = 1,2)$, such that*

$$Lu^i_N = \varepsilon^N Q^i_N(x,t,\varepsilon), \qquad u^1_N\big|_{t=0} = u^2_N\big|_{t=0}, \qquad \left|Q^i_N \Big/ \varepsilon\frac{\partial u^i_N}{\partial t}\right|_{u^i_N=0} \leqslant \text{const},$$

and constants ε_0, $\delta_0, \delta_1 > 0$, such that $\max_x |u^1_N - u^2_N| \geqslant \delta_1 > 0$, $\delta_1 = \text{const}$ for $t = \delta_0$ and $\varepsilon < \varepsilon_0$.

The proof of this theorem is based on the construction of functions u^i_N in the form (31), where $a = \infty$ in case (b) and $a = 0$ in case (c). In both cases (b) and (c), for $|b| > |b_{\min}|$, the solution of the form (31) satisfies conditions (33) without any additional conditions of the form (32), since, by the estimates

$$\frac{f(t,\tau)\chi'(\tau)}{VK(\chi)} \sim \tau^{2\alpha-1+b/K_1}, \qquad \tau \to 0, \qquad K_1 = \lim_{\tau\to 0}\frac{K(\chi)}{\tau},$$

in case (b) (here $b_{min}/K_1 = 2\alpha$) and by the estimates

$$\frac{f(t,\tau)\chi'(\tau)}{VK(\chi)} \sim \exp\left(\tau\sqrt{b^2 - 4|(FK)'_\chi|_{\chi=1}|}\right), \qquad \tau \to \infty.$$

in case (c), the integral (32) diverges and $\varphi^1(t)$ cannot be determined.

Formula (31) implies that for $\tau \gg 1$ the sum of the first N terms can be represented in the form (31),

$$u_N(x,t,\varepsilon) \cong 1 - \left[\exp(-l\tau)\sum_{n=0}^{N} \varepsilon^n P_2 n(t,\tau)\right]\Bigg|_{\tau=S/\varepsilon},$$

where $P_{2n}(t,\tau)$ is a polynomial of degree $2n$ with coefficients depending on t, $l = \big(-b + \sqrt{b^2 + 4|R'(1)|}\big)/2K(1)$ in case (b), and $l = \big(-b - \sqrt{b^2 + 4|R'(1)|}\big)/2K(1)$ in case (c). The next statement, proved in Chapter IV, allows to construct a global asymptotic solution and to calculate the coordinate of the wave front.

Lemma 2. *For $c_1\varepsilon^{1-\gamma_1} \leqslant (x + \varphi^0) \leqslant c_2\varepsilon^{1-\gamma_2}$ and for any $0 < \gamma_1 < \gamma_2 < 1/2$, $c_1, c_2 > 0$, the following relation holds:*

$$1 - u_N(x,t,\varepsilon) = \exp\big(Q_N(x + \varphi^0, t, \varepsilon)/\varepsilon\big)\big(1 + O(\varepsilon^N)\big). \tag{34}$$

where $Q_N(z,t,\varepsilon) \leqslant 0$ is a polynomial in z of degree N with coefficients smoothly depending on t, ε, $Q_N(0,t,\varepsilon) = O(\varepsilon)$.

Denoting $1-u_N = v_N$, taking into account that the function v_N is (exponentially) small in the domain given above, and using (28), we obtain the following equation for v_N:

$$\varepsilon\frac{\partial v_N}{\partial t} - \varepsilon^2 K(1)\frac{\partial}{\partial x}\left(\lambda\frac{\partial v_N}{\partial x}\right) - \gamma F'(1)v_N = 0. \tag{35}$$

Its solution, depending on one fast variable Φ_N/ε, $\Phi_N \geqslant 0$, has the form

$$v_N = \exp(-\Phi_N/\varepsilon), \qquad \Phi_N = \sum_{k=0}^{N} \varepsilon^k \Phi^k, \tag{36}$$

where the function Φ^0 satisfies the Hamilton–Jacobi equation

$$\frac{\partial\Phi^0}{\partial t} + K(1)\lambda\left(\frac{\partial\Phi^0}{\partial x}\right)^2 + \gamma F'(1) = 0. \tag{37}$$

If λ and γ are real analytical coefficients, then (36) and (37) imply that the function $Q_N(x + \varphi^0, t, 0)$ is a partial sum of the Taylor series for the function $\Phi^0 = \Phi^0(x,t)$. By construction, $\Phi^0|_{x=-\varphi^0} = 0$. The projections of trajectories of the Hamilton system corresponding to equation (37) are schematically depicted by solid lines in Figure 3 and Figure 4 for the cases (b) and (c) respectively; for increasing t, the direction of motion is shown by arrows.

Obviously, in case (b), the solution is not determined by the initial conditions for (37) and the function $\varphi^0(t)$ must be defined by some additional conditions. Then, the solution in the form (1.6) turns out to consist of an infinite number of simple waves $\chi = \chi(t,\tau)$ corresponding to the values of the parameter $b = b(t) = \varphi^0_t\beta/\gamma(-\varphi^0, t)$. Here $\varphi^0(t)$ must be such that $b(t) > b_{\min}$.

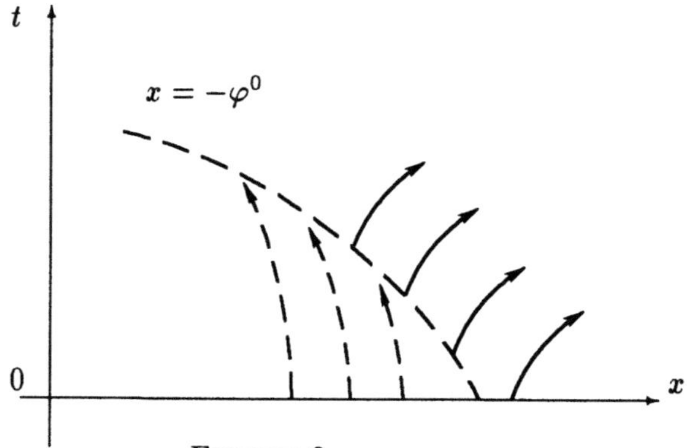

FIGURE 3

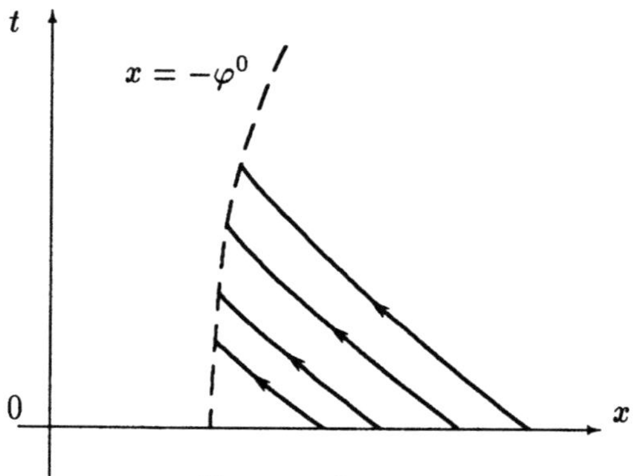

FIGURE 4

In case (c) the initial condition $\Phi^0|_{t=0} = \Phi^0_0(x)$ completely determines the solution (at least for small t), and the boundary of the solution support is given by the formula $\Phi^0(-\varphi^0, t) = 0$. Here the solution also consists of an infinite number of waves, for which the parameter $b(t)$ is calculated precisely as in case (b). Thus, in case (b), the solution of the Cauchy problem for (28) is asymptotically ill-posed, since the law of motion of the weak discontinuity front is arbitrary, while in case (c) this law is determined by a fixed exponential asymptotic behavior of the initial condition (just as in the theory of Kolmogorov–Petrovskii–Piskunov (KPP) equations, whose trajectories are shown in Figure 3 by dashed lines; the KPP equations combine cases (b) and (c)).

Now let us explain the last situation more precisely following [21]. Here we use constructions, similar to those developed for degenerate parabolic equations in [20], and apply them to semilinear parabolic equations of the form

$$Lu = \varepsilon\frac{\partial u}{\partial t} - \varepsilon^2\frac{\partial}{\partial x}\left(\lambda(x,t)K(u)\frac{\partial u}{\partial x}\right) - \gamma(x,t)F(u) = 0, \tag{38}$$

$$u\big|_{x\to-\infty} \to 0, \qquad u\big|_{x\to\infty} \to 1, \qquad 0 < u < 1. \tag{39}$$

We assume that $\lambda(x,t), \gamma(x,t) \geqslant \delta > 0 \in C^\infty$; $F \in C^\infty([0,1])$, $F(0) = F(1) = 0$, and consider two cases

(a) $F'(0) > 0$, $F'(0) > F'(u)$, $F(u) > 0$ for $u \in (0,1)$, $F(1) < 0$;
(b) $F'(0) = 0$, $F(1) < 0$, $F(u) > 0$ for $u \in (0,1)$.

It turns out that, precisely as in [21], in case (a) the Cauchy problem for equation (38) is ill-posed in the class of additive asymptotics (i.e., in the class of solutions u_N satisfying the equation $Lu_N = O(\varepsilon^N)$ for any given ε), but is well-posed in the class of multiplicative asymptotics (i.e., in the class of solutions u_N satisfying the equation $Lu_N = O(\varepsilon^N)\partial u_N/\partial t$, see [21]).

In case (b) the Cauchy problem is well-posed in the class of additive asymptotics.

The scheme of the proof of these statements is outlined below.

We recall that in the case $\lambda \equiv \gamma \equiv 1$ the problem (38), (39) has invariant solutions of the form $u = \chi(x + bt)$, where $\chi = \chi(\tau)$ satisfies the ordinary differential equation

$$\begin{aligned} &b\chi' - \chi'' - F(\chi) = 0, \qquad b = \text{const}, \\ &\chi\big|_{\tau\to-\infty} \to 0, \qquad \chi\big|_{\tau\to\infty} \to 1. \end{aligned} \tag{40}$$

In case (a) the solution of problem (40) exists for any $b \geqslant b_{\min} = 2\sqrt{F(0)}$ [107] and in case (b) the solution of problem (40) exists only for a single value of b [33, 39]. In both cases, for a fixed b, the solution of (40) is unique within arbitrary translations $\tau \to \tau + \text{const}$.

As $\tau \to -\infty$, we have $\chi = O(\exp(l^-\tau))$, where $l^- = \left(b - \sqrt{b^2 - 4F(0)}\right)/2$, $b > b_{\min}$ in case (a) and $l^- = b$ in case (b). As $\tau \to \infty$, we have $1 - \chi = O(\exp(-l^+\tau))$, $l^+ = \left(-b - \sqrt{b^2 - 4|F(1)|}\right)/2$. The asymptotic solution u_N satisfying the condition $Lu_N = O(\varepsilon^N)$ is given by the formula

$$u_N(x,t,\varepsilon) = \left[\chi(\tau) + \varepsilon\chi_\tau(\tau)\int_0^\tau \frac{V}{(\chi')^2}\int_a^{\tau'} \frac{f(t,\tau'')\chi'_\tau}{V}\,d\tau''d\tau' + \dots\right]\bigg|_{\tau=S/\varepsilon}, \tag{41}$$

where V and f denote the Wronskian and the right-hand side of the linearized equation (see [51]). The solutions of the form (41) have different properties in cases (a) and (b). In case (b) the inner integral in the second term in (41) converges as $\tau \to \pm\infty$. Therefore the function $u_N(\chi, t, \varepsilon)$ satisfies the boundary conditions (39) for $a = -\infty$,

$$\int_{-\infty}^{\infty} \frac{f(t,\tau)\chi'_\tau}{V}\,d\tau = 0. \tag{42}$$

The last relation defines the function $\varphi^1(t)$. Considering the omitted terms in (41) and using (39), we obtain equations that define the functions $\varphi^i(t)$, $i > 1$, and are similar to (42). The equation for $\varphi^0(t)$ is a consequence of the condition that the derivatives of the solution with respect to t vanish as $\tau \to \pm\infty$. In case (b) this equation has the form $b = \varphi^0_t\beta/\gamma(-\varphi^0, t)$, where $b = \text{const}$, $\beta = \sqrt{\gamma(-\varphi^0,t)/\lambda(-\varphi^0,t)}$. Thus in case (b) the boundary conditions and the relation $Lu_N = O(\varepsilon^N)$ imply equations for all functions involved in (41). In case (a), the estimate of the function χ shows that, as $\tau \to -\infty$, the inner integral in the second term in (41) does not converge as $\tau \to -\infty$. Therefore, setting $a = \infty$, we see that, for any arbitrary function $\varphi^1(t)(\varphi^i(t))$, the second term vanishes (the omitted terms also vanish). Moreover, the conditions that the derivatives with respect to t are bounded holds if

$\varphi_t^0\beta/\gamma(-\varphi^0,t) = b(t) \in C^\infty$, $b(t) > b_{\min}$. Thus, in case (a), the general asymptotic solution, satisfying the boundary conditions and the condition $Lu_N = O(\varphi^N)$ for any $N > 0$, has the form (41), where $\chi = \chi(\tau,t)$ is the solution of problem (40) with $b = b(t)$, $b(t) = b_{\min}$ being an arbitrary smooth function. We see that in this case the wave front position is not determined. Clearly, if we set the initial conditions of the form (41) (that is, if we fix $\varphi^i\big|_{t=0}$, $i = 0,\dots,N$) and take any arbitrary smooth continuations of these functions for $t > 0$ (so that $b(t) > b_{\min}$), we get a continuum of solutions defined by these continuations.

The solution becomes unique if we define it more precisely. Namely, to construct a multiplicative asymptotic solution, it is necessary to match the solution of form (41) with the solution of the linearized equation in the domains $x + \varphi^0 > O(\varepsilon^{1-\delta})$, $x + \varphi^0 < -O(\varepsilon^{1-\delta})$, $0 < \delta < 1/2$. From (41) we get that, as $\tau \to \pm\infty$, the sum of the first N terms can be written as follows

$$\begin{aligned} u_N^-(x,t,\varepsilon) &= \left[\exp(-l^-\tau)\sum_{n=0}^{N}\varepsilon^n P_{2n}^-(t,\tau)\right]\bigl(1 + O(\exp(-l^-\tau))\bigr)\Big|_{\tau=S/\varepsilon}, \quad \tau \to -\infty,\\ u_N^+(x,t,\varepsilon) &= \left[\exp(-l^+\tau)\sum_{n=0}^{N}\varepsilon^n P_{2n}^+(t,\tau)\right]\bigl(1 + O(\exp(-l^+\tau))\bigr)\Big|_{\tau=S/\varepsilon}, \quad \tau \to +\infty, \end{aligned} \tag{43}$$

where $P_{2n}^{\pm}(t,\tau)$ are polynomials in τ of degree $2n$ with coefficients belonging to $C^\infty([0,T])$.

Similar to [21], we have the following statement:

Theorem 3. a). *For any $c_1^+\varepsilon^{1-\gamma_1^+} \leqslant x + \varphi^0 \leqslant c_2^+\varepsilon^{1-\gamma_2^+}$ in the domain $0 < \gamma_1^+ < \gamma_2^+ < 1/2$, $c_1^+, c_2^+ > 0$, the relation*

$$1 - u_N(x,t,\varepsilon) = \exp\bigl(Q_N^+(x+\varphi^0,t,\varepsilon)/\varepsilon\bigr)\bigl(1 + O(\varepsilon^{N_1^+})\bigr) \tag{44}$$

holds.

b). *For any $-c_1^-\varepsilon^{1-\gamma_2^-} \leqslant x + \varphi^- \leqslant c_2^-\varepsilon^{1-\gamma_1^-}$ in the domain $0 < \gamma_1^- < \gamma_2^- < 1/2$, $c_1^-, c_2^- > 0$, the relation*

$$u_N(x,t,\varepsilon) = \exp\bigl(Q_N^-(x+\varphi^0,t,\varepsilon)/\varepsilon\bigr)\bigl(1 + O(\varepsilon^{N_1^-})\bigr) \tag{45}$$

holds. Here $Q_N^{\pm}(z,t,\varepsilon)$ are polynomials in z of degree N with coefficients smoothly depending on t, ε, $Q_N^{\pm}(0,t,\varepsilon) = O(\varepsilon)$, $N_1^{\pm} \to \infty$ as $N \to \infty$.

Setting $u_N^+ = 1 - u_N$ for $x + \varphi^0 \gg 1$ and $u_N^- = u_N$ for $x + \varphi^0 \ll -1$, we get $u_N^{\pm} \to 0$ as $|\tau| \to \infty$. Then the linearized equation for $u_N^{\pm}$ takes the form

$$\varepsilon\frac{\partial u_N^{\pm}}{\partial t} - \varepsilon^2\frac{\partial}{\partial x}\left(\lambda\frac{\partial u_N^{\pm}}{\partial x}\right) - \gamma F'(\sigma_\pm)u_N^{\pm} = 0, \qquad \sigma_+ = 1, \quad \sigma_- = 0. \tag{46}$$

The solutions of equation (46) depending on one fast variable $\Phi_N^{\pm}/\varepsilon$ have the form $\Phi_N^{\pm} \geqslant 0$, $u_N^{\pm} = \exp\bigl(-\Phi_N^{\pm}/\varepsilon\bigr)$, $\Phi_N^{\pm}/\varepsilon = (\Phi_\pm^0 + \varepsilon\Phi_\pm^1 + \dots)$, where the functions

$\Phi^0_\pm$ satisfy the Hamilton–Jacobi equations. The comparison of the expressions for $u^\pm_N$ with (44) and (45) shows that the polynomials $Q^\pm\_N(x+\varphi^0,t,\varepsilon)$ are N-jets of functions $\Phi^\pm_N$ at $x=-\varphi^0$. According to Theorem 3, this fact implies the boundary conditions for the functions $\Phi^0_\pm$, $\Phi^0_\pm|_{x=-\varphi^0}=0$. For $\Phi^0_\pm$, the Hamilton–Jacobi equation has the form

$$\frac{\partial\Phi^0_\pm}{\partial t}+\lambda\left(\frac{\partial\Phi^0_\pm}{\partial x}\right)^2+\gamma F'(\sigma_\pm)=0.$$

In case (a) of the Kolmogorov–Petrovskii–Piskunov equations, the projections of trajectories of the corresponding Hamilton system are shown in Figure 5 (for increasing t, the direction of motion is shown by arrows).

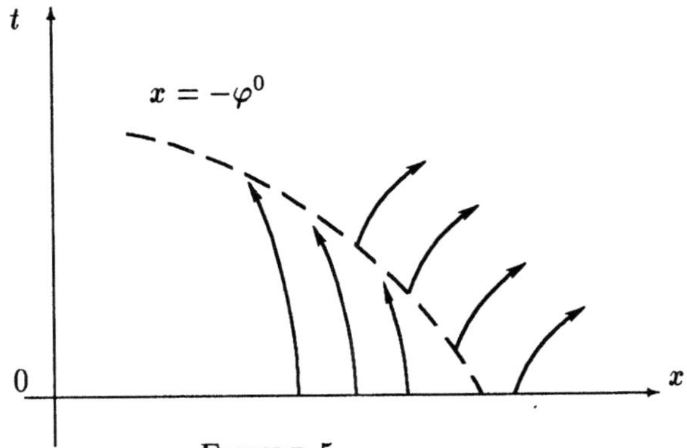

FIGURE 5

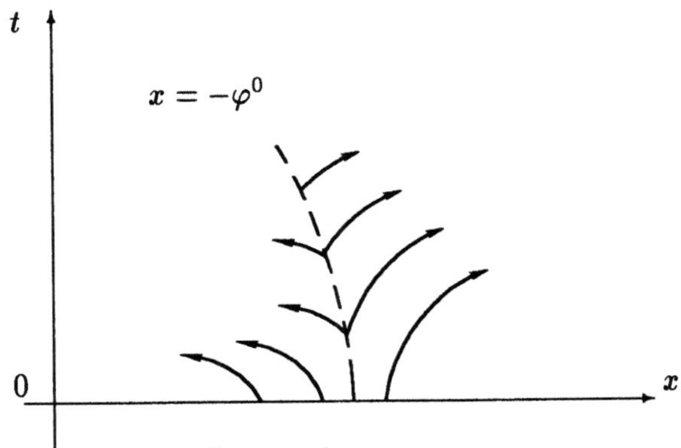

FIGURE 6

Clearly, the initial conditions for Φ^0_- at $t=0$ define the zero set $x=-\varphi^0(t)$ of the function Φ^0_- for $t>0$, and the initial-boundary-value problem for the function Φ^0_+ can be solved for any given function $\varphi^0(t)$. Thus, the principal term of the wave front coordinate $\varphi^0(t)$ is given by the equation $\Phi^0_-(-\varphi^0,t)=0$ and the initial data $\Phi^0_-|_{t=0}$ must be taken so that the inequality

$$\varphi^0_t\beta/\gamma(-\varphi^0,t)>b_{\min}$$

holds.

In case (b) Theorem 3 also holds. The projections of trajectories are shown in Figure 6. In this case, the initial data for $\Phi_{\pm}^{0}$ do not define the functions $\varphi^{0}(t)$, and the solutions $\Phi_{\pm}^{0}$ exist for $\varphi^{0}(t)$ defined by the equation $\varphi_{t}^{0} = b\sqrt{\gamma(-\varphi^{0},t)/\lambda(-\varphi^{0},t)}$, where b is the Zeldovich constant.

In fact, some formulas relating asymptotics of the initial data ahead the wave-front and the law of motion of the wave-front have been known in certain specific cases. For example, in [107] the case of a function Φ_{-}^{0} linear in x for $t = 0$ was considered. A more general situation, in which the equation for characteristics of the linearized problem has a nontrivial solution Φ_{0}^{0}, was considered by M. R. Booty, R. Haberman, and A. A. Minzoni [82].

In Chapter IV the above results are discussed in detail.

Note that this book deals with asymptotic solutions, i.e., with solutions approximately satisfying equations up to any (beforehand given) precision. There is a natural question, whether the exact solutions of problems considered (if they exist) and the asymptotic solutions of these problems (or solutions close to the exact solutions) have similar properties. This problem is beyond the scope of this book since it requires different methods.

It should be noted that the asymptotic solutions are of significant independent interest. The matter is that the equations considered in this book approximately model some actual physical processes and phenomena. There is no sense to construct solutions to these equations with precision greater than that with which they describe these processes and phenomena.

Moreover, the asymptotics, constructed with the help of methods presented in this book, are local attractors. Sometimes, this fact can be proved (see Appendix and [100]). The ideas proposed in [100] allow to solve the Cauchy problem by using the theorems about local attractors for equations with constant coefficients (theorems of the Kolmogorov-Petrovskii-Piskunov type and of [107]-type). In [113] an interesting approach based on probability methods was developed to investigate the wave front propagation.

Recently*, it was proved that the local attractor is an asymptotic solution of the phase field system (a mathematical model for the phase transition generalizing the classical Stefan problem).

---

*V. G. Danilov, G. A. Omel'yanov, and E. V. Radkevich, *Asymptotic solution of a phase field system and the modified Stefan problem*, Differential'nye Uravneniya, 1995, v. 31. (Russian.)

V. G. Danilov, G. A. Omel'yanov, and E. V. Radkevich, *Regularization of initial data for the modified Stefan problem*, Matem. Zametki, 1995. (Russian.)

CHAPTER I

PROPERTIES OF EXACT SOLUTIONS OF NONDEGENERATE AND DEGENERATE ORDINARY DIFFERENTIAL EQUATIONS

1.1. Standard equations

Examples considered in Introduction show that, in order to construct an asymptotics, it is necessary to obtain the exact solution of a certain (called *standard*) ordinary differential equation. In this section we discuss a method for investigating standard equations often met when localized asymptotic solutions are constructed. This method is based on reducing the standard equation to some well-known simple wave equations related to the Kolmogorov, Petrovskii, Piskunov [41] or Zeldovich [33, 34] equations. Their properties are considered below. Theorems 1.2–1.4 present the main results about solutions of the type that we shall need in our constructions.

Let us consider the ordinary differential equation with constant coefficients

$$b\frac{d\chi}{d\tau} - \frac{d}{d\tau}\left(\rho(\chi)\frac{d\chi^k}{d\tau}\right) - F(\chi) = 0, \tag{1.1}$$

where $\rho(\chi), F(\chi) > 0$ for $\chi \in]0,1[$, $\rho(0) > 0$, $\rho(1) > 0$, $k > 1$.

By $\theta(\xi)$ we denote the smooth monotone solution of the equation

$$b\frac{d\theta}{d\xi} - \frac{d^2\theta}{d\xi^2} - R(\theta) = 0, \qquad \theta \underset{\xi\to-\infty}{\longrightarrow} 0, \qquad \theta \underset{\xi\to\infty}{\longrightarrow} 1, \tag{1.2}$$

where $R(\theta) \in C^1([0,1])$. Recall that the following theorem was proved in [15, 51].

Theorem 1.1. *By the transformation*

$$\rho(\chi)\frac{d\chi^k}{d\tau} = \frac{d\theta}{d\xi}(\tau(\theta))\bigg|_{\theta=\chi}, \tag{1.3}$$

equation (1.2) *can be reduced to equation* (1.1), *where*

$$F(\chi) = \frac{R(\chi)\chi^{1-k}}{k\rho(\chi)}.$$

Further this transformation will be called a *localizing transformation.*

Proof. By (1.3), we have:

$$\frac{d\chi}{d\tau}(\xi) = \frac{1}{\rho(\chi)k\xi^{k-1}}\frac{d\theta}{d\xi}(\tau(\chi)),$$

$$\frac{d}{d\tau}\left(\rho(\chi)\frac{d\chi^k}{d\tau}\right) = \frac{d^2\theta}{d\xi^2}\left[\frac{d\theta}{d\xi}(\tau(\chi))\right]^{-1}\frac{d\chi}{d\tau} = \frac{d^2\theta}{d\xi^2}\left[\frac{d\theta}{d\xi}(\tau(\chi))\right]^{-1}\frac{1}{\rho(\chi)k\xi^{k-1}}\frac{d\theta}{d\xi}(\tau(\chi))$$

$$= \frac{1}{\rho(\chi)k\xi^{k-1}}\frac{d^2\theta}{d\xi^2}.$$

Expressing the derivatives of the function θ from the latter relations and substituting them into (1.2), we obtain (1.1). The theorem is proved.

The left-hand side of (1.3) has the physical meaning of the flow of a certain quantity carried by this flow. The localizing transformation was introduced in [15] and later in [92].

Remark. If, for $\chi = 0$, the function $R(\chi)\chi^{1-k}$ is not defined, then we must consider the equation

$$\rho(\chi)\chi^{k-1}\left[b\frac{d\chi}{d\tau} - \frac{d}{d\tau}\left(\rho(\chi)\frac{d\chi^k}{d\tau}\right)\right] + \frac{1}{k}R(\chi) = 0.$$

The monotone solutions of equations (1.2), where the function $R(\theta) \not\equiv 0$ has only two roots in the interval $[0,1]$, were studied in two situations: $R \in C^1([0,1])$, $R(0) = 0$, $R(1) = 0$,

$$\left.\frac{dR}{d\theta}\right|_{\theta=0} \cdot \left.\frac{dR}{d\theta}\right|_{\theta=1} < 0, \tag{1.4}$$

or

$$\left.\frac{dR}{d\theta}\right|_{\theta=0} = 0, \quad \left.\frac{dR}{d\theta}\right|_{\theta=1} < 0. \tag{1.5}$$

Equations (1.2), where the function $R(\theta)$ satisfies conditions (1.4), is known as the *Kolmogorov–Petrovskii–Piskunov* (KPP) *equation* after the authors who obtained and studied these equations in the now classical work [41].

This work attracted the attention of many researchers and provoked deep theoretical investigations. Here we show some results concerning the existence and properties of self-similar solutions of traveling-wave type satisfying the KPP equation. These solutions satisfy some boundary-value problems for equations (1.2).

Some of these results are obvious. For example, the monotone solution exists if and only if

$$b \geqslant b_{\min} = 2\sqrt{\left.\frac{dR}{d\theta}\right|_{\theta=0}}$$

This is due to the fact that, by the latter inequality, the characteristic exponents of the linearized ($\theta = 0$) equation (1.2) must be real. More delicate results are taken from [107].

Below we shall need the following definition of asymptotic equivalence often used in this book.

Definition 1.1. Continuous functions $\varphi(\tau)$ and $\psi(\tau)$ will be called *equivalent* as $\tau \to a$,

$$\psi \sim \varphi, \qquad \tau \to a,$$

if the limits $\lim_{\tau\to a}(\varphi/\psi)^{\pm 1}$ exist and

$$0 < \left| \lim_{\tau\to a} \frac{\varphi}{\psi} \right| < \infty.$$

Some important results are stated in the following theorem [107].

Theorem 1.2. (i) *There exists a constant $b_0 > 0$ such that the solution $\theta(\tau)$ of problem* (1.2) (*unique up to a translation*) *exists if and only if $b \geqslant b_0$.*

(ii) *The inequalities hold*:

$$2\sqrt{\frac{dR(\theta)}{d\theta}\bigg|_{\theta=0}} \leqslant b_0 \leqslant 2\sqrt{\sup_{\forall u\in[0,1]}\left(\frac{F(u)}{u}\right)}. \tag{1.6}$$

(iii) *For $b > b_0$,*

$$\theta(\tau) \sim \exp(l^-\tau), \qquad \tau \to -\infty, \tag{1.7}$$

where $l^- = \frac{1}{2}\left(b - \sqrt{b^2 - 4(dR/d\theta)\big|_{\theta=0}}\right)$.

For $b = b_0$,

$$\lim_{\tau\to-\infty} \frac{\theta'(\tau)}{\theta(\tau)} = l^+; \qquad l^+ = \frac{1}{2}\left(b + \sqrt{b^2 - 4\frac{dR}{d\theta}\bigg|_{\theta=0}}\right). \tag{1.8}$$

For $\tau \to +\infty$,

$$1 - \theta(\tau) \sim \exp(-l_0\tau), \qquad l_0 = \frac{1}{2}\left(-b + \sqrt{b^2 + 4|R'(1)|}\right).$$

Remark. If the strict inequality $b_0 > 2\sqrt{dR/d\theta\big|_{\theta=0}}$ holds, relation (1.8) is equivalent to $\theta(\tau) \sim \exp(l^+\tau)$, $\tau \to -\infty$.

Here we make comments about the proofs of these statements.

For the complete proof of Theorem 1.2, the reader is referred to work [107] and to a classical work [41]. One of the most interesting and important properties of solutions of problem (1.2), (1.4) is that if the inequality $R'(0) \geqslant R'(\theta)$ does not hold, then, first, generally speaking, no wave with velocity $b_{\min}$ exists and, second, the wave with minimal velocity $b_0 > b_{\min}$ is equivalent to $\exp(l^+\tau)$, $\tau \to -\infty$. This effect may seem to be strange, since the solution of the linearized equation (1.2) has the asymptotics

$$\theta \sim c_1 e^{l^+\tau} + c_2 e^{l^-\tau}, \qquad \tau \to -\infty,$$

and $l^+ > l^-$. So, in this case, the statement of theorem is equivalent to the statement that $c_2 = 0$.

Consider the function $G(\theta)$ defined and continuously differentiable in the region $0 \leqslant \theta \leqslant 1$ which satisfies the conditions [107]

$$G(0)=0, \qquad G(1)=0, \qquad \frac{dG(0)}{d\theta}>0, \qquad \frac{d^2G}{d\theta^2}=O(\theta^{-1}),$$
$$\left.\frac{dG}{d\theta}\right|_{\theta=0} \geqslant \frac{dG(\theta)}{d\theta} \quad \text{and} \quad G(\theta)>0 \quad \text{for} \quad 0<\theta<1. \tag{1.9}$$

We put $R(\theta)=G(\theta)H(\theta,k)$, where

$$H(\theta,k)=1+k^{-2}\Big(\frac{dG(0)}{d\theta}-\frac{dG(\theta)}{d\theta}\Big), \qquad k>0.$$

Let us define the function $\theta=\theta(\tau,k)$ by the integral (quadrature)

$$\tau=k\int_{1/2}^{\theta}\frac{du}{G(u)}. \tag{1.10}$$

This is a wave with velocity $b=k+\alpha/k$, $\alpha=G'(0)=R'(0)$. If $k=\sqrt{dR/d\theta|_{\theta=0}}=\sqrt{\alpha}$, then $b=2\sqrt{dR/d\theta|_{\theta=0}}$, and hence $b_0=b_{\min}$. Then, if k increases, the function $R(\theta)$ decreases and $b_{\min}$ does not grow.

For $k\geqslant\sqrt{\alpha}$ and increasing k, the values of $b=b(k)=k+\alpha/k$ grow, hence $b_{\min}=2\sqrt{\alpha}$.

If $k<2\sqrt{\alpha}$, then

$$\lim_{\tau\to-\infty}\frac{d\theta(\tau,k)/d\tau}{\theta(\tau,k)}=\frac{1}{k}\lim_{\theta\to-0}\frac{G(\theta)}{\theta}=\left.\frac{1}{k}\frac{dR}{d\theta}\right|_{\theta=0}=-\left(\frac{b}{2}+\sqrt{\frac{b^2}{4}-\left.\frac{dR}{d\theta}\right|_{\theta=0}}\right)=l^+.$$

Actually,

$$\left.\frac{1}{k}\frac{dR}{d\theta}\right|_{\theta=0}=\frac{\alpha}{k}=b-k=\frac{b\pm\sqrt{b^2-4\alpha}}{2}.$$

Since $b=2\sqrt{\alpha}$ for $k=\sqrt{\alpha}$ and $b=b(k)$ decreases with respect to k for $k<\sqrt{\alpha}$, it is sufficient to derive the trivial inequality

$$\frac{b-\sqrt{b^2-4\alpha}}{2}<\sqrt{\alpha}. \tag{1.11}$$

Then, we have $b=b(\alpha,k)>2\sqrt{\alpha}$.

Thus, for $k<\sqrt{\alpha}$, the quadrature given below generates a wave with minimal velocity and "strange" asymptotics.

Let us denote

$$R(\theta)=\theta\left.\frac{dR}{d\theta}\right|_{\theta=0}+\frac{1}{2}\zeta(\theta). \tag{1.12}$$

The two following theorems (see [107]) give asymptotics of self-similar waves.

Theorem 1.3. *Let* $b \geq b_0 > 2\sqrt{dR/d\theta|_{\theta=0}}$,

$$\int_{0+} |\zeta(\theta)|\theta^{-2}\,d\theta < \infty. \tag{1.13}$$

Then the function $\theta(\xi)$ *has the properties:*

$$\theta(\xi) = ae^{l\xi}(1+o(1)), \qquad \xi \to -\infty, \tag{1.14}$$

where

$$l = \begin{cases} l^+ = \frac{1}{2}\left(b + \sqrt{b_0^2 - 4\frac{dR}{d\theta}\Big|_{\theta=0}}\right), & b = b_0, \\ l^- = \frac{1}{2}\left(b - \sqrt{b^2 - 4\frac{dR}{d\theta}\Big|_{\theta=0}}\right), & b > b_0, \end{cases}$$

and a positive constant a *is determined as follows*

$$a = \begin{cases} \left(\sqrt{\frac{b^2}{4} - 4\frac{dR}{d\theta}\Big|_{\theta=0}}\right)^{-1} \\ \qquad \times\left\{ -l^-\theta(0) + \frac{d\theta}{d\tau}(0) - \int_{-\infty}^{0} e^{l^- s}\zeta(\theta(s))\,ds\right\} & \text{if } b > b_0, \\ -\left(\sqrt{\frac{b^2}{4} - 4\frac{dR}{d\theta}\Big|_{\theta=0}}\right)^{-1} \\ \qquad \times\left\{ -l^+\theta(0) + \frac{d\theta}{d\tau}(0) - \int_{-\infty}^{0} e^{l^+ s}\zeta(\theta(s))\,ds\right\} & \text{if } b = b_0. \end{cases}$$

Theorem 1.4. *Let* $b = 2\sqrt{dR/d\theta|_{\theta=0}}$ and

$$\int_{0+} |\zeta(\theta)|\theta^{-2}|\log\theta|\,d\theta < \infty.$$

Then the function θ *corresponding to the minimal velocity is estimated either by* (1.14) *with*

$$b = 2\sqrt{\frac{dR}{d\theta}\Big|_{\theta=0}}$$

or as

$$\theta = a_1\xi\exp\left(2\sqrt{\frac{dR}{d\theta}\Big|_{\theta=0}}\,\xi\right)(1+o(1)), \qquad \xi \to -\infty,$$

where a *and* a_1 *are positive constants determined by the relations*

$$a = \theta(0) + \int_{-\infty}^{0} s\exp\left(-2\sqrt{\frac{dR}{d\theta}\Big|_{\theta=0}}\,s\right)\zeta(\theta(s))\,ds,$$

$$a_1 = 2\sqrt{\frac{dR}{d\theta}\Big|_{\theta=0}}\,\theta(0) + \frac{d\theta(0)}{d\xi} - \int_{-\infty}^{0} \exp\left(-2\sqrt{\frac{dR}{d\theta}\Big|_{\theta=0}}\,s\right)\zeta(\theta(s))\,ds.$$

For $b \geqslant 2\sqrt{dR/d\theta|_{\theta=0}}$, *as* $\xi \to +\infty$, *the estimate holds*

$$\theta(\xi) = 1 - \Gamma \exp(-l_0\xi) + o(\exp(-l_0\xi)), \qquad l_0 = \frac{-b + \sqrt{b^2 + 4|R'(1)|}}{2},$$

where $\Gamma = \mathrm{const} > 0$.

Remark. If the sign of the function ζ is definite in a neighborhood of the point $\theta = 0$, then condition (1.13) is necessary for the estimate (1.14) to hold.

Obviously, the change $\theta = 1 - \widetilde{\theta}$ reduces the case $dR/d\theta|_{\theta=0} < 0$ to that considered above.

This theorem is proved in [107].

Now let us consider another class of equations (1.2). Namely, we assume that in (1.2) the function $R(\theta)$ satisfies the conditions

$$R(\theta) > 0, \quad \theta \in (0,1), \quad \left.\frac{dR}{d\theta}\right|_{\theta=1} < 0, \quad \left.\frac{dR}{d\theta}\right|_{\theta=0} = 0. \tag{1.15}$$

Equations, where the function $R(\theta)$ has such properties, arise when the combustion waves are modeled. These equations are called the *Zeldovich equations*. Ya. B. Zeldovich proved [33] that the solution of the boundary-value problem (1.2), satisfying the condition

$$\lim_{\xi \to -\infty} \theta^{-1} \frac{d\theta}{d\xi} = b > 0, \tag{1.16}$$

exists only for one value of the parameter b, say, for $b = b_0$. This number is called the *Zeldovich constant.*

By (1.16) and (1.15), as $\xi \to -\infty$, we have

$$\theta(\xi) \sim e^{b_0\xi}.$$

As $\xi \to \infty$, the estimates of the Zeldovich and Kolmogorov–Petrovskii–Piskunov equations coincide.

The equation

$$b_0 \frac{d\theta}{d\xi} - \frac{d^2\theta}{d\xi^2} - \theta^2(1-\theta) = 0$$

is a particular case of (1.2), namely, the function $R(\theta)$ satisfies conditions (1.15). This equation has the exact monotone solution

$$\theta(\xi) = \left(1 + e^{-\xi/\sqrt{2}}\right)^{-1}$$

and the constant b_0 (the Zeldovich constant) is equal to $1/\sqrt{2}$. As $|\xi| \to \infty$, we have the estimates

$$\begin{aligned} &\theta \sim e^{b_0\xi}, && \xi \to -\infty, \\ &1 - \theta \sim e^{-l_0\xi} + o(e^{-l_0\xi}), && \xi \to +\infty. \end{aligned}$$

These results can be generalized to the case of equations of the form (1.2), where the function $\theta^2(1-\theta)$ is replaced by a smooth function $R(\theta)$ satisfying conditions

(1.15). As is well-known [33], in this case the solution of traveling-wave type exists only for a single value of the constant $b = b_0$, and estimates of the form (1.16) hold as $|\xi| \to \infty$.

As pointed out above, the Zeldovich equations are studied in theory of combustion. By the physical meaning of these problems, the function $R(\theta)$ is nonnegative. However, the study of localized solutions of nonlinear equations based on Theorem 1.1 shows the necessity for considering the simple wave equations corresponding to the Zeldovich equation with nonpositive functions $R(\theta)$.

The equation

$$b\frac{d\theta}{d\xi} - \frac{d^2\theta}{d\xi^2} + \theta^2(1-\theta) = 0 \tag{1.17}$$

is a particular case of the Zeldovich equation with nonpositive $R(\theta)$.

Let us prove that the asymptotics of solutions, satisfying the conditions

$$\frac{d\theta}{d\xi} \neq 0, \qquad \theta\Big|_{\xi\to-\infty} \to 0, \qquad \theta\Big|_{\xi\to+\infty} \to 1, \tag{1.18}$$

is powerlike (not exponential) as $\xi \to -\infty$ (the existence of such solutions will be proved below).

Assume the contrary, i.e., assume that there are constants $\lambda_1 > 0$, $\lambda_2 > 0$ such that

$$\begin{aligned} \theta &= 1 - e^{\lambda_1\xi} + o(e^{\lambda_1\xi}), && \xi \to +\infty, \\ \theta &= o(e^{\lambda_2\xi}), && \xi \to -\infty. \end{aligned}$$

Then, by (1.17), we get

$$-b\lambda_1 + \lambda_1^2 + 1 = 0, \qquad b\lambda_2 - \lambda_2^2 = 0,$$

as $|\xi| \to \infty$. The real-valued solutions λ_1 exist only if $|b| \geqslant 2$. The second equation gives $b = \lambda_2$. Note that the following relations hold

$$b\int_{-\infty}^{\infty}\left(\frac{d\theta}{d\xi}\right)^2 d\xi = -\int_0^1 \theta^2(1-\theta)\,d\theta = -\frac{1}{12}.$$

Hence, we get $b < 0$ and $\lambda_2 < 0$, which contradicts the initial assumption. It is easy to verify that, as $|\xi| \to \infty$, the solutions of (1.17) satisfying (1.18) have the following asymptotics:

$$\begin{aligned} &\theta(\xi) \sim 1/|\xi|, && \xi \to -\infty, \\ &\theta(\xi) = 1 - \Gamma e^{\lambda_1\xi} + o(e^{\lambda_1\xi}), && \xi \to +\infty, \quad \Gamma = \text{const} > 0, \\ &\lambda_1 = \frac{b}{2} - \sqrt{\frac{b^2}{4} - 1}. \end{aligned} \tag{1.19}$$

This solution exists if $b < -2$.

Suppose that in equation (1.2) the function $R(\theta)$ is differentiable for $\theta \in [0,1]$. Then we have the following statement.

Lemma 1.1. *Let*

$$R(\theta) > 0, \quad \theta \in (0,1), \quad \left.\frac{dR}{d\theta}\right|_{\theta=1} < 0, \quad \left.\frac{dR}{d\theta}\right|_{\theta=0} = 0.$$

Then, for $b < -2\sqrt{dR/d\theta|_{\theta=1}}$, the solution of equation (1.2) *satisfies conditions* (1.18).

Obviously, the change of variables $\theta = 1-\vartheta$, $\xi \to -\zeta$ reduces (1.2) to the following simple wave equation:

$$|b|\frac{d\vartheta}{d\zeta} - \frac{d^2\vartheta}{d\zeta^2} + R(1-\vartheta) = 0.$$

The function $\hat{R}(\vartheta) = R(1-\vartheta)$ satisfies all but one of the conditions given in (1.4), namely, $\hat{R}'|_{\vartheta=1} \neq 0$ does not hold, but this is inessential for the proof of the existence theorem.

Let us consider the phase portrait of the first-order ordinary differential equation

$$\frac{dp}{d\theta} = \frac{bp - R(\theta)}{p}, \qquad b < 0,$$

to which equation (1.2) can be reduced (see Figure 1.1).

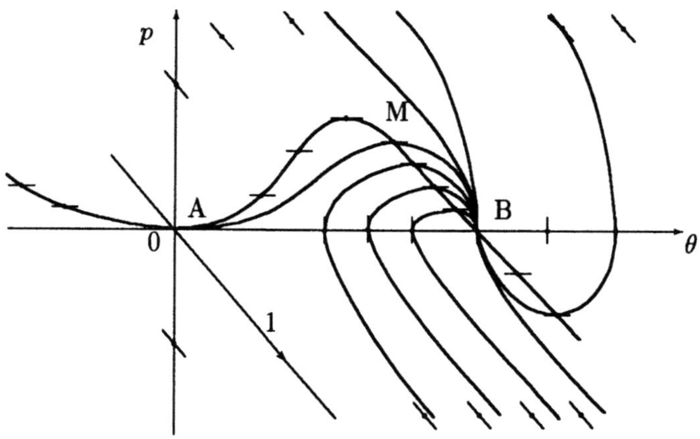

FIGURE 1.1

The singular point $A = (0,0)$ is a complicated singular point of saddle-node type with one direction distinguished (direction 1 in Figure 1.1). The singular point $B = (0,1)$ is a node for $|b| > 2\sqrt{dR/d\theta\,|_{\theta=1}}$ (a degenerate node for $|b| = 2\sqrt{dR/d\theta\,|_{\theta=1}}$). The eigenvalues and eigenvectors of the linearized equation have the form

$$\lambda_\pm = \frac{b}{2} \pm \sqrt{\frac{b^2}{4} - \left.\frac{dR}{d\theta}\right|_{\theta=1}}, \qquad v_- = (\lambda_-, 1), \qquad v_+ = (\lambda_+, 1)$$

The integral curve AMB corresponding to the solution of equation (1.17) with conditions (1.18) enters the singular point B along the direction of the eigenvector v_-. Lemma 1.1 is proved.

The change of variables $\theta = 1 - \vartheta$ used in the proof of Lemma 1.1 allows to apply all the results of Theorem 1. 1 to the case considered.

The following Lemma holds.

Lemma 1.2. *Let the function $R(\theta)$ in equation (1.2) satisfy conditions of Lemma 1.2 and $R'(1) \geqslant R'(\theta)$, $\theta \in (0,1)$, and let the point $\theta = 0$ be an algebraic branch point of this function, $R(\theta) \sim \theta^\beta$, $\beta > 1$. Then, as $|\xi| \to \infty$, the estimates:*

$$\theta(\xi) \sim |\xi|^{-1/(\beta-1)}, \qquad \xi \to -\infty,$$

$$\theta(\xi) = 1 - \Gamma e^{-\lambda\xi} + o(e^{-\lambda\xi}), \qquad \xi \to \infty, \quad \Gamma = \text{const} > 0,$$

$$\lambda = -\frac{b}{2} - \sqrt{\frac{b^2}{4} - \frac{dR}{d\theta}\bigg|_{\theta=1}} > 0.$$

hold for the solution $\theta(\xi)$ satisfying (1.18).

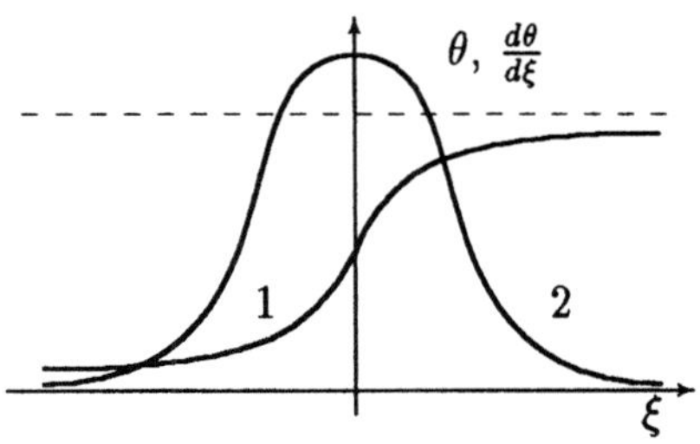

FIGURE 1.2

Now let us consider (1.1) and assume that in this equation the source-sink function F has the form $F = (k\chi^{k-1})^{-1}R(\chi)$, where the function R satisfies (1.5). Then the solution of (1.1) can be expressed in terms of solutions of the simple wave equation for the KPP equation by means of the method given in Theorem 1.1.

Let θ satisfy the simple wave equation

$$-b\frac{d\theta}{d\xi} + \frac{d^2\theta}{d\xi^2} + R(\theta) = 0. \tag{1.20}$$

In Figure 1.2, curve 1 is the plot of the function $\theta(\xi)$ satisfying (1.20). Curve 2 shows the function $d\theta/d\xi$. The graph of the function $f(\theta) = \frac{d\theta}{d\xi}(\theta^{-1}(\theta))$ is shown in Figure 1.3. Here $\theta^{-1}(\theta)$ is inverse to the function $\theta = \theta(\xi)$. Figure 1.2 and Theorem 1.1 imply that the function $\chi(\tau)$ with semibounded support (see Figure 1.4) satisfies equation (1.1) with $F(\chi)$ of the form $F = \left(\rho(\chi)k\chi^{k-1}\right)^{-1}R(\chi)$.

Let us consider the asymptotic behavior of $\chi(\tau)$ as $\chi \to 0$ and $\chi \to 1$. Without loss of generality, we can consider the case when the weak discontinuity front lies at the point $\tau = 0$.

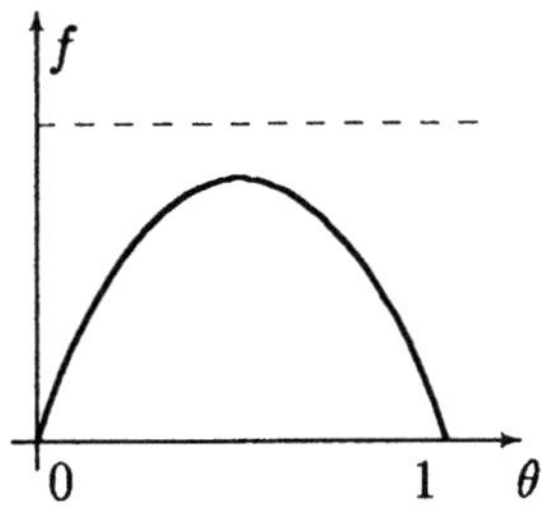

FIGURE 1.3

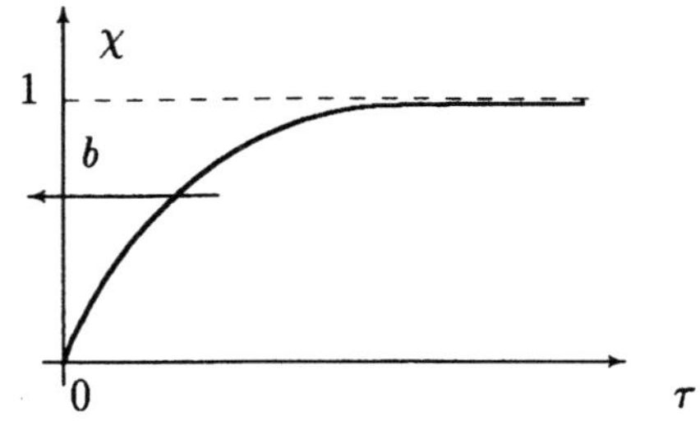

FIGURE 1.4

For $\tau \to 0$ we have

$$\chi(\tau) = \tau^{\alpha}\big(\gamma_1 + o(1)\big),$$

where $\gamma_1 = \gamma_1(\alpha)$ is a certain constant. Actually, by (1.3), it is sufficient to prove that the function $f(\theta)$ shown in Figure 1.3 satisfies the condition $df/d\theta\big|_{\theta=0} \neq 0$. To obtain the latter relation, we calculate the derivative

$$\frac{df(\theta)}{d\theta}\bigg|_{\theta=0} = \frac{d}{d\theta}\Big(\frac{d\theta}{d\xi}(\theta^{-1}(\theta))\Big)\bigg|_{\theta=0} = \frac{d^2\theta}{d\xi^2}\Big(\frac{d\theta}{d\xi}\Big)^{-1}\bigg|_{\xi=-\infty} = l.$$

By differentiating equation (1.20) with respect to ξ and applying L'Hospital's rule, we obtain

$$b = l + \frac{1}{l}\left|\frac{dR}{d\theta}\bigg|_{\theta=0}\right| \quad \text{as} \quad \xi \to -\infty. \tag{1.21}$$

Obviously, if $b > b_{\min}$ is a given constant, then $0 < |l| < \infty$ (see Figure 1.5A). If l belongs to the interval $[0, b/2]$, then the function $l(b)$ has the form

$$l^{-}(b) = \frac{b}{2} - \sqrt{\frac{b^2}{4} - R'(0)}.$$

If l belongs to the ray $[b/2, \infty]$, then the function $l(b)$ has the form

$$l^{+}(b) = \frac{b}{2} + \sqrt{\frac{b^2}{4} - R'(0)}.$$

The function $l(b)$ characterizes the asymptotic behavior of the function χ, which satisfies the nonlinear equation related to the semilinear equation (Theorems 1.2, 1.3).

Consider the asymptotic behavior of the function χ as $\chi \to 1 - 0$. Similarly to (1.21), as $\xi \to \infty$, we get

$$b = l_0 - \frac{1}{l_0}\left|\frac{dR}{d\theta}\bigg|_{\theta=1}\right|, \qquad l_0 = \lim_{\xi\to\infty} \frac{d^2\theta}{d\xi^2}\Big(\frac{d\theta}{d\xi}\Big)^{-1}, \qquad l_0 < 0.$$

This function is plotted in Figure 1.5B. Since $\theta \to 1 - 0$ as $\xi \to \infty$ and $b(l_0)$ is finite, we have

$$\lim_{\xi\to\infty} \frac{d^2\theta}{d\xi^2}\Big(\frac{d\theta}{d\xi}\Big)^{-1} = l_0 \neq 0.$$

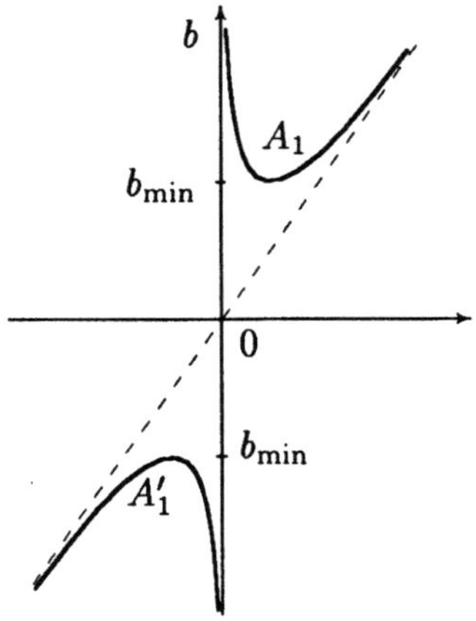

FIGURE 1.5A

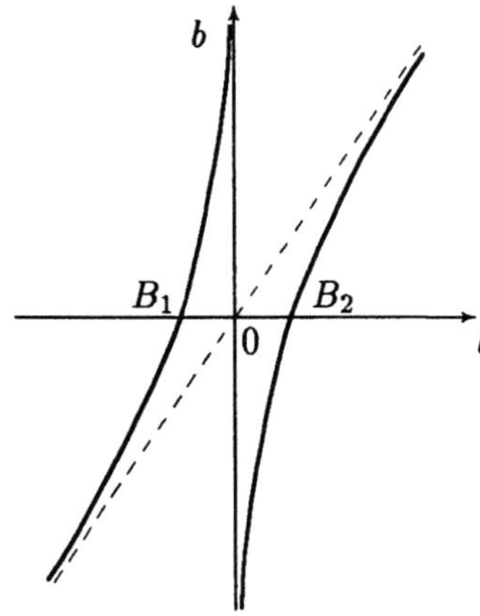

FIGURE 1.5B

This yields

$$\begin{aligned}\theta &= 1 - \Gamma \exp\{-l_0\xi\} + o(\exp\{-l_0\xi\}), \qquad \xi \to \infty,\\ \frac{d\theta}{d\xi} &= l_0\Gamma\exp\{-l_0\xi\} + o(\exp\{-l_0\xi\}) = -l_0(1-\Gamma\exp\{-l_0\xi\}) + l_2 + o(\exp\{-l_0\xi\})\\ &= -l_0\theta + l_0 + o(\exp\{-l_0\xi\}) = l_2(1-\theta) + o(\exp\{-l_2\xi\}).\end{aligned}$$

Then, taking into account (1.3), we see that, as $\tau \to \infty$, the function χ possesses the following asymptotics

$$\chi = 1 - \Gamma_1 \exp\Big\{-\frac{l_0\tau}{\rho(1)k}\Big\} + o\Big(\exp\Big\{-\frac{l_0\tau}{\rho(1)k}\Big\}\Big), \qquad \Gamma_1 = \text{const} > 0.$$

Theorem 1.5. *Let the function $F(\chi)$ in (1.1) be of the form $F(\chi) = \chi^q G(\chi)$, $k > 1$, $q > 0$, $k + q \geqslant 2$, the functions $\rho(\chi)$ and $G(\chi)$ be continuously differentiable for $\chi \geqslant 0$, $G(\chi) \neq 0$ for $\chi \in [0,1[$ and $G(1) = 0$, $dG/d\chi\big|_{\chi=1} \neq 0$, $\rho(\chi) > 0$ for $\xi \in [0,1]$. Then equation (1.1) possesses a monotone continuous nonnegative solution with semibounded support ($\chi(\tau) = 0$ for $\tau < 0$) satisfying the condition $d\chi^k/d\tau\,\big|_{\chi=0} = 0$ and such that*

(a) *if $k + q > 2$, $q \geqslant 1$, and $G(\chi) > 0$ for $\chi \in [0,1[$, then*

$$\begin{aligned}&\chi(\tau) \sim \tau^{1/(k-1)},\\ &\chi(\tau) = 1 - \gamma_1 \exp\Big\{-\frac{l_0\tau}{\rho(1)k}\Big\} + o\Big(\exp\Big\{-\frac{l_0\tau}{\rho(1)k}\Big\}\Big), \qquad \tau \to \infty,\\ &l_0 = -\frac{b_0}{2} + \sqrt{\frac{b_0^2}{4} - \frac{dR}{d\theta}\Big|_{\theta=1}},\end{aligned}$$

and b_0 is the Zeldovich constant for the Zeldovich equation

$$b_0\frac{d\theta}{d\xi} - \frac{d^2\theta}{d\xi^2} - R(\theta) = 0;$$

(b) *if* $k + q = 2$, $q < 1$, *and* $G(\chi) > 0$ *for* $\chi \in [0, 1[$, *then*

$$\chi(\tau) \sim \tau^{1/(k-1)} \sim \tau^{1/(1-q)}, \qquad \tau \to 0,$$
$$\chi(\tau) = 1 - \gamma_2 e^{-l_1\tau} + o(e^{-l_1\tau}), \qquad \tau \to \infty,$$

where

$$b \geqslant 2\sqrt{\left.\frac{dR}{d\theta}\right|_{\theta=0}}, \qquad l_1 = -\frac{b}{2} + \sqrt{\frac{b^2}{4} - \left.\frac{dR}{d\theta}\right|_{\theta=1}};$$

(c) *if* $k + q > 2$, $q > 1$, *and* $G(\chi) < 0$ *for* $\chi \in [0, 1[$, *then*

$$\chi(\tau) \sim \tau^{1/(1-q)}, \qquad \tau \to 0,$$
$$\chi(\tau) = 1 - \gamma_3 e^{-l_2\tau} + o(e^{-l_2\tau}), \qquad \tau \to \infty,$$
$$b < -2\sqrt{\left.\frac{dR}{d\theta}\right|_{\theta=1}}, \qquad l_2 = \frac{1}{k\rho(1)}\left(-\frac{b}{2} - \sqrt{\frac{b^2}{4} - \left.\frac{dR}{d\theta}\right|_{\theta=1}}\right),$$

where $R(\theta) = k\rho(\theta)\theta^{k+q-1}G(\theta)$, $\gamma_i = \text{const} > 0$.

The proof of this theorem follows from (1.3) and the properties of solutions of the corresponding equation (1.2).

Remark. If $q \geqslant 1$, $k + q > 2$, and $G(\chi) < 0$ for $\chi \in [0, 1[$, equation (1.1) has no monotone nonnegative solution $\chi \to \infty$ vanishing for $\tau < 0$ and tending to 1 as $\tau \to \infty$. In fact, by Theorem 1.5 this solution must satisfy the estimate

$$\chi(\tau) = o(\tau^{1/(k-1)}), \qquad \tau \to 0,$$

and thus, by (1.3), the solution of the corresponding equation (1.2) must satisfy

$$\theta(\xi) \sim \exp(\lambda_1 \xi), \qquad \lambda_1 = \text{const} > 0$$

as $\xi \to -\infty$, which contradicts Lemma 1.3.

1.2. Examples

In this section we consider some examples that illustrate the above statements. First, we show how important it is for the derivative of the source-sink function $dR/d\theta$ (see equation (1.1)) to be continuous for $\theta \in [0, 1]$.

Example 2.1. Consider the function

$$\theta(\xi) = \begin{cases} e^{-1/\xi}, & \xi \geqslant 0, \\ 0, & \xi < 0, \end{cases} \tag{2.1}$$

that is smooth in $\xi \in R$ and satisfies the simple wave equation

$$b\frac{d\theta}{d\xi} - \frac{d^2\theta}{d\xi^2} - \theta \ln^2\theta\,[b - \ln\theta(2 + \ln\theta)] = 0, \qquad \theta \in [0, 1]. \tag{2.2}$$

In this case the function R in (1.2) has the form

$$R = \theta \ln^2\theta\,[b - \ln\theta(2 + \ln\theta)],$$

and its first derivative is unbounded at the point $\theta = 0$. Therefore, the solution (2.1) of equation (2.2) can be localized.

Example 2.2. The localizing transformation

$$K(\chi)\frac{d\chi}{d\tau} = \frac{d\theta}{d\xi}(\theta^{-1}(\chi))$$

relates the monotone solution of the standard equation corresponding to the Bürgers type equation

$$b\frac{d\theta}{d\xi} - \frac{d^2\theta}{d\xi^2} + \frac{d\varphi(\theta)}{d\xi} - F(\theta)K(\theta) = 0$$

and the solution of the nonlinear standard equation

$$b\frac{d\chi}{dt} - \frac{d}{d\tau}\left(K(\chi)\frac{d\chi}{d\tau}\right) + \frac{d\varphi(\chi)}{d\tau} - F(\chi) = 0.$$

As an illustration, consider the well-known Bürgers equation

$$\frac{\partial\theta}{\partial t} + \frac{1}{2}\frac{\partial\theta^2}{\partial x} - D\frac{\partial^2\theta}{\partial x^2} = 0, \tag{2.3}$$

which can be reduced to a linear parabolic equation by the Bäcklund transformation

$$\frac{\partial \ln\rho(x,t)}{\partial x} = -\frac{\theta}{2D}, \qquad \frac{\partial \ln\rho(x,t)}{\partial t} = -\frac{1}{4D}\left(2D\frac{\partial\theta}{\partial x} - \theta^2\right).$$

There is another method for reducing the Bürgers equation for simple waves to a linear equation. This method illustrates Theorem 1.1 and can be applied to solutions invariant under the action of groups of translations. Consider the invariant of the translation group $\tau = x + bt$ and rewrite equation (2.3) as follows

$$b\frac{\partial\theta}{\partial\tau} + \frac{1}{2}\frac{d\theta^2}{d\tau} - D\frac{d^2\theta}{d\tau^2} = 0.$$

We shall calculate the function $\theta(z)$ satisfying the functional equation $z = b\theta + \theta^2/2$. We obtain $\theta(z) = -b \pm \sqrt{b^2 + 2z}$.

The equation for z has the form

$$\frac{dz}{d\tau} - D\frac{d^2\theta(z)}{d\tau^2} = 0.$$

By Theorem 1.1, its solution can be expressed in terms of the solution of the linear equation $d\mu/d\xi - Dd^2\mu/d\xi^2 = 0$:

$$\frac{d\theta(z)}{d\tau} = \frac{d\mu}{d\xi}(\mu^{-1}(z)). \tag{2.4}$$

By putting $\mu = e^{\tau/D}$, we get the explicit form of equation (2.4) related to our example:

$$\pm\frac{d}{d\tau}\sqrt{b^2 - 2z} = \frac{z}{D}.$$

By solving this equation for z, we obtain

$$\mp\frac{1}{b}\ln\left|\frac{\sqrt{b^2 - 2z} - b}{\sqrt{b^2 - 2z} + b}\right| = \tau + c_1, \qquad c_1 = \text{const}\,.$$

Substituting the expression for z in terms of θ, we get the solution of the Bürgers equation invariant under the translation

$$\theta = b\left[1 \pm \frac{1 + \exp\{\pm\tau b + c_1\}}{1 - \exp\{\pm\tau b + c_1\}}\right].$$

Example 2.3. Consider the standard equation for the nonlinear diffusion equation

$$b\frac{d\chi}{d\tau} - D\frac{d}{d\tau}\left(\chi^m \frac{d\chi}{d\tau}\right) - \chi^q(1-\chi^\nu) = 0 \qquad m, q > 0, \qquad m+q=1, \qquad \nu > 0. \tag{2.5}$$

Its localized solution is expressed in terms of the solution to the simple wave equation for the KPP type equation

$$b\frac{d\theta}{d\xi} - D\frac{d^2\theta}{d\xi^2} - \theta(1-\theta^\nu) = 0.$$

One of the solutions of this equation has the form

$$\theta(\xi) = \left(\frac{e^\xi}{1+e^\xi}\right)^{2/\nu}, \qquad b = \frac{\nu(4+\nu)}{2(2+\nu)}, \qquad D = \frac{\nu^2}{2(2+\nu)}.$$

Substituting the derivative

$$\frac{d\theta}{d\xi} = \frac{2}{\nu}\theta(1-\theta^{\nu/2})$$

into the expression on the right-hand side of (1.3), we arrive at the ordinary differential equation

$$\chi^m\frac{d\chi}{d\tau} = \frac{\nu}{2}\chi(1-\chi^{\nu/2}),$$

integrable for certain specific values of the parameters m, ν.

So, for $m = \nu/2$, we have the one-sided localized solution of equation (2.5):

$$\chi(\tau) = \begin{cases} \left(1 - e^{-\nu^2\tau/4}\right)^{2/\nu}, & \tau \geqslant 0, \\ 0, & \tau < 0, \end{cases}$$

for $m = \nu$, we have

$$-\chi^{\nu/2} - \ln|1-\chi^{\nu/2}| = \tau\nu^2/4 \quad \text{for} \quad \tau \geqslant 0.$$

Let us consider the equation close in form to (2.5):

$$(2\mu-1)\frac{d\chi}{d\tau} + \frac{d^2\chi^k}{d\tau^2} + \frac{1}{k}[\mu(\mu-1)\chi + \chi^\nu]^{1-k} = 0,$$

and the corresponding equation of the Kolmogorov–Petrovskii–Piskunov type

$$(2\mu-1)\frac{d\theta}{d\xi} + \frac{d^2\theta}{d\xi^2} + \mu(\mu-1)\theta + \theta^\nu = 0. \tag{2.6}$$

Equation (2.6) reduces to the first order system

$$\frac{d\theta}{d\xi} = p,$$
$$\frac{dp}{d\xi} = -(2\mu-1)p - \mu(\mu-1)\theta - \theta^\nu.$$

In the case $\nu = 5$, $\mu = \frac{1}{2}$, this system has the first integral

$$p^2 - \frac{\theta^2}{4} + \frac{\theta^6}{3} = \text{const}\,. \tag{2.7}$$

We shall consider only the solutions of equation (2.6) satisfying the conditions

$$\lim_{\tau\to+\infty} \theta = 0, \qquad \lim_{\tau\to\pm\infty} p = 0.$$

Then the constant on the right-hand side of (2.7) is equal to zero and thus we have

$$\frac{d\theta}{d\xi} = \sqrt{\frac{\theta^2}{4} - \frac{\theta^6}{6}}. \tag{2.8}$$

This equation can be easily integrated, and the solution (the function $\chi(\tau)$) can be calculated since it is not necessary to know the explicit form of θ.

By (2.8) and (1.3), we have the equation

$$\frac{d\chi^k}{d\tau} = \sqrt{\frac{\chi^2}{4} - \frac{\chi^6}{6}}$$

that can be easily solved. For example, for $k = 5$

$$\chi(\tau) = \begin{cases} \frac{3}{4}\left(1 - \left(\frac{2}{15}\right)^2 (c-\tau)^2\right)^{1/4}, & \tau \in \left[c - \frac{15}{2}, c + \frac{15}{2}\right], \\ 0, & \tau \notin \left[c - \frac{15}{2}, c + \frac{15}{2}\right], \end{cases} \qquad c = \text{const}\,.$$

The existence of a two-sided localized solution is based on the fact that the right-hand side of (2.8) is real not for all values of the variable ξ.

For $k = 3$, the localized solution has the form

$$\chi = \begin{cases} \sqrt[4]{\frac{3}{2}} \sin^{1/2}\left(\sqrt{\frac{2}{3}}\frac{\tau}{3} + c\right), & \tau \in \Omega_1, \\ 0, & \tau \notin \Omega_1, \end{cases}$$

$$\Omega_1 = \left[3\sqrt{6}\pi m - c_{10}, (2m-1)3\sqrt{\frac{3}{2}}\pi - c_1\right],$$

$$c_1 = 3c\sqrt{\frac{3}{2}}, \qquad m \in \mathbb{Z}$$

By choosing solutions of (2.6) with different properties (so as the function $\theta(\xi)$ be real and monotone on the whole line ξ), we can construct a one-sided localized solution of (2.5) for other values of the constant μ.

Example 2.4. Consider the equation simulating the flow of a non-Newtonian fluid in a transversal magnetic field. This is a particular case of the equation describing the turbulent filtration for $n = 1$, $x \in \mathbb{R}$, $k = 1$,

$$\frac{\partial u}{\partial t} - \frac{\partial}{\partial x}\left(\frac{\partial u}{\partial x}\right)^{\nu} + \gamma u = 0, \qquad \nu > 0, \qquad \frac{\partial u}{\partial x} \geqslant 0. \tag{2.9}$$

It is well-known that the change of variables

$$z = \frac{1 - \exp\{-\gamma(\nu - 1)t\}}{\gamma(\nu - 1)}, \qquad u(x,t) = v(z,x)e^{-\gamma t}$$

transforms (2.9) into the equation

$$\frac{\partial v}{\partial z} - \frac{\partial}{\partial x}\left(\frac{\partial v}{\partial x}\right)^{\nu} = 0.$$

The simple wave equation corresponding to the latter one, has the form

$$b\frac{dv}{d\tau} - \frac{d}{d\tau}\left(\frac{dv}{d\tau}\right)^{\nu} = 0, \qquad \tau = x + bz. \tag{2.10}$$

Its solution is constructed from the solutions of the equation

$$b\frac{d\theta}{d\xi} - \frac{d^2\theta}{d\xi^2} = 0$$

by using of a localizing transformation. The general solution of the linear equation is given by the formula

$$\theta(\xi) = c_1 e^{b\xi} + c_2, \qquad c_1, c_2 = \mathrm{const}\,.$$

Then the localized solution has the form

$$\frac{d}{dt}\left(\frac{dv^{\nu}}{d\tau}\right) = \frac{d\theta}{d\xi}\left(\theta^{-1}\left(\frac{dv}{d\tau}\right)\right) \equiv b\frac{dv}{d\tau}.$$

This is a consequence of Theorem 1.1 and the fact that equation (2.10) is reduced by the change of variables $\chi dv/d\tau$ to an equation of type (1.1).

The function v can be easily calculated:

$$v = \begin{cases} \dfrac{\nu^2 - 1}{b\nu^2}[A + b\tau]^{\nu/(\nu+1)} + B, & \tau > -\dfrac{A}{b}, \\ 0, & \tau \leqslant -\dfrac{A}{b}. \end{cases}$$

where $A > 0$, $B > 0$ are arbitrary constants.

CHAPTER II

DIRECT METHODS FOR CONSTRUCTING EXACT SOLUTIONS OF SEMILINEAR PARABOLIC EQUATIONS

In Chapter I some results about the qualitative properties of wave solutions to semilinear and quasilinear parabolic equations were presented. These properties are used in Chapters IV and V for constructing asymptotic solutions. The readers interested in applications of the preceding results and in asymptotic methods may omit this chapter which was included into the book to satisfy a natural desire of the authors and, perhaps, of the readers to consider some examples of solutions with properties described in Chapter I.

2.1. Preliminary notes

Recently some new methods were proposed for integrating nonlinear equations with sufficiently large number of laws of conservation. However these methods cannot be used to integrate semilinear parabolic equations describing the processes with dissipation. For such equations only explicit formulas for self-similar solutions have been well-known, i.e., the formulas for solutions $u(x,t)$ representable in the form

$$u(x,t) = \chi(\tau), \tag{1.1}$$

where τ is a self-similar variable, $\tau = x + pt$, $p = \text{const}$. First, the solutions of semilinear parabolic equations were investigated in the now classical work by Kolmogorov, Petrovskii, Piskunov [41] who considered self-similar solutions of the equation

$$u_t - u_{xx} - F(u) = 0, \tag{1.2}$$

taking values between the roots of the equation $F(u) = 0$ (we shall denote them by A and B, $A < B$ (see Figure 2.1), and suppose $F(A) = F(B) = 0$, $F'(A) \cdot F'(B) < 0$). In this case a self-similar solution is defined by the solution of the equation:

$$p\chi' - \chi'' - F(\chi) = 0, \tag{1.3}$$

satisfying the boundary conditions

$$\begin{aligned} &\text{a)} \quad \chi\big|_{\tau\to-\infty} \to A, \qquad \chi\big|_{\tau\to+\infty} \to B, \\ &\text{or} \\ &\text{b)} \quad \chi\big|_{\tau\to-\infty} \to B, \qquad \chi\big|_{\tau\to+\infty} \to A. \end{aligned} \tag{1.4}$$

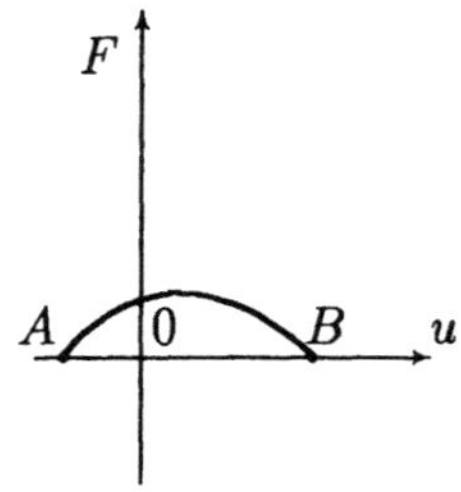

FIGURE 2.1A

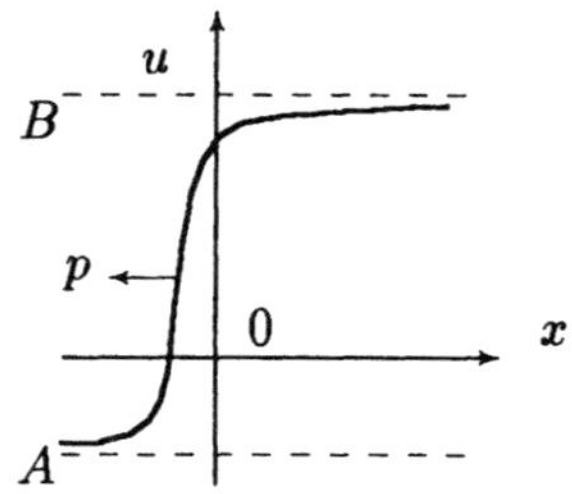

FIGURE 2.1B

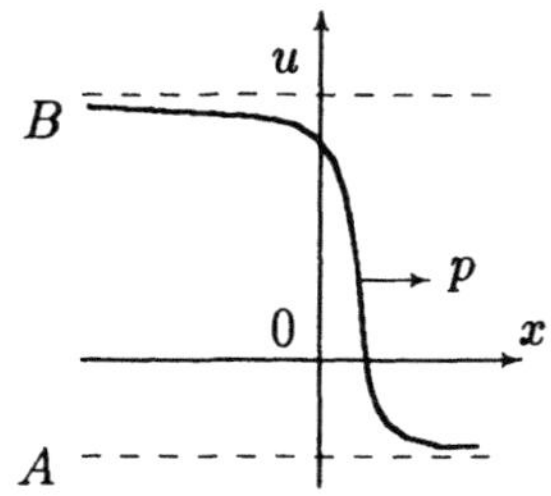

FIGURE 2.1C

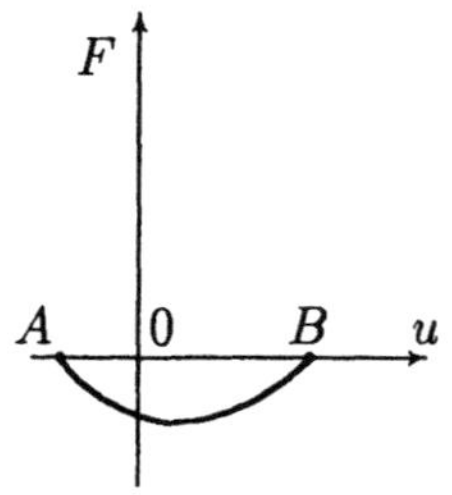

FIGURE 2.2A

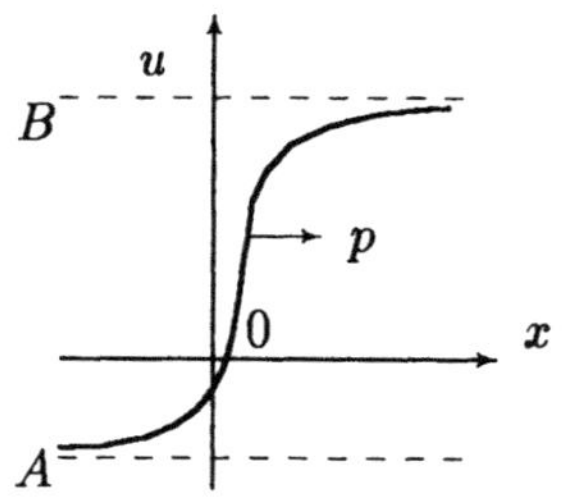

FIGURE 2.2B

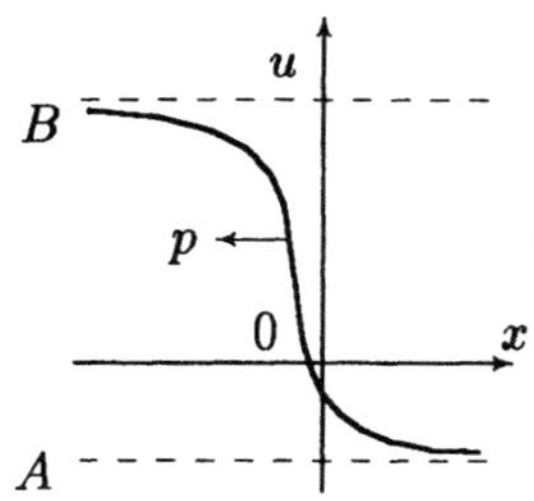

FIGURE 2.2C

In [41, 107] it was proved that the solution of problem (1.3)–(1.4) exists for $p \geqslant p_{\min} = 2\sqrt{F'(A)}$ (for $F(A) > 0$, see Figure 2.1A). The solution (1.1) of problem (1.2), (1.4a) has the form shown in Figure 2.1B. Since equation (1.2) is invariant under the change of variables $x \to -x$, the problem (1.2), (1.4b) has also wave solutions shown in Figure 2.1C.

For $F'(B) > 0$ (the function $F(u)$ changes its sign, Figure 2.2A), the solution of problem (1.3), (1.4) exists for $p \geqslant p_{\min} = 2\sqrt{F'(B)}$ and the wave solutions of problems (1.2), (1.4A) and (1.2), (1.4B) have the form shown in Figures (2.2.B) and (2.2.C) respectively.

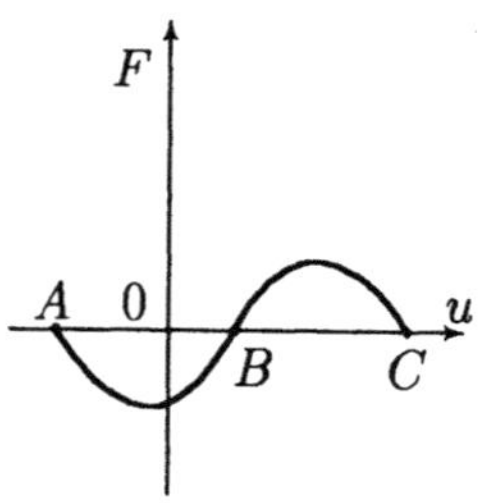

FIGURE 2.3A

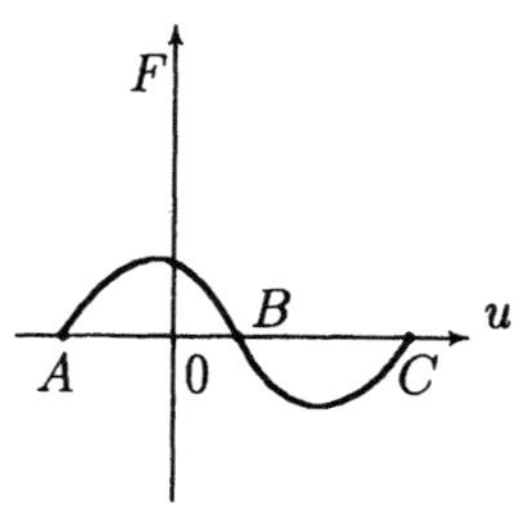

FIGURE 2.3B

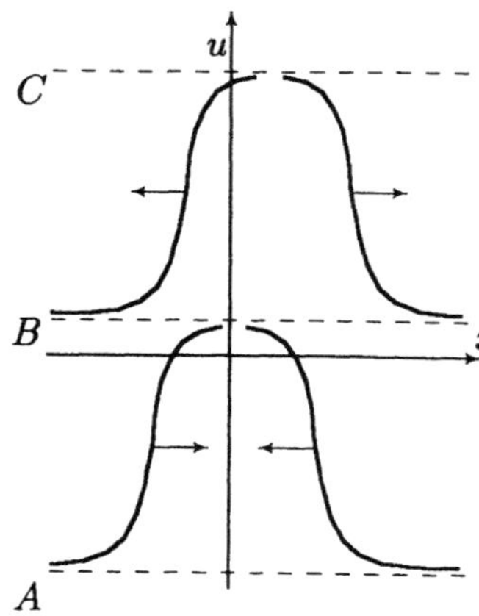

FIGURE 2.3C

Suppose the equation $F(u) = 0$ has three roots $F(A) = F(B) = F(C) = 0$ (Figures 2.3A and 2.3B). In this case the plot of the function $F(u)$ is the union of plots of functions $F(u)$ shown in Figures 2.1A and 2.2A. We shall consider only the function $F(u)$ shown in Figure 2.3A, since in case 2.3B the problem (1.3), (1.4) has no solutions between the zeros $u = A$ and $u = B$ of $F(u)$.

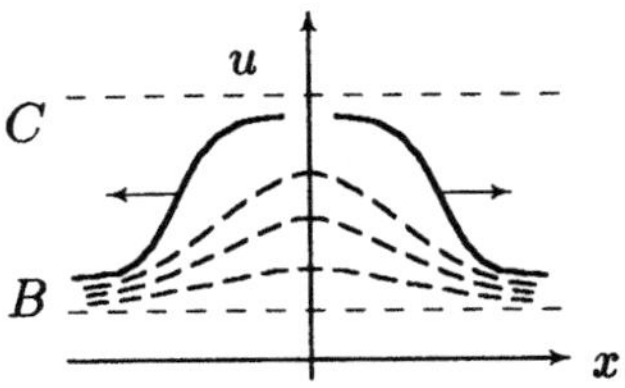

FIGURE 2.4A

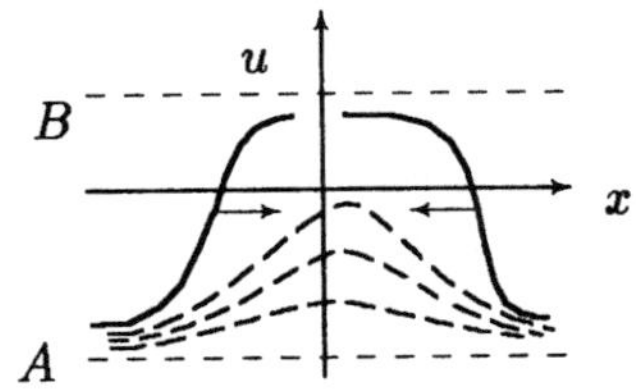

FIGURE 2.4B

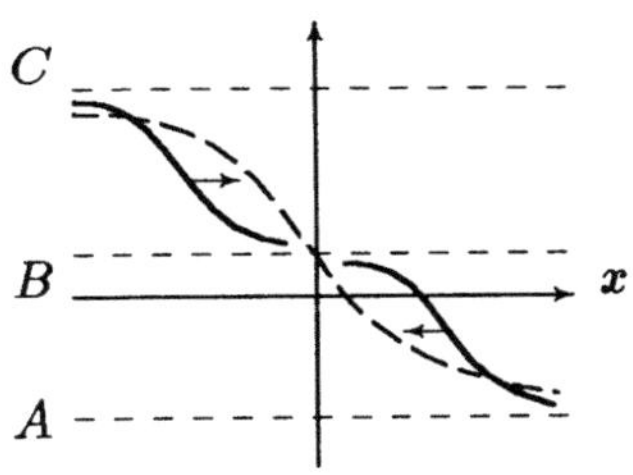

FIGURE 2.4C

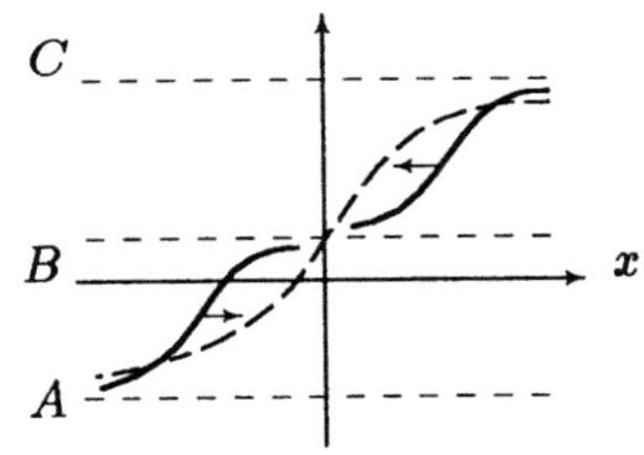

FIGURE 2.4D

Then the problem (1.2), (1.4) has wave solutions of the form (1.1) shown in Figure 2.3C. There are two types of wave interaction: the wave annihilation in a strip region (Figures 2.4A and 2.4B) and the confluence of waves propagating in different strip regions (Figures 2.4C and 2.4D). Obviously, to construct explicit formulas, first of all, it is necessary to construct explicit formulas describing the propagation of a single wave. For the simplest KPP equation

$$u_t - u_{xx} - u(1-u) = 0, \tag{1.5}$$

the well-known explicit formula describing a self-similar wave has the form

$$u = (1 + e^{-\tau/\sqrt{6}})^{-2}, \qquad \tau = x + pt, \qquad p = 5/\sqrt{6} > p_{\min} = 2. \tag{1.6}$$

For equations with cubic nonlinearity, formulas describing the wave interaction of confluence type were obtained for $F = u(1-u^2)$ and $F = u(1-u)(\nu + u)$ in [96]. Unfortunately, the authors were not familiar with these works (in [51] formulas from [96] were obtained independently as well as formulas for equations with other cubic nonlinearities of a particular form). In [25] there is a formula describing the wave interaction of this type for equations, where the nonlinearity is given by an arbitrary cubic polynomial $F(u)$ with a negative coefficient at its highest-degree term. It is necessary to point out that all the above formulas describing the wave interaction of confluence type are constructed under the assumption that each interacting wave propagates with a certain velocity, i.e., for each equation the explicit formula describing the confluence of two self-similar waves with certain velocities is known. This is due to the fact that for equation (1.5), among all existing waves, only one wave with velocity $p = 5/\sqrt{6}$ is given by the explicit formula (1.6). Ablowits and Zeppetella [76], by applying the Painlevé test to equation (1.5), obtained the same velocity. Moreover, the Painlevé test, applied to an equation with nonlinearity in the form of an arbitrary cubic polynomial, gives three possible values of velocity for the pole motion, and precisely for waves propagating with these velocities, formulas describing the confluence are constructed.

It turns out [25, 27] that this property is related to the expansion of the solution into a Dirichlet series. It is well-known that the solution of an ordinary differential equation for a self-similar wave can be represented as a Dirichlet series in a neighborhood of the roots of nonlinearity, i.e., in the domain where the equation can be linearized "in the main". Assume that we have series in terms of $e^{\lambda\tau}$ and $e^{-\lambda\tau}$ as $\tau \to -\infty$ and $\tau \to \infty$ respectively, $\lambda \in R_+$, $\tau = x + pt$. The set of such solutions for each equation is very poor (see [25, 27] and Theorem 2.1 below). For example, for equation (1.5) the constants λ and p are determined uniquely. On the other hand, the structure of Dirichlet series is suggested by the following representation of a self-similar solution:

$$\chi(\tau) = U\left(\frac{\varphi(e^{\lambda\tau})}{\psi(e^{\lambda\tau})}\right), \tag{1.7}$$

where $U = U(z)$ is an analytic function holomorphic near the real axis and at infinity, $\varphi(z)$ and $\psi(z)$ are polynomials. Obviously, the representation (1.7) is not unique, but the wave velocity is determined uniquely. Theorem 2.1 mentioned above implies a criterion for uniqueness of such solutions (with precision up to translations). Further, we note that representation (1.7) is closely related to the results obtained by using the Painlevé test, since the poles of the solution obviously coincide with the zero set of the polynomial $\psi(e^{\lambda\tau}) = \psi(e^{\lambda(x+pt)})$ and move with velocity $\dot{x} = -p$. Our restriction to the search of solutions in the form (1.7) is essential. For example, for equation (1.5) the self-similar wave propagating with minimal velocity cannot be represented in this form. The gain is that we can use constructive methods for obtaining solutions in the form (1.7), for example, the Hirota method and its modifications.

The interaction (confluence) of waves is described by solutions of the form (1.7), where polynomials in one variable are replaced by polynomials in two variables. Similar solutions describe the interaction of solitons and kinks for equations of the Sine-Gordon type. Apparently, there is a certain general yet unknown mechanism ensuring the existence of such solutions both for integrable and nonintegrable equations. This statement is confirmed in [84, 110], where the WTC method based on the Painlevé test is developed for constructing such solutions. The structure of solutions obtained by the WTC method is also birational, though this was not supposed in advance.

A separate problem is to construct an explicit formula for the wave interaction of annihilation or creation type. We present the explicit formula only for the principal term of the asymptotic solution describing the wave annihilation (creation) in the

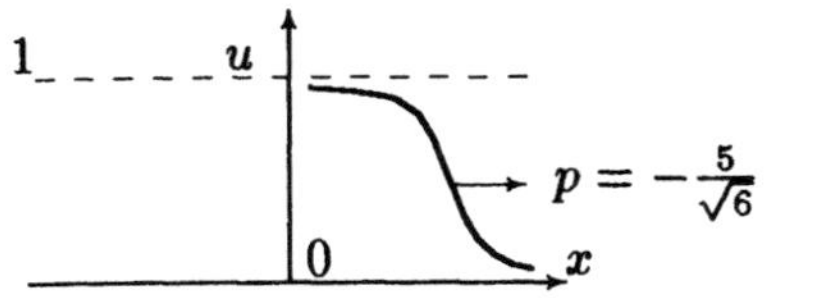

FIGURE 2.5A

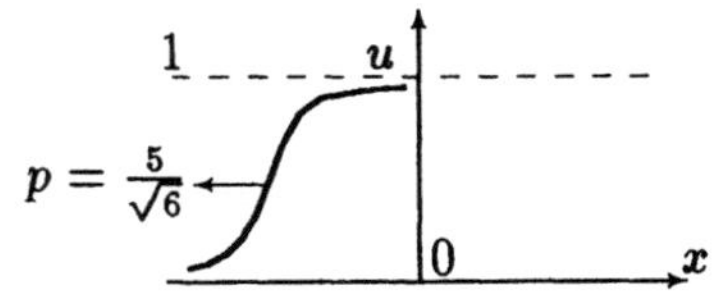

FIGURE 2.5B

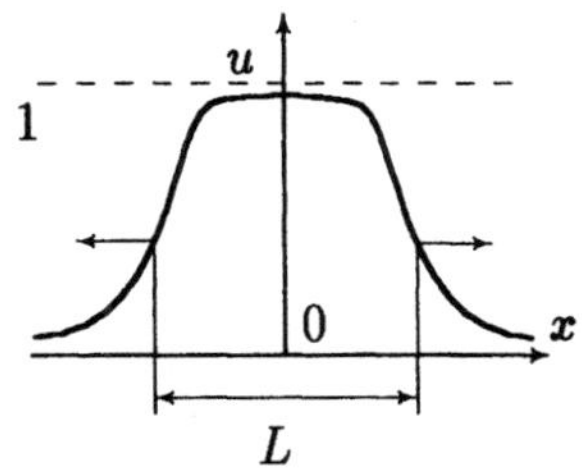

FIGURE 2.5C

case when the distance between the wave fronts is large. Here we have some interesting effects, which will be considered for equation (1.5). For this equation there exist two self-similar waves shown in Figures 2.5A and 2.5B. The product of these two functions is the function shown in Figure 2.5C. Obviously, this function satisfies the equation with precision up to $O(L^{-N})$, where $N > 0$ is an arbitrary number, L is the distance between the wave fronts.

Thus, in this case, the asymptotic law of superposition of waves is the product, and we may hope to construct an explicit formula for the principal term of the asymptotic solution by multiplying the explicit formulas for the solutions shown in Figures 2.5A and 2.5B in a way similar to that for solitons, kinks and the interaction of wave confluence type (in the latter case the asymptotic law of interaction is the sum). However, in the case of wave creation (annihilation), the principal term has the form

$$u = \left(1+\chi^{-1/4}\psi(\sqrt{\chi_2})\exp\left(-\frac{\tau_1}{\sqrt{6}}\right)\right)^{-2} \times \left(1+\chi^{-1/4}\psi(\sqrt{\chi_1})\exp\left(-\frac{\tau_2}{\sqrt{6}}\right)\right)^{-2}\left[1+O\left(\exp\{-L/\sqrt{6}\}\right)\right], \quad (1.8)$$

where $\chi_i = \chi(\tau_i)$, $i = 1,2$, are self-similar waves with plots given in Figures 2.5A and 2.5B, $\chi(\tau) = \left(1+\exp(-\tau/\sqrt{6})\right)^{-2}$, τ_i are self-similar variables, for example, $\tau_1 = x + \frac{5}{\sqrt{6}}t$, $\tau_2 = -x + \frac{5}{\sqrt{6}} + L$, $\psi(z)$ is the real solution of the problem

$$z^2(z-1)\psi'' + (9z-1)z\psi' - 8\psi = 0,$$
$$\psi(1) = 1, \qquad |\psi(0)| < \infty,$$

which can be expressed in terms of hypergeometric functions.

The asymptotic behaviour of the function $\psi(z)$ is considered in Section 2.5. The results of this section, in particular, imply that, due to the initial conditions, the function u is close to the product in the domain between the wave fronts. Outside this domain (ahead of the wave fronts) the functions $\psi(\sqrt{\chi_i})$ oscillate, thus, ahead of the wave fronts, the function (1.8) can be asymptotically represented in the form

$$u \cong U(\tau_1, \tau_2, i\tau_1, i\tau_2).$$

In Section 2.2 we give the conditions under which self-similar solutions of equation (1.2) can be represented as entire functions of the ratio of exponential polynomials. In Section 2.3 we show how to apply the Hirota method described in the preceding section to the Korteweg-de Vries (KdV) equation. Section 2.4 presents exact two-phase solutions of equation (1.2) with an arbitrary cubic nonlinearity. In Section 2.5 we consider the relation between the number of phases and the number of roots of equation $F(u) = 0$ and prove that N-phase solutions ($N \geq 3$) representable as the ratio of exponential polynomials (1.7) do not exist for equations with cubic nonlinearity. In Section 2.6 we derive formula (1.8). The main results of this chapter were first published in [24, 25, 27].

2.2. Representation of self-similar solutions in terms of rational functions

Let us assume that the solution of problem (1.3), (1.4a) can be expanded in a Dirichlet series as $\tau \to \pm\infty$ [30]:

$$\chi = A + \sum_{k=1}^{\infty} a_k e^{k\lambda\tau}, \qquad \tau \to -\infty, \tag{2.1}$$

$$\chi = B + \sum_{k=1}^{\infty} b_k e^{-k\lambda\tau}, \qquad \tau \to +\infty, \tag{2.2}$$

where $\lambda \in R_+$, $a_0 = A$, $b_0 = B$. Below we shall write the conditions imposed on the function $F(\chi)$ under which λ is uniquely determined by relations (2.1), (2.2). It turns out that, in this case, the number λ uniquely determines the parameter p in equation (1.3). By well-known theorems (see, for example, [41, 107]), the solution of problem (1.3), (1.4a) is uniquely (with precision up to translations) determined by the parameter p. Thus the uniqueness of λ means that the solution of problem (1.3), (1.4a) satisfying (2.1), (2.2) is unique (with precision up to translations).

In a neighborhood of the points $\chi = A$ and $\chi = B$ we represent the function $F(\chi)$ in the form

$$F(\chi) = \alpha(\chi - A) + (\chi - A)^2 \Phi(\chi; A), \quad \tau \to -\infty, \qquad \alpha = \left.\frac{dF}{d\chi}\right|_{\chi=A}, \tag{2.3}$$

$$F(\chi) = \beta(\chi - B) + (\chi - B)^2 \Psi(\chi; B), \quad \tau \to +\infty, \qquad \beta = \left.\frac{dF}{d\chi}\right|_{\chi=B} \tag{2.4}$$

The following recurrent relations for the coefficients of the series are the necessary conditions for the existence of expansions (2.1), (2.2) (compare with [76]):

$$a_k d_k = \widehat{\Phi}_k(a_1, \ldots, a_{k-1}), \qquad k \geq 1, \tag{2.5}$$

where $d_k = d_k(\lambda, p) = \lambda^2 k^2 - p\lambda k + \alpha$,

$$\widehat{\Phi}_k = -\frac{1}{k!}\left[\frac{d^k}{dz^k}\left(\left(\sum_{m\geq 1} a_m z^m\right)^2 \Phi\left(A + \sum_{m\geq 1} a_m z^m; A\right)\right)\right]\Bigg|_{z=0},$$

and

$$b_k D_k = \widehat{\Psi}(b_1, \ldots, b_{k-1}), \qquad k \geq 1, \tag{2.6}$$

where $D_k = D_k(\lambda, p) = \lambda^2 k^2 + p\lambda k + \beta$,

$$\widehat{\Psi}_k = -\frac{1}{k!}\left[\frac{d^k}{dz^k}\left(\left(\sum_{m\geq 1} b_m z^m_m\right)^2 \Psi\left(B + \sum_{m\geq 1} b_m z^m; B\right)\right)\right]\Bigg|_{z=0}.$$

The conditions (2.5), (2.6) are obtained by substituting the series (1.1) and (1.2) into (1.3), taking into account formulas (2.3), (2.4), and setting the coefficients at equal powers of the exponential equal to each other.

Proposition 2.1. *The relations hold*:

$$\widehat{\Phi}_k(\underbrace{0, \ldots, 0}_{k-1}) = 0, \qquad \widehat{\Psi}_k(\underbrace{0, \ldots, 0}_{k-1}) = 0, \qquad k \geq 1.$$

Hence, (2.5), (2.6) imply that for certain k', k'' the relations hold:

$$d_{k'}(\lambda, p) = 0, \qquad D_{k''}(\lambda, p) = 0.$$

Lemma 2.1. *Let $d_k(\lambda, p) = 0$ ($D_k(\lambda, p) = 0$) only for one value of $k > 1$. Then, by changing the order of numbering, one can set $k = 1$.*

Proof. Suppose $d_m = 0$ and $d_n \neq 0$ for $n \neq m$, $m \neq 1$ (here and below $i, j, k, l, m, n \in \mathbb{N}$). Then, by formulas (2.3) and (2.5), we have that only the coefficients a_k in (2.1) with indices $k = ml$, $l \in \mathbb{N}$, are nonzero. However, by setting $\lambda = \widehat{\lambda}/m$, the series (2.1) can be rewritten in the form

$$\chi(\tau) = A + \sum_{k\geq 1} \widehat{a}_k e^{k\widehat{\lambda}\tau},$$

where $\widehat{a}_k = a_l$, $l = m \cdot k$, $\widehat{a}_1 = a_m \neq 0$ and $d_1(\widehat{\lambda}, p) = d_m(\lambda, p) = 0$. The same arguments take place for the series (2.2). The lemma is proved.

The arbitrariness of the choice of $\widehat{a}_1$ corresponds to the translation invariance of the solution of problem (1.3), (1.4). Suppose $\widehat{a}_k$ are the "old" coefficients of the series (2.1), $\widetilde{a}_k$ are the "new" coefficients. Then

$$\widetilde{a}_k = \widehat{a}_k \left(\frac{\widetilde{a}_1}{\widehat{a}_1}\right)^k, \qquad k \geq 1.$$

Recall the notation:

$$\alpha = \left.\frac{dF}{du}\right|_{u=A}, \qquad \beta = \left.\frac{dF}{du}\right|_{u=B}.$$

Denote

$$c = \begin{cases} -\beta/\alpha; & \alpha > 0, \quad \beta \le 0, \\ -\alpha/\beta; & \alpha \le 0, \quad \beta > 0. \end{cases} \tag{2.7}$$

Theorem 2.1. *The following conditions are sufficient for the uniqueness (up to translations) of the solution (2.3), (2.4) satisfying (2.1), (2.2):*

1)

$$\alpha + \beta < 0 \tag{2.8}$$

or

2) $\alpha + \beta \ge 0$, *the pair of numbers* $n, m \in \mathbb{N}$ *such that*

$$n = \frac{m+1}{cm-1}, \qquad n, m > \frac{1}{c}$$

is unique.

Remark. If the pair of numbers from Theorem 2.1 does not exist, then there is no solution of problem (1.3), (1.4) satisfying (2.1), (2.2) (the proof see below).

Proof. As shown above, for a given function $F(\chi)$, the solution of problem (1.3), (1.4) is completely determined by the dispersion relations (2.5), (2.6). This implies the following possible cases:

a) $d_1(\lambda, p) = D_1(\lambda, p) = 0$,

b) $d_m = d_n = D_k = 0$,

c) $d_k = D_m = D_n = 0$,

d) $d_k = d_l = D_m = D_n - 0$.

Obviously, the versions $d_l = d_m = d_n = 0$ and $D_l = D_m = D_n = 0$ are impossible, since $d_k = 0$ and $D_k = 0$ are quadratic equations in k for given λ and p.

Consider case a):

$$d_1(\lambda, p) = \lambda^2 - p\lambda + \alpha = 0, \qquad D_1(\lambda, p) = \lambda^2 + p\lambda + \beta = 0.$$

By solving this system of equations with respect to λ and p, we obtain:

$$\lambda^2 = -\frac{\alpha+\beta}{2}, \qquad p = \frac{\alpha - \beta}{\sqrt{-2(\alpha+\beta)}}. \tag{2.9}$$

Since we are interested in real solutions, (2.9) gives us the first condition (1.8): $\alpha + \beta < 0$. If $\alpha = \beta < 0$, then $p = 0$, and in this case there is only a stationary solution of the initial equation (1.2).

In cases b) and c), according to Lemma 2.1, we can, without loss of generality, consider the relations $D_k = 0$ and $d_k = 0$ for $k = 1$.

Consider case b):

$$d_m(\lambda, p) = \lambda^2 m^2 - p\lambda m + \alpha = 0,$$
$$d_n(\lambda, p) = \lambda^2 n^2 - p\lambda n + \alpha = 0.$$

By solving this system of equations with respect to λ and p, we get:

$$\lambda^2 = \frac{\alpha}{m \cdot n}, \qquad p = \sqrt{\alpha}\frac{m+n}{\sqrt{m \cdot n}}. \tag{2.10}$$

The branch of the root is chosen so that the boundary conditions (1.4) hold. Since $\lambda, p \in \mathbb{R}$, this yields a necessary condition for α:

$$\alpha > 0. \tag{2.11}$$

The solution of the system

$$d_m(\lambda, p) = \lambda^2 m^2 - p\lambda m + \alpha = 0,$$
$$D_1(\lambda, p) = \lambda^2 + p\lambda + \beta = 0$$

has the form

$$\lambda^2 = -\frac{\alpha + \beta m}{m(m+1)}, \qquad p = \frac{\alpha - \beta m^2}{\sqrt{-(\alpha + \beta m)m(m+1)}}. \tag{2.12}$$

Now, by setting the right-hand sides in (2.10), (2.12) equal to each other, we get the equation that relates m and n:

$$\lambda^2 = \frac{\alpha}{m \cdot n} = -\frac{\alpha + \beta m}{m(m+1)}, \tag{2.13}$$
$$n = -\frac{\alpha(m+1)}{\alpha + \beta m} = \frac{m+1}{-\beta/\alpha\, m - 1} = \frac{m+1}{cm-1},$$

where $c = -\beta/\alpha$.

Since $m, n \in \mathbb{N}$, then $cm - 1 > 0$. This implies that $c > 1/m > 0$. By (2.11), we get $\alpha > 0$, $\beta < 0$, and c is a rational number. Formula (2.13) is symmetric with respect to m and n: by expressing m in terms of n, we obtain

$$m = \frac{n+1}{cn-1}. \tag{2.14}$$

Thus, without loss of generality, we assume below that $n > m$ in (2.13).

Let us consider equation (2.13). We formally calculate the derivatives:

$$\frac{\partial n}{\partial m} = -\frac{c+1}{(cm-1)^2} < 0, \tag{2.15}$$

$$\frac{\partial n}{\partial c} = -\frac{m(m+1)}{(cm-1)^2} < 0. \tag{2.16}$$

Thus the function $n(m,c)$ monotonically decreases in variables m and c in the whole domain.

First we assume that $c > 1$, and hence, $m > 1/c$. Then the monotonicity of the function $n(m,c)$ (2.15), (2.16) and the conditions $n > m$, $n, m \in \mathbb{Z}_+$, imply that the number m can take only one value, $m = 1$. Hence, there are two cases:

1) $d_1 = d_n = D_1 = 0$, if $n = \frac{2}{c-1} \in \mathbb{N}$.

2) $d_1 = D_1 = 0$, if $n = \frac{2}{c-1} \notin \mathbb{N}$.

In both cases the phase constants λ and p are uniquely determined by (2.9). The assumptions $c > 1$ imply condition (2.8).

Now we consider the case $c \leq 1$ and $m \geq 2$. In this case there exist pairwise different $k, l, m, n \in \mathbb{N}$ such that

$$\begin{aligned} &\text{a)} \qquad d_k = d_l = D_1 = 0, \\ &\text{b)} \qquad d_m = d_n = D_1 = 0. \end{aligned} \tag{2.17}$$

Let us find a number c for which such natural numbers exist. This number is determined by (2.15), (2.16) and the condition $m \geq 2$. Then k, m, n take values 2, 3, 4 respectively. By (2.17), we get:

$$c = \left. \frac{m+n+1}{m \cdot n} \right|_{\substack{m=3,\\ n=4}} = \frac{2}{3}. \tag{2.18}$$

By (2.13) and (2.17), we obtain

$$l = \left. \frac{k+1}{ck-1} \right|_{\substack{k=2,\\ c=2/3}} = 9.$$

In this case there are two solutions with phase constants

$$\lambda_1 = \frac{\sqrt{\alpha}}{3\sqrt{2}}, \qquad p_1 = \sqrt{\alpha}\frac{11}{3\sqrt{2}},$$

$$\lambda_2 = \frac{\sqrt{\alpha}}{2\sqrt{3}}, \qquad p_2 = \sqrt{\alpha}\frac{7}{2\sqrt{3}}; \qquad -\frac{\beta}{\alpha} = \frac{2}{3},$$

corresponding to the dispersion relations (2.17). Thus, for certain $c < 2/3$, the problem (1.3), (1.4) can possess several solutions satisfying conditions (2.1), (2.2). If, for a given c, there are no different $k, l, m, n > 1/c$, such that the conditions (2.17) hold simultaneously, then there exists a unique solution, and the phase constants λ, p are determined by formulas (2.10) or (2.12).

In case c) the line of reasoning is similar to that in case b) where $c = -\alpha/\beta$ and $\alpha < 0$, $\beta > 0$.

It remains to consider the case d).

Consider two systems of equations.

$$1) \qquad d_k = \lambda^2 k^2 - p\lambda k + \alpha = 0, \qquad d_l = \lambda^2 l^2 - p\lambda l + \alpha = 0$$

with solution

$$\lambda^2 = \frac{\alpha}{kl}, \qquad p = \sqrt{\alpha}\frac{k+l}{\sqrt{kl}}, \tag{2.19}$$

and

$$2) \qquad D_m = \lambda^2 m^2 + p\lambda m + \beta = 0, \qquad D_n = \lambda^2 n^2 + p\lambda n + \beta = 0.$$

The solution of system 2) has the form:

$$\lambda^2 = \frac{\beta}{m \cdot n}, \qquad p = -\sqrt{\beta}\frac{m+n}{\sqrt{m \cdot n}}. \tag{2.20}$$

By setting the right-hand sides in formulas (2.19), (2.20) equal to each other, we get

$$\lambda^2 = \frac{\alpha}{k \cdot l} = \frac{\beta}{m \cdot n}, \tag{2.21}$$

$$p = \sqrt{\alpha}\frac{k+l}{\sqrt{k \cdot l}} = -\sqrt{\beta}\frac{m+n}{\sqrt{m \cdot n}}. \tag{2.22}$$

Relation (2.21) implies:

$$kl = \frac{\alpha}{\beta}mn. \tag{2.23}$$

By substituting (2.23) into equation (2.22), we come to the contradiction:

$$k + l = -(m+n)$$

(since $k, l, m, n \in \mathbb{N}$). Thus, case d) does not take place. The theorem is proved.

The set $c \leq 1$, for which the solution is unique, is discrete. We give all the values of c lying between 2/3 and 1, for which the solution is unique.

k	l	c
2	3	1
2	5	4/3
3	7	5/7

This list can be extended: the closest value of $c < 2/3$ is equal to 7/11 for $k = 2$, $l = 11$.

Corollary. *The following problem for the Kolmogorov–Petrovskii–Fisher equation*

$$u_t - u_{xx} - u(1-u) = 0,$$

$$u\big|_{x \to -\infty} \to 0, \qquad u_{x \to +\infty} \to 1,$$

has a unique self-similar solution satisfying conditions (2.1), (2.2).

Proof. For the KPP-equation we have

$$\alpha = \left.\frac{dF}{du}\right|_{u=0} = 1, \qquad \beta = \left.\frac{dF}{du}\right|_{u=1} = -1, \qquad c = -\frac{\beta}{\alpha} = 1.$$

This problem corresponds to statement b) of the theorem. Formulas (2.13) imply $m = 2$ and $n = 3$, and thus

$$d_2 = d_3 = D_1.$$

Formulas (2.10) give single phase constants λ and p:

$$\lambda = \frac{1}{\sqrt{6}}, \qquad p = \frac{5}{\sqrt{6}}.$$

The corollary is proved.

The recurrent formulas (2.5), (2.6) for the coefficients a_k and b_k take the form:

$$a_k = \frac{\sum_{\substack{i+j=k,\\ i\geq 2, j\geq 2}} a_i a_j}{d_k}, \qquad k \geq 4, \tag{2.24}$$

where $a_1 = 0$, and a_2, a_3 are arbitrary,

$$d_k(\lambda, p) = \frac{1}{6}(k-2)(k-3), \qquad b_k = D_k^{-1} \sum_{\substack{i+j=k,\\ i\geq 1, j\geq 1}} b_i b_j, \qquad k \geq 2, \tag{2.25}$$

where k is arbitrary, and

$$D_k(\lambda, p) = \frac{1}{6}(k+6)(k-1).$$

In (2.29), (2.30) we choose

$$a_2 = 1, \qquad a_3 = -2, \qquad b_1 = -2.$$

Then for the coefficients a_k, b_k of the series (2.1), (2.2), corresponding to problem (2.24), we have $\left(\lambda = \frac{1}{\sqrt{6}}, p = \frac{5}{\sqrt{6}}\right)$:

$$a_k = (-1)^k (k-1), \qquad b_k(-1)^k(k+1), \qquad k \in \mathbb{N}. \tag{2.26}$$

By applying the Pade approximation to the series (2.1), (2.2) with coefficients (2.26), we obtain a well-known solution of the KPP-Fisher equation (compare with [76]):

$$u(x,t) = U(\tau) = \frac{e^{2\lambda\tau}}{(1+e^{\lambda\tau})^2},$$

where $\tau = x + 5t/\sqrt{6}$, $\lambda = 1/\sqrt{6}$.

In the subsequent section we present a method for constructing such solutions in a sufficiently general situation.

2.3. Construction of exact one-phase and two-phase solutions

1. To recall Hirota's method for obtaining exact solutions, let us first consider the following example of the KdV equation

$$\frac{\partial \mu}{\partial t} + 6\mu \frac{\partial \mu}{\partial x} + \frac{\partial^3 \mu}{\partial x^3} = 0.$$

The solutions of the KdV equations, satisfying the conditions

$$\mu \to 0, \qquad \frac{\partial \mu}{\partial x} \to 0 \quad \text{for} \quad |x| \to \infty,$$

will be called *soliton-like solutions*. By setting $\mu = \partial u/\partial x$, integrating the KdV equation with respect to x, and taking into account the conditions at infinity, we obtain

$$\frac{\partial u}{\partial t} + 3\left(\frac{\partial u}{\partial x}\right)^2 + \frac{\partial^3 u}{\partial x^3} = 0; \qquad \frac{\partial u}{\partial x} \to \infty, \quad |x| \to \infty. \tag{3.1}$$

Substituting $u = G(x,t)/F(x,t)$, we get

$$\begin{aligned} F^2\left(\frac{\partial G}{\partial t}F - G\frac{\partial F}{\partial t}\right) &+ 3\left(\frac{\partial G}{\partial x}F - G\frac{\partial F}{\partial x}\right)^2 \\ &+ F^2\left(f\frac{\partial^3 G}{\partial x^3} - 3\frac{\partial^2 G}{\partial x^2}\frac{\partial F}{\partial x} - 3\frac{\partial G}{\partial x}\frac{\partial^2 F}{\partial x^2} - G\frac{\partial^3 F}{\partial x^3}\right) \\ &+ 6\left(F\frac{\partial G}{\partial x}\left(\frac{\partial F}{\partial x}\right)^2 + FG\frac{\partial F}{\partial x}\frac{\partial^2 F}{\partial x^2} - G\left(\frac{\partial F}{\partial x}\right)^3\right) = 0. \end{aligned} \tag{3.2}$$

The characteristic property of this equation, common to linear differential equations, is its homogeneity. Equation (3.2) is homogeneous both in F and G (the degree of homogeneity is equal to 4).

In the method proposed by Hirota, the equation (3.2) is splitted into two equations by introducing an arbitrary constant λ. To do this, we rewrite equation (3.2) as follows

$$\begin{aligned} F^2\Bigg[&\frac{\partial G}{\partial t}F - G\frac{\partial F}{\partial t} + 3\lambda\left(\frac{\partial G}{\partial x}F - G\frac{\partial F}{\partial x}\right) \\ &+ \frac{\partial^3 G}{\partial x^3}F - 3\frac{\partial^2 G}{\partial x^2}\frac{\partial F}{\partial x} + 3\frac{\partial G}{\partial x}\frac{\partial^2 F}{\partial x^2} - G\frac{\partial^3 F}{\partial x^3}\Bigg] \\ &+ 3\left[\frac{\partial G}{\partial x}F - G\frac{\partial F}{\partial x}\right]\left[\frac{\partial G}{\partial x}F - G\frac{\partial F}{\partial x} - 2\left(F\frac{\partial^2 F}{\partial x^2} - \left(\frac{\partial F}{\partial x}\right)^2\right) - \lambda F^2\right] = 0, \end{aligned} \tag{3.3}$$

The following relations are sufficient for (3.3) to hold:

$$\begin{aligned} \frac{\partial G}{\partial t}F - G\frac{\partial F}{\partial t} + 3\lambda\left(F\frac{\partial G}{\partial x} - G\frac{\partial F}{\partial x}\right) + \frac{\partial^3 G}{\partial x^3}F& \\ -3\frac{\partial^2 G}{\partial x^2}\frac{\partial F}{\partial x} + 3\frac{\partial G}{\partial x}\frac{\partial^2 F}{\partial x^2} - G\frac{\partial^3 F}{\partial x^3} &= 0, \end{aligned} \tag{3.4}$$

$$2\left(F\frac{\partial^2 F}{\partial x^2} - \left(\frac{\partial F}{\partial x}\right)^2\right) + \lambda F^2 - \left(\frac{\partial G}{\partial x}F - G\frac{\partial F}{\partial x}\right) = 0. \tag{3.5}$$

Further we introduce the bidifferential operators:

$$D_t = \frac{\partial}{\partial t} - \frac{\partial}{\partial t'}, \qquad D_x = \frac{\partial}{\partial x} - \frac{\partial}{\partial x'}.$$

In Hirota's method the calculus of bidifferential operators is developed, which allows to simplify the process of solving the bilinear equations. For example, the action of a bidifferential operator D_t is given by the formula

$$D_t(G(x,t)\cdot F(x',t')) = \frac{\partial G}{\partial t}F - G\frac{\partial F}{\partial t'}.$$

Similarly,

$$\begin{aligned} D_x(G\cdot F) &= \frac{\partial G}{\partial x}F - G\frac{\partial F}{\partial x'}, \\ D_x^2(G\cdot F) &= \frac{\partial^2 G}{\partial x^2}F - 2\frac{\partial G}{\partial x}\frac{\partial F}{\partial x'} + G\frac{\partial^2 F}{\partial x'^2}. \end{aligned}$$

On the "diagonal", i.e., for $x = x'$, $t = t'$, the following formulas hold:

$$D_x^m(A\cdot 1) = \frac{\partial^m A}{\partial x^m}; \tag{3.6}$$

$$D_x^m(A\cdot B) = (-1)^m D_x^m(B\cdot A); \tag{3.7}$$

$$D_x^m(A\cdot A) = 0 \qquad \text{for all} \quad m; \tag{3.8}$$

$$D_x^m(A\cdot A) = 2D_x^{m-1}\left(\frac{\partial A}{\partial x}A\right) \quad \text{for even} \quad m; \tag{3.9}$$

$$D_x D_t(A\cdot A) = 2D_x\left(\frac{\partial A}{\partial t}A\right) = 2D_t\left(\frac{\partial A}{\partial x}A\right); \tag{3.10}$$

$$D_x\left(\exp\{p_1 x\}\exp\{p_2 x'\}\right) = (p_1 - p_2)^m \exp\{(p_1+p_2)x\}; \tag{3.11}$$

$$D_x^m(e^{px}A)(e^{px'}B) = E^{2px}D_x^m(AB). \tag{3.12}$$

By using the bidifferential operators, we rewrite (3.4), (3.5):

$$\begin{aligned} &\{(D_t + 3\lambda D_x + D_x^3)G(x,t)F(x',t')\}\big|_{x=x',\,t=t'} = 0, \\ &\{(D_x^2 + \lambda)(F(x,t)F(x',t')) - D_x(G(x,t)F(x',t'))\}\big|_{x=x',\,t=t'} = 0. \end{aligned} \tag{3.13}$$

The second equation in (3.13) can be written as follows

$$\lambda = \frac{\partial}{\partial x}\left(\frac{G}{F}\right) - 2\frac{\partial^2}{\partial x^2}(\ln F), \tag{3.14}$$

hence the value of the constant λ is determined by the boundary condition for $u = G/F$ and by the asymptotic expansion of the function F as $|x| \to \infty$. For example, for soliton-like solutions, we get $u \to 0$, $F = e^{\alpha x}$. Then, obviously, (3.14) implies $\lambda = 0$ and, by (3.14), we obtain

$$G = 2\frac{\partial F}{\partial x}. \tag{3.15}$$

Then

$$u = \frac{G}{F} = 2\frac{\partial \ln F}{\partial x}.$$

By substituting this expression into the first equation in (3.13), we get

$$(D_t - D_t^3)\left(\frac{\partial F}{\partial x} F\right) = 0. \tag{3.16}$$

By (3.10), equation (3.16) is equivalent to the equation

$$D_x(D_t + D_x^3)(FF) = 0. \tag{3.17}$$

To solve the latter, we expand the function F into a formal power series with respect to the parameter h:

$$F = 1 + hf_1 + h^2 f_2 + \dots. \tag{3.18}$$

Then the expression for the function $u(x,t)$ takes the form

$$u(x,t) = 2\frac{hf_{1x} + h^2 f_{2x} + \dots}{1 + hf_1 + h^2 f_2 + \dots}.$$

Obviously, the chosen representation of the solution implies that the function f_{1x} satisfies the linearized equation (3.1). By substituting the series (3.18) into equation (3.17) and setting the coefficients at equal powers of h equal to each other, we obtain the following system of equations

$$2D_x(D_t + D_x^3)(f_{1x} \cdot 1) = 0, \tag{3.19}$$

$$2D_x(D_t + D_x^3)(f_{2x} \cdot 1) = -d_x(D_t + D_x^3)(f_1 \cdot f_1), \tag{3.20}$$

$$2D_x(D_t + D_x^3)(f_{3x} \cdot 1) = -D_x(D_t + D_x^3)(f_2 f_1 + f_1 f_2), \tag{3.21}$$

Obviously, the operators on the left-hand sides of equations in system (3.19)–(3.21) have the form

$$2D_x(D_t + D_x^3)(f \cdot 1) = 2\frac{\partial}{\partial x}\left(\frac{\partial}{\partial t} + \frac{\partial^3}{\partial x^3}\right)f.$$

The first equation in the system is linear homogeneous, and the other equations are linear nonhomogeneous. We can cut this system at the ith equation by setting $f_i = 0$ and putting the right-hand side equal to zero. By solving the obtained system, we find a finite number of functions f_l $(l = 1, \dots, i-1)$. Then one can set $h = 1$ in the final expression. The system (3.19)–(3.21) has two types of solutions: polynomial and exponential ones.

The polynomial solutions correspond to the so-called rational solitons. For example, the solution of system (3.19)–(3.21) can be obtained by setting

$$f_1 = a_0 + a_1 x + a_2 x^2 + a_3 x^3 + bt, \qquad f_i = 0 \quad \text{for} \quad i > 1, \tag{3.22}$$

where $b = 12a_3$, $a_2^2 = 3a_1 a_3$, a_0, a_1, a_3 are arbitrary constants. Then

$$F = 1 + h[a_0 + a_1 x + (3a_1 a_3)^{1/2} x^2 + a_3(x^3 + 12t)].$$

If we require that the condition

$$u\Big|_{x=0} = 0 \tag{3.23}$$

holds, we get $a_1 = a_2 = 0$. By setting $h = 1$, we finally obtain

$$\begin{aligned} F &= a_3(x^3 + 12(t + t_0)), \qquad t = \text{const}, \\ u &= 6x^2/(x^3 + 12(t + t_0)). \end{aligned} \tag{3.24}$$

The function u is the solution of equation (3.1) with infinite discontinuity. This solution corresponds to a rational soliton for the KdV equation

$$\mu = \frac{\partial u}{\partial x} = -\frac{6x(x^3 - 24t)}{(x^3 + 12t)^2}$$

(here we put $t_0 = 0$).

The exponential solutions of (3.19)–(3.21) correspond to N-soliton solutions, which are very cumbersome in general, and we do not write them here. We only note that these solutions are the ratio of two exponential polynomials

$$u = \frac{Q(z_1, \ldots, z_N)}{P(z_1, \ldots, z_N)}, \qquad z_i = \exp\{b_i t + a_i x\}.$$

As an example consider the two-soliton solution of the KdV equation for $N = 2$:

$$\mu = 6\frac{\partial}{\partial x}\left(\frac{1}{\varphi}\frac{\partial \varphi}{\partial x}\right), \qquad \varphi = 1 + e^{\tau_1} + e^{\tau_2} + \frac{(\alpha_2 - \alpha_1)^2}{(\alpha_2 + \alpha_1)^2} e^{\tau_1 + \tau_2}, \tag{3.25}$$

where

$$\tau_1 = -2\alpha_i(x + x_i - 4\alpha_i^2 t) \qquad (i = 1, 2), \quad \alpha_i, x_i = \text{const}.$$

Now consider the KdV equation with a small parameter $0 < \varepsilon < 1$:

$$\frac{\partial v}{\partial t} + 6v\frac{\partial v}{\partial x} + \varepsilon^2 \frac{\partial^3 v}{\partial x^3} = 0. \tag{3.26}$$

Obviously, the two-soliton solution of this equation is the function obtained from (3.25) by substituting τ_i/ε instead of τ_i:

$$v = 6\varepsilon^2 \frac{\partial}{\partial x}\left(\frac{1}{\psi}\frac{\partial \psi}{\partial x}\right), \qquad \psi = 1 + e^{\tau_1} + e^{\tau_2} + \frac{(\alpha_2 - \alpha_1)^2}{(\alpha_2 + \alpha_1)^2} e^{\tau_1 + \tau_2}, \tag{3.27}$$

where

$$\tau_i = -2\alpha_1 \frac{x + x_i - 4\alpha_i^2 t}{\varepsilon} \qquad (i = 1, 2), \quad \alpha_1, x_1 = \text{const}.$$

Let, for definiteness, $x_1 > x_2$, $\alpha_2 > \alpha_1$. Then for $t < (x_1 - x_2)/(4(\alpha_1^2 - \alpha_1^2))$ $(\delta > 0)$ as $\varepsilon \to 0$ the two-soliton solution can be represented, up to any power of the small parameter ε, as the sum of one-soliton solutions:

$$v = v_1 + v_2 + O(\varepsilon^M), \tag{3.28}$$

where $M > 0$ is an arbitrary large number, v_1 and v_2 are one-soliton solutions

$$v_1 = 12\alpha_1 e^{\tau_1}(1 + e^{\tau_1})^{-2}, \qquad v_2 = 12\alpha_2 e^{\tau_2}(1 + e^{\tau_2})^{-2}, \tag{3.29}$$

and τ_1, τ_2 are defined above.

The functions v_1 and v_2 are ratios of exponential polynomials, and the exponents in formulas (3.29) are the same as in (3.27). Thus we have obtained the "law of conservation of phases": a two-soliton solution is the ratio of exponential polynomials with exponents from the formulas for one-soliton solutions. This fact together with the "asymptotic law of addition" for solitons (3.28) allows to construct a two-soliton solution by taking one-soliton solutions. Namely, we calculate the sum $v_1 + v_2$ by establishing the common denominator for the fractions in (3.29) and obtain the ratio of exponential polynomials. In this ratio we replace the obtained coefficients by arbitrary ones and substitute the ratio into the equation. Thus we obtain a system of equations for these coefficients. This system turns out to be solvable, and, by solving it, we get the exact formula for the two-soliton solution. Thus, the sum of two solitons is, in a sense, an ansatz for the formula of a two-soliton solution. Similarly, in the case of a semilinear parabolic equation, we can construct exact formulas for wave solutions using Hirota's method.

2.4. Formulas for solutions of semilinear parabolic equations with common cubic nonlinearity

In this section we shall construct exact formulas for one-phase (self-similar) and two-phase (not self-similar) solutions of a one-dimensional semilinear parabolic equation

$$u_t - u_{xx} - \Phi(u) = 0, \tag{4.1}$$

where $\Phi(u) = \sum_{k=0}^{3} \varphi_k u^k$ is a polynomial of degree three.

These solutions have the form

$$u(x,t) = U(\tau_1, \tau_2) = \frac{C_3 + C_1 e^{\tau_1} + C_2 e^{\tau_2}}{1 + e^{\tau_1} + e^{\tau_2}},$$

where $\tau_i = a_i x + b_i t + t_{0i}$, $i = 1, 2$, a_i, b_i are constants (they can be expressed in terms of the coefficients of the polynomial $\Phi(u)$), t_{0i} are the translation constants, C_i, $i = 1, 2, 3$, are the roots of the equation $\Phi(u) = 0$, $C_i \neq C_j$, $i \neq j$. Besides of this, equation (4.1) is studied for general exponential (powerlike) functions $\Phi(u)$.

In this case one-phase solutions are obtained and the form of the function $\Phi(u)$ is given, for which the method for constructing one-phase solutions can be applied.

First consider equation (4.1) with

$$\Phi(u)=\sum_{k=0}^{N}\varphi_k u^{q_k}, \qquad \varphi_N\neq 0, \tag{4.2}$$

where φ_k, $q_k\in\mathbb{R}$, $k=0,\dots,N$, and the sequence q_k is strictly ordered, $q_i>q_j$ for any $i>j$, $i,j=0,\dots,N$.

The solution of (4.1), (4.2) will be sought in the form:

$$u(x,t)=\left(\frac{G(x,t)}{F(x,t)}\right)^p, \tag{4.3}$$

where the exponent $p>0$ depends on the exponent q_N (nonlinearity) in $\Phi(u)$ and will be determined in the sequel.

By substituting the solution (4.3) into (4.1), (4.2), for the functions G and F we get

$$p\big[GF^2(G_t-G_{xx})-G_x^2F(F_t-F_{xx})-(p-1)F^2G_x^2 \\ +2pGFG_xF_x-(p+1)G^2F_x^2\big]=\widehat{\Phi}(G,F), \tag{4.4}$$

where

$$\widehat{\Phi}(G,F)=\sum_{k=0}^{N}\varphi_k G^{pq_k-p+2}F^{-pq_k+p+2}.$$

Equation (4.4) is underdetermined, since it contains two unknown functions. We shall construct the solutions of this equation by using the following method: we pass from equation (4.4) to an overdetermined system of equations, so that to solve this system be sufficient to solve equation (4.4). This reduction can be established in different ways. Here we follow the method proposed by Hirota. In this method, as explained in the preceding section, the functions G and F are represented as polynomials in a certain formal parameter ε, and after the substitution of these functions into equation (4.4), we obtain series in powers of the parameter ε. Obviously, in order to have a finite number of equations, which are the coefficients at powers of ε, it is necessary for the exponents in (4.4) to be nonnegative integer.

Let the functions G and F have the form

$$G=C+\varepsilon g, \qquad F=1+\varepsilon f, \tag{4.5}$$

where $g=g(x,t)$, $f=f(x,t)$, C is a certain constant, ε is a parameter.

Lemma 4.1. *Let the functions G and F be defined by (4.5). In order the number of equations for functions G and F be minimal, it is necessary that the nonlinearity be of the form:*

$$\Phi(u)=u\sum_{k=0}^{4}\varphi_k u^{(q-1)(k-2)/2}, \qquad \varphi_L\neq 0, \tag{4.6}$$

where $q = \max_k q_k = q_N$ *and* $p = 2/(q-1)$ *in* (2.3).

Proof. Denote

$$m(q_k) = pq_k - p + 2, \qquad l(q_k) = -pq_k + p + 2 = 4 - m(q_k). \tag{4.7}$$

The powers of G and F on the left-hand side of (4.4) are integer. Thus, by substituting the functions G and F (4.5) into the expression on the left-hand side of (4.4), we get a polynomial in ε. Then it is necessary that $m(q_k)$ and $l(q_k)$ take only nonnegative integer values. Hence, by using notation (4.7), we see that the integer-valued functions $m(q_k)$ and $l(q_k)$ must satisfy the following relations

$$l(q_k) + m(q_k) = 4, \qquad o \leqslant l(q_k) \leqslant 4, \qquad 0 \leqslant m(q_k) \leqslant 4.$$

Since $p > 0$, we have that $l(q_k)$ decreases monotonically, and $m(q_k)$ increases monotonically. Then, for $q_k = q = \max_i q_i$, the function $l(q_k)$ takes its minimal value that is equal to zero, an for $q_k = q$, respectively, $m(q_k)$ takes its maximum: $m(q) = 4$. Thus we get the parameter $p = 2/(q-1)$, and hence it is necessary that $q > 1$.

By (4.7), we obtain

$$q_k = 1 + \frac{(q-1)}{2}(k-2) = 1 + \frac{1}{p}(k-2), \qquad k = 0, \ldots, 4.$$

This formula for q implies that the function $\Phi(u)$ has the form (4.6). The lemma is proved.

The function $\Phi(u)$ can also be represented in the form:

$$\Phi(u) = pu^{1-2/p}\big(\alpha + \beta u^{1/p} + \gamma u^{2/p} + \lambda u^{3/p} + \mu u^{4/p}\big), \qquad \mu \neq 0. \tag{4.8}$$

By Lemma 4.1 and formula (4.8), we get the function

$$\widehat{\Phi}(G, F) = p\big(\alpha F^4 + \beta G F^3 + \gamma G^2 F^2 + \lambda G^3 F + \mu G^4\big).$$

By substituting this expression into (4.4), we obtain the following equation

$$\begin{aligned} &GF^2(G_t - G_{xx}) - G^2 F(F_t - F_{xx}) - (p-1)F^2 G_x^2 + 2pGFG_xF_x \\ &\qquad - (p+1)G^2F_x^2 = \alpha F^4 + \beta G F^3 + \gamma G^2 F^2 + \lambda G^3 F + \mu G^4. \end{aligned} \tag{4.9}$$

Lemma 4.2. *If the functions G an F have the form* (4.5), *then equation* (4.9) *is equivalent to the following systems of equations:*

a) *for* $C \neq 0$

$$\alpha + \beta C + \gamma C^2 + \lambda C^3 + \mu C^4 = 0, \tag{4.10a}$$

$$\begin{aligned} &C(Cf - g)(g_t - g_{xx}) - (p-1)Cg_x^2 + 2pC^2 g_x f_x - (p+1)C^3 f_x^2 \\ &\qquad + (2\beta + 3\gamma C + 3\lambda C^2 + 2\mu C^3)(Cf - g)^2 = 0, \end{aligned} \tag{4.10b}$$

$$\begin{aligned} &C^2(Cf - g)(f_t - f_{xx}) - (p-1)Cg_x^2 + 2pC^2 g_x f_x - (p+1)C^3 f_x^2 \\ &\qquad + (\beta + \gamma C - 2\mu C^3)(Cf - g)^2 = 0, \end{aligned} \tag{4.10c}$$

$$(p-1)C^2 g_x^2 + \alpha(Cf - g)^2 = 0, \tag{4.10d}$$

$$(p+1)f_x^2 + \mu(Cf - g)^2 = 0; \tag{4.10e}$$

b) *for* $C = 0$

$$g(g_t - g_{xx}) - (p-1)g_x^2 - \gamma g^2 = 0, \tag{4.11a}$$

$$g(f_t - f_{xx}) - 2pg_x f_x + \lambda g^2 = 0, \tag{4.11b}$$

$$(p+1)f_x^2 + \mu g^2 = 0. \tag{4.11c}$$

Remark. Obviously, the case $C = 0$ is possible only if $\alpha = 0$. Then it is necessary that $\beta = 0$ for $p \neq 1$.

This lemma is proved by substituting (4.5) into (4.9) and collecting the summands with equal powers of the parameter ε [25].

From equation (4.10a) we get that the constant C is the nonzero root of equation (4.8), $\Phi(C^p) = 0$.

Further we shall assume that $g \neq Cf$, since otherwise $u = (G/F)^p = ((C+\varepsilon g)/(1+\varepsilon f))^p = C^p = \text{const}$ and this solution is of no interest.

Lemma 4.3. *Let* $p \neq 1$ *and* C_1, C_2 *be the roots of equation* $\Phi(u) = 0$ (4.8). *Then the one-phase (self-similar) solution of equation* (4.1), (4.8) *has the form*

$$u(x,t) = \left(\frac{C_1 + C_2 e^{\tau}}{1+e^{\tau}}\right)^p, \tag{4.12}$$

where $\tau = ax + bt + t_0$, a, b, C_1, C_2 *are constants in terms of the coefficients of the function* $\Phi(u)$, $t_0 = \text{const}$.

a) *For* $\alpha \neq 0$

$$a = \pm(C_1 B - A), \qquad A = \sqrt{-\frac{\alpha}{(p-1)C_1^2}}, \qquad B = \sqrt{-\frac{\mu}{p+1}},$$

$$b = a\left(a + \frac{2\alpha + \beta C_1 - 2pC_1^3 AB}{C_1^2 A}\right) = a\left(a + \frac{\lambda + 2\mu C_1 - 2pAB}{B}\right),$$

where $\Phi(C_1) = 0$ *and the coefficients of the function* $\Phi(u)$ *necessarily satisfy the condition*

$$B(2\alpha + \beta C_1 - 2pC_1^3 AB) = C^2 A(\lambda + 2\mu C - 2pAB).$$

b) *For* $\alpha = 0$

$$a = \pm BC_1, \qquad b = -B^2 C_1(C_1 \pm (p+1)^2(C_1 - C_2))/\mu, \qquad B = \sqrt{-\frac{\mu}{p+1}}.$$

In this case, necessarily $\beta = 0$ *in* (4.8).

Proof. In the sequel the main goal is to prove that for $p \neq 1$ there are no other solutions of the form (4.3) except those given in the lemma. Assume that $\alpha \neq 0$ and equations (4.10d), (4.10e) can be reduced to the form

$$g_x = \pm A(Cf - g), \qquad f_x = \pm B(Cf - g), \tag{4.13}$$

where $A = \sqrt{-\frac{\alpha}{(p-1)C^2}} \neq 0$, $B = \sqrt{-\frac{\mu}{p+1}} \neq 0$.

This yields

$$g_x = \frac{A}{B} f_x. \tag{4.14}$$

By substituting (4.13) into (4.10b) and (4.10c), we get:

$$g_t - g_{xx} \pm \frac{K}{C^2 A} g_x = 0, \tag{4.15}$$

where $K = 2\alpha + \beta C - 2pC^3 AB$, and

$$f_t - f_{xx} \pm \frac{M}{B} f_x = 0, \tag{4.16}$$

where $M = \lambda + 2\mu C - 2pAB$.

By differentiating the first equation in (4.13) in x an taking (4.14) into account, we obtain

$$g_{xx} \pm (CB - A) g_x = 0.$$

This gives $g(x,t)$:

$$g = D \exp\{ax + \varphi(t)\} + \psi_1(t), \tag{4.17}$$

where $a = \pm(CB - A)$, $D = \text{const}$.

From (4.14) and (4.17) we obtain $f(x,t)$:

$$f = \frac{DB}{A} \exp\{ax + \varphi(t)\} + \psi_2(t). \tag{4.18}$$

By substituting (4.17) into (4.15), we find $\varphi(t)$ and $\psi_1(t)$:

$$\varphi(t) = a\left(a + \frac{K}{C^2 A}\right)t + t_{01} = b_1 t + t_{01}, \quad \psi_1(t) = \psi_{01} = \text{const}, \tag{4.19}$$

where t_{01} is an arbitrary constant.

Similarly, by (4.18) and (4.16), we get $\varphi(t)$ and $\psi_2(t)$:

$$\varphi(t) = a\left(a + \frac{M}{B}\right)t + t_{02} = b_2 t + t_{02}, \quad \psi_2(t) = \psi_{02} = \text{const}, \tag{4.20}$$

where t_{02} is an arbitrary constant.

By setting $b = b_1 - b_2$ and taking (4.13) into account, we get the condition for the coefficients of $\Phi(u)$ and the constants t_{01}, t_{02}, ψ_{01}, ψ_{02}:

$$BK = C^2 AM, \qquad \psi_0 = \psi_{02} = \frac{1}{C}\psi_{01}, \qquad t_0 = t_{01} = t_{02}.$$

Then the solution of problem (4.1), (4.8) has the form

$$u(x,t) = v(\tau) = \left(\frac{C(1+\psi_0) + \frac{A}{B}e^{\tau}}{1 + \psi_0 + e^{\tau}}\right)^p,$$

where $\tau = ax + bt + t_0$ can be obtained from (4.17), (4.19), (4.20).

Now assume that $\alpha = 0$, $\mu \neq 0$. Equations (4.13) imply that $g_x = 0$, i.e., the function g depends only on the time $g = g(t)$. Hence, by (4.10a), equation (4.10b) takes the form

$$Cg_t = \beta(Cf - g). \tag{4.21}$$

Recall that $Cf \neq g$ and $\mu \neq 0$. Then, by (4.13), $f_x \neq 0$. Hence, to have an equality, it is necessary that $\beta = 0$ and $g = C\psi_0 = \text{const}$.

Then the function $\Phi(u)$ (4.8) can be written as follows

$$\Phi(u) = u(\gamma + \lambda u^{1/p} + \mu u^{2/p}) = \mu u(C_1 - u^{1/p})(C_2 - u^{1/p}), \tag{4.22}$$

where $\gamma = \mu C_1 C_2$, $\lambda = -\mu(C_1 + C_2)$.

As mentioned above, the constant C must be equal to one of the roots C_1, C_2 of the equation $\Phi(C^p) = 0$. Equation (4.10c) takes the form

$$f_t - f_{xx} \pm (p+1)B(C_i - C_j)f_x = 0, \qquad i,j = 1,2, \quad i \neq j. \tag{4.23}$$

Equation (4.13) gives the function

$$f = e^{ax+\varphi(t)} + \psi_0, \qquad a = \pm BC_i, \qquad i = 1,2.$$

By substituting this solution into (4.13), we find the function $\varphi(t)$:

$$\varphi(t) = \big(a^2 \pm (p+1)B(C_i - C_j)a\big)t + t_0 = bt + t_0.$$

Hence, in this case, the solution of (4.1), (4.22) is given by the following formula

$$u_i(x,t) = U(\tau_i) = \left(\frac{C_i(1+\psi_0)}{1 + \psi_0 + \exp(\tau_i)}\right)^p, \tag{4.24}$$

where $\tau_i = a_i x + b_i t + t_{0i}$, $a_i = \pm BC_i$, $b_i = -B^2 C_i(C_i \pm (p+1)^2(C_i - C_j))/\mu$, $B = \sqrt{-\mu/(p+1)}$, $t_{0i} = \text{const}$, $i,j = 1,2$, $i \neq j$.

Consider the case $C = 0$ (the system of equations (4.11)). Then from (4.10b) we have $\beta = 0$ and the function $\Phi(u)$ is given by (4.22). The function g is completely determined by equation (4.11c)

$$g = \mp\frac{1}{B}f_x, \qquad B = \sqrt{-\frac{\mu}{p+1}}. \tag{4.25}$$

Differentiating equation (4.25) with respect to x and t and substituting the function g and its derivatives into equations (4.11a) and (4.11b), we get

$$f_x f_{xt} - f_x f_{xxx} - (p-1)f_{xx}^2 - \gamma f_x^2 = 0, \tag{4.26a}$$

$$f_t - (1+2p)f_{xx} \mp \frac{\lambda}{B}f_x = 0. \tag{4.26b}$$

Differentiating equation (4.26b) with respect to x and substituting f_{tx} into (4.26a), we obtain the following ordinary differential equation in x:

$$2pf_xf_{xxx}-(p-1)f_{xx}^2\pm\frac{\lambda}{B}f_xf_{xx}-\gamma f_x^2=0.$$

By substituting the values of the coefficients λ and γ from (4.22) into this equation, we get

$$\frac{2p}{p+1}f_xf_{xxx}-\frac{p-1}{p+1}f_{xx}^2\pm B(C_1+C_2)f_xf_{xx}+B^2C_1C_2f_x^2=0. \qquad (4.27)$$

The solution of (4.27) will be sought in the form $f=e^{ax+\varphi(t)}+\psi(t)$. Then the coefficient a is determined by the following equation

$$a^2\pm B(C_1+C_2)a+B^2C_1C_2=(a\pm BC_1)(a\pm BC_2)=0.$$

Hence $a=\mp BC_i$, $i=1,2$. The functions $\varphi(t)$ and $\psi(t)$ are given by (4.26b)

$$\varphi(t)=B\big((1+2p)C_i^2\mp(p+1)(C_1+C_2)\big)t+t_0=bt+t_0,$$
$$i,j=1,2, \qquad i\neq j \qquad t_0=\text{const}, \qquad \psi(t)=\psi_0=\text{const}.$$

By obtaining the function g from (4.25), we get the solution of equation (4.1), (4.22), corresponding to system (4.11) ($C=0$)

$$u_i(x,t)=U(\tau_i)=\left(\frac{C_i\exp(\tau_i)}{1+\psi_0+\exp(\tau_i)}\right), \qquad (4.28)$$

where $\tau_i=a_ix+b_it+t_{0i}$, $a_i=\pm BC_i$, $b_i=-B^2C_i(pC_i-(p+1)C_j)$, $B=\sqrt{-\mu/(p+1)}$, $\psi_0,t_{0i}=\text{const}$, $i,j=1,2$, $i\neq j$. The lemma is proved.

It is easy to see that if in (4.28) we divide the numerator and denominator by e^{τ_i}, the result is exactly formula (4.24). Thus (4.24) and (4.28) define the same solutions of equation (4.1), (4.22).

Now consider the case $p=1$.

Equation (4.10d) implies that $\alpha=0$. Then the function $\Phi(u)$ (4.8) is a cubic polynomial

$$\Phi(u)=\beta+\gamma u+\lambda u^2+\mu u^3=-\mu(C_1-u)(C_2-u)(C_3-u), \qquad (4.29)$$

where $\beta=-\mu C_1C_2C_3$, $\gamma=\mu(C_1C_2+C_2C_3+C_3C_1)$, $\lambda=-\mu(C_1+C_2+C_3)$.

Lemma 4.4. *Suppose equation $\Phi=0$ (see (4.29)) has three different roots $u=C_1$, $u=C_2$, $u=C_3$. Then equation (4.1), (4.29) has the exact two-phase (not self-similar) solutions*

$$u_k(x,t)=U(\tau_i^k,\tau_j^k;C_k)=\frac{C_k+C_i\exp(\tau_i^k)+C_j\exp(\tau_j^k)}{1+\exp(\tau_i^k)+\exp(\tau_j^k)}, \qquad (4.30)$$

where $\tau_i^k = a_i^k x + b_i^k t + t_{0i}^k$, $a_i^k = \pm B(C_k - C_i)$, $b_i^k = -B^2(C_k - C_i)(C_k + C_i - 2C_j)$, $B = \sqrt{-\mu/2}$, $t_{0i}^k = \text{const}$, $k, i, j = 1, 2, 3$, $k \neq i \neq j$.

Proof. Precisely as in Lemma 4.3, first consider the case $C \neq 0$. Recall that C is one of the roots of equation $\Phi(C) = 0$ (4.29). To begin, we put $C = C_3 \neq 0$.

We get the function g from (4.10e):

$$g = C_3 f \mp \frac{1}{B} f_x, \qquad B = \sqrt{-\frac{\mu}{2}}. \tag{4.31}$$

Differentiating (4.31) with respect to x and substituting the value of g_x into (4.10c), we get the following equation

$$f_t - 3f_{xx} \mp \frac{1}{B}(\lambda + 3\mu C_3) f_x = 0. \tag{4.32}$$

Similarly, equation (4.10b) reduces to the form

$$g_t \pm \frac{1}{B} f_{xxx} - 3C_3 f_{xx} \pm \frac{1}{BC_3}(2B^2 C_3^3 - \beta) f_x = 0. \tag{4.33}$$

Differentiating (4.31) with respect to t, we get

$$f_{xt} = \pm B(C_3 f_t - g_t),$$

and using (4.32), (4.33), we obtain

$$f_{xt} - f_{xxx} - (\gamma + 2\lambda C_3 + 3\mu C_3^2) f_x = 0. \tag{4.34}$$

Differentiating (4.32) with respect to t and substituting the obtained expression for f_{tx} into (4.34.), we get

$$2f_{xxx} \pm \frac{1}{B}(\lambda + 3\mu C_3) f_{xx} - (\gamma + 2\lambda C_3 + 3\mu C_3^2) f_x = 0.$$

By substituting the values of γ and λ from (4.29) into the latter equation, we obtain

$$f_{xxx} \mp B(2C_3 - C_1 - C_2) f_{xx} + B^2(C_3^2 - C_3 C_1 - C_3 C_2 + C_1 C_2 C_3) f_x = 0, \tag{4.35}$$

whose general solution has the form:

$$f = e^{a_1 x + \varphi_1(t)} + e^{a_2 x + \varphi_2(t)} + \psi(t), \tag{4.36}$$

where

$$a_i = \pm B(C_3 - C_i), \qquad i = 1, 2;$$

$a_1 \neq a_2$ since $C_1 \neq C_2$.

By substituting the solution (4.36) into (4.32), we find the functions $\varphi_1(t)$, $\varphi_2(t)$, $\psi(t)$:

$$\begin{aligned}\varphi_i(t) &= -(B^2(C_3 - C_i)(C_3 + C_i - 2C_j))t + t_{0i},\\ t_{0i} &= \text{const}, \quad j = 3 - i, \quad i = 1, 2; \quad \psi(t) = \psi_0 = \text{const}.\end{aligned} \tag{4.37}$$

From (4.31), we find the function $g(x, t)$:

$$g = C_3\psi_0 + C_1 e^{a_1x+b_1t+t_{01}} + C_2 e^{a_2x+b_2t+t_{02}},$$

where ψ_0, a_i, b_i, $i = 1, 2$, were defined in (4.36), (4.37).

The obtained solution shows that the parameters a_i and b_i are determined by the choice of the root of equation $\Phi(u) = 0$ (4.29).

Introduce the notation

$$a_j^i = a_j(C_i), \qquad b_j^i = b_j(C_i), \qquad i, j = 1, 2, 3, \qquad i \neq j.$$

By fixing an arbitrary nonzero root C_k, $k = 1, 2, 3$, of $\Phi(u) = 0$ (4.39), we get from (4.36):

$$\begin{gathered} a_i^k = \pm B(C_k - C_i), \qquad b_i^k = -B^2(C_k - C_i)(C_k + C_i - 2C_j), \\ k, i, j = 1, 2, 3, \qquad k \neq i \neq j, \end{gathered} \tag{4.38}$$

and the two-phase solutions of the equation (4.1), (4.29) have the form (4.30) for $C_k \neq 0$.

Now consider the system of equations (4.11), i.e., the case $C = 0$. The function $\Phi(u)$ (4.22) has the form of a cubic polynomial with one zero root (coincides with (4.29) for $\beta = 0$). Equations (4.25), (4.26b), (4.27) coincide with equations (4.31), (4.32), (4.25) respectively if we put $C_3 = 0$ in the latter equation. Thus the system of equations (4.11) determines the solution of equation (4.1), (4.22) in the form (4.30) for $C_k = 0$. Thus formula (4.30) holds. The lemma is proved.

Lemma 4.5. *Formulas* (4.30) *for* $k = 1, 2, 3$ *determine the same two-phase solution of equation* (4.1).

Proof. Actually, if in formula (4.30) for $u_i(x, t)$ we divide the numerator and denominator by $\exp(\tau_i^k)$, then we get $u_k(x, t)$, $k \neq i$. Let us show this.

Obviously, by (4.38), we have

$$a_i^k = -a_k^i, \qquad b_i^k = -b_k^i. \tag{4.39}$$

It is easy to verify that

$$\begin{aligned} a_j^k &= a_j^i - a_k^i, \qquad b_j^k = b_j^i - b_k^i,\\ a_j^i - a_k^i &= \pm B(C_i - C_j - C_i + C_k) = \pm B(C_k - C_j) = a_j^k,\\ b_j^i - b_k^i &= -B^2\left[(C_i - C_j)(C_i + C_j - 2C_k) - (C_i - C_k)(C_i + C_k - 2C_j)\right]\\ &= -B^2(C_k - C_j)(C_k + C_j - 2C_i) = b_j^k. \end{aligned} \tag{4.40}$$

Dividing the numerator and denominator in (4.30) by $\exp(\tau_i^k)$ and taking into account (4.39), (4.40), we transform the obtained expression (without loss of generality, we set $t_{0i} = 0$):

$$\begin{aligned} u_i(x,t) &= \frac{[C_i + C_j \exp[a_k^i x + b_k^i t] + C_k \exp[a_k^i x + b_k^i t]] \exp[-(a_k^i x + b_k^i t)]}{[1 + \exp(\tau_j^i) + \exp(\tau_k^i)] \exp[-(\tau_k^i)]} \\ &= \frac{C_k + C_i \exp[-(a_k^i x + b_k^i t)] + C_j \exp[(a_j^i - a_k^i)x + (b_j^i - b_k^i)t]}{1 + \exp(\tau_j^i) + \exp(\tau_k^i)} \\ &= u_k(x,t). \end{aligned}$$

The lemma is proved.

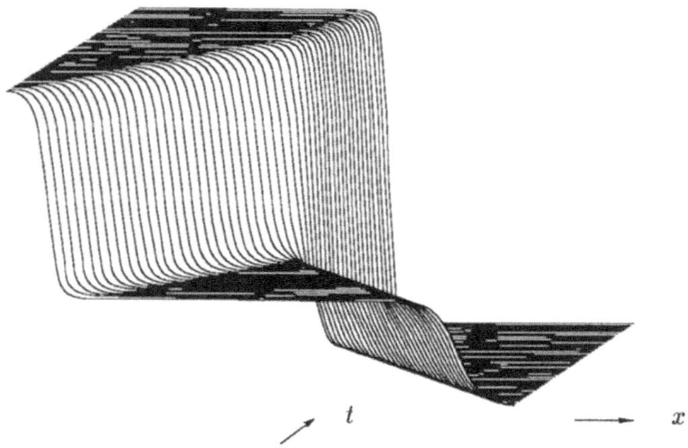

FIGURE 2.6

Example 4.1. Consider the equation

$$\frac{\partial u}{\partial t} - \frac{\partial^2 u}{\partial x^2} - 2(4-u)(2-u)(1-u) = 0. \tag{4.41}$$

In this case, $C_1 = 1$, $C_2 = 2$, $C_3 = 4$, $\mu = -2$, and the solution given by (4.30) is plotted in Figure 2.6 ($t = -4 \div -2.5$, $x = -20 \div 25$). It is clearly seen that after the interaction the velocity of the resulting wave decreases in contrast to the velocity of the wave coming from the left.

Example 4.2. Consider a particular case of equation (4.41)

$$\frac{\partial u}{\partial t} - \frac{\partial^2 u}{\partial x^2} - u(1-u^2) = 0, \qquad |u|\big|_{|x|\to\infty} \to 1. \tag{4.42}$$

This equation is called the Semenov equation or the Newell–Whithead equation [103].

This example has been examined earlier. Let us show that there exists a two-phase solution of equation (4.42) with singularities on a certain ray

$$u(x,t) = \frac{1+\exp(\tau_1)}{1-\exp(\tau_1)+\exp(\tau_2)}\Bigg|_{\substack{\tau_1=-\sqrt{2}x-C_1\\ \tau_2=(3t/2-x/\sqrt{2})-C_2}}$$

Actually, substitute a function of the form $u = G(x,t)/F(x,t)$ into the equation

$$F\left(\frac{\partial G}{\partial t}F - G\frac{\partial F}{\partial t}\right) - F\left(\frac{\partial^2 G}{\partial x^2}F - 2\frac{\partial G}{\partial x}\frac{\partial F}{\partial x} - G\frac{\partial^2 F}{\partial x^2}\right)$$
$$- 2FG\left(\frac{\partial F}{\partial x}\right)^2 - G(F^2 - G^2) = 0. \quad (4.43)$$

Choose the functions F and G in the form

$$F = 1 + hf(x,t), \qquad G = 1 + hg(x,t).$$

Substitute these expressions into (4.43) and set the coefficients at equal powers of the parameter h equal to zero. Then

$$h^1: \quad \left(\frac{\partial}{\partial t} - \frac{\partial^2}{\partial x^2} + 2\right)(g-f) = 0, \ldots, \quad (4.44)$$

$$h^2: \quad f\left(\frac{\partial}{\partial t} - \frac{\partial^2}{\partial x^2}\right)(g-f) + \left[\frac{\partial g}{\partial t}f - g\frac{\partial f}{\partial t} - \frac{\partial^2 g}{\partial x^2}f + 2\frac{\partial g}{\partial x}\frac{\partial f}{\partial x} + g\frac{\partial^2 f}{\partial x^2}\right]$$
$$- 2\left(\frac{\partial f}{\partial x}\right)^2 + 2g(g-f) + (g^2 - f^2) = 0, \quad (4.45)$$

$$h^3: \quad f\left[\frac{\partial g}{\partial t}f - g\frac{\partial f}{\partial t}\right] + f\left[-\frac{\partial^2 g}{\partial x^2}f + 2\frac{\partial g}{\partial x}\frac{\partial f}{\partial x} + g\frac{\partial^2 f}{\partial x^2}\right]$$
$$- 2g\left(\frac{\partial f}{\partial x}\right)^2 + g(g^2 - f^2) = 0. \quad (4.46)$$

By using (4.44), we transform (4.45) and obtain the equation

$$\frac{\partial f}{\partial t}(f-g) - \frac{\partial^2 f}{\partial x^2}(f-g) + 2\frac{\partial g}{\partial x}\frac{\partial f}{\partial x} - 2\left(\frac{\partial f}{\partial x}\right)^2 + 3(f-g)^2 = 0. \quad (4.47)$$

By using (4.44) and (4.47), we eliminate the time derivatives from (4.46) and get

$$\text{a)} \quad \sqrt{2}\frac{\partial f}{\partial x} - f = -g \qquad \text{or} \qquad \text{b)} \quad -\sqrt{2}\frac{\partial f}{\partial x} - f = -g. \quad (4.48)$$

Rewrite (4.44) in the form

$$\left(\frac{\partial}{\partial t} - \frac{\partial^2}{\partial x^2}\right)g = -2g, \qquad \left(\frac{\partial}{\partial t} - \frac{\partial^2}{\partial x^2}\right)f = -2f.$$

Then (4.45) implies

$$2(g-f)^2+(g^2-f^2)+2\left(\frac{\partial g}{\partial x}-\frac{\partial f}{\partial x}\right)\frac{\partial f}{\partial x}=0.$$

Equation (4.46) gives

$$2\left(f\frac{\partial g}{\partial x}-g\frac{\partial f}{\partial x}\right)\frac{\partial f}{\partial x}+g(g^2-f^2)=0.$$

Equation (4.47) implies

$$3(f-g)^2-2f(f-g)+2\left(\frac{\partial g}{\partial x}-\frac{\partial f}{\partial x}\right)\frac{\partial f}{\partial x}=0.$$

By combining the two last relations, we get (4.48) and one extra relation

$$\frac{\partial g}{\partial x}=\pm\sqrt{2}g,$$

which determines the function g:

$$g=-\exp\{\pm\sqrt{2}x+C(t)\}.$$

By substituting the latter expression into (4.44), we get

$$\alpha_1=\pm\sqrt{2},\qquad b_1=0,\qquad C\equiv 0.$$

Consider the solution of (4.44) of the form

$$g-f=C_1\exp(\tau_1)+C_2\exp(\tau_2),\qquad \tau_i=b_it+\alpha_ix,\quad i=1,2,\quad C_i=\text{const}.$$

Then, according to (4.44) and the latter relation, we get

$$f=\pm\frac{1}{\sqrt{2}}\left[\frac{C_1}{\alpha_1}\exp(\tau_1)+\frac{C_2}{\alpha_2}\exp(\tau_2)\right]+C_3,\qquad C_3=\text{const}. \tag{4.49}$$

We set the constants C_1 and C_2 equal to $C_i=\alpha_i\sqrt{2}$, $C_3=0$. Then, in cases a) and b) (see (4.48)), equation (4.47) takes the form :

$$\begin{aligned}&\text{a)}\qquad \frac{\partial f}{\partial t}-3\frac{\partial^2 f}{\partial x^2}+3\sqrt{2}\frac{\partial f}{\partial x}=0,\\ &\text{b)}\qquad \frac{\partial f}{\partial t}-3\frac{\partial^2 f}{\partial x^2}-3\sqrt{2}\frac{\partial f}{\partial x}=0,\end{aligned} \tag{4.50}$$

and the constants α_i, b_i are determined by the system

$$\begin{aligned}&\text{a)}\qquad b_i-\alpha_i^2+2=0,\qquad b_i-3\alpha_i^2+3\sqrt{2}\alpha_i=0,\\ &\text{b)}\qquad b_i-\alpha_i^2+2=0,\qquad b_i-3\alpha_i^2-3\sqrt{2}|\alpha_i|=0,\end{aligned}$$

and take the values

$$\begin{array}{lllll} \text{a)} & b_1 = 0, & b_2 = -3/2, & \alpha_1 = +\sqrt{2}, & \alpha_2 = +1/\sqrt{2}; \\ \text{b)} & b_1 = 0, & b_2 = -3/2, & \alpha_1 = -\sqrt{2}, & \alpha_2 = -1/\sqrt{2}. \end{array}$$

This solution was considered by Newell [103]. By choosing the sign minus in (4.49), one can construct a two-phase solution with discontinuity of the second kind:

$$u(x,t) = \frac{1+\exp(\tau_1)}{1-\exp(\tau_1)+\exp(\tau_2)}\Bigg|_{\substack{\tau_1=-\sqrt{2}x/\varepsilon - C_1 \\ \tau_2=(3t/2-x/\sqrt{2})/\varepsilon - C_2}} \tag{4.51}$$

Note that in Lemmas 4.3 and 4.4 the coefficient a in the exponential takes two opposite values (the functions (4.17), (4.24), (4.28), (4.30)). This is due to the invariance of equation (4.1) under the change $x \to -x$. Therefore, further we shall use only one value.

Let us return to equation (4.1), (4.29). The two-phase solution (4.30) of (4.1), (4.29) was obtained under the condition that the cubic polynomial (4.29) has no multiple roots.

Now consider the case when the function $\Phi(u)$ (4.29) has multiple roots.

Lemma 4.6. *Let equation* $\Phi(u) = 0$ (4.29) *possess two different roots, then equation* (4.1), (4.29) *admits a solution of the form:*

$$u(x,t) = U(\tau_2,\tau_2) = \frac{A_1 + C_1\exp(\tau_1) + C_2\exp(\tau_2)}{A_2 + \tau_1 + \exp(\tau_2)},$$

where $\tau_i = a_i x + b_i t + t_{0i}$, $i = 1,2$, C_i *are zeros of the function* $\Phi(u)$ (*the multiplicity of* C_1 *is 2*, $C_3 = C_1$), p_0, t_{0i} *are arbitrary constants.*

Proof. Here we can use the previous calculations. Since

$$\frac{d\Phi}{du}\bigg|_{u=C_1} = \gamma + 2\lambda C_1 + 3\mu C_1^2 = 0,$$

the equation (4.45) admits a solution of the form ($C_3 = C_1$):

$$f = p(t)x + q(t) + \exp(ax + bt + t_0). \tag{4.52}$$

By (4.32), (4.34), we find the unique values of the coefficients a and b:

$$a = \mp\frac{3}{2}B(C_1 - C_2), \qquad b = a^2, \qquad B = \sqrt{-\mu/2}.$$

Substituting (4.52) into equation (4.32), (4.34), we determine the functions $p(t)$ and $q(t)$ from the system

$$p'(t) = 0, \qquad q'(t) + 3B(C_1 - C_2)p(t) = 0,$$

and hence

$$p = p_0 = \text{const}, \qquad q = -2ap_0t + q_0,$$

where $q_0 = \text{const}$.

Equation (4.31) implies the function $g(x,t)$:

$$g = C_1q_0 - \frac{p_0}{B} + C_1p_0(x - 2at) + C_2\exp(a(x + at) + t_0).$$

Then the solution of equation (4.1), (4.23) (for $\Phi(u)$ with two different roots) has the form

$$u(x,t) = \frac{C_1q_0 - p_0/B + C_1p_0(x - 2at) + C_2\exp(a(x + at) + t_0)}{q_0 - p_0(x - 2at) + \exp(a(x + at) + t_0)}, \tag{4.53}$$

where $q^0 = 1 + q_0$.

If we introduce the notation

$$\tau_1 = p_0(x - 2at), \qquad \tau_2 = a(x + at) + t_0, \qquad A = C_1q^0 - \frac{p_0}{B},$$

formula (4.53) can be rewritten in the form

$$u(x,t) = U(\tau_2,\tau_2) = \frac{A_1 + C_1\exp(\tau_1) + C_2\exp(\tau_2)}{q^0 + \tau_1 + \exp(\tau_2)},$$

The lemma is proved.

Thus, for multiple roots of the equation $\Phi(u) = 0$ (4.29), the solution of equation (4.1), (4.29) remains to be a two-phase solution, but sharply changes its structure, this solutions ceases to be continuous and cannot be expanded into a Dirichlet series.

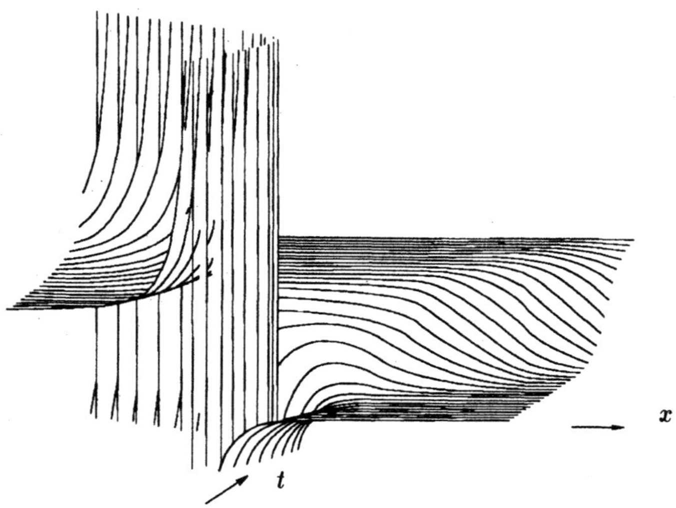

FIGURE 2.7

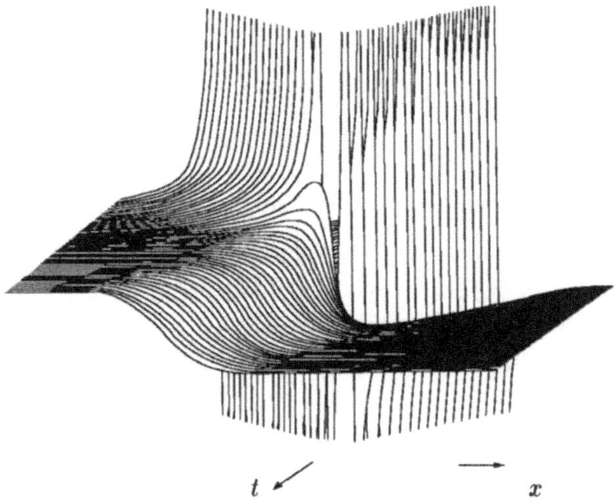

FIGURE 2.8

Example 4.3. For $C_1 = C_3 = 2$, $C_2 = 1$, $\mu = -2$, $a_+ = 1$, the solution u_+ given by (4.53) is plotted in Figure 2.7 ($t = -10 \div 5$, $x = -10 \div 10$).

Example 4.4. For $C_1 = C_3 = 2$, $C_2 = 1$, $\mu = -2$, $a_- = 1$, the solution u_- is plotted in Figure 2.8.

The motion of singularities (the curves of discontinuity) is shown in Figure 2.9 ($u = u_+$) and in Figure 2.10 ($u = u_-$)

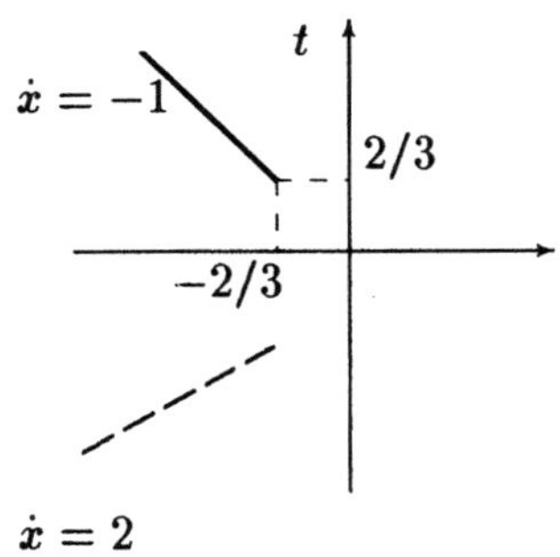

FIGURE 2.9

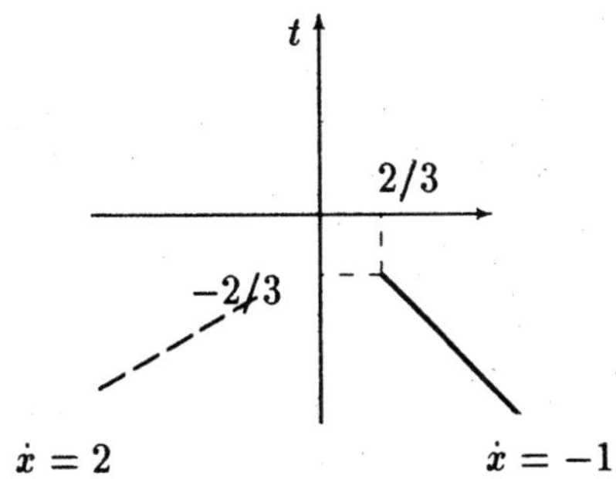

FIGURE 2.10

Lemma 4.7. *Let equation $\Phi(u) = 0$ (4.29) have a single root of multiplicity 3. Then the solution of equation (4.1), (4.29) has the form*

$$u(x,t) = C - \frac{1}{B}\frac{2p_0x + q_0}{p_0x^2 + q_0x + 6p_0t + r^0}, \tag{4.54}$$

where $B = \sqrt{-\mu/2}$, p_0, q_0, r^0 are arbitrary constants, C is the root of the equation $\Phi(u)x = 0$ of multiplicity 3.

Proof. Since the root C is multiple, the first and second derivatives of the function $\Phi(u)$ (4.29) are zero at the point $u = C$:

$$\left.\frac{d\Phi}{du}\right|_{u=C} = \gamma + 2\lambda C + 3\mu C^2 = 0, \qquad \left.\frac{1}{2}\frac{d^2\Phi}{du^2}\right|_{u=C} = \lambda + 3\mu C = 0, \tag{4.55}$$

where $\beta = -\mu C^3$, $\gamma = 3\mu C^2$, $\lambda = -3\mu C$.

By (4.55), equation (4.35) reduces to the form $f_{xxx} = 0$. This implies

$$f = p(t)x^2 + q(t) + r(t). \tag{4.56}$$

Substituting the solution (4.56) into equation (4.32), we get the functions $p(t)$, $q(t)$ and $r(t)$ from the system of equations

$$p'(t) = 0, \qquad q'(t) = 0, \qquad r'(t) = 6p(t).$$

This yields

$$p = p_0, \qquad q = q_0, \qquad r = 6p_0t + r_0.$$

Thus the function $f(x,t)$ has the form

$$f(x,t) = p_0x^2 + q_0x + 6p_0t + r_0,$$

where p_0, q_0, r_0 are constants of integration. Equation (4.31) determines the function $g(x,t)$

$$g = p_0Cx^2 + (q_0C - 2p_0/B)x + 6p_0C + Cr_0 - q_0/B.$$

This gives the solution (4.54).

For $t > t_0 = (q_0^2 - 4p_0r_0)/24p_0^2$, the function $u(x,t)$ defined by (4.54) is a continuous function, namely, a "rational two-phase soliton" with phases

$$\tau_1 = 2p_0x + q_0, \qquad \tau_2 = q_0(1 - p_0)x + 6p_0t.$$

The solution (4.54) can be represented in the form

$$u(x,t) = U(\tau_2, \tau_2) = C - \frac{1}{B}\frac{\tau_1}{q^0 + (\tau_1/2)^2 + \tau_2},$$

where $q^0 = r^0 - q_0^2/14$. The lemma is proved.

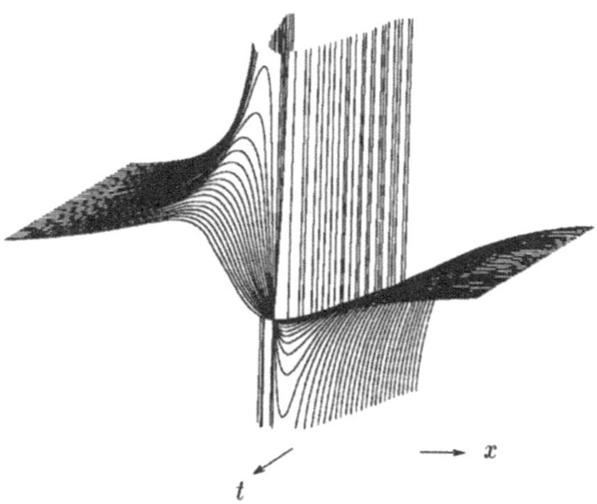

FIGURE 2.11

Example 4.5. For $C_1 = C_3 = 2$, $C_2 = 1$, $\mu = -2$, the solution given by (4.54) is plotted in Figure 2.11 ($t = -1 \div 0.5$, $x = -9 \div 9$) and in Figure 2.12 ($t = -4 \div 15$, $x = -9 \div 9$), the rear view. The singularity of the solution disappears for $t > -1/8$.

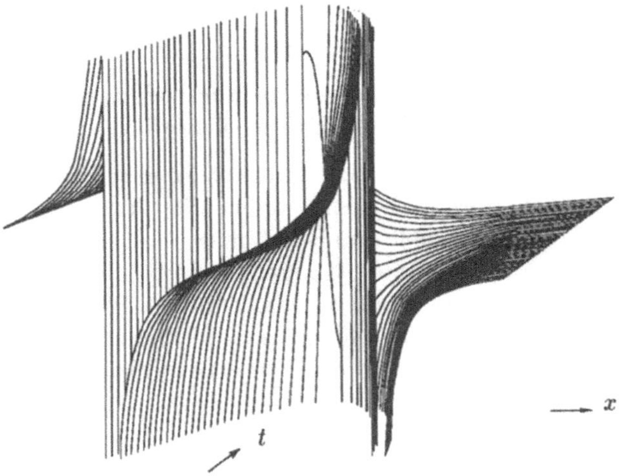

FIGURE 2.12

Now let us consider some applications of these results. Since the choice of parameters of the function $\Phi(u)$ (4.29) is arbitrary, complex solutions may exist for certain definite values of coefficients of this function. Represent the coefficients β, γ, λ, μ of this function in the form

$$\beta = \beta_R + i\beta_I, \qquad \gamma = \gamma_R + i\gamma_I, \qquad \lambda = \lambda_R + i\lambda_I, \qquad \mu = \mu_R + i\mu_I. \tag{4.57}$$

Then the solution of equation (4.1), (4.29) will be sought in the form

$$u(x,t) = v(x,y) + iw(x,t), \tag{4.58}$$

where $v(x,t)$ and $w(x,t)$ are real-valued functions.

Substituting (4.57) and (4.58) into equation (4.1), (4.29) and taking the real and imaginary parts, we get the system of equations

$$\begin{aligned} v_t - v_{xx} - \chi(v,w) &= 0, \\ w_t - w_{xx} - \Omega(v,w) &= 0, \end{aligned} \tag{4.59}$$

where

$$\begin{aligned} \chi(v,w) &= \operatorname{Re}\Phi(u) = \beta_R + v(\gamma_R + \lambda_R v + \mu_R v^2) \\ &\quad - w(\gamma_I + \lambda_R w + \mu_I w^2) - vw(2\lambda_I + 3\mu_I w + 3\mu_R w), \\ \Omega(v,w) &= \operatorname{Im}\Phi(u) = \beta_I + v(\gamma_I + \lambda_I v + \mu_I v^2) \\ &\quad - w(-\gamma_R + \lambda_I w + \mu_R w^2) + vw(2\lambda_R + 3\mu_R v + 3\mu_I w), \\ |\mu| &= \mu_R^2 + \mu_I^2 \neq 0. \end{aligned}$$

The solution of system (4.59) has the form

$$v = \operatorname{Re} u, \qquad w = \operatorname{Im} u,$$

where u is defined by (4.30), (4.53), (4.54). The choice of parameters (4.57) of the function $\Phi(u)$ (4.29) allows to find solutions of a large class of systems of equations of the form (4.59). In the same way, equation (4.1) (4.6) can be reduced to a similar system of equations.

2.5. Relation between the number of phases in the solution and the degree of nonlinearity

In the previous section we presented some formulas for exact two-phase solutions of semilinear parabolic equations with cubic nonlinearity. In this section we shall study the relation between the possible number of phases and the number of roots of the function $\Phi(u)$ (4.2), which determines the nonlinearity of the equation. We shall prove that, for polynomials of degree three, there exist no real N-phase solution ($N \geqslant 3$), which take values in the strip between the roots of the polynomial $\Phi(u)$ (4.29) and can be represented as an entire function of the ratio of exponential polynomials:

$$u = U\left(\frac{\varphi(\exp(\tau_1), \ldots, \exp(\tau_N))}{\psi(\exp(\tau_1), \ldots, \exp(\tau_N))}\right), \tag{5.1}$$

where $U(z)$ is an entire function, $\varphi(y_1, \ldots, y_N)$, $\psi(y_1, \ldots, y_N)$ are polynomials, $\tau_i = a_i x + b_i t$, $a_i, b_i = \text{const}$, $i-1, \ldots, N$. It is assumed that the phases τ_i are independent, i.e., linear combinations of phases with integer coefficients do not vanish.

Let us prove two preliminary lemmas.

Lemma 5.1. *If the function*

$$u(x,t) = V(\tau_1, \ldots, \tau_N), \tag{5.2}$$

where $\tau_i = a_i x + b_i t$, $i = 1, \ldots, N$, *satisfy equation* (4.1), *then the function* $\widehat{u}(x,t) = V(\tau_1 + C_1, \ldots, \tau_N + C_N)$, *where* C_i, $i = 1, \ldots, N$ *are arbitrary constants, also satisfies equation* (4.1).

Proof. If the function $V(\tau_1, \ldots, \tau_N)$ (5.2) is a solution of equation (5.1), then V satisfies the following relation

$$\sum_{k=1}^{N} b_k \frac{\partial V}{\partial \tau_k} - \sum_{\substack{k=2 \\ i+j=k \\ i\geqslant 1,\, j\geqslant 1}}^{2N} a_i a_j \frac{\partial^2 V}{\partial \tau_i \partial \tau_j} - \Phi(V) = 0. \tag{5.3}$$

Denote $\xi_i = \tau_i + C_i$. By substituting the function $\widehat{u}(x,t)$ (5.3) into equation (5.1), we obtain the relations

$$\begin{aligned} \widehat{u}_t - \widehat{u}_{xx} - \Phi(\widehat{u}) &= \sum_{k=1}^{N} b_k \frac{\partial V}{\partial \xi_k} - \sum_{\substack{k=2 \\ i+j=k \\ i\geqslant 1,\, j\geqslant 1}}^{2N} a_i a_j \frac{\partial^2 V}{\partial \xi_i \partial \xi_j} - \Phi(V) \\ &= \sum_{k=1}^{N} b_k \frac{\partial V}{\partial \tau_k} - \sum_{\substack{k=2 \\ i+j=k \\ i\geqslant 1,\, j\geqslant 1}}^{2N} a_i a_j \frac{\partial^2 V}{\partial \tau_i \partial \tau_j} - \Phi(V) = 0. \end{aligned} \tag{5.4}$$

The lemma is proved.

Lemma 5.2. *The following relations hold*

$$1) \quad \alpha_i + \alpha_j < 0, \qquad i,j = 1,2,3, \qquad i \neq j,$$
$$2) \quad \alpha_1\alpha_2 + \alpha_2\alpha_3 + \alpha_3\alpha_1 = 0,$$

where $\alpha_k = d\Phi/du\big|_{u=C_k}$, $k = 1,2,3$.

Proof. We represent the function $\Phi(u)$ in the form

$$\Phi(u) = -\mu(C_1 - u)(C_2 - u)(C_3 - u), \tag{5.5}$$

where $\mu < 0$, and calculate α_k from (5.4) and (5.5):

$$\alpha_k = \mu(C_i - C_k)(C_j - C_k); \qquad k,i,j, = 1,2,3; \quad k \neq i \neq j. \tag{5.6}$$

Substituting (5.6) into (5.4), we prove the lemma.

The main assertion of this section is the following theorem:

Theorem 5.1. *There does not exist any solution of equation* (4.1), (5.5) *of the form* (5.1) *for* $N \geqslant 3$.

Proof. We suppose that a solution of the form (5.1) exists, then, by Lemma 5.1, the function

$$\widehat{u} = U\left(\frac{\varphi(\exp(\tau_1 + C_1), \ldots, \exp(\tau_N + C_N))}{\psi(\exp(\tau_1 + C_1), \ldots, \exp(\tau_N + C_N))}\right)$$

satisfies equation (4.1), (5.5). Passing to the limit as $C_i \to -\infty$, $i = 1, \ldots, k-1, k+1, \ldots, N$, we see that N functions

$$\widehat{u}_k = U\left(\frac{\varphi(0, \ldots, \exp(\tau_k + C_k), \ldots, 0)}{\psi(0, \ldots, \exp(\tau_k + C_k), \ldots, 0)}\right)$$

are one-phase solutions of equation (4.1), (4.5) taking values in the strip region between the roots of the polynomial $\Phi(u)$. Since, by Lemma 5.2, the polynomial $\Phi(u)$ of degree three satisfies the conditions of Lemma 4.4, then, according to this lemma, there exist not more than three one-phase solutions of equation (4.1), (5.5). Thus, $N \leqslant 3$, i.e., there exist not more than three independent phases τ_1, τ_2, τ_3.

Let us show that there are only two independent phases. For each of them we write the dispersion relations:

$$\begin{aligned}
&1) \quad a_1^2 - b_1 + \alpha_1 = 0, \qquad a_1^2 + b_1 + \alpha_3 = 0,\\
&2) \quad a_2^2 - b_2 + \alpha_2 = 0, \qquad a_2^2 + b_2 + \alpha_1 = 0,\\
&3) \quad a_3^2 - b_3 + \alpha_2 = 0, \qquad a_3^2 + b_3 + \alpha_3 = 0.
\end{aligned}$$

These relations are based on the invariance of equation (4.1) with respect to reflection $x \to -x$. Thus we can find the values of the phase constants a_i, b_i, $i = 1,2,3$,

$$
\begin{aligned}
b_1 &= \frac{\alpha_1 - \alpha_3}{2}, & a_1^2 &= -\frac{\alpha_1 + \alpha_3}{2},\\
b_2 &= \frac{\alpha_2 - \alpha_1}{2}, & a_2^2 &= -\frac{\alpha_2 + \alpha_1}{2},\\
b_3 &= \frac{\alpha_2 - \alpha_3}{2}, & a_3^2 &= -\frac{\alpha_2 + \alpha_3}{2}.
\end{aligned}
\tag{5.7}
$$

Relations (5.7) imply that $b_3 = b_1 + b_2$. Let us show that $a_3 = a_1 + a_2$.
By Lemma 5.2, we have

$$\alpha_1 = \sqrt{\alpha_1^2 + \alpha_1\alpha_2 + \alpha_2\alpha_3 + \alpha_3\alpha_1} = \sqrt{(\alpha_1 + \alpha_2)(\alpha_1 + \alpha_3)} = 2a_1a_2.$$

Then

$$0 = -\frac{\alpha_2 + \alpha_1}{2} - \frac{\alpha_1 + \alpha_3}{2} + \alpha_1 + \frac{\alpha_2 + \alpha_3}{2} = a_2^2 + a_1^2 + 2a_1a_2 - a^3 = 0.$$

This yields

$$\tau_3 = \tau_2 + \tau_1,$$

where $\tau_i = a_i x + b_i t$, $i = 1,2,3$. The theorem is proved.

2.6. Asymptotics of wave creation for the KPP-Fisher equation

Suppose the function $u(x,t) = \chi(x + pt)$ is a self-similar solution of runnig wave type satisfying equation (1.2) (the function χ is the solution of the boundary-value problem (1.3), (1.4a)) (see Figure 2.1a). Then since equation (1.2) is invariant under the change of variables $x \to -x$, the function $u(x,t) = \chi(-x + pt + L)$, $L = \text{const}$, is the solution of equation (1.2) satisfying the boundary conditions (1.4b) (see Figure 2.1b).

The attempts to construct solutions describing the interaction of such waves in the form (5.1) (shown in Figures 2.4a, 2.4b) were unsuccessful. Here we derive the asymptotic formula given in Section 2.1, which describes this interaction. We shall consider only the waves for which exact formulas are known. For the KPP-equation with nonlinearity of the form $u(1-u)$, this is a unique pair of waves:

$$u_\pm(x,t) = (1 + \exp(-\tau_\pm/\sqrt{6}))^{-2},$$

corresponding to the phases $\tau_\pm = \pm x + \frac{5}{\sqrt{6}}t + L^\pm$, $L^+ = 0$, $L^- = L$.

As mentioned above, a natural ansatz U for solutions describing the wave generation is the product of self-similar solutions. Such solutions satisfy the equation with precision up to $\exp(-L)U$, but the inversion of the operator on this discrepancy gives a "correction" to the ansatz U. This is the effect of the solution instability with respect to (exponentially) small perturbations mentioned in [21].

To construct formulas describing the wave interaction, we represent the solution in the form of a Dirichlet series as $\tau_2 \to +\infty$

$$u(x,t) = \chi(\tau_1, \tau_2) = \chi_0(\tau_1) + \exp(-\gamma_1 \tau_2)\chi_1(\tau_1) + \exp(-\gamma_2 \tau_2)\chi_2(\tau_1) + \dots, \tag{6.1}$$

where

$$\chi_0(\tau_1) = \left(1 + \exp(-\tau_1/\sqrt{6})\right)^{-2},$$
$$\tau_1 = x + \frac{5}{\sqrt{6}}t, \qquad \tau_2 = -x + \frac{5}{\sqrt{6}}t + L; \qquad \gamma_2 > \gamma_1 = 1/6.$$

Here we set $L^+ = 0$, $L^- = L$. Equation (1.2) in variables τ_+, τ_- has the form,

$$-\frac{5}{\sqrt{6}}\left(\frac{\partial}{\partial \tau_+} + \frac{\partial}{\partial \tau_-}\right)\chi + \left(\frac{\partial^2}{\partial \tau_+^2} - 2\frac{\partial^2}{\partial \tau_+ \partial \tau_-} + \frac{\partial^2}{\partial \tau_+^2}\right)\chi + \chi(1-\chi) = 0. \tag{6.2}$$

Substituting the series (6.1) into this equation and collecting the terms at equal powers of $\exp(-\gamma_1 \tau_2)$, we get

$$-\frac{5}{\sqrt{6}}\frac{d}{d\tau_1}\chi_0 + \frac{d^2}{d\tau_1^2}\chi_0 + \chi_0(1-\chi_0) = 0, \tag{6.3a}$$

$$\exp(-\tau_2/\sqrt{6}): \quad \frac{d^2}{d\tau_1^2}\chi_1 - \frac{3}{\sqrt{6}}\frac{d}{d\tau_1}\chi_1 + 2(1-\chi_0)\chi_1 = 0. \tag{6.3b}$$

The function χ_0 (6.1) satisfies equation (6.3a) identically. Since the function χ_0 is monotonic and invariant under displacements, we can replace the variable τ_1 by a new independent variable χ_0 (similar transformations in a different situation were considered in [8]):

$$\begin{aligned} \varphi(\chi_0) &= \frac{d\chi_0}{d\tau_1} = \frac{2}{\sqrt{6}}\chi_0(1-\sqrt{\chi_0}), \\ \frac{d}{d\tau_1} &= \varphi\frac{d}{d\chi_0}, \qquad \frac{d^2}{d\tau_1^2} = \varphi^2\frac{d^2}{d\chi_0^2} + \varphi\frac{d\varphi}{d\chi_0}\frac{d}{d\chi_0}. \end{aligned} \tag{6.4}$$

In new variables equation (6.3a) has the form

$$-\frac{5}{\sqrt{6}}\varphi + \varphi\frac{d\varphi}{d\chi_0} + \chi_0(1-\chi_0) = 0. \tag{6.5}$$

Substituting (6.4) into equation (6.3b) and taking into account (6.5), we get

$$\varphi^2\frac{d^2\chi_1}{d\chi_0^2} + \left(\frac{2}{\sqrt{6}}\varphi - \chi_0(1-\chi_0)\right)\frac{d\chi_1}{d\chi_0} + 2(1-\chi_0)\chi_1 = 0. \tag{6.6}$$

By introducing a new variable $z = \sqrt{\chi_0}$ in this equation, we obtain an ordinary differential equation of the Riemann type

$$z^2(1-z)\frac{d^2\chi_1}{dz^2} - 2z(1+z)\frac{d\chi_1}{dz} + 12(1+z)\chi_1 = 0, \tag{6.7}$$

whose solution can be expressed in terms of hypergeometric functions.

We shall change the variables:

$$\chi_1 = -2z^{3/2}\Psi(z).$$

Substituting the function $\chi_1(z)$ into equation (6.7), we get the following equation for the function $\Psi(z)$

$$4z^2(z-1)\Psi'' + 4(5z-1)z\Psi' - 3(13+11z)\Psi = 0. \tag{6.8}$$

We have the following asymptotic behavior of the solution of (6.8) (as $z \to 0$):

$$\Psi(z) = z^{\pm i\sqrt{39}/2}(1+O(z)), \qquad z \to 0. \tag{6.9}$$

The general solution is defined by the formula:

$$\begin{aligned}\Psi(z) = C_+ z^{i2\sqrt{2}} F(i2\sqrt{2},\ 8+i2\sqrt{2};\ 1+i4\sqrt{2};\ z)\\ + C_- z^{-i2\sqrt{2}} F(-i2\sqrt{2},\ 8-i2\sqrt{2};\ 1-i4\sqrt{2};\ z),\end{aligned}$$

where $F(\alpha,\beta;\gamma;z)$ is a hypergeometric function, C_+ and C_- are constant.

Recall that we consider the real solutions of equation (1.2) and hence the function χ_1 has the form

$$\chi_1(\chi_0) = -2\chi_0^{3/4}\,\mathrm{Re}\,\Psi(\sqrt{\chi_0}) = -2\chi_0^{3/4}\psi(\sqrt{\chi_0}), \tag{6.10}$$

where $\psi(z) = \mathrm{Re}\,\Psi(z)$ is a real solution of equation (6.8). The asymptotic behavior of the function $\chi_1(\chi_0)$ as $\chi_0 \to 0$ ($\tau_1 \to -0$) is defined by formula (6.9):

$$\chi_1(\chi_0) = \chi_0^{3/4}[C_1\cos(\sqrt{2}\ln\chi_0) + C_2\sin(\sqrt{2}\ln\chi_0) + O(\chi_0^{1/4})], \qquad \chi_0 \to 0.$$

The boundary conditions imposed on the function χ_1 (and hence on the function $\psi(z)$) can be defined by expanding the ansatz U into a Dirichlet series in terms of $\exp(-\gamma_1\tau_2)$:

$$\begin{aligned}U(\tau_1,\tau_2) &= \big(1+\exp(-\tau_1/\sqrt{6})\big)^{-2}\big(1+\exp(-\tau_2/\sqrt{6})\big)^{-2}\\ &= \big(1+\exp(-\tau_1/\sqrt{6})\big)^{-2}\\ &\quad - 2\big(1+\exp(-\tau_1/\sqrt{6})\big)^{-2}\exp(-\tau_2/\sqrt{6}) + O\big(\exp(-2\tau_2/\sqrt{6})\big)\\ &= \chi_0(\tau_1) - 2\chi_0(\tau_1)\exp(-\tau_1/\sqrt{6}) + O\big(\exp(-2\tau_2/\sqrt{6})\big), \qquad \tau_2 \to +\infty.\end{aligned}$$

In the domain behind the wave front, the influence of small perturbations (see [21]) on the behavior of waves is weak, thus the solution must behave in the same way as the anzatz. This yields the boundary conditions for the function χ_1 as $\tau_1 \to +\infty$:

$$\begin{aligned}\chi_1\big|_{\tau_1\to+\infty} &= \chi_1\big|_{\chi_0=1} = -2,\\ \chi_1\big|_{\tau_1\to-\infty} &= \chi_1\big|_{\chi_0=0} = 0.\end{aligned}$$

Thus we get the conditions for the function $\psi(z)$:

$$\psi(1) = 1, \qquad |\psi(0)| < \infty.$$

Formulas (6.1) and (6.10) describe the asymptotic behavior of $\chi(\tau_1, \tau_2)$ as $\tau_2 \to +\infty$:

$$\begin{aligned}
\chi(\tau_1, \tau_2) &= \chi_0(\tau_1) + \chi_1(\tau_1) e^{-\tau_2/\sqrt{6}} (1 + O(\exp(-\tau_2/\sqrt{6}))) \\
&= \chi_0(\tau_1)[1 - 2\chi^{-1/4}\psi(\sqrt{\chi_0(\tau_1)}) e^{-\tau_2/\sqrt{6}}](1 + O(\exp(-\tau_2/\sqrt{6}))) \\
&= (1 + e^{-\tau_1/\sqrt{6}})^{-2}(1 + \chi^{-1/4}\psi(\sqrt{\chi_0(\tau_1)}) e^{-\tau_2/\sqrt{6}})^{-2} \\
&\quad \times (1 + O(\exp(-\tau_2/\sqrt{6}))), \qquad \tau_2 \to +\infty.
\end{aligned}$$

This asymptotic expansion was obtained expanding the function $\chi(\tau_1, \tau_2)$ with respect to the self-similar solution $\chi_0(\tau_1)$ of equation (1.2) (the left-hand wave). Since the ansatz V is symmetric, then, by expanding the function $\chi(\tau_1, \tau_2)$ with respect to the solution $\chi_0(\tau_2)$ of equation (1.2) (the right-hand wave), we obtain the following asymptotic expansion of $\chi(\tau_1, \tau_2)$:

$$\begin{aligned}
\chi(\tau_1, \tau_2) &= (1 + \chi^{-1/4}\psi(\sqrt{\chi_0(\tau_2)}) e^{-\tau_1/\sqrt{6}})^{-2}(1 + e^{-\tau_2/\sqrt{6}})^{-2} \\
&\quad (1 + O(\exp(-\tau_2/\sqrt{6}))), \qquad \tau_1 \to +\infty.
\end{aligned}$$

By combining the obtained formulas, we get the asymptotic solution $\chi(\tau_1, \tau_2)$ with respect to the parameter L (the distance between the fronts):

$$\begin{aligned}
\chi(\tau_1, \tau_2) &= (1 + \chi^{-1/4}\psi(\sqrt{\chi_0(\tau_2)}) e^{-\tau_1/\sqrt{6}})^{-2} \\
&\quad \times (1 + \chi^{-1/4}\psi(\sqrt{\chi_0(\tau_1)}) e^{-\tau_2/\sqrt{6}})^{-2}(1 + O(e^{-L/\sqrt{6}})),
\end{aligned}$$

where N is an arbitrary number. Thus the formula presented in Section 3.1 is proved.

CHAPTER III

SINGULARITIES OF NONSMOOTH SOLUTIONS TO QUASILINEAR PARABOLIC AND HYPERBOLIC EQUATIONS

3.1. Main definitions

The aim of this chapter is to introduce the main notions for the construction of localized asymptotic solutions to quasilinear parabolic equations with a small parameter at the derivatives (i.e., solutions that satisfy the equations up to sufficiently smooth functions tending to zero as a sufficiently great power of a small parameter).

Let us consider the quasilinear parabolic equation with a small parameter at the derivatives:

$$Lu = \varepsilon\frac{\partial u}{\partial t} - \varepsilon^2 \operatorname{div}\big(K(x,t,u)\nabla u\big) + \varepsilon\sum_{j=1}^{n}\frac{\partial\varphi(x,t,u)}{\partial x_j} + F(x,t,u) = 0, \tag{1.1}$$

where $K(x,t,u)$, $\varphi(x,t,u)$, $F(x,t,u) \in C^\infty$ for $u > 0$, $K > 0$ for $u > 0$; $K \sim u^{k-1}$, $k > 1$, $F \sim u^q$, $q > 0$, $\varphi \sim u^m$, $m > 0$ for $u \to 0$; $0 < \varepsilon \ll 1$ is a small parameter.

Definition 1.1. A nonnegative function $u(x,t,\varepsilon)$ defined for $x \in \mathbf{R}^n$, $t \in \times[0,T]$, is called a *localized asymptotic (up to $O(\varepsilon^N)$) solution (synerget)* to equation (1.1) if it satisfies the following conditions:

1. the function $u(x,t,\varepsilon)$ vanishes for all ε on the closure of a certain domain in $\mathbf{R}^n \times [0,T]$;

2. the maximum in x, t of the function $u(x,t,\varepsilon)$ is bounded below by a positive constant independent of ε;

3. the function $u(x,t,\varepsilon)$ is smooth for $(x,t,\varepsilon) \in \overset{\circ}{\operatorname{supp}}\, u$ and

$$Lu = \varepsilon^N Q_N(x,t,\varepsilon)$$

in the domain $\overset{\circ}{\operatorname{supp}}\, u$, the function $Q_N(x,t,\varepsilon)/u(x,t,\varepsilon)$ is uniformly bounded in this domain;

4. (the flow continuity condition): for $(x,t,\varepsilon) \in \overset{\circ}{\operatorname{supp}}\, u$, $(x,t,\varepsilon) \to (x_0,t_0,\varepsilon_0) \in \Gamma$ ($\Gamma = \operatorname{supp} u \setminus \overset{\circ}{\operatorname{supp}}\, u$),

$$\lim K(x,t,u)\nabla u = 0. \tag{1.2}$$

3.2. Asymptotic solutions bounded as $\varepsilon \to 0$

Example 2.1. Consider the quasilinear parabolic equation:

$$\varepsilon\frac{\partial u}{\partial t} - \varepsilon^2\frac{\partial^2(u^{2-q})}{\partial x^2} - m\frac{u^q}{(1-q)^2}\left(1-u^{2(1-q)}\right)\left[\frac{(1-q)b-2+q}{m} + u^{2(1-q)}\right] = 0,$$
$$1 > q > 0, \qquad b > \frac{2-q}{1-q}, \qquad m = (3-2q)(2-q). \tag{2.1}$$

The function

$$u(x,t,\varepsilon) = \begin{cases} \left[\tan\left(\frac{x+bt-x_0}{\varepsilon}\right)\right]^{1/(1-q)}, & x - x_0 + bt \geqslant 0, \\ 0, & x - x_0 + bt < 0, \end{cases} \tag{2.2}$$

is the solution of equation (2.1). For $\varepsilon = 1$ and $\varepsilon \ll 1$ the solution is plotted in Figure 3.1 (curves 1 and 2 respectively). One can see that the function (2.2) vanishes for $x \leqslant -bt + x_0$. There is a region of width of the order ε, where the solution varies from zero to magnitudes exponentially close to 1. Moreover, $\partial(u^{2-q})/\partial x\big|_{x=-bt+x_0} = 0$. This coincides with the flow continuity condition (1.2). Thus the solution (2.2) satisfies all assumptions listed in Definition 1.1.

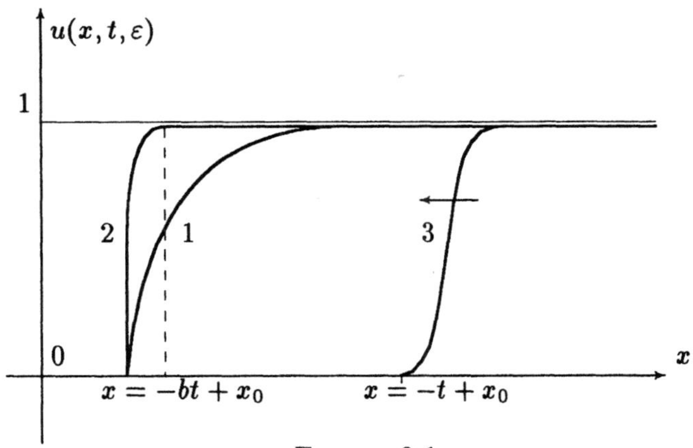

FIGURE 3.1

It is easy to see that,

$$u(x,t,\varepsilon) = \left(\frac{x+bt-x_0}{2\varepsilon}\right)^{1/(1-q)} + O\left(\left(\frac{x+bt-x_0}{2\varepsilon}\right)^{1/(1-q)+1}\right),$$

as $x - x_0 + bt \to 0_{+0}$, which means that the solution has a singularity of branching type with exponent $\alpha = 1/(1-q)$ in a neighborhood of the boundary of its support.

In this case, comparing equations (2.1) and (1.1), we get

$$K(x,t,u) = ku^{k-1}, \qquad k = 2-q, \qquad \varphi(x,t,u) \equiv 0,$$
$$F(u) = m\frac{\left(1-u^{2(1-q)}\right)}{(1-q)^2}\left[\frac{b(1-q)-2+q}{m} + u^{2(1-q)}\right]u^q.$$

Obviously, the functions $u \equiv 0$ and $u \equiv 1$ satisfy equation (2.1) since $F(0) = F(1) = 0$. The equation $F(u) = 0$ has the root $u = 1$, hence there exist solutions bounded as $\varepsilon \to 0$.

Example 2.2. The support of the solution considered in Example 2.1 is noncompact (semibounded). However, by using this solution, one can construct an asymptotic solution to (2.1), such that its compact support grows at the rate b.

For this purpose, in order to match the solution with unity, we introduce smooth functions E_1, E_2:

$$E_1(x,t) = \begin{cases} 0, & x < -bt - a_0, \\ 1, & x > -bt - a_0 + \delta_1, \quad a_0 = \text{const} > 0, \quad 1 > \delta_1 > 0, \end{cases}$$
$$E_2(x,t) = \begin{cases} 0, & x < bt + \gamma_1 - a_0, \\ 1, & x > bt + \gamma_1 - a_0 - \delta_0, \quad \gamma_1 = \text{const} > 0. \end{cases}$$

The equation (2.1) is invariant under the change of variables $x \to -x$, therefore the function $u(-x,t,\varepsilon)$ satisfies also (2.1) ($u(x,t,\varepsilon)$ is given by (2.2)). Hence the function

$$u(x,t,\varepsilon) = \Big[\theta(x + bt - x_0)\tan^{\alpha}\Big(\frac{x + bt - x_0}{\varepsilon}\Big)(1 - E_1) + E_1\Big]$$
$$\times \Big[\theta(-x + bt + \gamma_1 - x_0)\tan^{\alpha}\Big(\frac{-x + bt + \gamma_1 - x_0}{\varepsilon}\Big)(1 - E_2) + E_2\Big],$$
$$\alpha = \frac{1}{1-q}, \qquad \theta(\xi) = \begin{cases} 1, & \xi > 0, \\ 0, & \xi \leqslant 0, \end{cases}$$

satisfies (2.1) up to $O(\varepsilon^N)$ ($N > 0$ is an arbitrary number) and its support is compact. An "exponentially small" residual appears in the equation since the solution (2.2) is replaced by 1 in the domain where this solution is only exponentially close to 1 as $\varepsilon \to 0$ (see Figure 3.2).

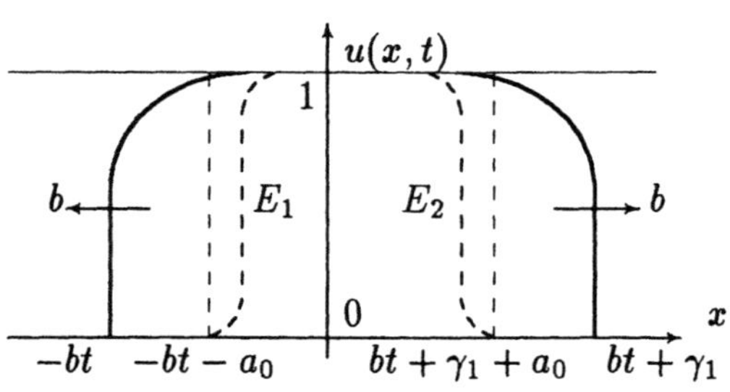

FIGURE 3.2

The properties of solutions given in Examples 1 and 2 are generalized in Definitions 2.1 and 2.2.

Definition 2.1. A localized asymptotic (up to $O(\varepsilon^N)$) solution $u(x,t,\varepsilon)$ is called a *localized cloud* (LC) if:

a) the function $u(x,t,\varepsilon)$ is uniformly bounded as $\varepsilon \to 0$;

b) for $(x,t,\varepsilon) \in \operatorname{supp} u(x,t,\varepsilon)$,

$$Lu = \varepsilon^N Q_N(x,t\varepsilon),$$

where Q_N is such that

$$\left| \frac{\partial^{|\delta|+\mu}}{\partial t^\mu \partial x^\delta} \frac{Q_N(v,t,\varepsilon)}{u(x,t,\varepsilon)} \right| \leqslant c_{\delta\mu}\varepsilon^{-(|\delta|+\mu)},$$

$$c_{\delta\mu} = \text{const} > 0, \quad \delta \in \mathbb{Z}^n_+, \quad |\delta| + \mu \leqslant N-1, \quad |\delta| = \sum_{i=1}^N \delta_i;$$

c) the flow continuity condition (1.2) holds.

In order to describe the class of LC-solutions, we need the following

Definition 2.2. A function $W(x,t,\tau)$ is said to belong to a class B^α, if

(1) $W(x,t,\tau) \in C^\infty(\mathbf{R}^n \times [0,T] \times (\mathbf{R} \setminus \{0\}))$;

(2) for any $(x,t) \in \mathbf{R}^n \times [0,T]$, the function $W(x,t,\tau)$ is continuous in τ, $W(x,t,\tau) > 0$ for $\tau > 0$, and $W(x,t,\tau) \equiv 0$ for $\tau \leqslant 0$;

(3) there exist functions $\psi_0(x,t) \in C^\infty(\mathbf{R}^n \times [0,T])$ and

$$\Psi(x,t,\tau) \in C^\infty[\mathbf{R}^n \times [0,T] \times (\mathbf{R}_+ \setminus \{0\})]$$
$$\Psi(x,t,\tau) = O(\tau^\omega), \quad \tau \to 0, \quad \omega > 0,$$

such that

$$W(x,t,\tau) = \tau^\alpha[\psi_0(x,t) + \Psi(x,t,\tau)]$$

for $\tau \to 0$;

(4) $\lim_{\tau\to\infty} W(x,t,\tau) = V(x,t)$, $V(x,t) \in C^\infty$ and the difference $W(x,t,\tau) - V(x,t)$ decreases faster than any negative power of τ as $\tau \to \infty$;

(5) for any multi-index $\beta \in \mathbb{Z}^n_+$ and any $\nu \in \mathbb{Z}_+$, the derivatives $\frac{\partial^{|\beta|+\nu}}{\partial x^\beta \partial t^\nu} W(x,t,\tau)$ satisfy the above conditions.

Example 2.3. The solution (2.2) to equation (2.1) has the form

$$u(x,t,\varepsilon) = W(x,t,\tau)\big|_{\tau=(x+bt-x_0)/\varepsilon},$$

where the function

$$W(x,t,\tau) = \begin{cases} [\tan \tau]^{1/(1-q)}, & \tau > 0, \\ 0, & \tau < 0, \end{cases}$$

is independent of x, t and belongs to B^α for $\alpha = 1/(1-q)$.

As shown in Chapter IV, LC-asymptotic solutions (similar to (2.2) up to $O(\varepsilon^N)$) can be written as follows

$$u(x,t,\varepsilon) = \sum_{j=0}^{N} \varepsilon^j W_j(x,t,\tau)\Big|_{\tau=S(x,t,\varepsilon)/\varepsilon}, \qquad (2.3)$$
$$S(x,t,\varepsilon) \in C^\infty, \quad \nabla S\big|_{S=0} \neq 0, \quad W_j \in B^{\alpha+\omega_j},$$

where $\omega_j \geqslant 0$ is a nondecreasing sequence of positive numbers, $\omega_0 = 0$.

The structure of LC-asymptotic solutions implies that formula (2.3) can be rewritten as follows

$$u(x,t,\varepsilon) = W\left(\tau + \sum_{j=0}^{N+1} \varepsilon^w h W_j(x,t,\tau), x, t\right)\Bigg|_{\tau=S(x,t,\varepsilon)/\varepsilon}, \qquad (2.4)$$
$$S(x,t,\varepsilon) \in C^\infty, \quad \nabla S\big|_{S=0} \neq 0, \quad W_j \in B^\alpha, \quad \widehat{W}_j \in B^{1+\omega_j}.$$

The functions in (2.3) and (2.4) are calculated by the method given in Chapter IV.

3.3. Asymptotic solutions unbounded as $\varepsilon \to 0$

Example 3.1. Consider the quasilinear parabolic equation:

$$\varepsilon\frac{\partial u}{\partial t} - \varepsilon^2 k\left(\frac{\partial}{\partial x}\left(u^{(k-1)}\frac{\partial u}{\partial x}\right)\right) = 0, \qquad k > 1, \qquad u \geqslant 0. \qquad (3.1)$$

The function

$$u(x,t,\varepsilon) = \begin{cases} \dfrac{1}{(1+4t)^{1/(k+1)}}\left[\frac{1}{\varepsilon}\left(1 - \frac{(k-1)2x^2}{k(k+1)(1+4t)^{2/(k+1)}}\right)\right]^{1/(k-1)}, & -x_f < x < x_f, \\ 0, & x \notin (-x_f, x_f), \end{cases} \qquad (3.2)$$

where $x_f^2 = \frac{k(k+1)(1+4t)^{2/(k+1)}}{2(k-1)}$, satisfies equation (3.1). For $\varepsilon = 1$ and $\varepsilon \ll 1$, the solution is plotted in Figure 3.3. In a neighborhood of the points $x = \pm x_f$ the function (3.2) has a branching type singularity:

$$u(x,t,\varepsilon) = \frac{1}{(1+4t)^{2/(k+1)}|x_f/2|^{1/(k-1)}}\left(\frac{x_f+x}{\varepsilon}\right)^{1/(k-1)} + O\left(\left(\frac{x_f+x}{\varepsilon}\right)^{k/(k-1)}\right), \qquad x \to -x_f + 0,$$

$$u(x,t,\varepsilon) = \frac{1}{(1+4t)^{2/(k+1)}|x_f/2|^{1/(k-1)}}\left(\frac{x_f-x}{\varepsilon}\right)^{1/(k-1)} + O\left(\left(\frac{x_f-x}{\varepsilon}\right)^{k/(k-1)}\right), \qquad x \to x_f - 0,$$

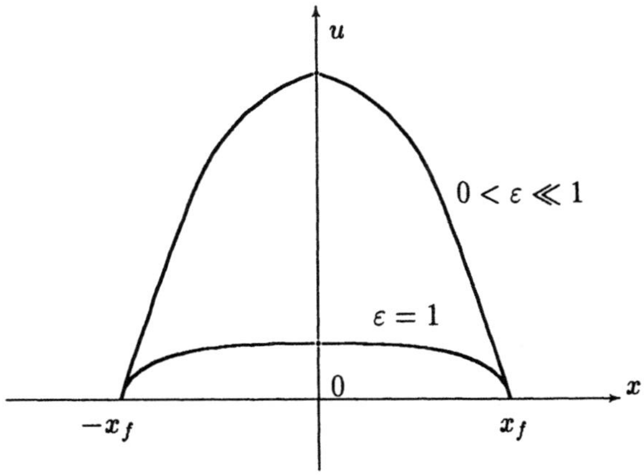

FIGURE 3.3

Moreover, $\frac{\partial(u^k)}{\partial x}\Big|_{x=\pm x_f} = 0$. This coincides with the flow continuity condition. Thus the solution (3.2) satisfies all assumptions listed in Definition 1.1.

It must be pointed out that at any interior point of the support, outside an ε-independent neighborhood of its boundary, we have $u(x,t,\varepsilon) = O(\varepsilon^{-1/(k-1)})$ as $\varepsilon \to 0$, i.e., the solution increases like a nonnegative power of the parameter ε as $\varepsilon \to 0$.

There exist localized asymptotic solutions increasing exponentially as $\varepsilon \to 0$.

Example 3.2. The solution of the quasilinear parabolic equation

$$\varepsilon\frac{\partial u}{\partial t} - \varepsilon^2\frac{\partial}{\partial x}\left(\frac{ku^{k-1}}{1+u^{k-1}}\frac{\partial u}{\partial x}\right) - \frac{\varepsilon u}{2(t+1)} = 0, \qquad k > 1 \tag{3.3}$$

has the form

$$u(x,t,\varepsilon) = \begin{cases} \left[\exp\left\{\frac{1}{\varepsilon}\left(1 - \frac{(k-1)x^2}{4k(t+1)}\right)\right\} - 1\right]^{1/(k-1)}, & x \in (-x_f, x_f), \\ 0, & x \notin (-x_f, x_f), \end{cases} \tag{3.4}$$

$$x_f = 2\sqrt{\frac{k}{k-1}(t+1)}$$

and satisfies all assumptions listed in Definition 1.1. However, in contrast to Example 3.1, as $\varepsilon \to 0$, the solution (3.4) increases exponentially at interior points of its support outside a certain neighborhood of the boundary of the support, where the solution has a branching type singularity with the same exponent as in the preceding example.

The rate of increase (as $\varepsilon \to 0$) is different at interior points of the support since the properties of the coefficients in (3.1) and (3.3) are different as $u \to \infty$. If we write the functions $K(x,t,u)$ and $F(x,t,u)$ (see (1.1)) for equations (3.1) and (3.3), then in the first case we have

$$K \sim u^{k-1}, \qquad u \to \infty, \qquad F = 0,$$

while in the second case

$$K = O(1), \qquad F = O(u), \qquad u \to \infty. \tag{3.5}$$

It turns out that (see Chapter V), in general, a relation of the form (3.5) provides the existence of solutions with properties similar to those of the solution (3.4).

The relations (3.5) are quite natural from the viewpoint of applications. In order to describe a class of functions, to which the asymptotic localized unbounded (as $\varepsilon \to 0$) solutions belong, we introduce the following definition. By $\beta(\tau)$ denote a certain fixed function belonging to $C^\infty(\mathbf{R}_+ \setminus \{0\})$ and increasing infinitely as $\tau \to \infty$.

Definition 3.1. A function $W(x,t,\tau)$ is said to belong to a class $\mathfrak{U}_\beta^\alpha$ if

(1) $W(x,t,\tau) \in C^\infty(\mathbf{R}^n \times [0,T] \times (\mathbf{R} \setminus \{0\}))$;

(2) for any $(x,t) \in \mathbf{R}^n \times [0,T]$, the function $W(x,t,\tau)$ is continuous in τ and

$$\begin{aligned} W(x,t,\tau) > 0 \qquad &\text{for} \quad \tau > 0,\\ W(x,t,\tau) = 0 \qquad &\text{for} \quad \tau \leqslant 0; \end{aligned}$$

(3) there exist functions $\psi(x,t) \in C^\infty(\mathbf{R}^n \times [0,T])$ and $\Psi(x,t,\tau) \in C^\infty(\mathbf{R}^n \times [0,T] \times (\mathbf{R} \setminus \{0\}))$ such that, for $\tau \to 0$, $\Psi(x,t,\tau) = O(\tau^\omega)$, $\omega > 0$ and

$$W(x,t,\tau) = \tau^\alpha[\psi(x,t) + \Psi(x,t,\tau)];$$

(4)

$$\lim_{\tau\to\infty} \frac{W(x,t,\tau)}{\beta(\tau)} = V(x,t) \in C^\infty[\mathbf{R}^n \times [0,T]]$$

uniformly in (x,t);

(5) for any multi-index $\delta \in \mathbb{Z}_+^n$ and any $\nu \in \mathbb{Z}_+$, the derivatives $\frac{\partial^{|\delta|+\nu}}{\partial x^\delta \partial t^\nu} W(x,t,\tau)$ satisfy assumptions (1)–(4) of this definition.

Example 3.3. The solutions (3.3) and (3.4) to equations (3.1) and (3.3) can be written in the form

$$u(x,t,\varepsilon) = W(x,t,\tau)\big|_{\tau=S(x,t)/\varepsilon},$$

where the function $W(x,t,\tau)$ is independent of x and, in the first case:

$$\begin{aligned} W(t,\tau) &= \begin{cases} \frac{1}{(1+4t))^{1/(k+1)}}\tau^{1/(k-1)}, & \tau \geqslant 0,\\ 0, & \tau < 0, \end{cases}\\ S(x,t) &= 1 - \frac{(k-1)2x^2}{k(k+1)(1+4t)^{2/(k+1)}}, \end{aligned} \tag{3.6}$$

and in the second case:

$$W(\tau) = \begin{cases} (e^\tau - 1)^{1/(k-1)}, & \tau \geqslant 0,\\ 0, & \tau < 0, \end{cases} \qquad S(x,t) = 1 - \frac{(k-1)x^2}{4k(t+1)}. \tag{3.7}$$

The functions (3.6) and (3.7) belong to the classes $\mathfrak{U}^{\alpha}_{\tau^\alpha}$ and $\mathfrak{U}^{\alpha}_{\exp\{\tau\alpha\}}$, $\alpha = 1/(k-1)$, respectively.

Note that in both examples there exist interior points of the solution support, where $\nabla_x S = 0$. This property is typical for the class of solutions under consideration. Outside a neighborhood of such points, the unbounded (as $\varepsilon \to 0$) localized asymptotic solutions, constructed in Chapter V, have the form:

$$u(x,t,\varepsilon) = \sum_{j=0}^{N+1} \varepsilon^j W_j(x,t,\tau)x,t,\tau\big|_{\tau=S(x,t,\varepsilon)/\varepsilon}, \qquad S(x,t,\varepsilon) \in C^\infty, \qquad W_j \in \mathfrak{U}^{\alpha}_{\beta_j} \tag{3.8}$$

The sequence of functions β_j is such that $\beta_0 > \beta_j \geqslant \beta_{j+1}$, $j = 1, \ldots,$ for sufficiently large τ. Near the points, where $\nabla_x S = 0$, the solution has another form (for details, see Chapter V).

3.4. The structure of singularities of solutions to quasilinear parabolic equations near the boundary of the solution support

In this section we shall study the structure of localized solutions to quasilinear parabolic equations in a neighborhood of the boundary of the solution support. The property of solutions to be localized is independent of whether the equation contains a small parameter or not. However, when asymptotic solutions are constructed, one must take into account the weak discontinuity of the solution on the boundary of its support and investigate the differential properties of the solution near the weak discontinuity front. This is a local problem, i.e., this investigation can be carried out in an arbitrarily small neighborhood of the point under consideration.

Since the small parameter in equation (1.1) can be excluded by means of the scale change of variables of the type $t' = t/\varepsilon$, $x' = x/\varepsilon$, and the size of the neighborhood can be taken arbitrarily, without loss of generality, we can study the differential properties of solutions to equations without a small parameter. Of course, the obtained estimates will not be uniform in ε when we return to the initial variables. Nevertheless, the structure of the solution, obtained as the result of these investigations, is preserved for asymptotic (with respect to the parameter) solutions studied in Chapters IV and V.

In this section we shall show that, under certain (very weak) restrictions, the solution may have singularities only of exponential branching type in a neighborhood of the boundary of its support.

Consider the quasilinear parabolic equation

$$Lu = \frac{\partial u}{\partial t} - \langle \nabla, \rho(x,t,u)\nabla u^k \rangle + \langle \nabla, \varphi_i(x,t,u) \rangle + F(x,t,u) = 0 \tag{4.1}$$

where $x \in \mathbf{R}^n$, $k > 1$, $t \in [0,T]$, and the functions $\rho(x,t,u)$, $\varphi_i(x,t,u)$, $j = 1, \ldots, n$, $F(x,t,u)$ are smooth for $u > 0$; for $u \to +0$, they can be expanded into series with smooth coefficients with respect to u:

$$F(x,t,u) = u^q\Big[G_0(x,t) + \sum_{i\geqslant 1} u^{m_i} G_i(x,t)\Big] = u^q G(x,t,u),$$

$$\varphi_j(x,t,u) = u^m\Big[\varphi_{j0} + \sum_{i\geqslant 1} u^{n_i}\varphi_{ji}(x,t)\Big] = u^m\Phi_j(x,t,u),$$

$$\rho(x,t,u) = \rho_0(x,t) + \sum_{i\geqslant 1} u^{l_i}\rho_i(x,t),$$

where $q > 0$, $m > 0$, $G_0(x,t) \neq 0$, $\varphi_{j0} \neq 0$ for $j = 1,\dots,n$, $\rho_0(x,t) > 0$, $\{m_i\}$, $\{n_i\}$, $\{l_i\}$ are increasing unbounded sequences of natural numbers.

To describe the class of solutions under consideration, we introduce the following definitions.

Let Ω be a certain domain in $\mathbf{R}^n \times \mathbf{R}_+$.

Definition 4.1. Denote by $V(\Omega)$ the set of functions χ in variables x, t, τ, $(x,t) \in \Omega$, $\tau \in \mathbf{R}$ such that:

(1) the function χ is infinitely differentiable in x, t; it is infinitely differentiable in τ if $\tau > 0$, equal to zero for $\tau < 0$, positive for $\tau > 0$, and uniformly continuous for $\tau \geqslant 0$;

(2) for $\tau = 0$, the differentiation with respect to x, t does not change the structure of singularities of χ with respect to τ, i.e., for any multi-indices $\beta \in \mathbb{Z}_+^n$ and any $\gamma \in \mathbb{Z}_+$, the limits

$$\lim_{\tau\to+0}\left[\frac{1}{\chi(x,t,\tau)}\frac{\partial^{|\beta|}}{\partial x^\beta}\frac{\partial^\gamma}{\partial t^\gamma}\chi(x,t,\tau)\right]$$

exist uniformly in x, t and belong to $C^\infty(\Omega)$;

(3) for any $\tau \in [0,\tau_0)$, where τ_0 is sufficiently small, there exists a constant $c = c(\tau_0) > 0$ such that $c\chi(x,t,\tau) \geqslant \chi(x,t,\tau')$ for $0 < \tau' < \tau$.

Example 4.1. For $\alpha > 0$, $\beta > 0$, the function $\tau^\alpha(\beta + \sin^2\tau^{-1})$ belongs to $V(\Omega)$, and $c \geqslant 1 + 1/\beta$ in item (3) of Definition 4.1. Therefore, the latter property means that the point $\tau = 0$ is not a point of accumulation for the zero set of the function $\chi(x,t,\tau)$ with respect to τ.

Definition 4.2. The function $u(x,t)$ is called a *local self-similar asymptotic* (*with respect to smoothness*) solution of equation (4.1) in the domain Ω, if the following conditions hold.

(1) The function $u(x,t)$ can be represented in the form

$$u(x,t) = \chi\big(x,t,S(x,t)\big) + g(x,t), \tag{4.2}$$

where $\chi(x,t,\tau) \in V(\Omega)$, $S(x,t)$ is a certain smooth function such that

$$\nabla_x S(x,t)\big|_{S=0} \neq 0, \qquad (x,t) \in \Omega,$$

the function $g(x,t)$ belongs to C^∞ for $(x,t) \notin \{(x,t)\colon S(x,t) = 0\}$, and the estimate

$$\frac{g(x,t)}{\chi(x,t,S(x,t))} = O(\chi)$$

is uniform for $(x,t) \in \Omega$.

(2) $Lu = g_1(x,t)$, where $g_1 \in C^\infty$ for $(x,t) \notin \{(x,t)\colon S(x,t) = 0\}$, and

$$\int_0^S g_1(x,t,S)\, dS = o\Big(\frac{d\chi^k}{d\tau}(x,t,S)\Big), \qquad S \to +0,$$

if the front does not move, i.e., $\partial S/\partial t = 0$ for $(x,t) \in \Gamma\{(x,t)\colon S(x,t) = 0\}$ ($m > 1$, $q \geqslant 1$) or $\big(\alpha\frac{\partial S}{\partial t} + \Phi(x,t,0)\frac{\partial S}{\partial x}\big) = 0$ for $(x,t) \in \Gamma$, $m = 1$, $q > 1$; otherwise

$$\int_0^S g_1(x,t,S)\, dS = o\big(\chi(x,t,S)\big), \qquad S \to +0.$$

(3) The flow satisfies the continuity condition (1.1) of the form

$$\lim_{\chi \to +0} \frac{\partial}{\partial \tau}(\chi + g)^k = 0.$$

The definition of a local self-similar solution implies that in a neighborhood of the weak discontinuity front (i.e., on the set $\{S(x,t) = 0\} \cap \Omega$) the structure of the solution is defined only by the differential properties of the function χ with respect to τ.

We shall say a few words about Definitions 4.1 and 4.2.

The notion "singularity" is often met in mathematics, but it has different meanings. The so-called singularities of general position, i.e., singularities that do not vary under the action of "small perturbations", play an essential role. Obviously, the objects of certain stability are important in applications, since stable phenomena are most often met in nature. Definitions 4.1 and 4.2 deal precisely with such "stable" singularities. Actually, the properties of the function χ imply that the structure of the "almost" solution (4.2) does not vary both in time and in space (when passing from point to point in a certain arbitrarily small domain). Below we shall show that "stable" singularities cannot be arbitrary. We also point out that the classes of functions $B^\alpha\big|_\Omega$ and $\mathfrak{U}^\alpha_\beta\big|_\Omega$, introduced in the previous section, are imbedded into $V(\Omega)$ for any Ω.

The notion of solutions with singularities also needs further comments. In mathematical literature, a function satisfying a partial differential equation (for example, a nonlinear equation in divergent form) in a certain space of distributions, e.g., in D', is called a solution with singularity. In physical literature, an equivalent definition based on the integral identity (its "density" is the initial equation) is conventional. Obviously, under our additional assumptions about the structure of the solution singularities, for $g_1 \equiv 0$, the definition of asymptotic (with respect to smoothness) solutions (see item (2) in Definition 4.2) is equivalent to the standard definition of solutions in the sense of distributions.

Without loss of generality, we can assume that the domain Ω is sufficiently small, so that $|\nabla_x S(x,t)| > 0$ for $(x,t) \in \Omega$. Below we shall consider two cases: $|\partial S/\partial t| > 0$ for $(x,t) \in \Omega$ or $\partial S/\partial t \equiv 0$ for $(x,t) \in \Omega$. This consideration implies: if in Ω the sign of the derivative $\partial S/\partial t$ varies, then the singularity does not preserve its type except the case when $m \geqslant 1$, $k + q = 2$.

The main statement of this section is*

*For the classification of singularities of solutions satisfying (4.1) see also [13, 72].

Theorem 4.1. *Let a local self-similar asymptotic solution of equation* (4.1) *exist in the domain* Ω. *Then, in a neighborhood of the point* $\tau = 0$, *for* $(x,t) \in \Omega$, *the function* $\chi(x,t,\tau)$ *can be represented in the form:*

$$\chi(x,t,\tau) = \begin{cases} c\tau^\alpha + o(\tau^\alpha), & \tau \to 0+0, \\ 0, & \tau < 0, \end{cases} \qquad (4.3)$$
$$\alpha \in \mathbf{R}_+, \qquad c = c(x,t) \in C^\infty, \qquad c > 0.$$

Note that this statement is inverse (in a sense) to that of D. G. Aronson [77, 79] about the solution structure near the points of degeneration. The difference is that we obtain the asymptotic (with respect to smoothness) solution under the assumption that the flow is continuous and the surface of degeneration is rigid. Besides of this, though a priori estimates are not established, our theorem shows how perturbations influence the smoothness of the solution. Namely, any perturbations of the right-hand side of the equation satisfying the conditions in item (2) of Definition 4.2, do not change the exponent of the principal term of the asymptotics with respect to smoothness.

Before we prove the theorem, we note that if the point $u = 0$ is not an exponential branching point for the coefficients, then the solution of (4.1) may also have singularities (not only of branching type).

Example 4.2. Consider the equation

$$\frac{\partial u}{\partial t} = \frac{\partial}{\partial x}\Big(2u\frac{\partial u}{\partial x}\Big) - u\big[1 - 2u(1 + 2\ln u)\ln u\big]\ln^2 u = 0.$$

Comparing this equation with (4.1), we get

$$\rho(x,t,u) \equiv 1, \qquad k = 2,$$
$$F(x,t,u) = -u\big[1 - 2u(1 + 2\ln u)\ln u\big]\ln^2 u.$$

The considered equation has an invariant solution $u = \chi(x+t)$, where the function $\chi = \chi(\tau)$ satisfies the ordinary differential equation

$$\frac{d\chi}{d\tau} - \frac{d^2}{d\tau^2}\chi^2 - F(\chi) = 0,$$

one of whose localized solutions

$$\chi(\tau) = \begin{cases} \exp\{-1/\tau\}, & \tau \geqslant 0, \\ 0, & \tau < 0, \end{cases}$$

is a smooth and uniformly bounded function.

If we introduce a small parameter ε into the this equation (compare with equation (1.1)), then the exact invariant solution takes the form $u = \chi((x+t)/\varepsilon)$. This solution belongs to the class of bounded (as $\varepsilon \to 0$) solutions, but, in contrast to those considered in this chapter, its derivatives have no exponential asymptotics as $\tau \to \infty$.

On the basis of Theorem 4.1 we can classify the principal singularities at $\tau = 0$.

TABLE I

	α	Existence conditions ($\partial S/\partial t \neq 0$)
1	$\dfrac{1}{k-1}$	$q \geqslant 1; \quad m \geqslant 1$
2	$\dfrac{1}{1-q}$ or $\dfrac{1}{k-1}$	$q < 1; \quad m \geqslant 1, \quad k+q \geqslant 2$
3	$\dfrac{1}{k-m}$	$0 < m \leqslant 1, \quad 2m \leqslant k+q, \quad q \geqslant m$
4	$\dfrac{1}{m-q}$	$0 < q < m, \quad 2m \leqslant k+q, \quad 0 \leqslant m \leqslant 1$
5	$\dfrac{2}{k-q}$	$k > q, \quad 0 \leqslant \dfrac{k+q}{2} \leqslant \min(1, m)$

TABLE II

	α	Existence conditions ($\partial S/\partial t = 0$)
1	$\dfrac{1}{k-m}$	$k > m, \quad m > 0, \quad 2m \leqslant \min(k+q, k+1)$
2	$\dfrac{2}{k-q}$	$\dfrac{2}{k-q} > 0 \quad q \leqslant 1, \quad 0 \leqslant \dfrac{k+q}{2} \geqslant m$
3	$\dfrac{1}{m-q}$	$q < m, \quad 2m \leqslant k+q, \quad q \leqslant 1$

Corollary 4.1. *Let $k > 1$, $m > 0$, $q > 0$. Then in a neighborhood of the weak discontinuity front the local self-similar solution of equation* (4.1) *can possess only the principal singularities which are given in Tables* I *and* II.

A more precise analysis shows that a local self-similar asymptotic (with respect to smoothness) solution can be expanded into an asymptotic series with respect to a system of functions from $V(\Omega)$. Generally speaking, this series may consist not only of powers of τ, though, by Theorem 4.1, the principal term of this series is a certain power. To formulate the corresponding results, we must define a local self-similar asymptotic solution more precisely.

By $\mathcal{P}(\{w_i\})$, $i \in N$, denote the set of functions obtained from the set $\{w_i\}$ by means of the following transformations:

A) taking a linear combination with smooth coefficients,

B) multiplication,

C) raising to a real power.

In general, the sequence of generating functions $\{w_i\}$ is not finite and can be chosen in different ways.

The sequence of generating functions $\{w_i\}$ will be called *minimal*, if $\mathcal{P}(\{w_i\})$ cannot be obtained from any subsequence of the sequence $\{w_i\}$.

Everywhere below, in the definition of the set $\mathcal{P}(\{w_i\})$, the sequence $\{w_i\}$ is supposed to be minimal.

Definition 4.3. The function $u(x,t)$ is called a *local self-similar asymptotic* (*with* *respect to smoothness) solution* to equation (4.1) in the domain Ω, if for $(x,t) \in \Omega$:

1. The function $u(x,t)$ can be written in the form

$$u(x,t) = \chi(x,t,S(x,t)) + g(x,t),$$

where $S(x,t)$ is a certain smooth function such that

$$S(x_0,t_0) = 0, \qquad \nabla S\Big|_{S=0} \neq 0, \qquad \frac{\partial S(x,t)}{\partial t}\Big|_{S=0} \neq 0, \qquad (x_0,t_0) \in \Omega.$$

As $\tau \to +0$, the function $\chi(x,t,\tau)$ can be expanded into an asymptotic series with respect to the functions $\{\psi_i(x,t,\tau)\}$ from the set $V(\Omega)$:

$$\chi(x,t,\tau) = \sum_i \psi_i(x,t,\tau), \qquad \lim_{\tau \to +0} \frac{\psi_{i+1}}{\psi_i} \to 0.$$

The function $g(x,t)$ belongs to C^∞ if $(x,t) \notin \{(x,t)\colon S(x,t) = 0\}$ and the estimate

$$\frac{g(x,t)}{\chi(x,t,S(x,t))} = O(S^N)$$

holds uniformly in $(x,t) \in \Omega$ for any $N > 0$.

2. $Lu = g_1(x,t)$, where $g_1 \in C^\infty$ for $(x,t) \notin \{(x,t)\colon S(x,t) = 0\}$ and $g_1(x,t,S)/\chi(x,t,S) = O(S^N)$, $S \to +0$, $\forall N > 0$.

3. The flow continuity condition holds:

$$\lim_{\chi \to +0} \frac{\partial \chi^k}{\partial \tau} = 0.$$

Definition 4.4. A local self-similar solution

$$u(x,t) = \chi(x,t,S(x,t)) + g(x,t),$$

where the function $\chi(x,t,\tau)$ can be expanded into an asymptotic series with respect to a system of functions $\{\psi_i(x,t,\tau)\}$ from the set $\mathcal{P}(\{\omega_i\})$, is called *structurally stable*, if for any sufficiently small $\delta > 0$, there exists a local self-similar solution

$$u_\delta(x,t) = \chi_\delta(x,t,S_\delta(x,t)) + g_\delta(x,t),$$

such that $S_\delta(x,t) \in C^\infty(V)$, $S(x,t_0) = S_\delta(x,t_0)$, $S_\delta(x,t) \to S(x,t)$ as $\delta \to 0$, and:

(1) $\displaystyle\lim_{\tau \to +0} \frac{\chi(x,t,\tau)}{\chi_\delta(x,t,\tau)} = 1 + r_\delta$, where r_δ is a continuous positive function, $r_\delta \to 0$ as $\delta \to 0$ in the maximum norm;

(2) the function $\chi_\delta(x,t,\tau)$ can be expanded into an asymptotic series with respect to a certain system of functions $\{\psi_i^\delta(x,t,\tau)\}$ from the set $\mathcal{P}(\{\omega_i\})$, which do not belong to any subset of $\mathcal{P}(\{\omega_i\})$ invariant under the transformations A, B, C.

The following statement is an analog of Theorem 4.1.

Theorem 4.1′. *A structurally stable local self-similar solution of equation* (4.1) *can be expanded into an asymptotic series with respect to a system of functions from the set* $\mathcal{P}(\{\tau_+\})$, $\tau_+ = \tau\theta(\tau)$.*

Further, we prove Theorem 4.1 and Theorem 4.1′ using the techniques developed in the proof of Theorem 4.1. At the end of this section we prove two lemmas needed to construct the asymptotic solutions in the subsequent chapter.

Proof of Theorem 4.1. To simplify the notation, we shall consider the case $x \in \mathbf{R}$. If, in general, $x \in \mathbf{R}^n$, $n > 1$, the proof is precisely the same.

Substitute the function (4.2) into (4.1). For $(x, t) \in \Omega$, we get

$$\begin{aligned}\frac{\partial S}{\partial t}\frac{\partial \chi}{\partial \tau} - \left(\frac{\partial S}{\partial x}\right)^2 \frac{\partial}{\partial \tau}\left(\rho(x,t,\chi)\frac{\partial \chi^k}{\partial \tau}\right) + \chi^q G_0 \\ = -\frac{\partial \chi}{\partial t} + \rho(x,t,\chi)\frac{\partial}{\partial x}\left(\frac{\partial S}{\partial x}\frac{\partial}{\partial \tau} + \frac{\partial}{\partial x}\right)\chi^k \\ + \left(\frac{\partial S}{\partial x}\frac{\partial}{\partial \tau} + \frac{\partial}{\partial x}\right)\rho(x,t,\chi)\frac{\partial \chi^k}{\partial x} - \frac{\partial g(x,t)}{\partial t} \\ + \left(\frac{\partial S}{\partial x}\right)^2 \frac{\partial}{\partial \tau}\left[\sum_{\substack{\mu \geqslant 1 \\ l \geqslant 1}} \frac{\partial^\mu}{\partial \chi^\mu}\rho(x,t,\chi) g^\mu(x,t) \frac{\partial}{\partial \tau}\left(\frac{\partial^l \chi^k}{\partial \chi^l}\right) g^l(x,t)\right] + \dots\end{aligned} \tag{4.4}$$

Here the arguments x, t, $\tau = S(x,t)$ of the functions χ, g are omitted. The dots here and below in (4.13) stand for omitted summands. These summands can be analyzed similarly to those written explicitly in formula (4.4).

On the left-hand side of this relation, we collect the most "nonsmooth" functions. Note that in a neighborhood of $\tau = 0$ the summand $\partial\chi(x,t,\tau)/\partial t$ possesses the same singularity with respect to τ as the function $\chi(x,t,\tau)$. This holds for the derivatives of any order in x, t.

The subsequent constructions can be divided into some lemmas describing all possible combinations of the parameters. Let us list them.

I. $k + 1 \geqslant 2\min(m, 1)$. (4.5)

II. $k + 1 < 2\min(m, 1)$. (4.6)

In case I, the following situations are possible:

I.1. $m \geqslant 1$, $q \geqslant 1$; (4.7)

I.2. $m \geqslant 1$, $0 < q < 1$; (4.8)

I.3. $0 < m \leqslant 1$, $q \geqslant m$ (4.9)

(in this case, obviously, $k > m$ since $k > 1$);

I.4. $0 < m \leqslant 1$, $0 < q < m$. (4.10)

Similarly, in case II we have:

II.1. $m \geqslant 1$;

II.2. $0 < m < 1$. (4.11)

---

*Recall that $\theta(\tau)$ is the Heaviside function.

Moreover, the case $k = q = m = 1$ is possible. Then, as is known, equation (4.1) has no localized solutions and this situation is not considered below. For the combination of parameters I.1–I.4, the statement of the theorem is proved in Lemmas 4.1–4.4 and for II.1–II.2 in Lemmas 4.5 and 4.6 respectively.

These lemmas are proved similarly. Therefore we prove only Lemmas 4.1–4.3 in detail and outline the other proofs.

Lemma 4.1. *Let the conditions of theorem be satisfied. Let either* $|\frac{\partial S}{\partial t}| \neq 0$ *and* $m > 1$, *or* $|\frac{\partial S}{\partial t} + \Phi(x,t,0)\frac{\partial S}{\partial x}| \neq 0$ *and* $m = 1$, *and let the parameters of equation* (4.1) *satisfy conditions* I.1..

Then in a neighborhood of the zero surface of the function $S(x,t)$ *either* $\frac{\partial S}{\partial t} > 0$ (*for* $m > 1$), *or* $\frac{\partial S}{\partial t} + \Phi(x,t,0)\frac{\partial S}{\partial x} > 0$ (*for* $m = 1$), *and the statement of Theorem* 4.1 *holds.*

Proof. The cases $m > 1$ and $m = 1$ will be considered separately. First, let $m > 1$. Note that, by conditions of Theorem 4.1, the relation (4.4) holds only for $\tau = S(x,t)$ and does not hold for any independent variables x, t, τ. Thus to derive the required equation, one must change the order of operations, i.e., first substitute $\tau = S(x,t)$ (this was the second operation) and then differentiate with respects to τ.

Let us change the variables: $(x,t) \to (\xi_1, \xi_2)$, $\xi_1 = S(x,t)$, $\xi_2 = S(x-t, x+t)$. By assumption, the Jacobian

$$J = \begin{vmatrix} \frac{\partial S}{\partial x} & \frac{\partial S}{\partial x} + \frac{\partial S}{\partial t} \\ \frac{\partial S}{\partial t} & -\frac{\partial S}{\partial x} + \frac{\partial S}{\partial t} \end{vmatrix} = -|\nabla_{x,t} S|^2$$

does not vanish. We put $\chi(x,t,S) = X(\xi_1, \xi_2, \xi_1)$. Note that, by Definition 4.1, the differentiation with respect to the first two arguments does not change the type of singularity of the function $X(\xi_1, \xi_2, \xi_1)$. Moreover, we have

$$\frac{\partial}{\partial \tau}\chi(x,t,\tau)\bigg|_{\tau=S} = \frac{\partial X}{\partial \xi_1}(\xi_1, \xi_2, \xi_1) - \frac{\partial}{\partial q}X(q, \xi_2, \xi_1)\bigg|_{q=\xi_1}.$$

Further, to simplify the notation, we shall use the operator ordering techniques [54]. We ascribe numbers to the arguments ξ_1, ξ_2, ξ_1 of the function X, i.e., $(\xi_1, \xi_2, \xi_1) = (\overset{1}{\xi_1}, \xi_2, \overset{3}{\xi_1})$, and fix this ordering. Then the last relation can be rewritten in the form

$$\frac{\partial}{\partial \tau}\chi(x,t,\tau)\bigg|_{\tau=S} = \left(\frac{\overset{4}{\partial}}{\partial \xi_1} - \frac{\overset{2}{\partial}}{\partial \xi_1}\right) X(\overset{1}{\xi_1}, \xi_2, \overset{3}{\xi_1}). \tag{4.12}$$

Here, following [54], we assume that the operator of differentiation acts only on the arguments whose numbers are less than the number of the operator itself. (In particular, the operator of differentiation $\overset{2}{\partial/\partial \xi_1}$ does not change the type of singularity).

Under this change of variables, the operators $\partial/\partial x$ and $\partial/\partial t$ turn into the operators D_x and D_t of the form:

$$D_x = a(\xi_1,\xi_2)\frac{\overset{2}{\partial}}{\partial\xi_1} + b(\xi_1,\xi_2)\frac{\partial}{\partial\xi_2},$$

$$D_t = c(\xi_1,\xi_2)\frac{\overset{2}{\partial}}{\partial\xi_1} + d(\xi_1,\xi_2)\frac{\partial}{\partial\xi_2},$$

where $a,b,c,d \in C^\infty$. It is easy to calculate the coefficients a, b, c, d explicitly, but we do not need their explicit form for further considerations.

In variables ξ_1,ξ_2, relation (4.4.) has the form (the wave over the function denotes the change of variables $x = x(\xi_1,\xi_2)$, $t = t(\xi_1,\xi_2)$):

$$\begin{aligned}
&\widetilde{\frac{\partial S}{\partial t}}\frac{\overset{4}{\partial}}{\partial\xi_1}X - \left(\widetilde{\frac{\partial S}{\partial x}}\right)^2\frac{\overset{4}{\partial}}{\partial\xi_1}\left[\widetilde{\rho}(\xi_1,\xi_2,0)\frac{\overset{4}{\partial}}{\partial\xi_1}X^k\right] + X^q\widetilde{G}_0\\
&\quad= -\widetilde{\frac{\partial S}{\partial t}}\frac{\overset{4}{\partial}}{\partial\xi_1}X + \left(\widetilde{\frac{\partial S}{\partial x}}\right)^2\left[-\frac{\overset{4}{\partial}}{\partial\xi_1}\left(\left(\widetilde{\rho}(\xi_1,\xi_2,0)\frac{\overset{2}{\partial}}{\partial\xi_1}\right)X^k\right)\right.\\
&\quad\left. -\frac{\overset{2}{\partial}}{\partial\xi_1}\widetilde{\rho}(\xi_1,\xi_2,0)\frac{\overset{4}{\partial}}{\partial\xi_1}X^k + \frac{\overset{2}{\partial}}{\partial\xi_1}\widetilde{\rho}(\xi_1,\xi_2,0)X^k\right]\\
&\quad - D_tX + \widetilde{\rho}(\xi_1,\xi_2,0)D_x\left(\widetilde{\frac{\partial S}{\partial x}}\left(\frac{\overset{4}{\partial}}{\partial\xi_1} - \frac{\overset{2}{\partial}}{\partial\xi_1}\right) + D_x\right)X^k\\
&\quad + \left[\widetilde{\frac{\partial S}{\partial x}}\left(\frac{\overset{4}{\partial}}{\partial\xi_1} - \frac{\overset{2}{\partial}}{\partial\xi_1}\right) + D_x\right]\widetilde{\rho}(\xi_1,\xi_2,0)D_xX^k - D_t\widetilde{g}\\
&\quad + \left(\widetilde{\frac{\partial S}{\partial x}}\right)^2\left(\frac{\overset{4}{\partial}}{\partial\xi_1} - \frac{\overset{2}{\partial}}{\partial\xi_1}\right)\sum_{\nu,l>0}\frac{\partial^\nu}{\partial X^\nu}\widetilde{\rho}(\xi_1,\xi_2,X)\bigg|_{X=0}\\
&\quad \times(\widetilde{g})^\nu\cdot\left(\frac{\overset{4}{\partial}}{\partial\xi_1} - \frac{\overset{2}{\partial}}{\partial\xi_1}\right)\left[\frac{\partial^lX^k}{\partial X^l}(\widetilde{g})^l\right] + \sum_{l\geqslant 1}\frac{\partial^l}{\partial X^l}(X^q\widetilde{G}(\xi_1,\xi_2,0))(\widetilde{g})^l\\
&\quad + D_x\cdot\sum_{\nu\geqslant 1}\frac{\partial}{\partial X}\widetilde{\rho}(\xi_1,\xi_2,X)\bigg|_{X=0}(\widetilde{g})^\nu\left(\widetilde{\frac{\partial S}{\partial x}}\left(\frac{\overset{4}{\partial}}{\partial\xi_1} - \frac{\overset{2}{\partial}}{\partial\xi_1}\right) + D_x\right)\sum_{l\geqslant 1}\frac{\partial^lX^k}{\partial X^l}(\widetilde{g})^l\\
&\quad - D_x(X^m\widetilde{\Phi}(\xi_1,\xi_2,0)) - \widetilde{\frac{\partial S}{\partial x}}\left(\frac{\overset{4}{\partial}}{\partial\xi_1} - \frac{\overset{2}{\partial}}{\partial\xi_1}\right)\sum_{l\geqslant 1}\frac{\partial^l}{\partial X^l}X^m\widetilde{\Phi}(\xi_1,\xi_2,0)(\widetilde{g})^l\\
&\quad + \widetilde{g}_1 + \dots
\end{aligned}\tag{4.13}$$

Here

$$\widetilde{\rho} = \widetilde{\rho}(\overset{1}{\xi_1},\xi_2,X(\overset{1}{\xi_1},\xi_2,\overset{3}{\xi_1})),\qquad \widetilde{g} = \widetilde{g}(\overset{3}{\xi_1},\xi_2),\qquad \widetilde{g}_1 = \widetilde{g}_1(\overset{3}{\xi_1},\xi_2),$$

$$\widetilde{\Phi} = \widetilde{\Phi}(\overset{1}{\xi}_1, \xi_2, X(\overset{1}{\xi}_1, \xi_2, \overset{3}{\xi}_1)), \qquad \widetilde{G} = \widetilde{G}(\overset{1}{\xi}_1, \xi_2, X(\overset{1}{\xi}_1, \xi_2, \overset{3}{\xi}_1)),$$

the dots stand for the sum of subsequent terms in the expansion by the Taylor formula,

$$\widetilde{\frac{\partial S}{\partial x}} = \frac{\partial S}{\partial x}\bigg|_{\substack{x=x(\xi_1,\xi_2)\\ t=t(\xi_1,\xi_2)}}, \qquad \widetilde{\frac{\partial S}{\partial t}} = \frac{\partial S}{\partial t}\bigg|_{\substack{x=x(\xi_1,\xi_2)\\ t=t(\xi_1,\xi_2)}}$$

Note that

$$\frac{\overset{4}{\partial}}{\partial \xi_1} X(\overset{1}{\xi}_1, \xi_2, \overset{3}{\xi}_1) \equiv \frac{\partial}{\partial \xi_1} X(\xi_1, \xi_2, \xi_1)$$

and rewrite (4.13) in the form

$$\widetilde{\frac{\partial S}{\partial x}} \frac{\partial X}{\partial \xi_1} - \left(\widetilde{\frac{\partial S}{\partial x}}\right)^2 \frac{\partial}{\partial \xi_1}\left(\widetilde{\rho}(\xi_1, xi_2, 0)\frac{\partial X^k}{\partial \xi_1}\right) = \widetilde{f}, \tag{4.13'}$$

where

$$\begin{aligned}\widetilde{f}(\xi_1, \xi_2, X) = &-\widetilde{\frac{\partial S}{\partial t}} \frac{\overset{2}{\partial}}{\partial \xi_1} X + \left(\widetilde{\frac{\partial S}{\partial x}}\right)^2 \frac{\overset{4}{\partial}}{\partial \xi_1}\left(\widetilde{\rho}(\xi_1, \xi_2, 0)\Big(-\frac{\overset{2}{\partial}}{\partial \xi_1}\Big) X^k\right)\\ &- \frac{\overset{2}{\partial}}{\partial \xi_1}\widetilde{\rho}(\overset{1}{\xi}_1, \xi_2, 0)\frac{\overset{4}{\partial}}{\partial \xi_1} X^k + \frac{\overset{2}{\partial}}{\partial \xi_1}\widetilde{\rho}(\overset{1}{\xi}_1, \xi_2, 0)\frac{\overset{2}{\partial}}{\partial \xi_1} X^k\\ &- \widetilde{\frac{\partial S}{\partial x}}\left(\frac{\overset{4}{\partial}}{\partial \xi_1} - \frac{\overset{2}{\partial}}{\partial \xi_1}\right)[X^m \eta \Phi(\overset{1}{\xi}_1, \xi_2, 0)] - X^q \widetilde{G}(0, \xi_2, 0) + \widetilde{f}_1.\end{aligned}$$

Here $\widetilde{f}_1$ denotes the right-hand side of (4.13).

Note that, for l, $\nu \geqslant 0$, $l + \nu \geqslant 1$, the L'Hospital rule implies

$$\begin{aligned}&\lim_{\xi_1 \to 0} \frac{\overset{4}{\partial}}{\partial \xi_1} \frac{\partial^\nu}{\partial X^\nu} \widetilde{\rho}(\xi_1, \xi_2, X)\bigg|_{X=0} (X)^\nu \frac{\overset{4}{\partial}}{\partial \xi_1}(X^{k-l}\widetilde{g}^l)\\ &\qquad \times \left(\frac{\overset{4}{\partial}}{\partial \xi_1}\widetilde{\rho}(\xi_1, \xi_2, X)\bigg|_{X=0} \frac{\overset{4}{\partial}}{\partial \xi_1} X^k\right)^{-1} = 0,\\ &\lim_{\xi_1 \to 0} \frac{\overset{2}{\partial}}{\partial \xi_1} \frac{\partial^\nu}{\partial X^\nu} \widetilde{\rho}(\overset{1}{\xi}_1, \xi_2, X)\bigg|_{X=0} (X)^\nu \frac{\overset{4}{\partial}}{\partial \xi_1}(X^{k-l}\widetilde{g}^l)\\ &\qquad \times \left(\frac{\overset{2}{\partial}}{\partial \xi_1}\widetilde{\rho}(\overset{1}{\xi}_1, \xi_2, 0)\frac{\overset{4}{\partial}}{\partial \xi_1} X^k\right)^{-1} = 0,\\ &\lim_{\xi_1 \to 0} \frac{\overset{2}{\partial}}{\partial \xi_1} \frac{\partial^\nu}{\partial X^\nu} \widetilde{\rho}(\overset{1}{\xi}_1, \xi_2, X)\bigg|_{X=0} (X)^\nu \frac{\overset{2}{\partial}}{\partial \xi_2}(X^{k-l}\widetilde{g}^l)\end{aligned}$$

$$\times \left(\frac{\overset{2}{\partial}}{\partial \xi_1} \widetilde{\rho}(\overset{1}{\xi_1}, \xi_2, 0) \frac{\overset{2}{\partial}}{\partial \xi_2} X^k \right)^{-1} = 0.$$

Now we integrate (4.13′) with respect to ξ_1. The summands in $\int_0^{\xi_1} \widetilde{f}\, d\xi_1$ can be divided into three groups with respect to the powers of k, q, m. As is easily seen, each successive summand in each group tends to zero (as $\xi_1 \to 0$) faster than the preceding summands.

Consider the first summands in each group and prove that, as $\xi_1 \to 0$, they tend to zero faster than X, except for the summand

$$\widetilde{\rho}(\overset{1}{\xi_1}, \xi_2, 0) \frac{\overset{4}{\partial}}{\partial \xi_1} X^k.$$

For example, let us consider a summand from the group "k" (see (4.13)). We have

$$\lim_{\xi_1 \to 0} \frac{1}{X} \int_0^{\xi_1} \frac{\overset{2}{\partial}}{\partial \xi_1} \widetilde{\rho}(0, \xi_2, 0) \frac{\overset{4}{\partial}}{\partial \xi_1} X^k$$
$$= \lim_{\xi_1 \to 0} \frac{1}{X} \int_0^{\xi_1} \frac{\partial \widetilde{\rho}}{\partial \xi_1} \frac{\partial X^k}{\partial \xi_1}\, d\xi_1 + \lim_{\xi_1 \to 0} \frac{1}{X} \int_0^{\xi_1} \widetilde{\rho} \frac{\overset{4}{\partial}}{\partial \xi_1} \frac{\overset{2}{\partial}}{\partial \xi_1} X^k.$$

Integrating by parts, we get

$$\lim_{\xi_1 \to 0} \frac{1}{X} \int_0^{\xi_1} \frac{\partial \widetilde{\rho}}{\partial \xi_1} \frac{\partial X^k}{\partial \xi_1}\, d\xi_1 = \lim_{\xi_1 \to 0} \frac{1}{X} \int_0^{\xi_1} \frac{\partial^2 \widetilde{\rho}}{\partial \xi_1^2} X^k\, d\xi_1 + \lim_{\xi_1 \to 0} \frac{\partial \widetilde{\rho}}{\partial \xi_1} X^{k-1}. \tag{4.14}$$

Obviously, the second summand vanishes. In the first summand, using condition (3) from Definition 4.2, we estimates the absolute value of the integral:

$$\left| \int_0^{\xi_1} \frac{\partial^2 \widetilde{\rho}}{\partial \xi_1^2} X^k\, d\xi_1 \right| \leqslant c \max \left| \frac{\partial^2 \widetilde{\rho}}{\partial \xi_1^2} \right| X^k(\xi_1, \xi_2, \xi_1) \xi_1.$$

This implies that the first summand also vanishes.

Now consider the first summand in group "m": $\partial X^m / \partial \xi_1$. Prove that, as $\xi_1 \to 0$, all successive summands vanish faster than the first summand. We have

$$\lim_{\xi_1 \to 0} \frac{\overset{4}{\partial}}{\partial \xi_1} \sum_{l \geqslant 1} \frac{\partial^l}{\partial X^l} (X^m \widetilde{\Phi}(\xi_1, \xi_2, 0))(X)^l \left(\frac{\overset{4}{\partial}}{\partial \xi_1} X^m \widetilde{\Phi}(\xi_1, \xi_2, 0) \right)^{-1} = 0,$$
$$\lim_{\xi_1 \to 0} \frac{\overset{2}{\partial}}{\partial \xi_1} \sum_{l \geqslant 1} \frac{\partial^l}{\partial X^l} (X^m \widetilde{\Phi}(\xi_1, \xi_2, 0))(X)^l \left(\frac{\overset{2}{\partial}}{\partial \xi_1} X^m \widetilde{\Phi}(\xi_1, \xi_2, 0) \right)^{-1} = 0.$$

Similarly, all other limits of this kind are equal to zero.

Thus relation (4.13) implies that

$$\widetilde{\frac{\partial S}{\partial t}}X - \left(\widetilde{\frac{\partial S}{\partial x}}\right)^2 (\widetilde{\rho}(0,\xi_2,0))\frac{\partial X^k}{\partial \xi_1} = X\Theta_1,$$

where $\Theta_1 = \Theta_1(\xi_1, \xi_2)$ is a continuous function tending to zero as $\xi_1 \to 0$. Since $|\partial S/\partial t| \neq 0$, the latter relation yields

$$\lim_{\xi_1 \to 0} \frac{\partial X^k}{\partial \xi_1} X^{-1} = c(\xi_2) = \operatorname*{const}_{\xi_1}. \tag{4.15}$$

Let us prove that the function $c(\xi_2)$ is positive. Assume the contrary. By (4.13), we have

$$\frac{\partial X^k}{\partial \xi_1} = (c+\theta)X, \qquad \theta \to 0 \quad \text{for} \quad \xi_1 \to 0.$$

Choose a sufficiently small δ so that the inequality $0 < \xi_1 < \delta$ holds for $c + \theta < 0$. By integrating the latter equation for $0 < \xi_1 < \delta$ and taking into account the flow continuity condition, we get

$$X^k = \int_0^{\xi_1} (c+\theta)X \, d\xi_1 < 0.$$

This is impossible since the function X is nonnegative. The obtained contradiction proves that $\partial S/\partial t > 0$.

Let us prove that

$$I = \lim_{\xi_1 \to 0} \frac{X}{\xi_1^\alpha} = \text{const} \neq 0, \qquad \alpha = \frac{1}{k-1}. \tag{4.16}$$

By the L'Hospital rule,

$$\lim_{\xi_1 \to 0} \frac{X}{\xi_1^\alpha} = \lim_{\xi_1 \to 0} \frac{X^k}{\xi_1^{\alpha k}} = \lim_{\xi_1 \to 0} \frac{\partial X^k}{\partial \xi_1}(\alpha k \xi_1^{\alpha k - 1})^{-1}.$$

For $\xi_1 \to 0$,

$$(I + o(1))\widetilde{\frac{\partial S}{\partial t}} - \left(\left(\widetilde{\frac{\partial S}{\partial x}}\right)^2 \widetilde{\rho}\right)\Bigg|_{\xi_1 = 0} k(I + o(1))^k = (I + o(1))\theta,$$

and we have (4.16) since the constant c is strictly positive.

Thus our lemma is proved for $m > 1$.

For $m = 1$, (4.13) can be reduced to the form:

$$\left(\widetilde{\frac{\partial S}{\partial t}} + \widetilde{\frac{\partial S}{\partial x}}\widetilde{\Phi}(0,\xi_2,0)\right)\Bigg|_{\xi_1=0} X - \left(\widetilde{\frac{\partial S}{\partial x}}\right)^2 \widetilde{\rho}(0,\xi_2,0)\frac{\partial X^k}{\partial \xi_1} = Xo(1).$$

We can deal with this relation precisely as with (4.11). Here and below the asymptotic expansion of the function χ with respect to powers of ξ_1 (or τ) can be constructed by using the methods of the theory of perturbations. Lemma 4.1 is proved.

Lemma 4.2. *Let the conditions of theorem be satisfied. Let either* $|\frac{\partial S}{\partial t}|_{S=0} \neq 0$ *for* $m > 1$ *or* $|\frac{\partial S}{\partial t} + \Phi(x,t,0)\frac{\partial S}{\partial x}|_{S=0} \neq 0$ *for* $m = 1$, *and let the parameters of equation* (4.1) *satisfy conditions* I.2. (*see* (4.8)).

Then in a neighborhood of the zero surface of the function $S(x,t)$ *the following relations hold:*

(1) *if* $m > 1$, $k+q > 2$, *then* $G(x,t,0)\frac{\partial S}{\partial t} > 0$,

(2) *if* $m = 1$, $k+q > 2$, *then* $G(x,t,0)(\frac{\partial S}{\partial t} + \Phi(x,t,0)\frac{\partial S}{\partial x}) > 0$,

(3) *if* $m > 1$, $k+q = 2$, *then* $G(x,t,0)\frac{\partial S}{\partial t} > 0$,

(4) *if* $m = 1$, $k+q = 2$, *then* $G(x,t,0)(\frac{\partial S}{\partial t} + \Phi(x,t,0)\frac{\partial S}{\partial x}) > 0$, *and the statement of Theorem* 4.1 *holds.*

Proof. The cases $m > 1$ and $m = 1$ are considered separately. For $m > 1$, similarly to Lemma 4.1, relation (4.6) has the form:

$$\widetilde{\frac{\partial S}{\partial t}}X - \left(\widetilde{\frac{\partial S}{\partial x}}\right)^2 \widetilde{\rho}(0,\xi_2,0)\frac{\partial X^k}{\partial \xi_1} - \int_0^{\xi_1} X^q \widetilde{G}(0,\xi_2,0)\,d\xi_1 = Xo(1). \tag{4.17}$$

Here one must consider three situations (the flow and function continuity conditions provide that the following limits exist):

$$\begin{aligned} I_1 &= \lim_{\xi_1\to\infty} \frac{\partial X^k}{\partial \xi_1}X^{-1} = \text{const},\\ I_2 &= \lim_{\xi_1\to 0}\int_0^{\xi_1} X^q\,d\xi_1 = \text{const},\\ I_3 &= \lim_{\xi_1\to 0}\frac{\partial X^k}{\partial \xi_1}\left(\int_0^{\xi_1} X^q\,d\xi_1'\right)^{-1} = \text{const}. \end{aligned} \tag{4.18}$$

Actually, dividing (4.17) by X, we get

$$\frac{\partial S}{\partial t} - \widetilde{\rho}(0,\xi_2,0)\frac{\partial X^k}{\partial \xi_1}X^{-1} - X^{-1}\int_0^{\xi_1} X^q \widetilde{G}\,d\xi_1' = o(1).$$

Thus, if the limits I_1 or I_2 are equal to $\pm\infty$, then we have (4.18). But if one of the limits I_1 or I_2 is finite, the other is also finite.

Assume that $I_1 = \text{const}$. Let us calculate the limit I_2 as $\xi_1 \to 0$, using the flow continuity condition $\frac{\partial X^k}{\partial \xi_1}|_{X=0} = 0$:

$$\begin{aligned} I_2 &= \lim_{\xi_1\to 0}\frac{1}{X^k}X^{k-1}\int_0^{\xi_1} X^q\,d\xi_1'\\ &= \lim_{\xi_1\to 0}(k-1)X^{k-2}\frac{\partial X}{\partial \xi_1}\left(\frac{\partial X^k}{\partial \xi_1}\right)^{-1}\int_0^{\xi_1} X^q\,d\xi + \lim_{\xi_1\to 0} X^q\frac{\partial X^k}{\partial \xi_1}\left(\frac{\partial X^k}{\partial \xi_1}\right)^{-1}\\ &= \lim_{\xi_1\to 0}\frac{k-1}{k}X^{-1}\int_0^{\xi_1} X^q\,d\xi + \lim_{\xi_1\to 0} X^q = \frac{k-1}{k}I_2. \end{aligned}$$

Hence, $I_2 = 0$. Thus this case reduces to the case considered in Lemma 4.1 and

$$X = O(\xi_1^\alpha), \qquad \alpha = 1/(k-1).$$

Consider the case when $I_2 = \text{const}$. Let us calculate the limit I_1 by using the L'Hospital rule:

$$I_1 = \lim_{\xi_1 \to 0} X^{q-1} \frac{\partial X^k}{\partial \xi_1} X^{-q} = \lim_{\xi_1 \to 0} k \int_0^{\xi_1} \frac{\partial X^{k+q-1}}{\partial \xi_1} d\xi_1 \left((k+q-1) \int_0^{\xi_1} X^q \, d\xi_1 \right)^{-1}$$
$$= \frac{k}{k+q-1} \lim_{\xi_1 \to 0} X^{k+q-1} \left(\int_0^{\xi_1} X^q \, d\xi_1 \right)^{-1} = \lim_{\xi_1 \to 0} X^{k+q-2} = 0.$$

Then (4.17) can be reduced to the form:

$$\frac{\partial S}{\partial t} X - \int_0^{\xi_1} X^q \widetilde{G}(0, \xi_2, 0) \, d\xi_1 = X\Theta, \tag{4.19}$$

where $\Theta(\xi_1 . \xi_2)$ is a continuous function tending to zero as $\xi_1 \to 0$. Precisely as in Lemma 4.1, when equation (4.15) was considered, we get

$$X = O(\xi_1^\alpha), \quad \xi_1 \to 0, \qquad \alpha = \frac{1}{1-q}, \qquad \widetilde{G}(0, \xi_2, 0) \widetilde{\frac{\partial S}{\partial t}} > 0.$$

Now consider (4.18):

$$\lim_{\xi_1 \to 0} \frac{\partial X^k}{\partial \xi_1} \left(\int_0^{\xi_1} X^q \, d\xi_1 \right)^{-1} = \text{const}. \tag{4.20}$$

We calculate the limit

$$I_2^{-1} = I_4 = \lim_{\xi_1 \to 0} X \left(\int_0^{\xi_1} X^q \, d\xi_1 \right)^{-1}.$$

Here, three different cases are also possible. First, let $I_4 = \infty$. Then $I_2 = 0$ and this case was considered above when condition (4.19) was examined. Hence, only two cases remain:

$$I_4 = 0, \tag{4.21}$$
$$I_4 = \text{const} \neq 0. \tag{4.22}$$

Consider the case $I_4 = 0$. Then, by assumption, (4.15) yields the equation:

$$-\left(\widetilde{\frac{\partial S}{\partial x}} \right)^2 \widetilde{\rho}(0, \xi_2, 0) \frac{\partial X^k}{\partial \xi_1} - \int_0^{\xi_1} X^q (\widetilde{G}(0, \xi_2, 0) + \Theta) \, d\xi_1 = 0.$$

Similarly to Lemma 4.1, this implies that $X = O(\xi_1^\alpha)$, $\xi_1 \to 0$, $\alpha = 2/(k-q)$. This estimate contradicts assumption (4.21) and condition (4.8). (This can be verified by direct substitution of the exponential function X into I_4). Hence, only the case (4.22) remains to be considered.

Assume that relations (4.20), (4.22) hold. Then the condition (4.17) is satisfied. Similarly to Lemma 4.1, we find

$$X = O(\xi_1^{\alpha}), \qquad \xi_1 \to 0, \qquad \alpha = \frac{1}{1-q} = \frac{1}{k-1}, \qquad k+q=2. \tag{4.23}$$

This estimate contradicts the conditions $m > 1$, $k+q > 2$, which hold by assumption. The lemma is proved for $m > 1$, $k+q > 2$.

Now let us consider the case $m = 1$, $k+q > 2$. Here the equation (4.13) yields:

$$\left(\widetilde{\frac{\partial S}{\partial t}} + \widetilde{\frac{\partial S}{\partial x}}\widetilde{\Phi}(0,\xi_2,0)\right)X - \left(\widetilde{\frac{\partial S}{\partial x}}\right)^2 \widetilde{\rho}(0,\xi_2,0)\frac{\partial X^k}{\partial \xi_1} - \int_0^{\xi_1} X^q \widetilde{G}(0,\xi_2,0)\,d\xi_1 = X\Theta, \tag{4.24}$$

where $\Theta = \Theta(\xi_1,\xi_2)$ is a continuous function tending to zero as $\xi_1 \to 0$. Here two situations are possible: $I_1 = \mathrm{const}$ or $I_2 = \mathrm{const}$ and they are considered as above (see (4.18)–(4.20)).

Now we shall examine the case $m > 1$, $k+q = 2$. In this situation, equation (4.13) has the form:

$$\widetilde{\frac{\partial S}{\partial t}}X - \left(\widetilde{\frac{\partial S}{\partial x}}\right)^2 \widetilde{\rho}(0,\xi_2,0)\frac{\partial X^k}{\partial \xi_1} - \int_0^{\xi_1} X^q G(0,\xi_2,0)\,d\xi_1 = X\Theta. \tag{4.25}$$

Here three different cases are also possible (see (4.18)), but they are treated in another way.

Actually, we assume that (4.18) holds and calculate I_4. Here also two different cases are possible: (4.21), (4.22). The calculation of I_4 for (4.21) results, as above, in a contradiction. For (4.22), we obtain the estimate (4.23) and the relation $G(\xi_1,\xi_2,0)\frac{\partial S}{\partial t} > 0$, thus our lemma is proved.

Let us consider the forth case, which is the last: $m = 1$, $k+q = 2$. Here the previous proof is repeated, but we get the estimate:

$$\widetilde{G}(0,\xi_2,0)\left(\widetilde{\frac{\partial S}{\partial t}} + \widetilde{\frac{\partial S}{\partial x}}\widetilde{\Phi}(0,\xi_2,0)\right) > 0.$$

Lemma 4.2 is proved.

Lemma 4.3. *Let the conditions of Theorem 4.1 be satisfied. Let the parameters of equation (4.1) satisfy conditions I.3 (see (4.9). Then, in a neighborhood of the zero surface of the function $S(x,t)$, the following inequalities:*

$$\Phi(x,t,0)\frac{\partial S}{\partial x} > 0, \qquad m < 1,$$

or

$$\Phi(x,t,0)\frac{\partial S}{\partial x}+\frac{\partial S}{\partial t}>0, \qquad m=1,$$

and the statement of Theorem 4.1 hold.

Proof. Here we must separate the cases $m<1$ and $m=1$. First let $m<1$. Similarly to Lemma 4.1, the relation (4.13) can be reduced to the form:

$$\widetilde{\frac{\partial S}{\partial t}}X+\widetilde{\Phi}(0,\xi_2,0)\widetilde{\frac{\partial S}{\partial x}}X^m-\left(\widetilde{\frac{\partial S}{\partial x}}\right)^2\widetilde{\rho}(0,\xi_2,0)\frac{\partial X^k}{\partial \xi_1}$$
$$-\int_0^{\xi_1}X^q\widetilde{G}(0,\xi_2,0)\,d\xi_1=X\Theta. \quad (4.26)$$

Considerations similar to those in Lemma 4.1 (see (4.15)) show that it is necessary to analyze the following situations:

$$I_1=\lim_{\xi_1\to 0}X^m=\operatorname*{const}_{\xi_1}, \quad (4.27)$$

$$I_2=\lim_{\xi_1\to\infty}\frac{\partial X^k}{\partial\xi_1}X^{-1}=\operatorname*{const}_{\xi_1}, \quad (4.28)$$

$$I_3=\lim_{\xi_1\to 0}X^{-1}\int_0^{\xi_1}X^q\,d\xi_1=\operatorname*{const}_{\xi_1}, \quad (4.29)$$

$$I_4=\lim_{\xi_1\to 0}\frac{\partial X^k}{\partial\xi_1}X^{-m}=\operatorname*{const}_{\xi_1}, \quad (4.30)$$

$$I_5=\lim_{\xi_1\to 0}\frac{\partial X^k}{\partial\xi_1}\left(\int_0^{\xi_1}X^q\,d\xi_1\right)^{-1}=\operatorname*{const}_{\xi_1}, \quad (4.31)$$

$$I_6=\lim_{\xi_1\to 0}X^{-m}\left(\int_0^{\xi_1}X^q\,d\xi_1\right)^{-1}=\operatorname*{const}_{\xi_1}. \quad (4.32)$$

The flow continuity condition and the fact that the function X is continuous imply that these limits exist.

The assumption that (4.27) holds immediately leads to a contradiction. Actually, by using the L'Hospital rule, we get

$$I_1=\lim_{\xi_1\to\infty}X^{m-1}=\infty, \qquad m<1.$$

Hence

$$\lim_{\xi_1\to\infty}X^{1-m}=0. \quad (4.33)$$

Assume that (4.28) holds. Calculate I_3:

$$I_3=\lim_{\xi_1\to 0}X^{-1}\int_0^{\xi_1}X^q\,d\xi_1$$
$$=\lim_{\xi_1\to 0}(k-1)X^{k-2}\frac{dX}{d\xi_1}\left(\frac{dX^k}{d\xi_1}\right)^{-1}\int_0^{\xi_1}X^q\,d\xi_1+\lim_{\xi_1\to 0}X^q\frac{dX^k}{d\xi_1}\left(\frac{dX^k}{d\xi_1}\right)^{-1}$$
$$=\lim_{\xi_1\to 0}\frac{k-1}{k}X^{-1}\int_0^{\xi_1}X^q\,d\xi_1+\lim_{\xi_1\to 0}X^q=\frac{k-1}{k}I_3+0.$$

Since $k > 1$, this implies $I_3 = 0$.

Further, calculate I_4:

$$I_4 = \lim_{\xi_1 \to 0} \frac{\partial X^k}{\partial \xi_1} X^{-m} = \lim_{\xi_1 \to 0} X^{1-m} \cdot \underset{\xi_1}{\text{const}} = 0.$$

Calculate I_5:

$$I_5 = \lim_{\xi_1 \to 0} \frac{\partial X^k}{\partial \xi_1} \left(\frac{\partial X}{\partial \xi_1} \right)^{-1} \lim_{\xi_1 \to 0} \frac{\partial X}{\partial \xi_1} \left(\int_0^{\xi_1} X^q \, d\xi_1 \right)^{-1}$$
$$= \text{const} \lim_{\xi_1 \to 0} \frac{\partial X^k}{\partial \xi_1} (kX^{k+q-1})^{-1} = \text{const} \lim_{\xi_1 \to 0} (kX^{k+q-2})^{-1} = \infty.$$

Hence

$$\lim_{\xi_1 \to 0} \left(\frac{\partial X^k}{\partial \xi_1} \right)^{-1} \int_0^{\xi_1} X^q \, d\xi_1 = 0.$$

Calculate I_6:

$$I_6^{-1} = \lim_{\xi_1 \to 0} X^{-m} \int_0^{\xi_1} X^q \, d\xi_1 = \lim_{\xi_1 \to 0} (k+m) \frac{dX^k}{d\xi_1} \left(k \frac{dX^{k+m}}{d\xi_1} \right)^{-1} \int_0^{\xi_1} X^q \, d\xi_1$$
$$= \lim_{\xi_1 \to 0} (k+m) X^k (kX^{k+m})^{-1} \int_0^{\xi_1} X^q \, d\xi_1$$
$$- \lim_{\xi_1 \to 0} (k+m)(kX^{k+m})^{-1} \int_0^{\xi_1} X^{k+q} \, d\xi_1$$
$$= \left(1 + \frac{m}{k}\right) I_6^{-1} - \lim_{\xi_1 \to 0} (k+m)(kX^{k+m})^{-1} \int_0^{\xi_1} X^{k+q} \, d\xi_1.$$

Here, as above, we integrate by parts. The second summand is zero according to item (3) in Definition 4.1. This yields

$$I_6^{-1} - \left(1 + \frac{m}{k}\right) I_6^{-1} = 0, \qquad I_6^{-1} = 0.$$

Hence

$$\lim_{\xi_1 \to 0} X^{-m} \int_0^{\xi_1} X^q \, d\xi_1 = 0.$$

Thus we obtain the contradiction: (4.32) implies that only one summand $\widetilde{\Phi} \frac{\partial X^m}{\partial \xi_1}$ possesses the principal singularity and nothing can compensate this singularity.

Further, consider the next case. Assume that condition (4.30) is satisfied. As shown above (see (4.33), $I_1^{-1} = 0$. Calculate I_2:

$$I_2 = \underset{\xi_1}{\text{const}} \lim_{\xi_1 \to 0} mX^{m-1} = \infty.$$

Hence

$$\lim_{\xi_1 \to 0} \frac{\partial X}{\partial \xi_1} \left(\frac{\partial^2 X^k}{\partial \xi_1^2} \right)^{-1} = 0.$$

Calculate I_5:

$$I_5 = \lim_{\xi_1\to 0} \frac{\partial X^m}{\partial \xi_1} X^{-q} \underset{\xi_1}{\text{const}} = \underset{\xi_1}{\text{const}} \lim_{\xi_1\to 0} \left(\frac{\partial X^k}{\partial \xi_1} X^{-m}\right) \lim_{\xi_1\to 0} \frac{1}{X^{k+q-1-m}} = \infty.$$

Hence

$$\lim_{\xi_1\to 0} X^q \left(\frac{\partial X^m}{\partial \xi_1}\right)^{-1} = 0.$$

Further, calculate I_6:

$$I_6 = \underset{\xi_1}{\text{const}} \lim_{\xi_1\to 0} \frac{1}{X^{k+q-1-m}} = \infty.$$

Hence

$$\lim_{\xi_1\to 0} X^q \frac{\partial X^m}{\partial \xi_1} = 0.$$

Thus, (4.26) takes the form:

$$\widetilde{\frac{\partial S}{\partial t}} \widetilde{\Phi}(0, \xi_2, 0) X^m - \frac{\partial X^k}{\partial \xi_1} (\widetilde{\rho}(0, \xi_2, 0) + \Theta_1) = 0.$$

By analogy to Lemma 4.2, this yields the estimate

$$X = O(\xi_1^\alpha), \qquad \xi_1 \to 0, \qquad \alpha = \frac{1}{k-m}, \qquad \widetilde{\Phi}(0, \xi_2, 0) \widetilde{\frac{\partial S}{\partial x}} > 0.$$

Further, similarly to Lemma 4.1, we consider the situations (4.30) and (4.31). Both of them lead to contradictions. Thus Lemma 4.3 is proved for $m > 1$.

Let us consider the case $m = 1$. We return to the situation studied in Lemma 4.2 (see (4.24)). Thus Lemma 4.3 is proved completely.

Lemma 4.4. *Let the conditions of Theorem* 4.1 *be satisfied. Let the parameters of equation* (4.1) *satisfy conditions* I.4 *(see* (4.10)*). Then in a neighborhood of the zero surface of the function* $S(x, t)$ *the following inequalities hold:*
if $0 < q < m$, $0 < m \leqslant 1$, $k + q > 2m$, *then*

$$\left(G\Phi \frac{\partial S}{\partial x}\right) < 0,$$

if $0 < q < m$, $m = 1$, $k + q = 2$, *then*

$$\frac{\partial S}{\partial t} + \frac{\partial S}{\partial x}\Phi < 0,$$

and the statement of Theorem 4.1 *holds.*

The *proof* is similar to that of Lemma 4.2.

Lemma 4.5. *Let the conditions of Theorem 4.1 be satisfied. Let the parameters of equation (4.1) satisfy conditions II (see (4.6). Then in a neighborhood of the zero surface of the function $S(x,t)$ the statement of Theorem 4.1 holds.*

Proof. In condition II (4.6) there are two cases: II.1 and II.2. Similarly to the preceding lemmas, we consider the case of rigorous inequalities in (4.6), (4.10), (4.11). Then the proofs of the preceding lemmas are repeated.

There is one more lemma, in which the first situation from Table 3.4.2 is considered.

Lemma 4.6. *Let the conditions of Theorem 4.1 be satisfied. Let the parameters of equation (4.1) satisfy the conditions, $q \leqslant 1$, $k > q$, $k+q \leqslant 2m$ and $\partial S/\partial t = 0$. Then in a neighborhood of the zero surface of the function $S(x,t)$ the inequality holds: $G(x,t,0) < 0$.*

Proof. Here we must consider the cases $q < 1$ and $q = 1$ separately. First consider $q < 1$, $k + q < 2m$. The relation (4.13) reduces to the form:

$$\widetilde{\Phi}(0,\xi_2,0)\widetilde{\frac{\partial S}{\partial x}}X^m - -\left(\widetilde{\frac{\partial S}{\partial x}}\right)^2\widetilde{p}(0,\xi_2,0)\frac{\partial X^k}{\partial \xi_1} - \int_0^{\xi_1} X^q\widetilde{G}(0,\xi_2,0)\,d\xi_1 = X^m o(1).$$

We must analyze the following situations

$$I_1 = \lim_{\xi_1\to 0} X^m\left(\int_0^{\xi_1} X^q\,d\xi_1\right)^{-1} = \underset{\xi_1}{\text{const}}. \tag{4.34}$$

$$I_2 = \lim_{\xi_1\to 0}\frac{\partial X^k}{\partial \xi_1}X^{-m} = \underset{\xi_1}{\text{const}}, \tag{4.35}$$

$$I_3 = \lim_{\xi_1\to 0}\frac{\partial X^k}{\partial \xi_1}\left(\int_0^{\xi_1} X^q\,d\xi_1\right)^{-1} = \underset{\xi_1}{\text{const}}, \tag{4.36}$$

Assume that (4.34) holds. Thus we get a contradiction. Actually, calculate I_2:

$$I_2 = \lim_{\xi_1\to 0}\frac{\partial X^k}{\partial \xi_1}\left(\int_0^{\xi_1} X^q\,d\xi_1\right)^{-1}\lim_{\xi_1\to 0} X^{-m}\int_0^{\xi_1} X^q\,d\xi_1 = \underset{\xi_1}{\text{const}}\, I_3. \tag{4.37}$$

On the other hand, by the L'Hospital rule, we get

$$\begin{aligned} I_2 &= \lim_{\xi_1\to 0} X^{q-m}\frac{\partial X^k}{\partial \xi_1}X^{-q} \\ &= \lim_{\xi_1\to 0}\frac{k}{k+q-m}X^{k+q-2m}\lim_{\xi_1\to 0}\left(X^m\left(\int_0^{\xi_1} X^q\,d\xi_1\right)^{-1}\right) \\ &= \text{const}\lim_{\xi_1} X^{k+q-2m} = \infty, \end{aligned}$$

since $k + q < 2m$. Hence $I_3 = \infty$, $I_3^{-1} = 0$.

Dividing the initial equation by $\int_0^{\xi_1} X^q\, d\xi_1$, we come to the contradiction: as $\xi_1 \to 0$, only the ratio $\frac{\partial X^k}{\partial \xi}\left(\int_0^{\xi_1} X^q\, d\xi_1\right)^{-1}$ increases infinitely, the other summands are finite.

Let relation (4.35) hold. Then, precisely as above, $I_1^{-1} = \infty$, hence relation (4.35) is impossible.

Further, consider the case when (4.36) holds. Calculate I_1:

$$I_1 = \operatorname*{const}_{\xi_1} \lim_{\xi_1 \to 0} X^m \left(\frac{\partial X^k}{\partial \xi}\right)^{-1} = \operatorname*{const}_{\xi_1} \lim_{\xi_1 \to 0} \xi X^{m-k} = 0.$$

Thus, by (4.36) and the condition $\widetilde{\rho} > 0$, we get the estimate

$$X = O(\xi_1^\alpha) \quad \text{for} \quad \xi_1 \to 0, \quad \alpha = \frac{2}{k-q},$$

and the condition $\widetilde{G}(0, \xi_2, 0) < 0$.

The other cases are considered in the same way. Lemma 4.6 is proved completely.

The forthcoming lemmas are proved by the same methods as above.

Lemma 4.7. *Let the conditions of Theorem* 4.1 *be satisfied. Let the parameters of equation* (4.1) *satisfy the conditions* $k > m$, $m > 0$, $2m \leqslant \min(k+q, k+1)$ *and* $\partial S/\partial t = 0$. *Then in a neighborhood of the zero surface of the function* $S(x,t)$ *the inequality holds:*

$$\Phi(x,t,0)\frac{\partial S}{\partial x} > 0.$$

Lemma 4.8. *Let the conditions of Theorem* 4.1 *be satisfied. Let the parameters of equation* (4.1) *satisfy the conditions* $q < m$, $2m \leqslant k + q$, $q \leqslant 1$ *and* $\partial S/\partial t = 0$. *Then in a neighborhood of the zero surface of the function* $S(x,t)$ *the inequality holds:*

$$\Phi(x,t,0)G(x,t,0)\frac{\partial S}{\partial x} > 0.$$

Theorem 4.1 is completely proved.

Proof of Theorem 4.1′ [16]. Precisely as in Theorem 4.1, we restrict ourselves to the case $x \in \mathbf{R}^1$ and put $m > 1$. As above, the proof reduces to the consideration of the following equation (compare with (4.17)):

$$b\frac{\partial X}{\partial \xi_1} - D^2 \frac{\partial^2 X^k}{\partial \xi_1^2} - X^q \widetilde{G} = \widetilde{f}, \tag{4.38}$$

where

$$b = \left.\frac{\partial \widetilde{S}}{\partial t}\right|_{\xi_1=0}, \qquad D^2 = \left(\frac{\partial \widetilde{S}}{\partial x}\right)^2 \rho(0, \xi_2, 0).$$

The right-hand side $\widetilde{f}$ is defined in (4.13). The proof of Theorem 4.1 shows that, for $k + q > 2$, the summand $X^q G$ is "less" singular (i.e., smoother) than the other summands on the left-hand side of (4.38) and thus all solutions are structurally stable. The structural instability (actually, this means that the asymptotic expansion of the solution contains functions other than certain powers of their arguments) is caused by resonances, for which to occur, it is necessary that the equation with zero right-hand side possess more than one nontrivial solution.

This is possible if all three summands on the left-hand side of (4.38) are of the same order of smallness as $\xi_1 \to 0$. In turn, this is possible (see Theorem 4.1) only for $k+q=2$.

First we consider the equation with zero right-hand side, i.e., $\tilde{f} = 0$. By the transformation

$$\frac{\partial X^k}{\partial \xi_1} = \frac{\partial \theta}{\partial \zeta}(\theta^{-1}(X)) \tag{4.39}$$

(θ^{-1} is inverse to θ), equation (4.38) can be reduced to a linear ordinary differential equation (see Theorem 1.1 of Chapter I):

$$b\frac{\partial \theta}{\partial \zeta} - D^2\frac{\partial^2 \theta}{\partial \zeta} - k\tilde{G}\theta = 0, \qquad k+q=2. \tag{4.40}$$

Note that the localizing transformation (4.39) takes the point $\xi_1 = 0$ to the point $\zeta = -\infty$ (for $\tilde{G} < \infty$, to the point $\zeta = \infty$). The principal term of the asymptotic series for the solution of equation (4.38) is expressed in terms of solutions of the linear equation (4.40). These solutions have the form:

$$\begin{gathered} \theta = c_1 e^{\lambda_1 \zeta} + c_2 e^{\lambda_2 \zeta} \\ \lambda_{1,2} = \frac{b \pm \sqrt{b^2 - 4D^2\tilde{G}k}}{2D^2}, \qquad b \geqslant 2D\sqrt{\tilde{G}k}. \end{gathered} \tag{4.41}$$

Consider the case $b = b_{\min} = 2D\sqrt{\tilde{G}k}$, then the general solution of equation (4.40) has the form

$$\theta = (c_1 + c_2\zeta)e^{\lambda\zeta}, \qquad \lambda = \lambda_1 = \lambda_2 = \frac{\sqrt{\tilde{G}k}}{D} \tag{4.42}$$

and is related to the solution of equation (4.38)

$$\begin{gathered} X = \left(\frac{\lambda_1(k-1)}{k}\right)^{1/(k-1)} \left(\xi_1^{1/(k-1)} + (B\xi_1)^{1/(k-1)-1} \int_0^{\xi_1} \frac{d\tau}{\ln(B\tau)}\right) + \dots, \\ B = \frac{(k-1)\sqrt{\tilde{G}k}}{Dk} > 0 \end{gathered} \tag{4.43}$$

(dots stand for the summands tending to zero as $\xi_1 \to +0$ faster than those written out in this formula).

Actually, we express the derivative of the function θ in terms of the function itself, i.e.,

$$\frac{d\theta}{d\zeta} = \lambda\theta + \frac{c_2\theta}{c_2\zeta + c_1}. \tag{4.44}$$

For sufficiently large ζ, we have:

$$\ln\theta = \lambda\zeta + \ln c_1 + \ln\delta\zeta + 2\Big[\frac{1}{2\delta\zeta+1} + \frac{1}{3(2\delta\zeta+1)^3} + \dots\Big], \qquad \frac{c_2}{c_1} = \delta < 0.$$

Therefore,

$$\frac{d\theta}{d\zeta} \sim \lambda\theta\Big(1 + \frac{1}{\ln\theta}\Big) \qquad \text{for} \quad \zeta \to -\infty.$$

By (4.39) and (4.44), we get the equation for the function X:

$$\frac{dX^k}{d\xi_1} \sim \lambda X\Big(1 + \frac{1}{\ln X}\Big) \qquad \text{for} \quad \xi_1 \to 0. \tag{4.45}$$

The function X is sought in the form

$$X = X_0(\xi_1) + X_1(\xi_1),$$

where $X_1/X_0 \to 0$ as $\xi_1 \to 0$. Then the equation (4.39) takes the form:

$$kX_0^{k-2}\Big[1 + \frac{(k-2)X_1}{X_0} + o\Big(\frac{X}{X_0}\Big)\Big]\frac{d(X_0+X_1)}{d\xi_1}$$
$$= \lambda\Big(1 + \frac{1}{\ln X_0 + \ln(\frac{X_1}{X_0}+1) + \text{const}}\Big).$$

Collecting the principal terms in this equation, we obtain the equation for X_0:

$$kX_0^{k-2}\frac{dX_0}{s\xi_1} = \lambda,$$

whose solution has the form

$$X_0 = \Big(\frac{\lambda(k-1)}{k}\Big)^{1/(k-1)}\xi_1^{1/(k-1)}. \tag{4.46}$$

Similarly, we get the equation for X_1:

$$\frac{dX_1}{d\xi_1} + \frac{X_1}{X_0}\Big(\frac{dX_0}{d\xi_1}(k-2)\Big) = \frac{\lambda X_0^{2-k}}{l\ln X_0},$$

whose solution has the form:

$$X_1 = \frac{k-1}{k}(B\xi_1)^{\alpha-1}\int_0^{\xi}\frac{d\xi}{\ln(B\xi)}. \tag{4.47}$$

Formula (4.43) is obtained.

In the same way, in the case of constant coefficients, the transformation (4.39) can reduce (4.38) to an equation of the form (4.40) with the right-hand side containing nonlinear summands (powers of the function θ: θ^{k+l}, θ^{2k-1+l_i}, θ^{m_i+1}, $\theta^{m+k-1+n_i}$,

θ^{m-1+n_i}). In general, when an asymptotic solution is constructed in the form of exponential series, a nonlinear right-hand side may cause resonances. The conditions that no resonances occur are given by the following inequalities:

$$\frac{\lambda_2}{\lambda_1} \neq mN_1 + kN_2 + N_3, \tag{4.48}$$

where N_1, N_2, N_3 are natural numbers from the set $N \cup \{0\}$.

If the inequality (4.48) holds, the solution of the equation (4.40) with right-hand side can be expanded into an asymptotic exponential series, and in the resonance case into an exponential series with polynomial coefficients. Respectively, by (4.39), the solution of (4.38) can be expanded into an asymptotic series in powers of ξ_1 if there are no resonances, and similarly to (4.43), in the resonance case, the asymptotic expansion contains summands with logarithms*.

If the resonance occur, the first nonexponential summand in the asymptotic expansion has the form

$$c_n(B\xi_1)^{\alpha-1} \int_0^{\xi_1} \ln(\xi B)(B\xi)^{\lambda_2/\lambda_1-1}\, d\xi, \tag{4.49}$$

where $C_n = \text{const} > 0$, $B = (\lambda(k-1)/k)$.

The resonance condition (4.48) means that the parameter b belongs to the sequence $\{b_i\}$,

$$b_i = \frac{(\beta+1)D\sqrt{\widetilde{G}k}}{\sqrt{\beta}}, \tag{4.50}$$

where the parameter β takes the values

$$\beta \in \{mN_1 + kN_2 + N_3\}, \qquad N_j \in N \cup \{0\}, \quad j = 1,2,3.$$

As will be shown below, only the asymptotic solutions of (4.38) that can be expanded into an asymptotic series with respect to a system of functions from $\mathcal{P}\{\xi_1\theta(\xi_1)\}$ are structurally stable.

First, note that, for any value of b, the principal term of the asymptotic expansion of the solution to (4.38) has the form (4.46).

Let us consider an asymptotic local self-similar solution of equation (4.38) satisfying assumption (1) of Definition 4.4. If the solution χ corresponds to $b \in \{b_i\}$, then, by (4.46), the solution χ_δ corresponds to $b = b_\delta \notin \{b_i\}$ for a sufficiently small δ, since the sequence $\{b_i\}$ has no accumulation points. Hence the solution can be expanded in an asymptotic series with respect to a system of functions from $\mathcal{P}\{\xi_1\theta(\xi_1)\}$ and, for $b \in \{b_i\}$, the asymptotic local self-similar solution of equation (4.1) is structurally unstable. This fact and the nondegeneracy of the change of variables $(x,t) \to (\xi_1,\xi_2)$, $\xi_1 = S(x,t)$, $\xi_2 = S(x-t, x+t)$, immediately prove our theorem in the case of constant coefficients.

*It is easy to see that in the resonance case the sequence of generating functions (see above the definition of a minimal sequence) related to the domain, on which an asymptotic solution is constructed, is infinite.

If the coefficients of the initial equation (1.1) depend on the variables (x,t), it is convenient to construct an asymptotic expansion of the solution X by using (4.13). Let $b > b_{\min}$ (the case $b = b_{\min}$ was considered above). Then the leading asymptotic solution has the form

$$X = M\xi_1^\alpha + X_1, \quad M = B^\alpha, \quad \alpha = \frac{1}{k-1}, \quad X_1/\xi_1^\alpha \to 0 \quad \text{for} \quad \xi_1 \to 0. \tag{4.51}$$

By (4.38), the correction X_1 satisfies the equation:

$$b\frac{\partial X_1}{\partial \xi_1} - D^2 k M^{k-1}\frac{d^2(\xi X_1)}{d\xi_1^2} - \widetilde{G}(2-k)M^{1-k}X_1/\xi_1 = f, \tag{4.52}$$

where

$$\begin{aligned} f_1 = {} & A_1 X + \sum_{j\geqslant 0} A_{2j}X^{k+l_j} + \sum_{j\geqslant 0} A_{3j}X^{l_j} + \sum_{j\geqslant 0} A_{4j}X^{m+n_j} + \sum_{j\geqslant 0} A_{5j}X^{n_i}\frac{dX^m}{d\xi_1} \\ & + \sum_{j\geqslant 0} A_{6j}X^{q+m_i} + D^2\frac{d^2}{d\xi_1^2}[(M\xi_1^\alpha + X_1)^k - M^k\xi_1^{\alpha k} - M^{k-1}k\xi_1 X_1] \\ & + \widetilde{G}(M\xi_1^\alpha + X_1)^{2-k} - M^{2-k}\xi_1^{\alpha(2-k)} - (2-k)M^{1-k}X_1/\xi_1, \end{aligned}$$

and $A_{\mu j} = A_{\mu j}(\xi_1, \xi_2)$ are smooth functions that can be expanded into series in ξ_1.

Denote by L the linear operator on the left-hand side of the last equation. Obviously, we have

$$L\xi_1^{\gamma+1} = L(\gamma)\xi_1^\gamma.$$

In this case the above condition (saying that the right-hand side is not in resonance) turns into the following one: the absence of resonance is equivalent to the inequality $L(\gamma) \neq 0$. The latter inequality is a condition imposed on the parameter b:

$$\begin{aligned} & L(\gamma) = 0 \Leftrightarrow b = b(\gamma) = D^2(\alpha+1)M^{k-1} + M^{1-k}\widetilde{G}/\alpha, \\ & M^{k-1} = D\sqrt{\widetilde{G}.k(\gamma+1)}\sqrt{\frac{\gamma+1-2\alpha+\alpha k}{\alpha(\alpha-\gamma-2)}}. \end{aligned} \tag{4.53}$$

Obviously, when an asymptotic solution is constructed for the equation (4.40), the exponents of γ that appear on the right-hand side of (4.52) take the following values

$$\gamma \in \{\alpha m N_1 + \alpha N_2 + N_3\}, \qquad N_i \in N \cup \{0\}, \qquad i = 1,2,3. \tag{4.54}$$

Therefore the resonance set $\{b_i\}$ of values of the parameter b is given by formula (4.53), where γ belongs to the set (4.54).

Further, all considerations literally coincide with the analysis of equations with constant coefficients. Theorem 4.1′ is proved.

The next statement is required to construct asymptotic solutions in the subsequent chapter.

Consider the ordinary differential equation

$$b\frac{\partial \chi}{\partial \tau} - \frac{\partial}{\partial \tau}\left(\rho \frac{\partial \chi^k}{\partial \tau}\right) - \chi^q G = 0, \qquad \chi\big|_{\tau=0} = 0, \tag{4.55}$$

where $k > 1$, $q > 0$, $k + q \geqslant 2$, $b = b(\xi)$, $\rho = \rho(\xi, \chi)$, $G = G(\xi, \chi)$ are functions depending on a certain parameter $\xi \in \mathbf{R}^m$ and satisfying the conditions $G(\xi, \chi)$, $b(\xi)$, $\rho(\xi, \chi) \in C^\infty$ for $\chi \geqslant 0$,

$$b(\xi), \qquad \rho(\xi, \chi), \qquad G(\xi, \chi) > 0.$$

Denote

$$K_1 = \lim_{\tau \to 0} \frac{k\chi^{k-1}\rho}{\tau}.$$

Lemma 4.9. *For any solution of equation* (4.55) *that is monotone in a certain neighborhood* $\Omega \in (\mathbf{R}^1_+ \times \mathbf{R}^1_+)$ *of the point* $(\chi = 0, \tau = 0)$ *the estimate*

$$b \leqslant K_1(2\alpha + 1), \qquad \alpha = 1/(k-1) \tag{4.56}$$

holds for $k + q > 2$.

Proof. The proof of this lemma is based on the representation

$$\chi(\xi, \tau) = c_0(\xi)\tau^\alpha + o(\tau^\alpha), \qquad \tau \to 0. \tag{4.57}$$

For solutions of partial differential equations, the existence of this representation was proved in Theorem 4.1 (in this case, the parameters ξ have the meaning of the coordinates on the surface $S = 0$). Therefore, to justify (4.57), one can use Theorem 4.1 with appropriate restrictions imposed on the properties of the solution. However, in our case, (4.57) can be immediately obtained from Theorem 1.1 of Chapter I by repeating the calculations performed when equation (4.40) was analyzed.

By (4.57), we have

$$K_1 = k\rho(\xi, 0)c_0^{k-1}.$$

The equation (4.55) implies

$$b(\xi) = \alpha\rho(\xi, 0)kc_0^{k-1}.$$

Thus, dividing the right-hand side by K_1, we get

$$\frac{b(\xi)}{K_1} = \alpha \leqslant 2\alpha + 1.$$

The lemma is proved.

In the case $k + q = 2$, the inequality given in this lemma does not, generally speaking, hold. As an example consider the equation with constant coefficients and nonlinearity of special form:

$$b\frac{d\chi}{d\tau} - \rho\frac{d^2\chi^k}{d\tau^2} - \chi^{2-k}(1 - \chi) = 0,$$
$$\rho, b = \text{const}, \qquad k > 1.$$

By Theorem 1.1 of Chapter I, this equation corresponds to the following semilinear equation:

$$b\frac{d\theta}{d\xi} - \frac{d^2\theta}{d\xi^2} - k\rho\theta(1-\theta) = 0.$$

Putting $\xi\sqrt{k\rho} = \zeta$, $\widehat{b} = b/\sqrt{k\rho}$, we obtain

$$\widehat{b}\frac{d\theta}{d\zeta} - \frac{d^2\theta}{d\zeta^2} - \theta(1-\theta) = 0.$$

For $\widehat{b} = 5/\sqrt{6}$, we get

$$\theta = (1 + \exp\{-\zeta/\sqrt{6}\})^{-2}.$$

Thus, by Theorem 1.1 of Chapter I, the equation (4.58) reduces to the following one:

$$\rho k\chi^{k-1}\frac{d\chi}{d\tau} = \frac{2\sqrt{\rho k}}{\sqrt{6}}(1-\sqrt{\chi})\chi.$$

The latter relation easily gives the constant c_0 (see (4.57)):

$$c_0 = \left(\frac{2}{\alpha\sqrt{6\rho k}}\right)^{1/(k-1)}.$$

By analogy to Lemma 4.9, we have

$$K_1 = k\rho c_0^{k-1}.$$

Thus

$$\frac{b}{K_1} = \frac{5\sqrt{\rho k}}{\sqrt{6}}\frac{1}{k\rho}\left(\frac{2}{\alpha\sqrt{6\rho k}}\right)^{-1} = \frac{5\alpha}{2}.$$

Therefore, in this case, the inequality $b/K_1 \leqslant 2\alpha + 1$ does not hold, for example, for $k < 3/2$ (i.e., for $\alpha > 2$).

Lemma 4.10. *The statement of Lemma 4.9 holds for $k + q = 2$:*

$$b = b_{\min} = 2\sqrt{k\rho(\xi,0)G(\xi,0)}.$$

Proof. By (4.55), we have for $k + q = 2$:

$$b(\xi) = \alpha\rho(\xi,0)kc_0^{k-1} + \frac{G(\xi,0)}{\alpha c_0^{k-1}}.$$

Obviously, this relation holds only for real values of c_0, such that $b \geqslant b_{\min}$. In particular, for $b \geqslant b_{\min}$, we have

$$c_0^{k-1} = \alpha^{-1}\sqrt{\frac{G(\xi,0)}{\alpha c_0^{k-1}}}.$$

On the other hand, as above,

$$K_1 = \rho(\xi,0)kc_0^{k-1}.$$

Thus

$$b_{\min}/K_1 = 2\alpha < 2\alpha + 1$$

Lemma 4.10 is proved.

3.5. The structure of singularities of nonsmooth self-similar solutions to quasilinear hyperbolic equations

In this section we expose the results obtained when the structure of nonsmooth self-similar solutions to systems of quasilinear hyperbolic equations was studied. In contrast to the previous section, here the type of singularities is not so closely related to the property of being nonlinear. In [50] the following result was stated: for hyperbolic equations, there are three types of singularities, concentrated on a submanifold of codimension 1:

- 1. strong discontinuity of θ-type (θ is the Heaviside function),
- 2. powerlike solutions with exponent 1/2,
- 3. infinitely narrow solitons

(the exact statement is given below).

Following [26, 50], let us consider the system of equations of the first order in the divergence form:

$$L[u] \equiv \frac{\partial}{\partial t}F_0(x,t,u) + \sum_{i=1}^{N} \frac{\partial}{\partial x_i}F_i(x,t,u) = 0, \tag{5.1}$$

where $(x,t) \in \Omega \subset \mathbf{R}^N \times \mathbf{R}_+$, $u \in Q \subset C^M$, $F_i(x,t,u)$ are vector-functions, whose components are entire functions of their arguments, infinitely differentiable in x and t for $(x,t) \in \Omega$ and taking values in C^M. The domain Ω is assumed to be sufficiently small.

To system (4.1) we associate the matrix:

$$A(x,t,p_0,p,u) = p_0\frac{\partial F_0}{\partial u}(x,t,u) + \sum_{i=1}^{N} \frac{\partial F_i}{\partial u}(x,t,u), \tag{5.2}$$

where $p_0 \in \mathbf{R}$, $p = (p_1,\dots,p_N) \in S^{N-1}$, $(x,t) \in \Omega$.

By $\lambda_j = \lambda_j(x,t,p_0,p,u)$, $j = 1,\dots,M$, we denote the eigenvalues of the matrix $A(x,t,p_0,p,u)$ and assume that $\lambda_j \in C^\infty(\overline{\Omega} \times \mathbf{R} \times S^{N-1} \times \Omega)$ and the equations

$$\lambda_j(x,t,p_0,p,u) = 0, \qquad j = 1,\dots,N$$

have different real solutions $p_{0j} = p_{0j}(x,t,p,u)$ belonging to $C^\infty(\overline{\Omega} \times S^{N-1} \times \Omega)$.

By $q_j = q_j(x,t,p,u)$ and $r_j = r_j(x,t,p,u)$ respectively, we denote the right and left zero-vectors of the matrix A corresponding to the eigenvalue λ_j for $p_0 = p_{0j}$.

Assume that for $(x,t) \in \Omega$, $u \in Q$, $p \in S^{N-1}$, the inequalities hold:

$$\left\langle r_j, \frac{\partial F_0}{\partial u}q_j \right\rangle \neq 0, \qquad j = 1,\dots,M, \tag{5.3}$$

where, as usual, $\langle\cdot,\cdot\rangle$ denotes the scalar product of vectors.

Our assumptions about the system (5.1) are similar to conditions of strong hyperbolity, and given the functions $F_i(x,t,u)$ are linear in u, our assumptions turn into the latter conditions. Following [26], we introduce the necessary definitions.

Definition 5.1. The system of equations (5.1) is called *essentially quasilinear* in $\Omega \times Q$ if, for all $j = 1, \ldots, M$, there exists a constant $\delta > 0$ such that the inequalities hold:

$$\left| \sum_{k=1}^{M} q_j^k \frac{\partial \lambda_j}{\partial u_k} \right| \geqslant \delta |q_j|, \tag{5.4}$$

where q_j^k is the kth component of the vector $q_j(x, t, p, u)$.

The condition (5.4) is similar to the Lax condition of essential quasilinearity and, by (5.3), is equivalent to the following inequality

$$\left| \sum_{k=1}^{M} q_j^k \frac{\partial p_{0_j}}{\partial u_k} \right| \geqslant \delta_1 |q_j|, \qquad \delta_1 = \text{const} > 0, \quad j = 1, \ldots, M. \tag{5.5}$$

In addition, later we assume that if $|u| \to \infty$, $u \in Q$ and $\lim_{|u| \to \infty} \frac{u}{|u|} = \vec{l}$, the limits

$$\lim_{|u| \to \infty} \frac{q_j}{|q_j|} = \mathcal{L}_j, \qquad j = 1, \ldots, M,$$

also exist, i.e., at infinity the eigenvectors of the matrix $A(x, t, p_0, p)$ "behave regularly" with respect to u.

Suppose $v = |u|$. Then $u = \vec{l}v$ as $u \to \infty$, and $\langle u, l \rangle = v$. Hence, for an arbitrary function $f(u)$ and a vector $a \in C^M$, the inequality holds (for sufficiently large u):

$$\left\langle a, \frac{\partial f}{\partial u} \right\rangle = \langle a, l \rangle \frac{\partial f}{\partial v}.$$

Thus (5.5) implies that the functions $p_{0j} = p_{0j}(x, t, p, u)$ increase infinitely (with respect to their absolute values) if the argument u tends to infinity along a certain direction $\vec{l}$ (depending, in general, on x, t) so that $|\langle \mathcal{L}_j, \vec{l} \rangle| > 0$ and $|A(x, t, p_0, p, u)l| \to 0$.

Hence, if $u \to \infty$ so that $\lim_{|u| \to \infty} u/|u| = \vec{l}$, then, under the condition $|p_0| \leqslant$ const, one can find a number R such that $|A(x, t, p_0, p, u)l| \geqslant \text{const} > 0$ for $|u| > R$.

Definition 5.2. The system of equations (5.1) is called *regular* if, for $(x, t) \in Q$ and for any vector $\vec{l} \in \mathbf{R}^M$ lying in Q, $|l| = 1$, there exist p_0, p, $|p_0| \leqslant$ const, $p \in S^{N-1}$ and constants $c_1 = c_1(p_0, p)$, $c_2 = c_2(p_0, p)$, such that as $u \to \infty$, $u \in Q$:

1. $$|\langle A(x, t, p_0, p, u)l, A(x, t, \widehat{p}_0, \widehat{p}, u)l \rangle| \leqslant c_1 |A(x, t, p_0, p, u)l|^2 \tag{5.6}$$

for any $|\widehat{p}_0| \leqslant$ const, $\widehat{p} \in S^{N-1}$;

2. $$\left| \left\langle A(x, t, p_0, p, u)l, \frac{\partial F_0}{\partial t}(x, t, \vec{l}z) + \sum_{i=1}^{N} \frac{\partial}{\partial x_i} F_i(x, t, \vec{l}z) \right\rangle \right| \leqslant c_2 |z| |A(x, t, p_0, p, u)\vec{l}|^2.$$

Note that the inequality in item 1 always holds for $N = 1$, $F_0(u) \equiv u$.

The above regularity conditions are sufficient for any solution of system (5.1), with singularities on a submanifold, to be bounded in a neighborhood of this submanifold. However, if beforehand we restrict our consideration to bounded solutions, no regularity conditions given in Definition 5.2 are needed for the statement about the structure of solutions.

Example 5.1. Let us consider the system of gas dynamics equations in the one-dimensional case:

$$\begin{aligned} &\frac{\partial \rho}{\partial t} + \frac{\partial}{\partial x}(\rho u) = 0, \\ &\frac{\partial u}{\partial t} + \frac{\partial}{\partial x}\Big(\frac{u^2}{2} + F(\rho)\Big) = 0, \qquad F(\rho) = \int \frac{1}{\rho}\frac{\partial p}{\partial \rho}\, d\rho. \end{aligned} \tag{5.7}$$

The matrix $A(x, t, p_0, p, u)$, related to this system, has the form

$$A = \begin{bmatrix} p_0 + p_1 u & p_1 \rho \\ p_1 F_1 & p_0 + p_1 u \end{bmatrix}, \qquad F_1 = \frac{dF}{d\rho},$$

its eigenvalues $\lambda_{1,2}$ are equal to

$$\lambda_{1,2} = p_0 + p_1(u \pm \sqrt{\rho F'}).$$

The corresponding right an left null eigenvectors are:

$$q_{1,2} = \begin{pmatrix} 1 & \pm\sqrt{\frac{F'}{\rho}} \end{pmatrix}, \qquad r_{1,2} = \begin{pmatrix} 1 \\ \pm\sqrt{\frac{\rho}{F'}} \end{pmatrix}.$$

Obviously, the strong hyperbolity condition (5.3) is satisfied for system (5.7). Assume that the condition (5.4) does not hold and the expression on the left-hand side of inequality (5.4) vanishes identically for the system (5.7). In this case, as is easy to see, the function $F(\rho)$ necessarily has the form $F(\rho) = A - C/\rho^2$, where A and C are constant. A medium, in which the pressure depends on the density in this way, is called the *Chaplygin gas*. It is well-known that in this case system (5.7) becomes linear:

$$\begin{aligned} &\Big(\frac{\partial}{\partial \tau} - \sqrt{2C}\frac{\partial}{\partial y}\Big) v = 0, \\ &\Big(\frac{\partial}{\partial \tau} + \sqrt{2C}\frac{\partial}{\partial y}\Big) w = 0, \end{aligned}$$

where

$$v(y, \tau) = u(x, t) + \frac{\sqrt{2C}}{\rho(x, t)}, \qquad w(y, \tau) = u(x, t) - \frac{\sqrt{2C}}{\rho(x, t)},$$

$$\tau = t, \qquad y = \int_a^x \rho(S, t)\, dS - \int_b^t \rho(a, S) u(a, S)\, dS, \qquad a, b, = \text{const}\,.$$

The system of equations (5.7) is regular in the sense of Definition 5.2. The proof of this statement for (5.7) and the system of equations considered in Example 5.2 will be given at the end of this section.

Example 5.2. Now let us consider the system of gas dynamics equations in the two-dimensional case:

$$\begin{aligned}
&\frac{\partial \rho}{\partial t}+\frac{\partial \rho u}{\partial x}+\frac{\partial \rho v}{\partial y}=0,\\
&\frac{\partial \rho u}{\partial t}+\frac{\partial}{\partial x}(\rho u^2+P(\rho))+\frac{\partial}{\partial y}(\rho uv))=0,\\
&\frac{\partial \rho v}{\partial t}+\frac{\partial}{\partial x}(\rho uv))+\frac{\partial}{\partial y}(\rho v^2+P(\rho))=0,
\end{aligned} \tag{5.8}$$

the matrix $A(x,y,t,p_0,p_1,p_2,u,v)$, related to this system, has the form

$$A=\begin{bmatrix} \widetilde{p} & p_1\rho & p_2\rho \\ \widetilde{p}u+p_1P' & \widetilde{p}\rho+p_1\rho u & p_1\rho u \\ \widetilde{p}v+p_2P' & p_1\rho v & \widetilde{p}\rho+p_2\rho v \end{bmatrix}, \qquad \widetilde{p}=p_0+p_1u+p_2v \tag{5.9}$$

its eigenvalues $\lambda_{1,2,3}$ are equal to

$$\begin{aligned}
&\lambda_1=(p_0+p_1u+p_2v)\rho,\\
&\lambda_{2,3}=\frac{(p_0+p_1u+p_2v)+(p_0+2p_1u+2p_2v)\rho\pm\sqrt{D}}{2},
\end{aligned} \tag{5.10}$$

where $D=[\widetilde{p}+(p_0+2p_1u+2p_2v)\rho]^2-4\rho[(p_0+p_1u+p_2v)-(p_1^2+p_2^2)P'(\rho).]$.

The equations $\lambda_j(x,y,t,p_0,p_1,p_2,u,v)=0$, $j=1,2,3$, have solutions of the form

$$\begin{aligned}
&p_{01}(x,y,t,p_0,p_1,p_2,u,v)=-(p_1u+p_2v),\\
&p_{0j}(x,y,t,p_0,p_1,p_2,u,v)=-(p_1u+p_2v)\pm\sqrt{P'(\rho)},\\
&\qquad j=2,3, \qquad p_1^2+p_2^2=1.
\end{aligned}$$

The left and right null eigenvectors corresponding to zero eigenvalues (5.10) have the form:

$$q_1=\begin{pmatrix}0\\-p_2\\p_1\\1\end{pmatrix}, \qquad r_1=\left(\frac{p_2u-p_1v}{p_1},-\frac{p_2}{p_1},1\right),$$

$$q_{2,3}=\begin{pmatrix}\pm\rho\\-p_1\sqrt{P'}\\-p_2\sqrt{P'}\end{pmatrix}, \qquad r_{2,3}=(p_1u+p_2v\pm\sqrt{P'},-p_1,-p_2),$$

The system (5.8) satisfies the strong hyperbolity condition (5.3) only for $\rho>0$. The system of equations (5.7) is not essentially quasilinear in the sense of Definition 5.1, since the inequality (5.5) does not hold for p_{01}. Therefore the results concerning the structure of nonsmooth solutions stated below hold only for the solutions of system (5.8) taking values in the subspace spanned on the vectors q_2, q_3.

By analogy to previous considerations, we shall describe the structure of non-smooth self-similar solutions in this situation.

Denote by $\mathfrak{A}_{F_0,\dots,F_N}$ the set of functions $\chi(x,t,\tau)$ locally integrable in variables $(x,t) \in \Omega$, $\tau \in \mathbf{R}$, such that

1. The functions $F_i(x,t,U(x,t)\chi(x,t,\tau))$, $i = 0,\dots,N$, belong to the space $L_1(\Omega \times \mathbf{R}_\tau)$ for any vector-function $U(x,t) \in C^\infty\overline{\Omega}$,

2. $\mathfrak{A}_{F_0,\dots,F_N} \subset C^\infty(\overline{\Omega} \times (R_\tau \setminus \{0\}))$.

3. For any function $\chi(x,t,\tau) \in \mathfrak{A}_{F_0,\dots,F_N}$, as $\tau \to \pm 0$, $x \to x_0$, $t \to t_0$, the limit $\lim\limits_{\substack{\tau\to\pm0\\ x\to x_0\\ t\to t_0}} \chi$ exists in all variables (possibly, infinite), $(x_0,t_0) \in \Omega$ is an arbitrary point.

We introduce the functions $\gamma^\pm(\chi;x,t)$, $(x,t) \in \Omega$, $\chi \in \mathfrak{A}_{F_0,\dots,F_N}$:

$$\gamma^\pm(\chi;x,t) \stackrel{\text{def}}{=} \begin{cases} \lim_{\tau\to\pm0}\chi(x,t,\tau), & |\lim_{\tau\to\pm0}\chi(x,t,\tau)| < \infty, \\ 0 & \text{otherwise,} \end{cases}$$

and assume that $\gamma^\pm(\chi;x,t) \in C^\infty(\overline{\Omega})$.

4. For arbitrary $\chi(x,t,\tau) \in \mathfrak{A}_{F_0,\dots,F_N}$, any multi-indices ν and nonnegative integer α, the limits

$$\lim_{\tau\to\pm0} \frac{\dfrac{\partial^{|\nu|+\alpha}}{\partial x^\nu \partial t^\alpha}\chi(x,t,\tau z) - \gamma^\pm\left(\dfrac{\partial^{|\nu|+\alpha}}{\partial x^\nu \partial t^\alpha}\chi;x,t\right)}{\chi(x,t,\tau) - \gamma^\pm(\chi;x,t)} = \mu_{\nu,\alpha}(x,t)$$

exist and $\mu_{\nu,\alpha}(x,t) \in C^\infty(\overline{\Omega})$.

Denote by $\varphi(\xi,x,t,h) \in C^\infty(\mathbf{R}_\xi \times \Omega \times [0,1])$ a one-parametric family of vector-functions, such that

$$\varphi(\xi,x,t,0) \in S^{-3}_{0,0}(\mathbf{R}_\xi \times \Omega), \qquad \frac{\partial}{\partial h}\varphi(\xi,x,t,0) \in S^0_{0,0}(\mathbf{R}_\xi \times \Omega) \tag{5.11}$$

(as usually, $\varphi(\xi,x,t) \in S^m_{\rho,\delta}$ means that, for any nonnegative integer γ, α and any multi-indices β, we have

$$\frac{\partial^{\gamma+|\beta|+\alpha}}{\partial\xi\partial x^\beta\partial t^\alpha}\psi(\xi,x,t) \leqslant C_{\gamma,\beta,\alpha}(1+|\xi|)^{m-\rho\gamma+\delta(|\beta|+\alpha)},$$

where $C_{\gamma,\beta,\alpha} = \text{const}$).

Definition 5.3. A vector-function $\varphi(\xi,x,t,h)$ will be called *consistent* with the nonlinearity of system (5.1), if for any $i = 0,\dots,N$ and any vector-functions $\Phi(x,t) \in C^\infty(\overline{\Omega})$, the following conditions hold

$$\begin{aligned} &1. \quad \int_{-\infty}^{\infty} \left[F_i(x,t,\Phi+\varphi(\xi,x,t,0)) - F_i(x,t,\Phi)\right] d\xi \leqslant A_1, \\ &2. \quad \int_{-\infty}^{\infty} \left[\frac{\partial}{\partial u}F_i(x,t,\Phi+\varphi(\xi,x,t,0)) - \frac{\partial}{\partial u}F_i(x,t,\Phi)\right]|\xi\, d\xi \leqslant A_2, \end{aligned} \tag{5.12}$$

where $A_i = \text{const} > 0$, $i = 1, 2$.

Below we shall assume that the vector-function $\varphi(\xi, x, t, h)$ is consistent.

Given $S(x,t,) \in C^\infty(\overline{\Omega})$, the equation $S(x,t) = 0$ determines a certain submanifold in Ω of codimension 1, $\nabla_x S(x,t)|_{S=0} \neq 0$.

Consider the one-parametric family of vector-functions

$$u(x,t,h) = U(x,t)\chi(x,t,S(x,t)) + \varphi\Big(\frac{S(x,t)}{h}, x, t, h\Big) + V(x,t), \qquad (5.13)$$

where $\chi(x,t,\tau) \in \mathfrak{A}_{F_0,\ldots,F_N}$, $\chi(x,t,S(x,t)) \ni C^1(\Omega)$, $U(x,t)$ is a vector-function with components $U_i(x,t) \in C^\infty(\overline{\Omega})$, $i = 1, \ldots, M$, $V(x,t)\big|_{S=0} \neq 0$, $V(x,t) \in C^{\mathbf{R}}(\overline{\Omega})$ for sufficiently large R.

Definition 5.4. A one-parametric family of vector-functions $u(x,t,h)$ of the form (5.13), such that $L[u] = O(h^2)$ in $D'(\Omega)$, is called a *self-similar solution* to the system of equations (5.1).

Let $\chi_k(x,t,\tau) \in \mathfrak{A}_{F_0,\ldots,F_N}$, $k = 0, 1, \ldots$. Suppose that a set of functions $u(x,t,h)$ satisfies the conditions

$$\lim_{\tau\to\pm 0} \frac{\chi_{k+1}(x,t,\tau) - \gamma^\pm(\chi_{k+1}; x,t)}{\chi_k(x,t,\tau) - \gamma^\pm(\chi_k; x,t)} = 0, \qquad k = 0, 1, \ldots,$$

where

$$\gamma^\pm(\chi_k; x,t) \overset{\text{def}}{=} \begin{cases} \lim_{\tau\to\pm 0} \chi_k(x,t,\tau), & |\lim_{\tau\to\pm 0} \chi_k(x,t,\tau)| < \infty, \\ 0, & |\lim_{\tau\to\pm 0} \chi_k(x,t,\tau)| = \infty, \end{cases}$$

Consider a family of vector-functions $u(x,t,h)$ that, for any $K > 0$, can be represented in the form:

$$u(x,t,h) = U(x,t)\sum_{k=0}^{K} \chi_k(x,t,S(x,t)) + \varphi\Big(\frac{S(x,t)}{h}, x, t, h\Big) \mod (C^{\mathbf{R}}(\Omega)), \quad (5.14)$$

where $\chi_k(x,t,\tau) \in \mathfrak{A}_{F_0,\ldots,F_N}$, $k \geqslant 0$, the functions $S(x,t)$ and $\varphi(\xi,x,t,h)$ satisfy the same conditions as the functions from (5.13) and $\chi_0(x,t,S(x,t)) \notin C^1(\Omega)$, $U(x,t)\big|_{S=0} \neq 0$.

It is easy to verify that the set of functions of the form (5.13) contains the set of functions defined by (5.14).

Let us introduce the notion of stability for self-similar generalized solutions to problem (5.1) having the form (5.14).

Definition 5.5. A self-similar generalized solution of the form (5.14), satisfying (5.1), is called *stable* if 1) there exists a small $\varepsilon > 0$ such that for any function $f(x,t) \in C^\infty(\overline{\Omega})$, $|f(x,t)| < \varepsilon$, there exists a generalized self-similar solution $u_f(x,t)$ of the form (5.14), satisfying the system

$$L[u_f] = f + O(h^2)$$

in $D'(\Omega)$ and such that, for any K,

$$u_f(x,t,h) = U(x,t;f)\sum_{k=0}^{K} \chi_k(x,t,S_f(x,t);f) + \varphi_f\Big(\frac{S_f(x,t)}{h},x,h,t\Big) \mod (C^K(\Omega)),$$

where the vector-function $\varphi_f(\xi,x,t,h)$ satisfies the same conditions as the function $\varphi(\xi,x,t,h)$ from (5.11) $S_f(x,t)\in C^\infty$, $\nabla_x S_f\big|_{S_f=0}\neq 0$, $\{S_f=0\}\cap\Omega\neq\varnothing$;

2) the limits

$$\lim_{\chi\to\pm 0}\frac{\chi_k(x,t,\tau)-\gamma^{\pm}(\chi_k;x,t)}{\chi_k(x,t,\tau;f)-\gamma^{\pm}(\chi_k(x,t,\tau;f);x,t)} = \varkappa_k^{\pm}(x,t), \qquad k=0,1,\dots,$$

exist and $\varkappa_k^{\pm}(x,t)\in C^\infty(\overline{\Omega})$.

Similarly, the definition of stability for a solution of the form (5.13) is weaker, since, by this definition, only the highest term of the asymptotic expansion of this solution is stable.

Further we assume that

$$\chi(x,t,\tau) = \sum_{k=0}^{K}\chi_k(x,t,\tau).$$

Let us formulate the main result of this section.

Theorem 5.1. *If* $u(x,t,h) = U(x,t)\chi(x,t,S(x,t)) + \varphi\big(\frac{S(x,t)}{h},x,h,t\big) + V(x,t)$, *is a real stable self-similar solution with values in* Q, *satisfying the regular, essentially quasilinear system* (5.1), *then either*

1. $\chi(x,t,\tau) = \sum_{i=0}^{K} d_i(x,t)\theta(\tau)\tau^i \mod C^K(\Omega)$, $d_i(x,t)\in C^N(\Omega)$, $d_0(x,t)=1$, *where* $(x,t)\in\Omega$, $\varphi\equiv 0$, $\theta(\tau)$ *is the Heaviside function, or*
2. $\chi(x,t,\tau) = 0 \mod(C^K(\Omega))$, *or*
3. $\varphi(\xi,x,t,h) = 0$, $\chi(x,t,\tau) = \sum_{i=1}^{2K} C_i(x,t)\theta(\tau)\tau^{(i-1)/2} \mod (C^K(\Omega))$, $C_i(x,t)\in C^\infty(\overline{\Omega})$, $C_1(x,t)=1$ *for* $(x,t)\in\overline{\Omega}$, *provided* $|\nabla_{x,t}\chi(x,t,S(x,t))|\to\infty$ as $x\to x_0$, $t\to t_0$, (x_0,t_0) *is an arbitrary point on the surface* $S(x,t)=0$, $K>0$ *is a natural number.*

As is known, the necessary conditions for the existence of solutions of shock wave type are the Hugoniot conditions, i.e., relations that must hold on the sets of discontinuity. Let us formulate the necessary conditions for the existence of solutions with singularities listed in Theorem 5.1.

Corollary 5.1. *In item 1 of Theorem 5.1 it is necessary that*

$$\Big[S_t\big(F_0(x,t,U+V)-F_0(x,t,U)\big) + \sum_{i=1}^{N} S_{x_i}\big(F_i(x,t,U+V)-F_i(x,t,U)\big)\Big]_{S(x,t)=0} = 0,$$

$$\int_{-\infty}^{\infty}\Big[S_t\big(F_0(x,t,V+\theta(\xi)U+\varphi(\xi,x,t,0))-F_0(x,t,V+\theta(\xi)U)\big) + \sum_{i=1}^{N} S_{x_i}\big(F_i(x,t,V+\theta(\xi)U)-F_i(x,t,V+\theta(\xi)U)\big)\Big]\,d\xi\Big|_{S(x,t)=0} = 0,$$

where the vector-function $V(x,t)$ (with smooth components) satisfies the equation outside the surface $S(x,t)=0$.

In item 2 of Theorem 5.1 it is necessary that

$$\int_{-\infty}^{\infty}\Big[S_t\big(F_0(x,t,V+\varphi(\xi,x,t,0))-F_0(x,t,V)\big) + \sum_{i=1}^{N} S_{x_i}\big(F_i(x,t,V+\varphi(\xi,x,t,0))-F_i(x,t,V)\big)\Big]\,d\xi\Big|_{S(x,t)=0} = 0.$$

In item 3 of Theorem 5.1 it is necessary that

$$\Big[S_t\big(F_0(x,t,U+V)-F_0(x,t,U)\big) + \sum_{i=1}^{N} S_{x_i}\big(F_i(x,t,U+V)-F_i(x,t,U)\big)\Big]_{S(x,t)=0} = 0,$$

$$\Big[S_t\frac{\partial}{\partial u}F_0(x,t,U+V) + \sum_{i=1}^{N} S_{x_i}\frac{\partial}{\partial u}F_i(x,t,U+V)\Big]U\Big|_{S(x,t)=0} = 0,$$

where the vector-function $V(x,t)$ (with smooth components) satisfies the equation outside the surface $S(x,t)=0$.

In Theorem 5.1 we present some versions of the structure of nonsmooth self-similar solutions and the conditions necessary for the existence of a solution with this structure. In this case we write only the "leading" necessary conditions corresponding to (or coinciding with) the Rankine–Hugoniot conditions concerning the shock wave front, which are well-known in mechanics. Note that, using the techniques developed here, we can write the whole infinite chain of conditions of the Rankine–Hugoniot type in all cases mentioned in this theorem.

Before proving the theorem, we note that solutions of the form $u(x,t,h) = U(x,t)\sum_{i=1}^{2K} C_i(x,t)\theta(S)S^{(i-1)/2}$ are similar to solutions well-known in the theory of detonation waves.

The solutions of the form $u(x,t,h) = U(x,t)\sum_{i=1}^{K} d_i(x,t)\theta(S)S^i$ describe the shock wave propagation in nonviscous fluids and gases. The solutions $u(x,t,h) =$

$\varphi(S/h, x, t, h)$ are infinitely narrow solitons (see [55, 56, 58, 59]). Theorem 5.1 implies that a soliton-like solution can propagate only on a smooth or a shock wave background.

The scheme of the proof is the following. First, using the regularity condition, we prove that the generalized self-similar solutions to system (5.1) are uniformly bounded for $(x,t) \in \Omega$. (Note that if we assume that the domain Ω is compact, the theorem holds without the regularity conditions (see Definition 5.2) and inequality (5.4)). For bounded self-similar generalized solutions, the algebraic function theorem implies that the function $\chi(x,t,\tau)$ possesses a singularity of the form $C_1\tau^\alpha + C_2\theta(\tau)\tau^\beta$ + smoother functions. Further, by applying the condition of essential quasilinearity to our system (see Definition 5.1), we complete the proof.

First of all, let us prove some auxiliary statements.

Without loss of generality, assume that Ω is an open ball with small radius ν. Denote

$$\Omega_\pm = \{(x,t) \in \Omega;\ S(x,t) \underset{(<)}{>} 0\}, \qquad \Gamma = \{(x,t) \in \Omega;\ S(x,t) = 0\}.$$

Lemma 5.1. *Let a strongly hyperbolic system of equations* (5.1) *be regular and let the condition* (5.1) *hold. Then a generalized self-similar solution* $u(x,t,h)$ *of the form* (5.13) *is uniformly bounded.*

Proof. Calculate the expression $L[u]$:

$$L[u] = \Big\{\frac{\partial}{\partial\tau}\Big[\frac{\partial S}{\partial t}F_0(x,t,u) + \sum_{i=1}^N \frac{\partial S}{\partial x_i}F_i(x,t,u)\Big] + \frac{\partial F_0}{\partial t}(x,t,u) + \sum_{i=1}^N \frac{\partial F_i}{\partial x_i}(x,t,u)\Big\}_{\tau=S(x,t)},$$

where $u = u(x,t,\tau,h) = U(x,t)\chi(x,t,\tau) + \varphi(\tau/h,x,t,h) + V(x,t)$.

To prove Lemma 5.1, it is sufficient to prove that the function $\chi(x,t,\tau)$ is uniformly bounded for $(x,t) \in \overline{\Omega}$. Obviously, as $h \to 0$, $\varphi(S/h,x,t,h) = O(h)$ in D'. Then the function $\widehat{\chi}(x,t,\tau) = U(x,t)\chi(x,t,\tau) + V$ is the pointwise limit of the function $u(x,t,\tau,h)$ as $h \to 0$, $(x,t) \notin \Gamma$. Taking into account that $\nabla_x S\big|_{S=0} \neq 0$, we obtain

$$\Big\{\frac{\partial S}{\partial t}\frac{\partial}{\partial\tau}F_0(x,t,\widehat{\chi}) + \sum_{i=1}^N \frac{\partial S}{\partial x_i}\frac{\partial}{\partial\tau}F_i(x,t,\widehat{\chi}) + \frac{\partial F_0}{\partial t}(x,t,\widehat{\chi}) + \sum_{i=1}^N \frac{\partial F_i}{\partial x_i}(x,t,\widehat{\chi})\Big\}_{\tau=S(x,t)} = 0 \qquad (5.15)$$

as $h \to 0$ in $D'(\Omega)$.

Assume that the statement of this lemma does not hold. For definiteness, let $|\chi(x,t,S)| \to \infty$ for $(x,t) \in \Omega_+$, $x \to x_0$, $t \to t_0$, $(x_0,t_0) \in \Gamma$. Then there exists an open ball $B_d \in \Omega_+$ such that $|U(x,t)\chi(x,t,S(x,t))| \geqslant d > 0$ for $(x,t) \in \bar{B}_d$ and arbitrary $d \in R_+$. Without loss of generality assume that $\nabla_x S \neq 0$ for $(x,t) \in \Omega$. Taking into account that the differentiation with respect to x and t does not change the structure of singularities with respect to τ, we rewrite (5.15):

$$\Big\{ |\nabla_x S| (A(x,t,p_0,p,\widehat{\chi}))U\chi_\tau + M(A(x,t,\widehat{p}_0,\widehat{p},\widehat{\chi})U(x,t))|\chi| \\ + \left(\frac{\partial F_0}{\partial t}(x,t,U(x,t)Z) + \sum_{i=1}^{N} \frac{\partial F_i}{\partial x_i}(x,t,U(x,t)Z) \right)_{Z=\chi(x,t,\tau)} \Big\}_{\tau=S(x,t)} = 0,$$

where $p_0 = \frac{\partial S}{\partial t}/|\nabla_x S|$, $p_i = \frac{\partial S}{\partial x_i}/|\nabla_x S|$, $\widehat{p}_0 = \frac{\partial \chi}{\partial t}/|\nabla_x S|$, $\widehat{p}_i = \frac{\partial \chi}{\partial x_i}/|\nabla_x \chi|$, $i = 1,\dots,N$, $M = M(x,t,S)$ is a scalar function bounded for $(x,t) \in \overline{\Omega}_+$.

By the regularity conditions given above (see Definition 5.2), taking the scalar product of the left-hand side and the vector $A(x,t,p_0,p,\widehat{\chi})U$, from the latter inequality for $(x,t) \in B_d$ we get:

$$\left| \frac{\partial}{\partial \tau}\chi(x,t,S(x,t)) \right| \leqslant C \big| \chi(x,t,S(x,t)) \big|, \tag{5.16}$$

where $C > 0$ is a certain constant. Thus the inequality (5.16) holds also for $(x,t) \in \bar{B}_d$. In $\overline{\Omega}_+$ we choose a system of coordinates $(\xi_0,\xi_1,\dots,\xi_N)$ putting $\xi_0 = S(x,t)$. Let $\chi(x,t,\tau) = \chi(\xi,\tau)$. Then (see the notation in [54]):

$$\begin{aligned} \frac{\partial}{\partial \tau}\chi(x,t,\tau)\Big|_{\tau=S(x,t)} &= \frac{\partial^1}{\partial \xi_0}\widehat{\chi}(\xi_0^2,\dots,\xi_N,\xi_0^1) \\ &= \frac{\partial \widehat{\chi}}{\partial \xi_0}(\xi_0,\dots,\xi_N,\xi_0) - \frac{\partial \chi}{\partial q}(q,\xi_1,\dots,\xi_N,\xi_0)\Big|_{q=\xi_0}. \end{aligned} \tag{5.17}$$

Since $\chi(x,t,\tau) \in \mathfrak{A}_{F_0,\dots,F_N}$, (5.16) and (5.17) imply that in the ball $\bar{B}$ we have the inequality

$$\left| \frac{\partial \widehat{\chi}}{\partial \xi_0}(\xi,\xi_0) \right| \leqslant \text{const}\, |\widehat{\chi}(\xi,\xi_0)|.$$

There is a contradiction: by assumption, the function $\chi(x,t,S(x,t))$ is unbounded. The lemma is proved.

Lemma 5.2. *Let the function $\chi(x,t,\tau)$ belong to $\mathfrak{U}_{(F_0,\dots,F_N)}$ and satisfy the equation*

$$\big(D_\tau \chi(x,t,\tau)\big)_{\tau=S(x,t)} = G(x,t,S(x,t)),$$

where $G(x,t,\tau) \in C^\infty(\overline{\Omega} \times (\mathbf{R} \setminus \{0\}))$.

Then

$$\chi(x,t,S) = \int_0^S \big\{ G + \big(l\chi(x,t,\tau)\big)_{\tau=S} \big\}\, dn + \gamma^\pm(\chi;x,t),$$

where the integral is calculated along the curve $\nabla S/|\nabla S|$, l is a first-order operator in variables (x,t) with smooth coefficients.

Proof. In the domain Ω we take an orthogonal system of coordinates $\xi = (\xi_0, \dots \xi_N)$ so that $\xi_0 = S(x,t)$. We put $\chi(x,t,\tau) = \widetilde{\chi}(\xi_0, \dots, \xi_N, \xi_0)$, $G(x,t,\tau) = \widetilde{G}(\xi_0, \dots, \xi_N, \xi_0)$. Then

$$\left(\frac{\partial}{\partial \tau}\chi(x,t,\tau)\right)_{\tau=S(x,t)} = \frac{\overset{2}{\partial}}{\partial \xi_0}\widetilde{\chi}(\overset{3}{\xi_0}, \dots, \xi_N, \overset{2}{\xi_0}) = \frac{\partial}{\partial \xi_0}\widetilde{\chi}(\xi_0, \dots, \xi_N, \xi_0) - \left(\frac{\partial}{\partial q}\widetilde{\chi}(q, \xi_1, \dots, \xi_N, \xi_0)\right)\bigg|_{q=\xi_0}.$$

This yields

$$\widetilde{\chi}(\xi_0, \dots, \xi_N, \xi_0) = \int_0^{\xi_0} \widetilde{G}(\xi_0', \dots, \xi_N, \xi_0')\, d\xi_0' + \int_0^{\xi_0} \left(\frac{\partial}{\partial q}\widetilde{\chi}(q, \xi_1, \dots, \xi_N, \xi_0')\right)\bigg|_{q=\xi_0'} d\xi_0' + \widetilde{\chi}\bigg|_{\xi_0=\pm 0}.$$

By passing to the coordinates (x,t) in this formula, we obtain the statement of our lemma. In this case, the operator l gives the expression for the derivative $\partial/\partial\xi_0$ in the coordinates (x,t).

Lemma 5.3. *Let $u(x,t,h) = U(x,t)\chi(x,t,S(x,t)) + \varphi(S/h, s, t, h)$ be a real stable generalized self-similar solution of the essentially regular quasilinear system of equations (5.1) and let this solution be bounded for $(x,t) \in \Omega$, $h \in [0,1]$. Then*

$$\chi(x,t,S) = X(x,t) + \sum_{k=1}^{K} U_k(x,t) S^{\alpha_k} + \sum_{k=1}^{K} \omega_k(x,t)\theta(S) S^{\beta_k} \bmod (O(S^{L_K})),$$

where $\{\alpha_k\}$, $\{\beta_k\}$ are monotonically increasing sequences of nonnegative real numbers without accumulation point, $K > 0$ is an integer number, the functions $X(x,t)$, $U_k(x,t)$, $\omega_k(x,t)$, $k \geqslant 1$, belong to $C^\infty(\overline{\Omega})$, $L_K = \max([\alpha_K], [\beta_K])$.

Proof. By the definition of stable generalized self-similar solutions of system (5.1), for $S(x,t,) \neq 0$ we have (the dependence of functions S and χ on f is omitted):

$$\lim_{h\to 0} L[u] = \left\{\frac{\partial}{\partial \tau}\left(\frac{\partial S}{\partial t}\widehat{F}_0(x,t,\chi(x,t,\tau)) + \sum_{i=1}^{N}\frac{\partial S}{\partial x_i}\widehat{F}_i(x,t,\chi(x,t,\tau))\right) + \frac{\partial \widehat{F}_0}{\partial t}(x,t,\chi(x,t,\tau)) + \sum_{i=1}^{N}\frac{\partial \widehat{F}_i}{\partial x_i}(x,t,\chi(x,t,\tau))\right\}_{\tau=S(x,t)} = f(x,t), \quad (5.18)$$

where $\widehat{F}_i(x,t,\chi) \overset{\text{def}}{=} F_i(x,t,\widehat{\chi}) = F_i(x,t,U\chi + V)$, $f(x,t)$ is an arbitrary smooth function, $|f(x,t)| \leqslant \varepsilon$, $(x,t) \in \overline{\Omega}$, $\varepsilon > 0$ is arbitrary small.

Let, for definiteness, $(x,t)\in\Omega_+$. Consider the first equation in (5.18). By Lemma 5.2, we get

$$\frac{\partial S}{\partial t}\widehat{F}_0^{(1)}(x,t,\chi)+\sum_{i=1}^{N}\frac{\partial S}{\partial\xi_i}\widehat{F}_i^{(1)}(x,t,\chi)=\int_0^S\left\{f-\frac{\partial\widehat{F}_0^{(1)}}{\partial t}-\sum_{i=1}^{N}\frac{\partial\widehat{F}_i^{(1)}}{\partial x_i}\right.$$
$$\left.+l\left(\frac{\partial S}{\partial t}\widehat{F}_0^{(1)}+\sum_{i=1}^{N}\frac{\partial S}{\partial x_i}\widehat{F}_i^{(1)}\right)\right\}\bigg|_{\tau=S}dn+G_+(x,t),\qquad(5.19)$$

where $\widehat{F}_i^{(1)}$ are the first components of the vector-function $\widehat{F}_i$ and

$$\widehat{G}_+(x,t)=\left[\frac{\partial S}{\partial t}\widehat{F}_0^{(1)}(x,t,V+U\gamma^+)-\sum_{i=1}^{N}\frac{\partial S}{\partial x_i}\widehat{F}_i^{(1)}(x,t,V+U\gamma^+)\right]_{S=0}.$$

The function χ is given by the equation

$$\frac{\partial S}{\partial t}\widehat{F}_0^{(1)}(x,t,\chi)-\sum_{i=1}^{N}\frac{\partial S}{\partial x_i}\widehat{F}_i^{(1)}(x,t,\chi)-\lambda\overset{\text{def}}{=}\mathcal{F}(\chi,\lambda)=0,$$

where x and t are considered as parameters.

By assumption, $\widehat{F}_i^{(1)}(x,t,\chi)$ are entire functions of χ, and hence $\mathcal{F}(\chi,\lambda)\not\equiv 0$, but $\mathcal{F}(\chi,\lambda_0)=0$, where λ_0 is the right-hand side of (5.19).

By the theorem on algebraic functions, in a small neighborhood of ω, $\lambda_0\in\omega$, we get

$$\chi=\gamma^+(\chi;x,t)\sum_{m\geqslant 1}C_m(\lambda-\lambda_0)^{m/p},$$

where p are certain natural numbers, $C_m=C_m(x,t)$ are coefficients depending on x, t as parameters and, under the assumption that the differentiation with respect to x and t does not change the type of singularities with respect to τ, the functions $C_m(x,t)$ belong to $C^\infty(\overline{\Omega}_+)$.

Thus the following relation holds

$$\chi(x,t,S)=\widehat{\gamma}^+(\chi;x,t)+\sum_{m\geqslant 1}[C_m]_K$$
$$\times\left(\left[\frac{\partial S}{\partial t}\widehat{F}_0^{(1)}(x,t,\chi)-\sum_{i=1}^{N}\frac{\partial S}{\partial x_i}\widehat{F}_i^{(1)}(x,t,\chi)\right]_K-[\lambda_0]_K\right)^{m/p},\qquad(5.20)$$

where $\widehat{\gamma}^+=\gamma^+\big|_{S=0}$, $[\Psi(x,t,\tau)]_K$ is the Taylor polynomial (of degree K) of the function $\Psi(x,t,\tau)$ with respect to S in a neighborhood of $S(x,t)=0$, $\Psi(x,t,\tau)-[\Psi(x,t,\tau)]_K=O(S^{K+1})$, the function $\lambda_0=\lambda_0(x,t)$, as defined above, is the right-hand side of (5.19). Note that the first summand in the sum over m in parentheses can be expanded into a formal series in powers of S, $\chi(x,t\tau)\big|_{S=0}$, the derivatives of the latter function with respect to x, t and the second summand $[\lambda_0]_K$ can be expanded into series in powers of S, whose coefficients depend on the coefficients of the expansion of the function f. This implies that for $(x,t) \in \Omega$ the function $\chi(x,t,S)$ can be represented in the form

$$\chi(x,t,s) = \widehat{\gamma}^{+} + \sum_{j=0}^{K} u_j(x,t) S^{(m_0+j)/p} + O\big(S^{(m_0+K)/p}\big),$$

where K is arbitrary integer, $m_0 \geqslant 1$. By using the theory of perturbations, one can prove this relation, i.e., calculate the function χ from (5.20). As the initial approximation, one can take $\widehat{\gamma}^{+}$, then, under the condition (this condition is satisfied for a certain f)

$$f - \frac{\partial \widehat{F}_0^{(1)}}{\partial t} - \sum_{i=1}^{N} \frac{\partial \widehat{F}_i^{(1)}}{\partial x_i} + l\Big(\frac{\partial S}{\partial t}\widehat{F}_0^{(1)} + \sum_{i=1}^{N} \frac{\partial S}{\partial x_i}\widehat{F}_i^{(1)}\Big)\Big|_{\substack{S=0\\ \chi=\widehat{\gamma}^{+}}} \neq 0,$$

the next summand in the expansion of χ is of order $O(S^{m_0/p})$, where m_0 is the number of the first nonzero coefficient in the expansion of χ with respect to $\lambda - \lambda_0$.

Similarly, in the domain Ω_-, we have the asymptotic expansion

$$\chi(x,t,S) = X_-(x,t) + \widetilde{u}_1(x,t)S^{\alpha_1} + \widetilde{u}_2(x,t)S^{\alpha_2} + \cdots + O(S^{K/p}),$$

where, since the function χ is real, $\widetilde{u}_j(x,t) \in C^{\infty}(\overline{\Omega})$, $\{\alpha_j\}$ is a monotonically increasing sequence of rational numbers.

The two last relations prove our lemma.

Lemma 5.4. *Let $q_m(x,t,p_0,p,u)$ and $r_m(x,t,p_0,p,u)$ be respectively the right and left eigenvectors of the matrix $A(x,t,p_0,p,u)$ (5.2), corresponding to the eigenvalue $\lambda_m(x,t,p_0,p,u)$. Then, for any $m = 1,\dots,M$:*

$$\sum_{l=1}^{M} r_m^l \sum_{k=0}^{N} \Big\langle p_k \frac{\partial^2 F_k^l}{\partial u^2} q_m, q_m \Big\rangle = \langle r_m, q_m\rangle \sum_{l=1}^{M} q_m^l \frac{\partial \lambda_m}{\partial u_l}, \tag{5.21}$$

where F_k^i, r_m^i, q_m^i, u_i are the components of the vectors F_k, r_m, q_m, u respectively.

Proof. Below the index m will be omitted. The following inequalities are obvious:

$$\sum_{k=0}^{N}\sum_{j=1}^{M} p_k \frac{\partial F_k^l}{\partial u_j} q^j = \lambda q^l, \qquad \sum_{k=0}^{N}\sum_{i=1}^{M} p_k \frac{\partial F_k^i}{\partial u_l} r^i = \lambda r^l. \tag{5.22}$$

We transform the scalar product on the left-hand side of (5.21):

$$\begin{aligned}
\Phi &= \sum_{k=0}^{N} \Big\langle p_k \frac{\partial^2 F_k^l}{\partial u^2} q, q\Big\rangle = \sum_{k=0}^{N} p_k \sum_{i,j=1}^{M} q^i \frac{\partial^2 F_k^l}{\partial u_i \partial u_j} q^j \\
&= \sum_{k=0}^{N} p_k \Big(\Big(\sum_{i=1}^{M} q^i \frac{\partial}{\partial u_i} \sum_{j=1}^{M} q^j \frac{\partial}{\partial u_j}\Big) F_k^l - \sum_{i,j=1}^{M} q^i \frac{\partial q^j}{\partial u_i}\frac{\partial F_k^l}{\partial u_j}\Big) \\
&= \sum_{i=1}^{M} q^i \frac{\partial}{\partial u_i}\Big(\sum_{k=0}^{N} p_k \sum_{j=1}^{M} \frac{\partial F_k^l}{\partial u_j} q^j\Big) - \sum_{i,j=1}^{M} q^i \frac{\partial q^i}{\partial u_i} \sum_{k=0}^{N} p_k \frac{\partial F_k^l}{\partial u_j}.
\end{aligned}$$

By using the first relation from (5.22), we get

$$\Phi = q^l \sum_{i=1}^{M} q^i \frac{\partial \lambda}{\partial u_i} + \lambda \sum_{i=1}^{M} q^i \frac{\partial q^l}{\partial u_i} - \sum_{i,j=1}^{M} q^i \frac{\partial q^l}{\partial u_i} \sum_{k=0}^{N} p_k \frac{\partial F_k^l}{\partial u_j}. \tag{5.23}$$

Multiplying the left- and right-hand sides by r^l and taking the sum from 1 to M, we obtain

$$\sum_{l=1}^{M} r^l \sum_{k=0}^{N} \left\langle p_k \frac{\partial^2 F_k^l}{\partial u^2} q, q \right\rangle = \langle r, q\rangle \sum_{i=1}^{M} q^i \frac{\partial \lambda}{\partial u_i} + \lambda \sum_{i,j=1}^{M} r^l \frac{\partial q^l}{\partial u_i} - \sum_{i,j=1}^{M} q^i \frac{\partial q^j}{\partial u_i} \sum_{k=0}^{N} \sum_{l=1}^{M} p_k \frac{\partial F_k^l}{\partial u_j} r^l. \tag{5.24}$$

By using the second relation in (5.22), we finally get the statement of our lemma.

Lemma 5.5. *Let $U(x,t)$, $V(x,t)$ and $W(x,t)$ be vector-functions with components from $C^\infty(\overline{\Omega})$, $\varphi(\xi, x, t, h)$ be vector-functions satisfying the conditions (5.11). Then the vector-function*

$$F_i\big(x, x_0, U + \theta(S)V + W\theta(S)\sqrt{S} + \varphi(S/h, x, x_0, h)\big), \qquad i = 0, \dots, N,$$

can be expanded into the following asymptotic series in $D'(\Omega)$:

$$\begin{aligned} F_i\big(x, x_0, U + V\theta + W\theta\sqrt{S} + \varphi\big) &= \theta(S)\Phi_i\big(x, x_0, U + V + W\sqrt{S}, U\big) + F_i(x, x_0, U) \\ &+ h\delta(S)\bigg[\int_{-\infty}^{\infty} \{\Phi_i(x, x_0, U + V + \varphi(\xi, x, x_0, 0), U + V)\theta(\xi) \\ &+ \Phi_i(x, x_0, U + \varphi(\xi, x, x_0, 0), U)(1 - \theta(\xi))\}\bigg|_{S=0} d\xi\bigg] \\ &+ h^{3/2}\delta(S) \int_0^\infty \Phi(x, x_0, U + V + \varphi(\xi, x, x_0, 0), U + V)\sqrt{\xi}\, d\xi \\ &+ O(h^2) + O(S\theta(S)), \end{aligned}$$

where $x_0 = t$, $\Phi_0(x, x_0, A, B) = F_i(x, x_0, A) - F_i(x, x_0, B)$, $O(h^2)$ denotes a distribution such that $|(O(h^2), \psi)| \leqslant \text{const}\, h^{3/2}$ for any base function ψ, $O(S\theta(S))$ denotes a regular function with singularity $\theta(S)S$ at $S = 0$ or a smoother function.

Proof. For any generating function $\psi(x,t)$, we have (the index i is omitted):

$$\begin{aligned} &\big(F(x, x_0, U + V\theta(S) + W\theta(S)\sqrt{S} + \varphi), \psi\big) \\ &= \int_{S \geqslant 0} F(x, x_0, U + V + W\sqrt{S} + \varphi)\psi\, dx + \int_{S \leqslant 0} F(x, x_0, U + \varphi)\psi\, dx. \end{aligned}$$

We transform the first integral on the right-hand side as follows

$$\int_{S\geqslant 0} F(x,x_0,U+V+W\sqrt{S}+\varphi)\psi\,dx$$
$$= \int_{S\geqslant 0} \Phi(x,x_0,U+V+W\sqrt{S}+\varphi,U+V+W\sqrt{S})\psi\,dx$$
$$+ \int_{S\geqslant 0} F(x,x_0,U+V+W\sqrt{S})\psi\,dx. \tag{5.25}$$

The integrand in the first summand on the right-hand side decreases rapidly as $S/h \to \infty$. Obviously, after the change of variables $(x,x_0) \to (\alpha, S/h)$, this summand turns into an integral over the surface $S=0 \pmod{O(h)}$, i.e., this summand corresponds to the a part of the coefficient at $\delta(S)$. The first summand contributes to the coefficient at $\theta(S)$. The summands that were omitted in formula (5.25) appear when the integral is taken over the domain $S \leqslant 0$. The detailed calculations are left to the reader.

Proof of Theorem 5.1. By Lemma 4, we have

$$u(x,t) = V_0(x,t) + \sum_{k\geqslant 1} V_k(x,t)S^{\alpha_k} + \sum_{k\geqslant 1} W_k(x,t)\theta(S)S^{\beta_k},$$

as $h \to 0$ in $D'(\Omega)$, where $V_0(x,t)$, $V_k(x,t)$, $W_\alpha(x,t)$, $k \geqslant 1$, are vector-functions with components $C^\infty(\overline{\Omega})$, $\alpha_k \in N$, $\beta_1 \geqslant 0$. Without loss of generality, consider the solution $u(x,t)$ in the domain Ω_+ and assume that $\gamma_k \notin \mathbb{Z}$ for all values of K:

$$u(x,t) = \check{V}_0(x,t) + \sum_{k\geqslant 1} V_k(x,t)S^{\gamma_k},$$

where $\{\gamma_k\}$ is an ordered subsequence of $\{\beta_k\}$. Note that, by assumption, $\gamma_1 < 1$.

Let us calculate the operator $L[u]$:

$$L[u] = A_0(x,t) + \sum_{k\geqslant 1} A_k(x,t)S^{\gamma_k - 1} + \sum_{k\geqslant 1} B_k(x,t)S^{\gamma_k}.$$

By assumption, $L[u]=0$ in D'. Hence

$$A_0(x,t)=0, \quad A_k(x,t)=B_k(x,t)=0, \qquad k\geqslant 1 \quad \text{for} \quad S(x,t)=0. \tag{5.26}$$

We shall prove that the sequence $\{\gamma_k\}$ is a subsequence of the sequence $\{(2k-1)/2\}_{k\geqslant 1}$. Suppose γ_m is a minimal number in the subsequence $\{\gamma_k\}$ which cannot be represented in the form $(2k-1)/2$, $k \in N$ and $U_m(x,t) \neq 0$. Calculate $A_m(x,t)$:

$$A_m(x,t) = \gamma_m\left(\frac{\partial S}{\partial t}\frac{\partial F_0}{\partial u}(x,t,\check{V}_0) + \sum_{i=1}^{N} \frac{\partial S}{\partial x_i}\frac{\partial F_i}{\partial u}(x,t,\check{V}_0)\right)V_m. \tag{5.27}$$

Since $A_m(x,t)\big|_{S=0} = 0$, then $V_m(x,t)$ is the right null eigenvector of the matrix $A(x,t,p_0,p,U_0)$ (5.2), where $p_0 = \frac{\partial S}{\partial t}/|\nabla_x S|$, $p_i = \frac{\partial S}{\partial x_i}/|\nabla_x S|$. Hence the matrix $A(x,t,p_0,p,U_0)$ is degenerate and possesses the left null vector $Z_m(x,t)$.

First we assume that $\gamma_m < 1$. Put $\gamma_{m'} = 2\gamma_m$. Taking into account that $2\gamma_m - 1 < \gamma_m$, we have

$$A_{m'} = \gamma_{m'}|\nabla_x S|A(x,t,p_0,p,V_0)V_{2m} + 2\gamma_m|\nabla_x S|\sum_{k=0}^{N} p_0 \sum_{i,j=1}^{M} \frac{\partial^2 F_k}{\partial u_i \partial u_j}(x,t,V_0)V_m^i V_m^j,$$

where V_m^i is the ith component of vectors V_m,

$$p_0 = \frac{\partial S}{\partial t}\Big/|\nabla_x S|, \qquad p_i = \frac{\partial S}{\partial x_i}\Big/|\nabla_x S|, \qquad i = 1,\dots,N. \tag{5.28}$$

By (5.26), $A_{m'}(x,t)\big|_{S=0} = 0$. Multiplying (5.27) by $Z_m(x,t)$ from the left, according to Lemma 5.4 and inequality (5.4), we get $U_m(x,t)\|_{S=0} = 0$. This contradicts our assumption. Hence, $\gamma_1 = 1/2$, $V_1(x,t) \neq 0$. Now let $\gamma_m > 1/2$ be the minimal exponent that does not belong to the sequence $\{(2l-1)/2\}$. Suppose $\gamma_n = \gamma_m + 1/2$. Then $\gamma_n \ni \{(2k-1)/2\}$, $k \in N$, and we have:

$$A_n(x,t) = \gamma_n|\nabla_x S|A(x,t,p_0,p,U_0)U_n + (\gamma_m+\gamma_1)|\nabla_x S|\sum_{k=0}^{N} p_k \sum_{i,j=1}^{M} \frac{\partial^2 F}{\partial u_i \partial u_j}U_m^i U_1^j,$$

where p_0, p_i, $i = 1,\dots,N$ are defined in (5.28).

By the same argument as in the case $\gamma_m < 1$, we obtain $U_m(x,t)\big|_{S=0} = 0$.

Thus it is proved that the sequence $\{\gamma_k\}$ is a subsequence of the sequence $\{(2k-1)/2\}$, $k \in N$.

Now we show that, in the case $\gamma_k \in \{(2k-1)/2\}$, the function $\varphi(S(x,t)/h, x, t, 0)$ is equal to zero. Substitute the solution $u(x,t,h) = U_0 + \sum_{k\geqslant 1} U_k \theta(S) S^{()k-1/2} + \varphi(S(x,t)/h, x, t)$ into equation (5.1). By Lemma 5.5, the Rankine–Hugoniot conditions have the form:

$$\begin{aligned}
&\delta(s): \left[\sum_{i=0}^{N} \frac{\partial S}{\partial x_i}\Phi_i(x,x_0,U+V,U)\right]\Bigg|_{S=0} = 0,\\
&\frac{\theta(s)}{\sqrt{s}}: \left[\sum_{i=0}^{N} \frac{\partial S}{\partial x_i}\sum_{k=1}^{N} \frac{\partial F_i(U+V)}{\partial u_k}W_k\right]\Bigg|_{S=0} = 0,\\
&h\delta'(s): \sum_{i=0}^{N} \frac{\partial S}{\partial x_i}\int_0^{\infty} \left[\Phi_i(x,x_0,U+V+\varphi(\xi,x,x_0,0),U+V)\right]_{S=0} d\xi\\
&\qquad + \sum_{i=0}^{N} \frac{\partial S}{\partial x_i}\int_{-\infty}^{0} \left[\Phi_i(x,x_0,U+\varphi(\xi,x,x_0,0),V)\right]_{S=0} d\xi = 0,
\end{aligned} \tag{5.29}$$

$$h^{3/2}\delta'(s) : \sum_{i=0}^{N} \frac{\partial S}{\partial x_i} \int_0^\infty \left[\sum_k \frac{\partial \Phi_i}{\partial u_k}(x, x_0, U + V + \varphi(\xi, x, x_0, 0), U + V) W_k \right]_{S=0} \times \xi^{1/2}\, d\xi = 0.$$

For convenience, to the left of the colon we write the singularity whose coefficient is set equal to zero. This system of equations must be completed by the following equations

$$\begin{aligned} &\sum_{i=0}^{N} \frac{\partial}{\partial x_i} F_i(x, x_0, U + V + \sqrt{s}W) = 0, \qquad s > 0, \\ &\sum_{i=0}^{N} \frac{\partial}{\partial x_i} F_i(x, x_0, U) = 0, \qquad s < 0, \end{aligned} \tag{5.30}$$

If $W = 0$, $\varphi = 0$, the system (5.29), (5.30) is standard and describes the propagation of shock waves. In this case only the first relation in (5.29) is nontrivial. This is the well-known Rankine–Hugoniot condition.

If $W \neq 0$, $\varphi = 0$, the Rankine–Hugoniot conditions must be completed by the second relation from (5.29). Then W is the null vector of the matrix A (depending only on $U+V_j$, but not on W). Therefore, for a certain j, we have $\lambda_j\big(x, t, \frac{\partial S}{\partial t}, \frac{\partial S}{\partial x}, U + V\big)\big|_{S=0} = 0$. Thus, the velocity of shock wave propagation must coincide with the velocity of weak discontinuity motion.

If $W \neq 0$, $\varphi \neq 0$, the two relations that hold on the discontinuity surface are completed by another two relations, the system (5.29), (5.30) becomes overdetermined.

If $W = 0$, $\varphi \neq 0$, there are only two relations in (5.29), the first and the third. This system together with (5.30) can be noncontradictory.

The theorem is proved.

Let us consider an example, where the condition $\nabla_{x,t}\chi \to \infty$ as $(x,t) \to \Gamma$ is necessary for the solution to be of the structure given in item 3 of Theorem 5.1.

Consider the equation

$$\frac{\partial u}{\partial t} + \frac{\partial}{\partial x}(u^2 + u^3) = 0. \tag{5.31}$$

The solution will be sought in the form

$$u(x,t) = u_0(x,t) + u_1(x,t)\theta(s) + \sum_{k\geqslant 2} u_k(x,t)\theta(S) S^{(k+2)/3}, \tag{5.32}$$

where k is an arbitrary natural number.

Substitute the series (5.32) into equation (5.31), collect the terms at equal powers of the function $S(x,t)$ and put them equal to zero:

$$\begin{aligned} &\frac{\partial u}{\partial t} + \frac{\partial}{\partial x}(u_0^2 + u_0^3) = 0, \\ &(S_t + S_x(2u_0 + u_1 + 3u_0^2 + 3u_0u_1 + u_1^2))u_1 = 0, \\ &(S_t + S_x(2u_0 + 2u_1 + 3u_0^2 + 6u_0u_1 + 3u_1^2))u_2 = 0, \end{aligned}$$

$$
\begin{aligned}
&\big(S_t + S_x(2u_0 + 2u_1 + 3u_0^2 + 6u_0u_1 + 3u_1^2)\big)u_3 = 0,\\
&\big(S_t + S_x(2u_0 + 2u_1 + 3u_0^2 + 6u_0u_1 + 3u_1^2)\big)u_4 = 0,\\
&\big(S_t + S_x(2u_0 + 2u_1 + 3u_0^2 + 6u_0u_1 + 3u_1^2)\big)u_5\\
&\qquad + \frac{\partial u_2}{\partial t} + \frac{\partial}{\partial x}(2((u_0 + u_1) + 3(u_0 + u_1)^2)u_2 = 0,\\
&\big(S_t + S_x(2u_0 + 2u_1 + 3u_0^2 + 6u_0u_1 + 3u_1^2)\big)u_6 + S_x(1 + 3u_0 + 3u_1)u_2^2\\
&\qquad + \frac{\partial u_3}{\partial t} + \frac{\partial}{\partial x}(2((u_0 + u_1) + 3(u_0 + u_1)^2)u_3 = 0,\\
&\dots\dots\dots\dots\dots\dots\dots\dots\dots\dots\dots\dots
\end{aligned}
$$

where S_t and S_x are partial derivatives of the function $S(x,t)$ with respect to t and x. All equations, beginning with the second one, are written for $S(x,t) = 0$. Obviously, the obtained chain of equations (corresponding to conditions of the Rankine–Hugoniot type) is solvable with respect to the functions

$$S(x,t), \quad u_k(x,t), k \geqslant 0,$$

provided that $(2u_1 + 3u_0)\big|_{S=0} = -1$. The latter relation is the consistency condition for our system. However, in contrast to the proof of Theorem 5.1, this condition is formal for a system of conditions of the Rankine–Hugoniot type.

In conclusion, we prove that the system of gas dynamics equations is regular in the two-dimensional case (see Example 5.2).

Introduce the notation

$$
\begin{aligned}
&a = Al = A(x, y, t, p_0, p_1, p_2, \rho, u, v)l,\\
&\hat{a} = \widehat{A}l = A(x, y, t, \hat{p}_0, \hat{p}_1, \hat{p}_2, \rho, u, v)l, \quad w = (\rho, u, v),
\end{aligned}
$$

where the matrix $A(x, y, t, p_0, p_1, p_2, \rho, u, v)$ has the form (5.9), $l = l(x, y, t)$ is a spatial vector from a conic neighborhood of the domain Q, and $l(x,t) \in C^\infty$.

Write the vector a (a_i are its components):

$$
\begin{aligned}
&a_1 = p_0l_1 + (p_1l_2 + p_2l_3)\rho + p_1l_1u + p_2l_1v,\\
&a_2 = p_0l_2\rho + p_0l_1u + p_1l_1p' + (2p_1l_2 + p_2l_3)\rho u + p_2l_2\rho v + p_1l_1u^2 + p_2l_1uv, \quad (5.33)\\
&a_3 = p_0l_3\rho + p_0l_1v + p_2l_1p'p_1l_3\rho u + (p_1l_2 + 2p_2l_3)\rho v + p_1l_1uv + p_2l_1v^2.
\end{aligned}
$$

The components $\widehat{a}_i$, $i = 1, 2, 3$, of the vector $\widehat{a}$ coincide with formulas (5.33), where the parameters p_j are replaced by $\widehat{p}_j$, $j = 0, 1, 2$.

Obviously, one can find p_0, p_1, p_2, $|p_0| \leqslant$ const, such that, as $|w| \to \infty$, the inequalities $|\hat{a}_i| \leqslant C_1|a_i|$, $i = 1, 2, 3$, hold for any $\hat{p}_0$, $\hat{p}_1$, $\hat{p}_2$, $|\hat{p}_0| \leqslant$ const, $\hat{p}_1^2 + \hat{p}_2^2 = 1$. Hence

$$
\begin{aligned}
&|\langle \widehat{A}l, \widehat{A}l\rangle| = |\langle a, \hat{a}\rangle| \leqslant |a_1, \hat{a}_1| + |a_2, \hat{a}_2| + |a_3, \hat{a}_3|\\
&\qquad \leqslant C_1|a_1^2 + a_2^2 + a_3^2| = C_1|\langle a, a\rangle| = C_1|\langle Al, Al\rangle| \quad \text{as} \quad |w| \to \infty. \quad (5.34)
\end{aligned}
$$

Now denote by b the right-hand factor in the scalar product on the left-hand side of the inequality given in item 2 of Definition 5.2:

$$b = \frac{\partial F_0}{\partial t}(x,y,t,lz) + \frac{\partial F_1}{\partial x}(x,y,t,lz) + \frac{\partial F_2}{\partial y}(x,y,t,lz),$$

where the vector $l = l(x,y,t)$ satisfies the above conditions.

The components b_i, $i = 1,2,3$, of the vector b have the form:

$$\begin{aligned} b_1 &= z\big(l_{1t} + (l_{2x} + l_{3y})\rho + l_{1x}u + l_{1y}v\big), \\ b_2 &= z\big(l_{2t}\rho + l_{1t}u + l_{1x}p' + (2l_{2x} + l_{3y})\rho u + l_{2y}\rho v + l_{1x}u^2 + l_{1y}uv\big), \\ b_3 &= z\big(l_{3t}\rho + l_{1t}v + l_{1y}p' + l_{3x}\rho u + (l_{2x} + 3l_{3y})\rho v + l_{1x}uv + l_{1y}v^2\big), \end{aligned}$$

where l_{it}, l_{ix}, l_{iy} are partial derivatives of the ith component of the vector $l(x,t)$ in t, x and y respectively.

Suppose the domain Q has the form

$$Q = \{w = (\rho,u,v) \colon \rho > 0, u \neq 0, v \neq 0\}. \tag{5.35}$$

By using the properties of $l(x,t)$, one can find a constant C_2 such that, as $|w| \to \infty$, $|b_i| \leqslant C_2|z||a_i|$, $i = 1,2,3$, in the domain Q (5.35). This yields

$$|\langle Al, b\rangle| \leqslant C_2|z||\langle Al, Al\rangle| \quad \text{as} \quad |W| \to \infty. \tag{5.36}$$

Therefore, the inequalities (5.34) imply that the system of gas dynamics equations is regular in the domain Q (5.36).

In the one-dimensional case, the regularity of this system is proved similarly.

Note that if $V = W = 0$, we consider a soliton-like solution of the initial system. It is well-known that the soliton motion essentially depends on the form of soliton, which, in its turn, is determined by the dispersion terms [58, 59] absent in equation (5.1). The problem is to find the proportionality coefficients $A = A(F,\varphi)$ in

$$\delta(s) = \lim_{h\to 0} \frac{\varphi(S/h,x,t,h)}{h} = A(F,\varphi) \lim_{h\to 0} \frac{F(\varphi(S/h,x,t,h))}{h},$$

where $F(z)$, for example, is a polynomial (the limit is taken in D'). Clearly, if the function φ is not given, it is impossible to determine the coefficient $A(F,\varphi)$. On the other hand, it is clear that for (5.1) conditions of the Hugoniot type cannot define the function $\varphi(\xi,x,t,h)$. However, one can write some necessary conditions for the existence of soliton-like solutions of system (5.1) without the function $\varphi(\xi,x,t,h)$. For simplicity, we consider only the simplest example, namely, the Hopf equation:

$$\frac{\partial u}{\partial t} + \frac{\partial}{\partial x}(u^2) = 0.$$

We have

$$\begin{aligned} u &= u_0(x,t) + \varphi\Big(\frac{S}{h},x,t,h\Big) \\ &= u_0(x,t) + hu_1(x,t)\delta(s) + hu_2(x,t)\theta(s) + O(h^2), \end{aligned}$$

where the estimate $O(h^2)$ is an estimate in D'. Then, in D', we have

$$u^2 = h\big(2u_0u_1\delta(s) + u_1^2A\delta(s) + 2u_0u_2\theta(s)\big) + O(h^2). \tag{5.37}$$

Substituting the expansions (5.37) and (5.36) into the Hopf equation and setting the coefficients at the leading singularities equal to zero, we obtain

$$\begin{aligned}\delta' &: \big[S_tu_1 + S_x(2u_0u_1 + Au_1^2)\big]_{S=0} = 0,\\ \delta &: \big[u_2 + u_{1t} + 2(u_0u_1)_x + (Au_1^2)_x\big]_{S=0} = 0.\end{aligned}$$

Putting $S = t - \psi(x)$ (in this case $u_1 = u_1(x)$), we get

$$\frac{d}{dx}\left[\left(\frac{d\psi}{dx}\right)^{-1} u_1 + u_2\right]\bigg|_{t=\psi(x)} = 0.$$

This relation is necessary for the Hopf equation to have a soliton-like solution. The pointwise limit (as $h \to 0$) of this solution is the function $u_0(x,t)$ for $t \neq \psi(x)$ and the function $u_0(x) + \varphi(0,x,t,0)$ for $t = \psi(x)$. In the general situation, a similar relation can be derived in the same way.

CHAPTER IV

WAVE ASYMPTOTIC SOLUTIONS OF DEGENERATE SEMILINEAR PARABOLIC AND HYPERBOLIC EQUATIONS

4.1. Self-stabilizing asymptotic solutions

This chapter presents a method for constructing nonsmooth asymptotic solutions uniformly bounded as a small parameter ε tends to zero. These solutions have semi-bounded (or compact) supports and week singularities on the boundary of supports. This is the difference from well-known asymptotic solutions describing the propagation of shock waves in media with low viscosity, or soliton-like waves in weakly dispersive media [55, 58, 59, 94]. The solutions constructed in this chapter generalize simple diffusion waves, known since the publication of the classical work by Kolmogorov, Petrovskii and Piskunov [41] (see also [107]).

An asymptotic solution of the form

$$u = W(x,t,S/\varepsilon,\varepsilon), \qquad S = S(x,t,\varepsilon) \in C^\infty, \qquad W(x,t,\tau,\varepsilon) \in C^\infty, \qquad \tau > 0,$$

is a *self-stabilizing* asymptotic solution if

$$|\tau|^N \frac{\partial^\gamma W}{\partial \tau^\gamma} \to 0 \quad \text{as} \quad \tau \to \pm\infty \qquad \forall \gamma, N > 0. \tag{1.1}$$

This type of solutions includes soliton-like asymptotic solutions of equations with weak dispersion [55, 58, 59, 94] and asymptotic solutions of shock wave type. The characteristic feature of asymptotic solutions is that the operator of multiplication by the function $S(x,t,\varepsilon)$ is "small" in the class of self-stabilizing asymptotic solutions. Actually, for all $\gamma, N > 0$,

$$\left| S^N \frac{\partial^\gamma W(x,t,S/\varepsilon,\varepsilon)}{\partial \tau^\gamma} \right| \leqslant \varepsilon^N \max \left| \tau^N \frac{\partial^\gamma W(x,t,\tau,\varepsilon)}{\partial \tau^\gamma} \right| = O(\varepsilon^N). \tag{1.2}$$

In some cases, the position of the weak discontinuity front ($x = -\varphi(t,\varepsilon)$ for $x \in \mathbf{R}^1$) can be determined locally. This means (see [20] and Section 4.3) that the position of the weak discontinuity front can be obtained in construction of an *additive* asymptotics, i.e., of a solution satisfying the equation with precision $O(\varepsilon^N)$ without any additional conditions on the behavior of the discrepancy $O(\varepsilon^N)$ outside a neighborhood of the weak discontinuity. However, if the standard equation (see Chapter I) reduces to the traveling wave equation of the Kolmogorov–Petrovskii–Piskunov type, the position of the weak discontinuity front is determined by the global structure of asymptotic solution. For example, in the one-dimensional case, the function $\varphi(t,\varepsilon)$

can be obtained only by matching the asymptotic solution in a neighbourhood of the weak discontinuity front (globally satisfying the equation up to a power of a small parameter) with the solution of the linearized equation far away from the weak discontinuity front.

This chapter is organized as follows. In Section 4.1, by using the method given in [15, 51], we construct asymptotic solutions for semilinear equations near the weak discontinuity front, and clarify when these solutions may be arbitrary to a certain extent. Section 4.2 covers the case when the solution is not uniquely determined by local considerations and, following [20], we match the solution with the solution of the linearized equation.

As shown above, in contrast to the previously known self-stabilizing asymptotic solutions, the majority of solutions considered in this chapter have the compact (or semibounded) support and a weak discontinuity on the boundary of the support.

Thus, we shall speak about the construction of an asymptotic solution (with respect to a small parameter) and, at the same time, about the asymptotic propagation of a weak discontinuity. In this chapter the problem is solved by using the general nonlinear WKB method, which also allows to construct nonsmooth solutions. On the weak discontinuity front, the higher terms of the asymptotic expansion with respect to a small parameter, obtained by the method given below, have the same singularity as the principal term (or even weaker).

4.2. Construction of nonsmooth asymptotic solutions. Derivation of basic equations

The equation considered in this section describes the processes in certain biological and chemical systems (see [,] for example, 5, 6, 8, 36, 101, 104 where the physical statement of such problems is given in detail). In these problems, there exists a natural small parameter $\varepsilon \in (0,1]$, and the following semilinear parabolic equation is often considered

$$\varepsilon\frac{\partial u}{\partial t} - \varepsilon^2\frac{\partial}{\partial x}\left(K(u)\lambda(x,t)\frac{\partial u}{\partial x}\right) - \gamma^2(x,t)F(u) = 0, \qquad x \in R, \quad t \in [0,T], \tag{2.1}$$

where $u = u(x,t)$ is an unknown function with values in $[0,1]$. The functions $K(u)$, $F(u)$, $\gamma(x,t)$ and $\lambda(x,t)$ are positive and infinitely differentiable (the first two of them for $u > 0$). The functions $\gamma(x,t)$ and $\lambda(x,t)$ describe slowly changing properties of the medium.

According to the definitions given in Chapter III, the functions $K(u)$ and $F(u)$ have the form

$$K(u) = u^{k-1}\rho(u), \qquad k > 1, \qquad F(u) = u^q G(u), \qquad q > 0,$$

where $\rho(0) > 0$, $G(0) \neq 0$, $G(1) = 0$, the functions $\rho(u)$ and $G(u)$ are smooth for $u \geqslant 0$, and $\rho(u) > 0$, $G(u) > 0$ for $u \in [0,1)$.

Let us consider the problem of constructing an asymptotic solution of (2.1) satisfying the conditions:

$$u(x,t,\varepsilon)\big|_{x=-\varphi(t,\varepsilon)} = 0, \qquad u(x,t,\varepsilon) \to 1, \qquad x \to \infty, \tag{2.2}$$

$$K(u)\frac{\partial u}{\partial x}\bigg|_{x=-\varphi(t,\varepsilon)} = 0, \tag{2.3}$$

where $\varphi(t,\varepsilon)$ is an unknown function to be determined (the asymptotic solutions bounded as $\varepsilon \to 0$ are defined in Chapter III).

Problem (2.1)–(2.3) describes the propagation of a highly intensive wave of the quantity transferred [33, 34]. The wave front is given by the relation $x = -\varphi(t,\varepsilon)$. The natural continuity of the this quantity and of its flux on the wave front implies the conditions:

$$u\big|_{x=-\varphi(t,\varepsilon)} = 0, \qquad K(u)\frac{\partial u}{\partial x}\bigg|_{x=-\varphi(t,\varepsilon)} = 0.$$

We recall that the necessary condition for the flux continuity was obtained in Chapter III, namely, $k+q \geqslant 2$. It turns out that the method for constructing the asymptotic solution essentially depends on the sum $k+q$.

As shown below, for $F(u) > 0$, the inequality $k+q > 2$ implies that the dependence of the asymptotic solution on the fast variable is given by an equation of the Zeldovich type [33, 35]. On the other hand, if $k+q = 2$, then the dependence of the asymptotic solution on the fast variable is defined by an equation of the Kolmogorov–Petrovskii–Piskunov type [41]. The required results about the properties of these equations are given in Chapter I.

For $K(u) = 1$, $\lambda, \gamma = \text{const}$, problem (2.1)–(2.3) has been thoroughly investigated (see [107]). The solutions of this problem are simple waves (i.e., functions of the argument $x + bt$), which were first studied in the classical work [41] (see Figure 4.1).

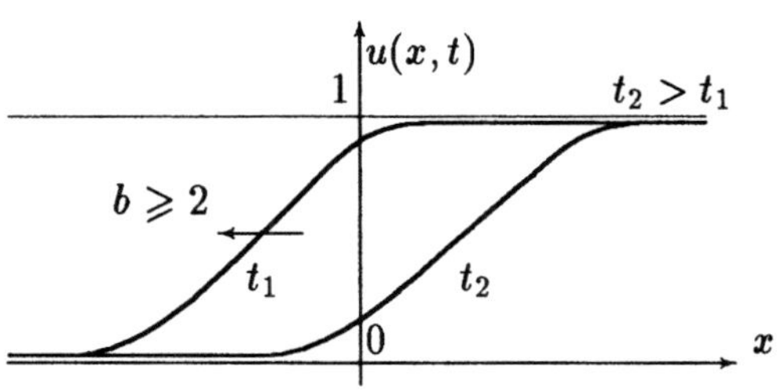

FIGURE 4.1

Since the transport coefficient $K(u)$ depends on the transferred quantity ($K(0) = 0$), the function $u(x,t,\varepsilon)$ is localized only from one side (see Figure 4.2), while the existence of an asymptotic solution of problem (2.1)–(2.3) for $\gamma \neq \text{const}$, $\lambda \neq \text{const}$ means, in particular, that the wave remains stable when perturbations arise.

A large number of exact localized solutions of equation (2.1) with constant coefficients λ, γ, or with coefficients of a special form, has been studied in many previous works. These solutions have the form

$$u(x,t,\varepsilon) = \begin{cases} \mu(x,t)\chi\big(S(x,t)/\varepsilon\big), & S \geqslant 0, \\ 0, & S < 0. \end{cases}$$

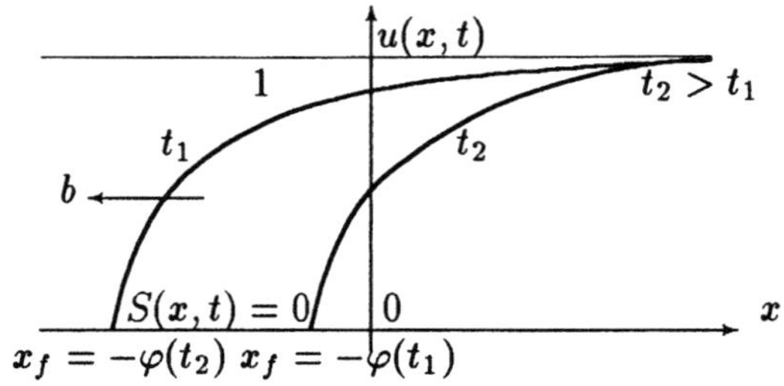

FIGURE 4.2

The functions μ and S are usually related to each other and to the invariant of the transformation group. In the general case of variable coefficients in (2.1), we seek the solution of problem (2.1)–(2.3) in the form

$$u(x,t,\varepsilon)=\begin{cases}\sum\limits_{i=0}^{N}\varepsilon^i W_i(x,t,\tau)\big|_{\tau=S(x,t,\varepsilon)/\varepsilon}+\varepsilon^{N+1}Q_N(x,t,S/\varepsilon,\varepsilon), & S\geqslant 0,\\ 0, & S<0.\end{cases} \tag{2.4}$$

Here the function $S(x,t,\varepsilon)$ is infinitely differentiable in its arguments and $\partial S(x,t,0)/\partial x \neq 0$ for $S(x,t,0)=0$. The functions $W_i(x,t,\tau)$ are infinitely differentiable in t, x and in τ for $\tau>0$, uniformly bounded as $\tau\to\infty$, and as $\tau\to 0$ possess an algebraic branching point (a powerlike singularity) $\sim\tau^\alpha$, $\alpha>0$, so that the differentiation with respect to t, x does not increase the singularity in τ (see Chapter III), while the derivatives in τ decrease faster than any negative power of τ as $\tau\to\infty$.

The exponent α depends on the properties of the functions $K(u)$, $F(u)$ (see Chapter III):

$$\alpha=\frac{1}{k-1},\qquad \text{if}\quad q\geqslant 1,\quad k+q>2,\quad F(u)>0,\quad u\in(0,1);$$

$$\alpha=\frac{1}{1-q}=\frac{1}{k-1},\qquad \text{if}\quad q<1,\quad k+q=2,\quad F(u)>0,\quad u\in(0,1).$$

The function $Q_N=Q_N(x,t,\tau,\varepsilon)$ is uniformly bounded together with its derivatives with respect to x, t, τ up to the order N.

Theorem 2.1. *The asymptotic solution of problem* (2.1)–(2.3) *has the form* (2.4), *and the function* W_0 *satisfies the simple wave equation*

$$\begin{gathered} b\frac{dW_0}{d\tau}-\frac{d}{d\tau}\left(K(W_0)\frac{dW_0}{d\tau}\right)-F(W_0)=0,\\ W_0(\tau)\big|_{\tau=0}=0,\qquad W_0(\tau)\big|_{\tau\to\infty}\to 1-0. \end{gathered} \tag{2.5}$$

The function $S(x,t,\varepsilon)$ *has the form*

$$S(x,t,\varepsilon)=\beta(t,\varepsilon)(x+\varphi(t,\varepsilon))+\beta_1(t,\varepsilon)(x+\varphi(t,\varepsilon))^2,$$

where $\beta^0(t) = \beta(t,0)$, $\varphi^0(t) = \varphi(t,0)$ *satisfy the system of equations*

$$\begin{aligned} \beta^0(t)\sqrt{\lambda(-\varphi^0(t),t)} &= \gamma(-\varphi^0(t),t), \\ \beta^0(t)\frac{d\varphi^0}{dt} &= \gamma^2(-\varphi^0(t),t)b. \end{aligned} \tag{2.6}$$

The function W_0 *satisfies the estimates*

$$\begin{aligned} &W_0(\tau) = O(\tau^\alpha), \qquad \tau \to 0, \\ &W_0(\tau) = 1 - \Gamma \exp\{-|l|\tau/\rho(1)\} + o(\exp\{-|l|\tau/\rho(1)\}), \qquad \tau \to \infty, \end{aligned} \tag{2.7}$$

where $\Gamma = \text{const} > 0$. *The following conditions hold:*

a) *if* $k + q > 2$, $q \geqslant 1$, $F(u) > 0$ *for* $u \in]0,1[$, *then*

$$\alpha = \frac{1}{k-1}, \qquad l = -\frac{b_0}{2} + \sqrt{\frac{b_0^2}{4} + \left|\frac{dR}{du}\right|_{u=1}},$$

where b_0 *is the Zeldovich constant for the Zeldovich equation:*

$$b_0\frac{d\theta}{d\xi} - \frac{d^2\theta}{d\xi^2} - R(\theta) = 0;$$

b) *if* $k + q = 2$, $q < 1$, $F(u) > 0$, $R'(0) > R'(u)$ *for* $u \in]0,1[$, *then*

$$\alpha = \frac{1}{k-1} = \frac{1}{1-q}, \qquad l = -\frac{b}{2} + \sqrt{\frac{b^2}{4} + \left|\frac{dR}{du}\right|_{u=1}}, \qquad b \geqslant 2\sqrt{\left.\frac{dR}{du}\right|_{u=0}},$$

where $R(u) = \rho(u)^{k+q-1}G(u) = K(u)F(u)$.

The functions W_1 *and* $\varphi^1 = \partial\varphi/\partial\varepsilon\big|_{\varepsilon=0}$ are defined in Lemma 2.1.

Remark. In this and subsequent theorems, as a rule, the construction of an asymptotic solution to the initial equation is reduced to solving an ordinary differential equation or a system of ordinary differential equations (system (2.6) in the last theorem). We assume that this system of equations is solvable. The solvability conditions are rather simple and we do not formulate them in theorems, except when these conditions are not obvious enough.

Before describing the structure of the correction W_1 we introduce the following notation:

$$\begin{aligned} \varphi(t,\varepsilon) &= \varphi^0(t) + \varepsilon\varphi^1(t) + \dots, \\ \beta(t,\varepsilon) &= \beta^0(t) + \varepsilon\beta^1(t) + \dots, \\ \beta_1(t,\varepsilon) &= \beta_1^0(t) + \varepsilon\beta_1^1(t) + \dots. \end{aligned}$$

Lemma 2.1. *The function $W_1(t,\tau)$ has the form*

$$W_1 = \frac{dW_0}{d\tau}\int_0^{\tau}\left(V(\tau')\left(\frac{dW_0}{d\tau}\right)^{-2}\int_{\varkappa}^{\tau'}\frac{f(t,\xi)}{V(\xi)K(W_0(\xi))}\frac{dW_0}{d\xi}\,d\xi\right)d\tau'. \tag{2.8}$$

where $\varkappa = 0$ for $k+q>2$, $\varkappa = +\infty$ for $k+q=2$, and

$$\begin{aligned}
V(\tau) &= \frac{1}{K^2(W_0)}\exp\left\{b\int\frac{d\tau}{K(W_0)}\right\},\\
f &= (\beta^0\lambda(-\varphi^0,t))^{-1}\left\{\frac{dW_0}{d\tau}+\beta^0\varphi^1_t+(\beta^0)^2\frac{\partial\lambda}{\partial x}(-\varphi^0,t)\varphi^1\frac{d}{d\tau}\left(K(W_0)\frac{dW_0}{d\tau}\right)\right.\\
&\quad-\frac{\partial\gamma^2}{\partial x}(-\varphi^0,t)\varphi^1F(W_0)-\beta^0\frac{\partial\lambda}{\partial x}\bigg|_{x=-\varphi^0}\tau\frac{d}{d\tau}\left(K(W_0)\frac{dW_0}{d\tau}\right)\\
&\quad\left.-\beta^0\frac{\partial\lambda}{\partial x}\bigg|_{x=-\varphi^0}K(W_0)\frac{dW_0}{d\tau}+\beta^0_t\frac{\tau}{\beta^0}\frac{dW_0}{d\tau}-\frac{\partial\gamma^2}{\partial x}\bigg|_{x=-\varphi^0}\frac{\tau}{\beta^0}F(W_0)\right\}.
\end{aligned} \tag{2.9}$$

(a) *If $k+q>2$ and*

$$\int_0^{\infty}\frac{f(t,\tau)}{V(\tau)K(W_0(\tau))}\frac{dW_0}{d\tau}\,d\tau = 0, \tag{2.10}$$

then the function $W_1(t,\tau)$ satisfies the estimates:

$$\begin{aligned}
W_1 &= O(\tau^{\alpha}), && \tau\to 0,\\
W_1 &= O\big(\tau^2\exp\{-|l|\tau/K(1)\}\big), && \tau\to\infty.
\end{aligned}$$

Condition (2.10) is an equation for the function φ^1:

$$\begin{aligned}
&-\beta^0\varphi^1_t I_5+\varphi^1(\beta^0)^2\frac{\partial\lambda}{\partial x}(-\varphi^0,t)I_4+\frac{\partial\gamma^2}{\partial x}(-\varphi^0,t)\varphi^1 I_6\\
&\qquad+\beta^0\frac{\partial\lambda}{\partial x}(-\varphi^0,t)(I_0+I_2)+\frac{\partial\gamma^2}{\partial x}(-\varphi^0,t)\frac{1}{\beta^0}I_3-\frac{1}{\beta^0}\frac{d\beta^0}{dt}I_1=0,
\end{aligned}$$

where

$$I_0 = \int_0^{\infty}N\tau\frac{d}{d\tau}\left(K(W_0)\frac{dW_0}{d\tau}\right)d\tau, \qquad I_1 = \int_0^{\infty}N\tau\frac{dW_0}{d\tau}d\tau,$$

$$I_2 = \int_0^{\infty}NK(W_0)\frac{dW_0}{d\tau}\,d\tau, \qquad I_3 = \int_0^{\infty}N\tau F(W_0)d\tau,$$

$$I_4 = \int_0^{\infty}N\frac{d}{d\tau}\left(K(W_0)\frac{dW_0}{d\tau}\right)d\tau,$$

$$I_5 = \int_0^{\infty}N\frac{dW_0}{d\tau}\,d\tau, \qquad I_6 = \int_0^{\infty}NF(W_0)\,d\tau,$$

$$N = K(W_0)\frac{dW_0}{d\tau}\exp\Big\{-b\int\frac{d\tau}{K(W_0)}\Big\}.$$

The functions $\beta_1^0(t)$ and $\beta^1(t)$ remain undetermined.

(b) *If $k+q=2$, then, for arbitrary smooth functions $\varphi^1(t)$, $\beta_1^0(t)$, and $\beta^1(t)$, the function $W_1(t,\tau)$ satisfies the estimates:*

$$\begin{aligned}
&W_1 = O(\tau^\alpha \ln\tau), && b = b_{\min}, && \tau\to 0;\\
&W_1 = O(\tau^\alpha), && b > b_{\min}, && \tau\to 0;\\
&W_1 = O(\tau^2\exp\{-|l|\tau/K(1)\}), && && \tau\to\infty.
\end{aligned}$$

Remark. For $k+q>2$ it can be shown that, with precision up to $O(\varepsilon^2)$, the sum $u(x,t,\varepsilon) = W_0(S/\varepsilon) + \varepsilon W_1(t,S/\varepsilon)$ is independent of the functions β^0 and β^1. Actually, this is a consequence of the fact that the function $\varphi^1(t)$ is independent of β_1^0 and β^1.

Proof. Let us verify the above statements. The main equations, obtained by substituting (2.4) into equation (2.1), have the form

$$\begin{aligned}
&\beta^0\frac{d\varphi^0}{dt}\frac{\partial W_0}{\partial\tau} - (\beta^0)^2\lambda(-\varphi_0,t)\frac{\partial}{\partial\tau}\Big(K(W_0)\frac{\partial W_0}{\partial\tau}\Big) - \gamma^2(-\varphi^0,t)F(W_0) = 0,\\
&\beta^0\frac{d\varphi^0}{dt}\frac{\partial W_1}{\partial\tau} - (\beta^0)^2\lambda(-\varphi^0,t)\frac{\partial^2}{\partial\tau^2}(K(W_0)W_1) - \gamma^2(-\varphi^0,t)\frac{dF}{dW_0}W_1\\
&\quad = -\Big\{\frac{dW_0}{d\tau}(\beta^1\varphi_t^0 + \beta^0\varphi_t^1) - \Big(2\lambda(-\varphi^0,t)\beta^0\beta^1 - (\beta^0)^2\frac{\partial\lambda}{\partial x}(-\varphi^0,t)\varphi^1\Big)\\
&\qquad\times\frac{d}{d\tau}\Big(K(W_0)\frac{dW_0}{d\tau}\Big) - \frac{\partial\gamma^2}{\partial x}(-\varphi^0,t)\varphi^1F(W_0)\\
&\qquad - \beta^0\frac{\partial\lambda}{\partial x}(-\varphi^0,t)\tau\frac{d}{d\tau}\Big(K(W_0)\frac{dW_0}{d\tau}\Big) - \beta^0\frac{\partial\lambda}{\partial x}(-\varphi^0,t)K(W_0)\frac{dW_0}{d\tau}\\
&\qquad - 2\beta_1^0\lambda(-\varphi^0,t)K(W_0)\frac{dW_0}{d\tau} - 4\tau\beta_1^0\lambda(-\varphi^0,t)\frac{d}{d\tau}\Big(K(W_0)\frac{dW_0}{d\tau}\Big)\\
&\qquad + (2\beta_1^0\varphi_t^0 + \beta_t^0)\frac{\tau}{\beta^0}\frac{dW_0}{d\tau} - \frac{\partial\gamma^2}{\partial x}(-\varphi^0,t)\frac{\tau}{\beta^0}F(W_0)\Big\}.
\end{aligned}\tag{2.11}$$

We have used the following formulas:

$$\begin{aligned}
S &= \beta(x+\varphi) + \beta_1(x+\varphi)^2 = \beta^0(x+\varphi^0) + \beta_1^0(x+\varphi^0)^2 + \varepsilon(\beta^0\varphi^1 + \beta^1(x+\varphi^0))\\
&\quad + O(\varepsilon^2 + (x+\varphi^0)^3 + \varepsilon(x+\varphi^0)),
\end{aligned}\tag{2.12}$$

$$\begin{aligned}
\frac{\partial S}{\partial t} &= \beta^0\frac{d\varphi^0}{dt} + \Big(\frac{d\beta^0}{dt} + 2\beta_1^0\frac{d\varphi^0}{dt}\Big)(x+\varphi^0) + \varepsilon\frac{d(\beta^0\varphi^1)}{dt} + \dots\\
&= \Big[\beta^0\frac{d\varphi^0}{dt} + \Big(\frac{d\beta^0}{dt} + 2\beta_1^0\frac{d\varphi^0}{dt}\Big)(\tau - \varphi^1\beta^0) + O(\varepsilon^2\tau^2)\Big]_{\tau=S(x,t,\varepsilon)/\varepsilon},
\end{aligned}$$

$$\frac{\partial S}{\partial x}=\left[\beta^0+\varepsilon 2\frac{\beta_1^0}{\beta^0}(\tau-\beta^0\varphi^1)+\varepsilon\beta^1+O(\varepsilon^2\tau^2)\right]_{\tau=S/\varepsilon},$$

$$x+\varphi^0=\left[\frac{\varepsilon(\tau-\beta^0\varphi^1)}{\beta^0}+O(\varepsilon^2\tau^2)\right|_{\tau=S/\varepsilon}.$$

By setting

$$\beta^0(t)\frac{d\varphi^0}{dt}=\gamma^2(-\varphi^0(t),t)b,\qquad \sqrt{\lambda(-\varphi^0(t),t)}\beta^0(t)=\gamma(-\varphi^0(t),t),$$

we see that the function $W_0(\tau)$ is the solution of problem (2.5). (The necessity of these conditions is shown in the Introduction to this book).

The properties of solutions satisfying equations of the form (2.5) are studied in detail in Chapters I and III. In particular, it is shown that the solution of (2.5) vanishes for $\tau=0$ and tends to unity faster than any negative power of τ as $\tau\to\infty$. The proof is based on the relation between the solution of problem (2.5) and the monotone solution of the equation

$$b\frac{d\theta}{d\xi}-\frac{d^2\theta}{d\xi^2}-R(\theta)=0,\qquad R=F(\theta)K(\theta),\tag{2.13}$$

namely, on the relation

$$K(W_0)\frac{dW_0}{d\tau}=\frac{d\theta}{d\xi}(\xi(\theta))\bigg|_{\theta(\xi)=W_0(\xi)},\tag{2.14}$$

where $\xi(\theta)$ is the inverse of the function $\theta(\xi)$, and $\xi(\theta(\xi))\equiv\xi$.

The mapping (2.14) transforms nonlocalized solutions of equation (2.1) (the traveling waves considered in Chapter I) into solutions localized from one side (see Figure 4.2).

Let us examine the behavior of the function W_0 as $W_0\to 0$ and $W_0\to 1$. The theorem on singularities, presented in Chapter III, implies that

$$W_0(\tau)=C_0\tau^\alpha+o(\tau^\alpha),\qquad \tau\to+0,\tag{2.15}$$

where $\alpha=1/(k-1)$ for $q\geqslant 1$ and for $q<1$, $k+q=2$, $F(u)>0$. For $k+q=2$, equation (2.13) is the Kolmogorov–Petrovskii–Piskunov equation for simple waves and has a wave-like solution under the condition (see Theorem 1.2. in Chapter I):

$$b\geqslant 2\sqrt{\frac{d}{dW_0}(F(W_0)K(W_0))\bigg|_{W_0=0}}.$$

Now we rewrite the equation for functions W_1 as follows

$$\frac{\partial^2 W_1}{\partial\tau^2}-\frac{\partial W_1}{\partial\tau}\left[\frac{1}{K(W_0)\beta^0}\left(\frac{d\varphi^0}{dt}\lambda^{-1}(-\varphi^0,t)-2\beta^0\frac{\partial K(W_0)}{\partial\tau}\right)\right]$$

$$+ W_1 \frac{1}{K(W_0)(\beta^0)^2}\left[\frac{\partial^2 K(W_0)}{\partial\tau^2}(\beta^0)^2 + \gamma^2(-\varphi^0,t)\lambda^{-1}(-\varphi^0,t)\frac{dF(W_0)}{dW_0}\right]$$
$$= -\frac{f\lambda^{-1}(-\varphi^0,t)}{(\beta^0)^2 K(W_0)}, \qquad W_1\big|_{\tau=0} = 0, \qquad W_1\big|_{\tau\to\infty} = 0, \tag{2.16}$$

where the function $f = f(t,\tau)$ is defined above.

Equation (2.15) is an ordinary differential equation with respect to the variable τ, its Wronskian has the form

$$V = \frac{1}{K^2(W_0)}\exp\left\{\frac{1}{\beta^0\lambda(-\varphi^0,t)}\frac{d\varphi^0}{dt}\int^\tau \frac{d\tau}{K(W_0)}\right\} = \frac{1}{K^2(W_0)}\exp\left\{b\int\frac{d\tau}{K(W_0)}\right\}.$$

By using the relation $W_0 \sim \tau^\alpha$ as $\tau \to 0$ (see (2.7)), it is possible to estimate the Wronskian as $\tau \to 0$:

$$V = O(\tau^{b/K_1-2}), \qquad q \geqslant 1, \quad k+q>2, \quad q<1, \quad k+q=2,$$

where

$$K_1 = \lim_{\tau\to+0}\frac{K(W_0)}{\tau}.$$

Now let us calculate the second solution W_{12} of the homogeneous equation (2.15) and estimate its singularities as $\tau \to 0$:

$$W_{12} = cW_{11}\int_0^\tau \frac{V}{W_{11}^2}\,d\tau, \qquad W_{12} = O(\tau^{b/K_1-\alpha}), \qquad \tau \to 0.$$

Here $c = \text{const}$, $W_{11} = dW_0(\tau)/d\tau$.

The right-hand side of (2.16) satisfies the estimate $f = O(\tau^{\alpha-1})$ as $\tau \to 0$. The general solution of (2.16) has the form

$$W_1 = c_1 W_{11} + c_2 W_{12} - \frac{W_{11}}{(\beta^0)^2\lambda(-\varphi^0,t)}\int_0^\tau\left(\int_0^{\tau'}\frac{fW_{11}}{VK(W_0)}\,d\tau''\right)\frac{V}{W_{11}^2}\,d\tau'. \tag{2.17}$$

In the variable τ, the function $W_1(x,t,\tau)$ must be smooth not less than the function W_0. Consequently, the singularity of the function W_1 as $\tau \to 0$ must be not greater than τ^α, and therefore $c_1 \equiv 0$.

Now consider the function W_1 as $\tau \to 0$. The following estimates hold

$$\begin{aligned} V &= O(\tau^{b/K_1-2}), \qquad W_{11} = O(\tau^{\alpha-1}), \qquad f = O(\tau^{\alpha-1}),\\ W_{12} &= O(\tau^{b/K_1-\alpha}), \qquad K(W_0) = O(\tau^{\alpha(k-1)}), \end{aligned} \tag{2.18}$$

Consider the case $k+q>2$. As proved in Chapter III (Lemma 4.10), we have

$$b/K_1 = \alpha. \tag{2.19}$$

This implies that the inner integral in (2.17) converges within the limits 0 and ∞. Therefore, putting $c_1 = c_2 = 0$, we see that the condition

$$\int_0^\infty \frac{fW_{11}}{VK(W_0)}\,d\tau = 0 \tag{2.20}$$

is necessary and sufficient for the function W_1 to decrease as $\tau \to \infty$. By using the explicit formula for f, we calculate (2.20) and prove that this expression does not contain $\beta^1(t)$ and $\beta_1^0(t)$. One can easily see that (2.20), written out in detail, has the form

$$\begin{aligned} &-(\varphi_t^0\beta^1 + \beta^0\varphi_t^1)I_5 + \Big(2\lambda(-\varphi^0,t)\beta^0\beta^1 - (\beta^0)^2\frac{\partial\lambda}{\partial x}(-\varphi^0,t)\Big)I_4 \\ &\quad + \frac{\partial\gamma^2}{\partial x}(-\varphi^0,t)\varphi^1 I_6 + \beta^0\frac{\partial\lambda}{\partial x}(-\varphi^0,t)I_0 + \Big(\beta^0\frac{\partial\lambda}{\partial x}(-\varphi^0,t) + 2\beta_1^0\lambda(-\varphi^0,t)\Big)I_2 \\ &\quad + 4\beta_1^0\lambda(-\varphi^0,t)I_0 - (2\beta_1^0\varphi_t^0 + \beta_t^0)\frac{1}{\beta^0}I_1 + \frac{\partial\gamma^2}{\partial x}(-\varphi^0,t)\frac{1}{\beta^0}I_3 = 0, \end{aligned} \tag{2.21}$$

where the constants I_j, $j = 1, \dots, 6$, are defined in Lemma 2.1.

We denote by $\sum_1$ the sum of expressions containing the factor β_1^0. By (2.21), we have

$$\sum\nolimits_1 = \beta_1^0\lambda(-\varphi^0,t)[2I_2 + 4I_0 - 2bI_1].$$

Integrating I_2 by parts (this integral converges due to (2.19)), we obtain

$$2I_0 = bI_1 - I_2,$$

therefore $\sum_1 \equiv 0$.

Similarly, denoting by $\sum_2$ the sum of terms with β^1, we get

$$\sum\nolimits_2 = \beta^1\gamma(-\varphi^0,t)\sqrt{\lambda(-\varphi^0,t)}[bI_5 - 2I_4].$$

Integrating I_4 by parts and taking into account (2.19), we obtain

$$\sum\nolimits_2 \equiv 0.$$

Obviously, $I_5 \not\equiv 0$, therefore the terms remaining in (2.21) form an ordinary linear differential equation with respect to the function $\varphi^1(t)$ given in the lemma. Of course, the relations between the integrals I_k hold when the upper limit (∞) is replaced by a variable limit (τ). This statement yields

$$W_1 = W_1\Big|_{\substack{\beta^1=0\\ \beta_1^0=0}}.$$

For $c_1 = c_2 = 0$, estimates (1.18) and formula (2.17) give

$$W_1 = O(\tau^\alpha), \qquad \tau \to 0.$$

As $\tau \to \infty$, condition (2.20) and the estimates

$$V = O\left(\exp\left\{\frac{b\tau}{K(1)}\right\}\right), \qquad W_{11} = O\left(\exp\left\{-\frac{|l|\tau}{K(1)}\right\}\right),$$
$$W_{12} = O\left(\exp\left\{\frac{b\tau}{K(1)} + \frac{|l|\tau}{K(1)}\right\}\right), \qquad f = O\left(\tau\exp\left\{-\frac{|l|\tau}{K(1)}\right\}\right), \qquad \tau \to \infty,$$
$$\lim_{\tau\to\infty} K(W_0) = K(1) = O(1),$$

obviously imply the estimate for W_1 given in the statement of this lemma.

Now consider the case $k+q=2$. It is known that the solution $W_0(\tau)$ of the standard equation exists for $b \geqslant b_{\min}$. Moreover,

$$b_{\min}/K_1 = 2\alpha$$

as proved in Chapter III, and consequently,

$$b/K_1 \geqslant 2\alpha. \tag{2.22}$$

Let us consider the asymptotic behavior of the integrand of the inner integral in (2.17) as $\tau \to 0$. We have*

$$\frac{fW_{11}}{VK(W_0)} \sim \tau^{2\alpha-1-b/K_1}.$$

Thus, in this case, the inner integral in (2.17) diverges near the lower limit. By choosing the constant c_2, we can reduce (2.17) to the form

$$W_1 = W_{11}\int_0^\tau V(W_{11})^{-2}\int_\infty^{\tau'} \frac{fW_{11}}{VK(W_{11})}\,d\tau''d\tau'. \tag{2.23}$$

(Here we put $c_1 = 0$ for the same reason as above).

By (2.22), as $\tau \to 0$, the inner integral can be asymptotically estimated for arbitrary functions φ^1, β^1 and β_1^0:

$$\int_\infty^\tau \frac{fW_{11}}{VK(W_{11})}\,d\tau' \sim \begin{cases} \ln\tau, & b = b_{\min}, \\ \tau^{2\alpha-b/K_1}, & b > b_{\min}. \end{cases}$$

This gives the required estimates for the function W_1 defined in (2.23). In conclusion, we note that for $b = b_{\min}$, as $\tau \to 0$, the "correction" $W_1(t,\tau)$ turns out to be less smooth than the principal term of the asymptotic expansion of $W_0(\tau)$. However, if $b > b_{\min}$, in the subsequent section, by analysing the global structure of the solution, we obtain formulas for the functions φ^1, β^1 and β_1^0 and thus estimate the global solution:

$$W_1(t,\tau) \sim \tau^\alpha, \qquad \tau \to 0.$$

*$\forall\varphi,\psi$: $\varphi \sim \psi$, $\tau \to a \Leftrightarrow \lim_{\tau\to a}|\varphi/\psi| = \text{const} \neq 0$.

Theorem 2.1 and Lemma 2.1 are proved.

Finally, below in Theorems 2.2 and 2.3, we present formulas for the asymptotic solution of localized wave type, satisfying (2.1) provided that

$$q < 1, \qquad k + q > 2, \qquad F(u) < 0, \qquad u \in (0,1).$$

Then, by Theorem 3.1 in Chapter III, the functions W_i in (2.4) can be estimated as follows:

$$W_i = O(\tau^{1/(1-q)}), \qquad \tau \to +0,$$

and, for $i = 0$, this estimate is exact, i.e.,

$$W_0 = c_0 \tau^{1/(1-q)} + O(\tau^{1/(1-q)+\nu}), \qquad \nu = \text{const} > 0.$$

As already noted, in a neighborhood of $u = 1$, the properties of the considered equation are similar to those of the standard equation of the Kolmogorov–Petrovskii–Piskunov type in a neighborhood of $u = 0$. In particular, the solution of the ordinary differential equation

$$b\frac{dW_0}{d\tau} - \frac{d}{d\tau}\Big(K(W_0)\frac{dW_0}{d\tau}\Big) - F(W_0) = 0, \tag{2.24}$$
$$W_0\big|_{\tau=0}, \qquad W_0\big|_{\tau\to\infty} \to 1,$$

can have the following asymptotics as $\tau \to \infty$:

$$\begin{aligned}
&\text{(a)} \quad 1 - W_0 \sim \exp\Big\{-\frac{\lambda_1 \tau}{K(1)}\Big\} + o\Big(\exp\Big\{-\frac{\lambda_1\tau}{K(1)}\Big\}\Big),\\
&\qquad \lambda_1 = -\frac{b}{2} - \sqrt{\frac{b^4}{4} - \frac{dR}{d\theta}\Big|_{\theta=1}}, \qquad b < -2\sqrt{\frac{dR}{d\theta}\Big|_{\theta=1}},\\
&\text{(b)} \quad 1 - W_0 \sim \exp\Big\{-\frac{\lambda_2 \tau}{K(1)}\Big\} + o\Big(\exp\Big\{-\frac{\lambda_2\tau}{K(1)}\Big\}\Big),\\
&\qquad \lambda_1 = -\frac{b}{2} + \sqrt{\frac{b^4}{4} - \frac{dR}{d\theta}\Big|_{\theta=1}}, \qquad b < -2\sqrt{\frac{dR}{d\theta}\Big|_{\theta=1}},
\end{aligned} \tag{2.25}$$

where $R(\theta) = \theta^{k+q-1}\rho(\theta)G(\theta) = K(\theta)F(\theta)$.

Under the condition

$$\frac{dR}{d\theta} \leqslant \frac{dR}{d\theta}\Big|_{\theta=1}, \tag{2.26}$$

we have the case (a) (due to results presented in Chapter I).

In this case the asymptotic solution is described in the following theorem.

Theorem 2.2. *For $q < 1$, $k + q > 2$, under the assumption (2.26), the problem (2.1)–(2.3) has an asymptotic solution of the form (2.4). The function $S(x, t, \varepsilon)$ has the form $S(x,t,\varepsilon) = \beta(t)(x + \varphi(t,\varepsilon))$, where β^0 and φ^0 satisfy the following system of relations*

$$\beta^0(t)\sqrt{\lambda(-\varphi^0,t)} = \gamma(-\varphi^0,t),$$
$$\beta^0(t)\frac{d\varphi^0(t)}{dt} = \gamma^2(-\varphi^0,t)b(t),$$

where $b < -2\sqrt{dR/du\big|_{u=1}}$. The function W_0 satisfies the estimates

$$W_0(\tau) = O(\tau^{1/(1-q)}), \qquad \tau \to 0,$$
$$W_0(\tau) = 1 - \Gamma_3 \exp\left\{-\frac{\lambda_1\tau}{K(1)}\right\} + o\left(\tau\exp\left\{-\frac{\lambda_1\tau}{K(1)}\right\}\right), \qquad \tau \to \infty,$$

where $\Gamma_3 = \mathrm{const}$, $b = -2\sqrt{\frac{dR}{du}(1)}$, $\lambda_1 = -\frac{b}{2} - \sqrt{\frac{b^2}{4} - \frac{dR}{du}\big|_{u=1}}$. The function W_1 is defined in Lemma 2.2.

Lemma 2.2. *The function $W_1(t,\tau)$ has the form*

$$W_1 = -\frac{1}{(\beta^0)^2\lambda(-\varphi^0,t)}\frac{dW_0}{d\tau}\int_0^\tau\left(V(\tau')\left(\frac{dW_0}{d\tau'}\right)^{-2}\times\int_0^{\tau'}\frac{f(t,\xi)}{V(\xi)K(W_0(\xi))}\frac{dW_0}{d\xi}\,d\xi\right)d\tau', \quad (2.27)$$

where V and f are defined as follows

$$V(\tau) = \frac{1}{K^2(W_0)}\exp\left\{b\int\frac{d\tau}{K(W_0)}\right\},$$
$$f = \left\{\frac{dW_0}{d\tau}\beta^0\varphi_t^1 + (\beta^0)^2\frac{\partial\lambda}{\partial x}(-\varphi^0,t)\varphi^1\frac{d}{d\tau}\left(K(W_0)\frac{dW_0}{d\tau}\right)\right.$$
$$-\frac{\partial\gamma^2}{\partial x}(-\varphi^0,t)\varphi^1F(W_0) - (\beta^0)^2\frac{\partial\lambda}{\partial x}(-\varphi^0,t)\tau\frac{d}{d\tau}\left(K(W_0)\frac{dW_0}{d\tau}\right)$$
$$\left.-\beta^0\frac{\partial\lambda}{\partial x}(-\varphi^0,t)K(W_0)\frac{dW_0}{d\tau} + \beta_t^0\tau\frac{dW_0}{d\tau} - \frac{\partial\gamma^2}{\partial x}(-\varphi^0,t)\tau F(W_0)\right\}. \quad (2.28)$$

The estimates

$$W_1 \sim (\tau^{1+1/(1-q)}), \qquad \tau \to 0,$$
$$W_1 = O\left(\tau^2\exp\left\{-\frac{\lambda_1\tau}{K(1)}\right\}\right), \qquad \tau \to \infty,$$

hold if $b < -2\sqrt{\frac{dR}{du}\big|_{u=1}}$ for any $\beta^0(t)$, $\varphi^0(t)$, $\varphi^1(t) \in C^\infty$ satisfying the relations given in Theorem 2.2.

Proof. In this proof we find unknown functions $W_0(\tau)$ and $W_1(t,\tau)$ and establish their properties. Substituting (2.4) into (2.1), we obtain the main equations of the form (2.11). For $F(u) < 0$, $u \in (0,1)$, the properties of (2.11) have been studied in Theorem 1.5.

Now let us examine the ordinary differential equation (2.16). As $\tau \to 0$, the Wronskian of this equation can be estimated as follows

$$V = O(\tau^{-2(k-1)\alpha}) \exp\left\{ -\frac{b\tau^{\alpha-\alpha k+1}}{(\alpha k-\alpha-1)c_{11}} \right\}, \qquad c_{11} = \rho(0)c_0^{k-1},$$

where $\alpha k - \alpha - 1 > 0$ for $q < 1$, $k + q > 2$. As $\tau \to 0$, we have the estimates:

$$W_{11} = O(\tau^{\alpha-1}), \qquad K(W_0) = O(\tau^{\alpha(k-1)}),$$
$$f = O(\tau^{\alpha-1}), \qquad W_1 = O(\tau^{\alpha}).$$

The first three estimates are obvious and completely coincide with those obtained in Theorem 2.2.

Now let us estimate the function W_1. First we estimate the ratio W_1/W_{11} as $\tau \to 0$ (as before, $c_1 = c_2 = 0$). From (2.16) we have

$$\frac{W_1}{W_{11}} = -\frac{1}{(\beta^0)^2\lambda(-\varphi^0,t)} \int_0^{\tau} \left(\int_0^{\tau'} \frac{fW_{11}}{VK(W_0)}\, d\tau'' \right) \frac{V}{W_{11}^2}\, d\tau'. \tag{2.29}$$

As $\tau \to 0$, the integrand in the inner integral has the asymptotics

$$\frac{fW_{11}}{VK(W_0)} \sim \tau^{\alpha(k+q)-1} \exp\left\{ \frac{b\tau^{\alpha-\alpha k+1}}{(\alpha k-\alpha-1)c_{11}} \right\}, \qquad b < 0.$$

Since $\alpha k - \alpha - 1 > 0$, the inner integral on the right-hand side of (2.29) converges for $\tau \to 0$.

Thus, as $\tau \to 0$,

$$\int_0^{\tau} \frac{W_{11}f}{VK(W_0)}\, d\tau' \sim \tau^{\alpha(2k+q)-\alpha} \exp\left\{ \frac{b\tau^{\alpha-\alpha k+1}}{(\alpha k-\alpha-1)c_{11}} \right\}.$$

This implies that the integrand in the expression for W_1 in (2.29) behaves asymptotically as follows

$$\frac{V}{W_{11}^2} \int_0^{\tau} \frac{W_{11}f}{VK(W_0)}\, d\tau' \sim \tau, \qquad \tau \to 0.$$

The latter relation gives us the required estimate for W_1:

$$W_1 \sim \tau^{1+1/(1-q)}, \qquad \tau \to 0.$$

Let us estimate the function W_1 at infinity. As $\tau \to \infty$, we have

$$W_{11} = O\left(\exp\left\{ -\frac{\lambda_1 \tau}{K(1)} \right\} \right), \qquad K(W_0) = O(1),$$

$$\frac{fW_{11}}{VK(W_0)} \sim \tau \exp\left\{\frac{b^2 - 4\frac{dR}{du}\big|_{u=1}}{K(1)}\right\}\Big), \qquad V \sim \exp\left\{-\frac{|b|\tau}{K(1)}\right\}.$$

Therefore, if the inner integral in the formula for W_1 diverges within the limits 0 and ∞ (for $|b| > |b_{\min}|$), we have the estimate

$$W_1 \sim \tau^2 \exp\left\{-\frac{\lambda_1 \tau}{K(1)}\right\}.$$

Thus Theorem 2.2 and Lemma 2.2 are proved.

Note that here, precisely as in Theorem 2.1 for $k+q=2$, the functions φ^0, φ^1 and β^0 depend on an undetermined parameter b taking values on the ray $b \leqslant -2\sqrt{\frac{dR}{d\theta}\big|_{\theta=1}}$. This follows from the fact that the kernel of the operator adjoint to the linearized operator of the equation for W_1 is trivial in the class of functions considered.

Now let us consider the case when the inequality (2.26) does not hold. By Theorem 1.2 of Chapter I, the solution of the equation for W_0 exists, generally speaking, for $b \leqslant b_* < -2\sqrt{\frac{dR}{d\theta}\big|_{\theta=1}}$. For $b = b_*$, the asymptotics of the function W_0 (as $\tau \to \infty$) has the form (b), for $b < b_*$, the form (a). Therefore, for $b < b_*$, we have the situation of Theorem 2.2, and for $b = b_*$, as is easy to see, the situation of Theorem 2.1 for $k+q>2$.

Theorem 2.3. *Let $q<1$, $k+q>2$, $F(u)<0$ for $u \in (0,1)$ and $b_* < -2\sqrt{\frac{dR}{d\theta}\big|_{\theta=1}}$ be the minimal value of the parameter b, for which, as $\tau \to \infty$, there exists a smooth monotone solution W_0 of problem (2.24) with asymptotic expansion (2.25) (b). Then the asymptotic solution of problem (2.1)–(2.3) has the form (2.4), where $W_0(\tau)$ is the solution of problem (2.24), $S(x,t,\varepsilon) = \beta^0(t)(x + \varphi^0(t) + \varepsilon\varphi^1(t))$, the functions β^0 and φ^0 satisfy the system of equations*

$$\beta^0(t)\sqrt{\lambda(-\varphi^0,t)} = \gamma(-\varphi^0,t),$$
$$\beta^0(t)\frac{d\varphi^0(t)}{dt} = b_*\gamma^2(-\varphi^0,t).$$

The functions W_1 and φ^1 are determined in Lemma 2.3.

Lemma 2.3. *The function $W_1(t,\tau)$ has the form*

$$W_1(t,\tau) = \frac{dW_0}{d\tau}\int_0^\tau \left(V(\tau')\left(\frac{dW_0}{d\tau}\right)^{-2}\int_0^{\tau'} \frac{f(t,\xi)}{V(\xi)K(W_0(\xi))}\frac{dW_0}{d\xi}\,d\xi\right)d\tau,$$

where the functions V and f are given in (2.28). The estimate holds:

$$W_1 \sim \tau^{1/(1-q)}, \qquad \tau \to 0.$$

The estimate

$$W_1 \sim \tau^2 \exp\left\{-\frac{\lambda_2 \tau}{K(1)}\right\}$$

holds under the condition (necessary and sufficient):

$$\int_0^\infty \frac{f(t,\xi)}{V(\xi)K(W_0(\xi))}\frac{dW_0}{d\xi}\,d\xi = 0. \tag{2.30}$$

The function $\varphi^1(t)$ is determined by equation (2.30) *coinciding with equation* (2.10) *in Lemma* 2.1.

Proof is similar to that of Lemma 2.2. The distinction is that here the asymptotic expansion (as $\tau \to \infty$) of the function W_0 implies the following asymptotic expansion (as $\tau \to \infty$) of the integrand in the inner integral:

$$\frac{f(t,\xi)}{V(\xi)K(W_0(\xi))} \sim \tau \exp\left\{-\frac{(2\lambda_2+b_*)\tau}{K(1)}\right\} = \tau \exp\left\{-\frac{\tau\sqrt{\frac{dR}{d\theta}\big|_{\theta=1}}}{K(1)}\right\}.$$

Therefore the inner integral converges in τ' within the limits 0 and ∞, and (2.30) is the necessary and sufficient condition for the function W_1 to decrease as $\tau \to \infty$. The theorem is proved.

Let $q < 1$, $k+q > 2$, $F(u) < 0$ for $u \in (0,1)$. Let conditions (2.26) hold. Consider the solution of the boundary-value problem (2.24) for $b = b_{\min}$. In this case, the properties of the functions $R(u) = K(u)F(u)$ as $\tau \to \infty$ yield two situations:

$$(\mathrm{c}):\quad 1 - W_0 \sim \exp\left\{\frac{b_{\min}\tau}{2K(1)}\right\},$$

$$(\mathrm{d}):\quad 1 - W_0 \sim \tau\exp\left\{\frac{b_{\min}\tau}{2K(1)}\right\},$$

where $b_{\min} = -2\sqrt{\frac{dR}{d\theta}\big|_{\theta=1}}$.

In both cases the statements of Theorem 2.3 and Lemma 2.3 hold except the estimates for the function W_1 as $\tau \to \infty$. Namely, for the case (c),

$$W_1 \sim \tau^2 \exp\left\{\frac{b_{\min}\tau}{2K(1)}\right\},$$

and for the case (d),

$$W_1 \sim \tau^3 \exp\left\{\frac{b_{\min}\tau}{2K(1)}\right\}.$$

Now, let us discuss the following fact. It turns out that it is possible to construct solutions more general than those considered in Theorems 2.1 and 2.2 for $k+q=2$ and $k+q>2$, $q<1$, $F(u)<0$, $u\in(0,1)$.

Actually, under the assumptions of Theorems 2.1 and 2.2, the position of the weak discontinuity front cannot be obtained by local examination even "in the principal term" (see the subsequent section of this chapter). Moreover, in these cases we can put $b = b(t)$, $|b(t)| \geqslant |b_{\min}|$, $b(t) \in C^\infty$. Indeed, the functions $\partial^i W_0/\partial t^i(t,\tau)$ and W_0 must satisfy the same conditions as $\tau \to \infty$ and $\tau \to 0$, i.e.,

$$\frac{\partial^i W_0}{\partial t^i} \sim W_0, \qquad \tau \to 0,$$
$$\frac{\partial^i W_0}{\partial t^i} = O(\tau^{-N}), \qquad \tau \to \infty,$$

then the inequality $\partial b/\partial t \neq 0$ does not lead to a contradiction. However, in this case, the position of the weak discontinuity front can be found by determining $b(t)$, not by solving the problem locally as in the above theorems.

Lemma 2.4. *Let $|b(t)| > |b_{\min}|$, $b(t) \in C^\infty$. Then, for $k+q=2$, $F(u) > 0$, $R'(0) > R(u)$, $u \in (0,1)$ or, for $k+q>2$, $q<1$, $F(u)<0$, $R'(1) > R(u)$, $u \in (0,1)$. The estimates hold:*

$$\frac{\partial W_0}{\partial t}(t,\tau) \sim \tau W_0, \qquad \tau \to +0,$$
$$1 - \frac{\partial W_0}{\partial t}(t,\tau) \sim \exp\Big(-\frac{l\tau}{K(1)}\Big), \qquad \tau \to \infty,$$

where

$$l = \begin{cases} \frac{-b+\sqrt{b^2+4|F'(1)|K(1)}}{2}, & k+q=2, \quad F \geqslant 0, \\ \frac{-b+\sqrt{b^2+4F'(1)K(1)}}{2}, & k+q>2, \quad F \leqslant 0. \end{cases}$$

In the case $k+q>2$, $F \geqslant 0$, the statement of this lemma holds if and only if $\partial b/\partial t \equiv 0$.

Proof. The function W_0 satisfies the equation

$$b\frac{\partial W_0}{\partial \tau} - \frac{\partial}{\partial \tau}\Big(K(W_0)\frac{\partial W_0}{\partial \tau}\Big) - F(W_0) = 0.$$

Denoting $U = \partial W_0/\partial t$, we obtain the following equation:

$$b\frac{\partial U}{\partial \tau} - \frac{\partial}{\partial \tau}\Big(K(W_0)\frac{\partial U}{\partial \tau}\Big) - \frac{\partial F}{\partial u}\Big|_{u=W_0} U = -\frac{\partial b}{\partial t}\frac{\partial W_0}{\partial \tau} \stackrel{\text{def}}{=} G.$$

This is the first variation equation considered in Theorems 2.1 and 2.2. This equation has the following solution:

$$U = -\int_0^\tau \frac{V}{(W_0')^2}\int_a^\tau \frac{GW_0'}{VK(W_0)}\,d\tau''d\tau', \qquad a = \begin{cases} \infty, & k+q=2, \quad F \geqslant 0, \\ 0, & \begin{cases} k+q>2, & F<0; \\ k+q=2, & F \geqslant 0. \end{cases} \end{cases}$$

Here, the estimates given in Theorem 2.1 for $k+q>2$ imply that the inner integral converges within the limits 0 and ∞, and the equality

$$\int_0^\infty \frac{GW_0'}{VK(W_0)}\,d\tau = 0$$

is necessary and sufficient for the estimates in Lemma 2.4 to hold. We have the latter equality only if $\partial b/\partial t \equiv 0$.

In the cases $k+q=2$, $F \geqslant 0$ (Theorem 2.1) and $k+q>2$, $F \leqslant 0$ (Theorem 2.2), the inner integral in the formula for U diverges between the limits 0 and ∞. Here, precisely as in the proofs of Lemmas 2.1 and 2.2, we obtain the required estimates for the function $\partial W_0/\partial t$ without any additional solvability conditions. The lemma is proved.

In conclusion, we estimate the discrepancies arising when the constructed asymptotic solutions are substituted into the equation. For $\varepsilon^2 Q_2(x,t,\varepsilon)$, we have

$$\begin{aligned} Q_2 = {} & \frac{\partial W_1}{\partial t} + \left(\beta^2\varphi_t^0 + \frac{\beta_t^1}{\beta^0}(\tau - \beta^0\varphi^1) + \frac{(\beta_1^0)_t}{(\beta^0)^2}(\tau - \beta^0\varphi^1)^2\right)\frac{\partial W_0}{\partial \tau} \\ & - (\beta^0)^2\lambda\frac{\partial}{\partial \tau}\left[(2K''_W(W_0)W_1^2 + O(W_1^3))\frac{\partial W_0}{\partial \tau}\right. \\ & \left. - \left(K'(W_0)\frac{\partial W_1}{\partial \tau}W_1 + O\left(W_1^2\frac{\partial W_1}{\partial \tau}\right)\right) - 2\beta^2(\beta^0)^{-1}K(W_0)\frac{\partial W_0}{\partial \tau}\right] \\ & + \frac{(\beta^0)^{-2}}{2}\frac{\partial^2\gamma^2}{\partial x^2}(\tau - \beta^0\varphi^1)^2 F(W_0) + (F''_W(W_0)W_1^2 + O(W_1^3)) + \ldots \end{aligned}$$

Here the dots stand for omitted terms, but all of them possess the same properties. The function $S(x,t,\varepsilon)$ is given in the general form (2.35). Now we shall show that, by choosing the function β^N, we can control the smoothness of discrepancies, but this choice does not essentially change the magnitude of the solution (2.4). Thus, for $\tau \to 0$, $k+q \geqslant 2$, $F(u) > 0$, $u \in (0,1)$, we have the estimate $Q_2 = O(\tau^{\alpha-1})$. The same estimate also holds for Q_N, $N > 2$.

If the asymptotics of the functions W_0, W_1 were of the form $W_i \sim \tau^\alpha + O(\tau^{\alpha+1})$, $\tau \to 0$, choosing the function β^2 (for $N=2$), we could obtain the estimate $Q_2 = O(\tau^\alpha)$, $\tau \to 0$ (similarly, choosing the function β^N, we could obtain estimates for any N). This statement follows from the fact that the discrepancy Q_N contains the function β^N multiplied by $2\beta^0\lambda\frac{\partial}{\partial\tau}(K(W_0)\frac{dW_0}{d\tau})$, which does not vanish and has the order $O(\tau^{\alpha-1})$ as $\tau \to +0$. Thus, for example, taking $\beta_k^i = 0$ (this is possible since the solution does not depend on the choice of these functions) and using the condition $Q_N = O(\tau^\alpha)$, $\tau \to 0$, we obtain an equation for the function β^N, too cumbersome to be written here.

On the other hand, obviously, a change in the function $\beta^2(\beta^N)$ contributes $O(\varepsilon^2)$ $(O(\varepsilon^N))$ to the solution.

Thus, the smoothness of discrepancies can be improved (up to that of the solution) by adding small terms to the solution. As pointed out above, this can be done, if the functions W_0 and W_1 have the following asymptotics as $\tau \to 0$:

$$W_i \sim \tau^\alpha + O(\tau^{\alpha+1}).$$

Indeed (see Chapter III and Theorem 2.1), we have the asymptotics

$$W_0 \sim \tau^\alpha + O(\tau^{\alpha+1}), \qquad \tau \to +0, \qquad \alpha = \frac{1}{k-1},$$

for $k+q>2$, $q \geqslant 1$ and for $k+q=2$, $b>b_{\min}$. For $b=b_{\min}$, $k+q=2$, we have

$$W_0 \sim \tau^\alpha + O(\tau^{\alpha+\nu}), \qquad \tau \to +0, \qquad \alpha = \frac{1}{k-1}, \qquad \nu > 0, \tag{2.31}$$

if the corresponding nonlocalized wave $\theta(\xi)$ (see Theorem 1.4 in Chapter I) has the asymptotics

$$\theta(\xi) = \exp(b_{\min}\xi) + o(\exp(b_{\min}\xi)), \qquad \xi \to -\infty. \tag{2.32}$$

If

$$\theta(\xi) = O(-\xi \exp(b_{\min}\xi)), \qquad \xi \to -\infty,$$

the solution of the corresponding nonlinear equation takes the form

$$W_0 \sim \tau^\alpha + c_1 \frac{\tau^\alpha}{\ln \tau} + O\left(\frac{\tau^\alpha \ln\ln\tau}{\ln^2 \tau}\right), \qquad \tau \to 0, \quad \alpha = \frac{1}{k-1}. \tag{2.33}$$

If $k+q>2$, $q<1$, $F(u)<0$ for $u \in (0,1)$, then

$$W_0 \sim \tau^\alpha + O\big(\tau^{1+(k+q-2)/(1-q)}\big), \qquad \alpha = \frac{1}{1-q}, \qquad \tau \to +0. \tag{2.34}$$

Thus, for $q \geqslant 1$, $k+q=2$, $b>b_{\min}$ and for $b=b_{\min}$ in the case (2.32) for $\nu \geqslant 1$, one can choose the function β^2 so that $Q_2 = O(\tau^\alpha)$ or

$$|Q_2/W_0| = \text{const} \qquad \text{as} \quad \tau \to 0.$$

In the next section we shall discuss the case, when an asymptotics of the form (2.33) can be obtained.

In the case $k+q>2$, $q<1$, $F(u)<0$, $u \in (0,1)$, the estimate for W_1 (W_N) always implies that $Q_2 = O(\tau^\alpha)$, $\tau \to 0$, see Lemma 2.2.

4.3. Global localized solutions and regularization of ill-posed problems

In the preceding section asymptotic solutions of degenerate semilinear parabolic equations were obtained. Generally speaking, these solutions are not uniform outside a neighborhood of the weak discontinuity front, which means that the ratio $\varepsilon W_1/(1-W_0)$ is unbounded outside this neighborhood. For example, for $1/2<\nu\leqslant 1$, we have

$$\varepsilon W_1/(1-W_0)\big|_{x+\varphi^0 \sim \varepsilon^{1-\nu}} \sim \varepsilon^{1-2\nu} \to \infty, \qquad \varepsilon \to 0.$$

It turns out that $O(\varepsilon^2)$-solutions cannot be constructed uniquely in all possible cases (and the correction $\varphi^1(t)$ to the coordinate of the weak discontinuity front cannot be determined).

Namely, for $k+q=2$, $F(u)>0$, $u \in (0,1)$ and for $k+q>2$, $q<1$, $F(u)<0$, $u \in (0,1)$ (see Theorems 2.1 and 2.2 and Lemma 2.4) there is no additional (orthogonality) condition that is necessary in other cases for the correction W_1 to satisfy the boundary conditions $W_1\big|_{\tau=0} = W_1\big|_{\tau\to\infty} = 0$. This means that the correction to the coordinate of the front $\varphi^1(t)$ may be chosen arbitrarily. Obviously, this additional condition also does not appear when the subsequent terms W_i, $i>1$ of the asymptotic expansion are constructed, since then the operator in variations is invertible. This means that degenerate semilinear parabolic equations are asymptotically ill-posed in the cases listed at the beginning of this section. Namely, the following statement holds [20].

Theorem 3.1. *Let the conditions of Theorem 2.1 be satisfied and let $k+q=2$. Then, for any $N>0$ equation (2.1) has two solutions, $u_N^1(x,t,\varepsilon)$ and $u_N^2(x,t,\varepsilon)$, satisfying the boundary conditions (2.2) and (2.3), such that:*

$$\text{(a)}\qquad Lu_N^i=\varepsilon^{N+1}Q_N^i(x,t,\varepsilon),\quad \left|Q_N^i\Big/\varepsilon\frac{\partial u_N^i}{\partial t}\right|\leqslant \text{const},\ u_N\to 0;$$

$$Q_n^i\to 0\quad \text{for}\quad u_N^i\to 1;\qquad u_N^1\big|_{t=0}=u_N^2\big|_{t=0};$$

(b) *there exists a constant $\delta>0$ such that, for $t=\delta$ and a sufficiently small ε, the estimate holds:*

$$\max|u_N^1-u_N^2|\geqslant\delta_1>0,\qquad \delta_1=\text{const}>0.$$

Proof. According to the algorithm given in Theorem 2.1, the solutions u_N^i, whose existence is stated, can be constructed in the form

$$u_N^i(x,t,\varepsilon)=W_0\left(\frac{S^i}{\varepsilon}\right)+\sum_{k=1}^{N}W_k^i\left(\frac{S^i}{\varepsilon},t\right),\tag{3.1}$$

where

$$S^i=\beta^0(x+\varphi^0)+\varepsilon\beta^0\varphi^i+\varepsilon^N\beta_i^N(x+\varphi^0),$$
$$\beta^0=[\lambda(-\varphi^0(t),t)]^{1/2}/\gamma(-\varphi^0(t),t),$$

the function $\varphi^0(t)$ satisfies the ordinary differential equation

$$\varphi_t^0=b\gamma(-\varphi^0,t)/\beta^0,\qquad \varphi^0\big|_{t=0}=\varphi_0^0,$$

and φ^i, $i=1,2$, are arbitrary smooth functions, such that

$$\left.\frac{\partial^j\varphi^1}{\partial t^j}\right|_{t=0}=\left.\frac{\partial^j\varphi^2}{\partial t^j}\right|_{t=0},\qquad j=0,\dots,N.$$

The function $W_0=W_0(\tau)$ is the solution of the boundary-value problem for the ordinary differential equation (2.5) for $b>b_{\min}$. The functions $W_k^i(\tau,t)$ satisfy the first variation equations with right-hand sides depending on W_0, W_j^i, $0<j<k$. As $\tau\to\infty$, the functions $W_k^i(\tau,t)$ decrease faster than any power of τ^{-1}, belong to C^∞ for $\tau>0$, $t\geqslant 0$, and satisfy the estimates

$$W_k^i(\tau,t)=O(\tau^\alpha),\qquad \tau\to+0,\qquad \alpha=\frac{1}{k-1}.$$

Now let us discuss a statement similar to statement (b) of the theorem. The assumption that the parameter ε is sufficiently small permits to consider only the principal term of the asymptotic expansion

$$u_N^i(x,t,\varepsilon)\sim W_0\left(\frac{S^i}{\varepsilon}\right)=W_0\left(\frac{\beta^0(x+\varphi^0)}{\varepsilon}+\beta^0\varphi^i\right).$$

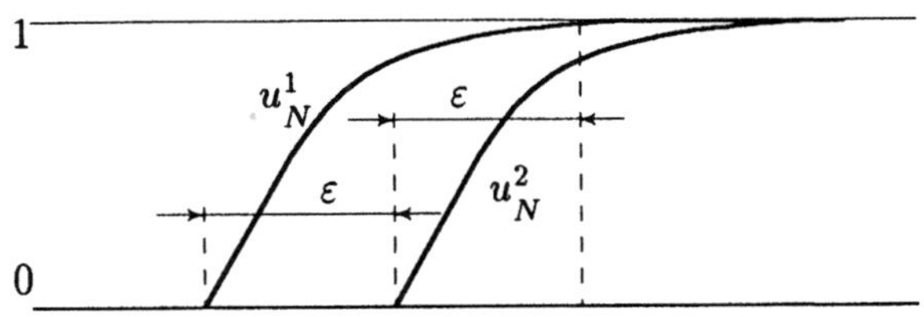

FIGURE 4.3

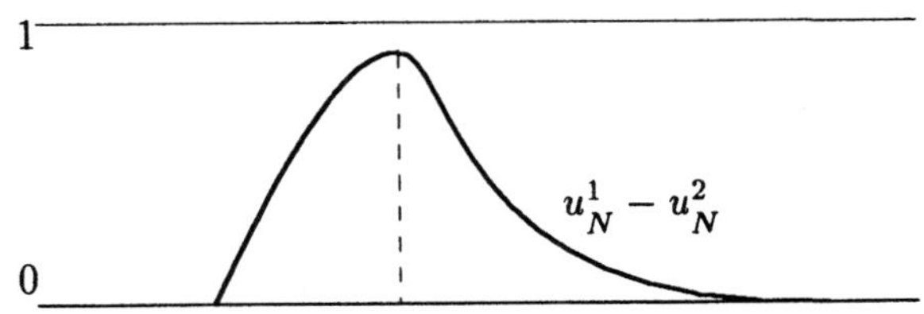

FIGURE 4.4

The plots for the principal terms are shown in Figure 4.2 (here, for definiteness, $\varphi^2 < \varphi^1$ at $t = \delta$)

Obviously, the difference $u_N^1 - u_N^2$ can be plotted as follows:

Thus, statement (b) of the theorem is proved. A similar statement can be formulated, also, for $k + q > 2$, $q < 1$, $F(u) < 0$, $u \in (0, 1)$.

Theorem 3.2. *Let the conditions of Theorem* 2.2 *be satisfied. Then the statement of Theorem* 3.1 *holds.*

The proof of this theorem consists in constructing the solutions $u_N^i(x, t, \varepsilon)$ according to the algorithm given in the preceding section.

Note that, by Lemma 2.4, under the conditions of Theorems 2.1 and 2.2, it is possible to construct solutions, for which the coordinates of the weak discontinuity front $\varphi^0(t)$ coincide at $t = 0$ and are different for $t > 0$.

Now we shall clarify the conditions, under which the constructions of Theorems 3.1 and 3.2 cannot be used. First, consider the case when the conditions of Theorem 2.1 are satisfied and $k + q = 2$. Then for the equation under consideration there is a family of asymptotic solutions having the form of distorted simple waves, and this family depends on the parameter b, i.e. on the velocity of a simple wave. In the preceding section (see Theorem 2.1) it is shown that, as $\tau \to +0$, the correction W_1 to the asymptotic solution corresponding to $b = b_{\min}$, is less smooth than the principal term.

It is natural to require that the next terms of the asymptotic solution be "not worse" (in the sense of smoothness) than the preceding ones. Thus the correction $\varphi^1(t)$ to the coordinate of the front must be chosen uniquely and, consequently, the solution must be unique.

Theorem 3.3. *Let the conditions of Theorem* 2.1 *be satisfied. Then, the asymptotic $O(\varepsilon^2)$-solution of the boundary-value problem* (2.1)–(2.3) *has the form of the* *distorted simple wave related to* $b = b_{\min}$:

$$u_2(x,t,\varepsilon) = W_0\left(\frac{S}{\varepsilon}\right) + \varepsilon W_1\left(\frac{S}{\varepsilon}, t, \varepsilon\right),$$

where $W_0 = W_0(\tau)$ *is the solution of the boundary-value problem* (2.5) *for* $b = b_{\min}$, W_1 *is defined in Lemma* 2.1,

$$S = S(x,t,\varepsilon) = \beta^0(x+\varphi^0) + \varepsilon\beta^0\varphi^1 + \varepsilon^2\beta^2(x+\varphi^0),$$
$$\beta^0 = \gamma(-\varphi^0(t),t)/[\lambda(-\varphi^0,t)]^{1/2},$$

the functions $\varphi^0(t)$ *and* $\varphi^1(t)$ *satisfy the following ordinary differential equations*:

$$\varphi_t^0 = b_{\min}\gamma^2(-\varphi^0,t)/\beta^0, \tag{3.2}$$

$$\beta^0\varphi_t^1 + \varphi^1(\beta^0)^2\frac{\partial\lambda}{\partial x}(-\varphi^0,t)J_1 + \frac{\partial\gamma^2}{\partial x}(-\varphi^0,t)\varphi^1 J_2 = 0, \tag{3.3}$$

where

$$J_1 = \lim_{\tau\to 0}\left(K(W_0)\frac{dW_0}{d\tau}\right)\Big/ W_0,$$
$$J_2 = \lim_{\tau\to 0} F(W_0)\Big/\frac{dW_0}{d\tau}.$$

The following estimates hold for asymptotics of the form (2.31):

$$W_1(\tau,t) \sim \tau^\alpha, \qquad \left|\frac{Lu_2}{u_2}\right| \leqslant \text{const}, \qquad \tau \to 0.$$

Proof. First, consider the case when the asymptotic behavior of the function $W_0(\tau)$ as $\tau \to 0$ has the form (2.36). Equation (3.3) is equivalent to the condition $\lim_{\tau\to 0} f(\tau,t)/\tau^{\alpha-1} = 0$. Hence, under the condition (3.3), the formula for $f(\tau,t)$ and relations (2.31) imply the estimate $f(\tau,t) \sim O(\tau^\alpha)$, $\tau \to +0$.

We recall that, in the case considered, W_1 is determined by the relation

$$W_1 = \frac{dW_0}{d\tau}\int_0^\tau V\left(\frac{dW_0}{d\tau'}\right)^{-2}\int_\infty^{\tau'} \frac{f\frac{dW_0}{d\tau''}}{VK(W_0)}\,d\tau''d\tau'. \tag{3.4}$$

Then we have the asymptotic behavior of the integrand of the inner integral in (3.4):

$$\frac{f dW_0/d\tau}{VK(W_0)} \sim \tau^{2\alpha - b/K_1}.$$

Thus, for $b/K_1 < 2\alpha+1$, we get

$$\int_\infty^\tau \frac{f dW_0/d\tau}{VK(W_0)}\,d\tau \sim \text{const}, \qquad \tau \to +0.$$

Consider the asymptotic behavior of the product $V(dW_0/d\tau)^{-2}$ for $b = b_{\min}$, $\tau \to 0$:

$$V\left(\frac{dW_0}{d\tau}\right)^{-2} \sim \tau^{-2}\tau^{b/K_1}\tau^{-2\alpha+2} \sim \tau^{b/K_1-2\alpha} \sim \tau^0$$

(recall that $b_{\min}/K_1 = 2\alpha$). This yields the sought asymptotic dependence $W_1 \sim \tau^\alpha$, $\tau \to +0$. By (2.31), precisely as at the end of the preceding section, one can choose the function β^2 (β^N for an $O(\varepsilon^N)$-asymptotic solution) so as to satisfy the relation

$$\left|\frac{Lu_2}{u_2}\right| \leqslant \text{const}\,. \tag{3.5}$$

The theorem is proved.

Below we propose an algorithm for calculating $\varphi^i(t)$ (when $|b| > |b_{\min}|$) by matching the solution constructed in the preceding section with the solution "at infinity". This solution has the form

$$u = 1 - v, \tag{3.6}$$

where $v = v(x,t,\varepsilon)$ is a small correction satisfying the linear parabolic equation

$$\varepsilon\frac{\partial v}{\partial t} - \varepsilon^2 K(1)\frac{\partial}{\partial x}\left(\lambda(x,t)\frac{\partial x}{\partial x}\right) - \gamma^2(x,t)F'(1) = 0. \tag{3.7}$$

The function $v(x,t,\varepsilon)$ plays the part of a background for the construction of asymptotic solutions in the form of infinitely narrow solitons [55, 56] or shock waves. In contrast to the cases mentioned above, here the background is exponentially small and, as is easily seen, its dependence on the small parameter ε is not smooth.

Besides the problem of matching the asymptotic solution of (3.7), satisfying the condition

$$v(x,t,\varepsilon) \to 0 \qquad \text{as} \quad (x+\varphi^0)/\varepsilon \to \infty, \tag{3.8}$$

with the asymptotic solution given in Theorem 2.1, there is a problem of constructing a "correct" asymptotic solution for equation (3.7). The point is that, as stated above, according to Theorem 2.1, an asymptotic solution of equation (3.7) must necessarily contain the factor $\exp(f/\varepsilon)$, where $f > 0$ is a certain smooth function. Only recently the problem has been well-posed for asymptotics of this type and some methods for constructing such asymptotic expansions have been proposed [49].

It turns out that it makes sense to consider such problems only for an extremely narrow class of equations called *tunnel equations* in [49]. Equation (3.7) belongs to this class and, as follows from the general construction [22] presented at the end of this section, the exponential ("WKB") asymptotic solution of (3.7) describes the exponential asymptotic behavior of its exact solution.

By matching the solution given by Theorem 2.1 with the exponential ("WKB") asymptotic solution of (3.7), we get the position of the weak discontinuity front of the solution.

First consider the case when the conditions of Theorem 2.1 are satisfied and $k + q = 2$. We put $\beta^k \equiv 0$, $k \geqslant 1$, $\beta_j^k \equiv 0$. Following the algorithm of Theorem

2.1 and taking into account Lemma 2.4 for $b(t) > b_{\min}$, we construct an $O(\varepsilon^{N+1})$ asymptotic solution $u_N(x,t,\varepsilon)$ in the form

$$u_N(x,t,\varepsilon) = \left[W_0(t,\tau) + \sum_{k=1}^{N} \varepsilon^k \frac{\partial W_0}{\partial \tau} \int_0^\tau \frac{V}{\left(\frac{\partial W_0}{\partial \tau'}\right)^2} \int_\infty^{\tau'} \frac{f_k(t,\tau'')\frac{\partial W_0}{\partial \tau''}}{VK(W_0)}\, d\tau'' d\tau'\right]_{\tau=S/\varepsilon}, \tag{3.9}$$

where $f_k = f_k(t,\tau)$ is the right-hand side of the first variation equation in construction of the kth approximation. The formula for $f_1 \equiv f(t,\tau)$ is given in Lemma 2.1.

Lemma 3.1. *For $\tau \gg 1$ the relation*

$$W_0(t,\tau) + \sum_{k=1}^{N} \varepsilon^k \frac{\partial W_0}{\partial \tau} \int_0^\tau \frac{V}{\left(\frac{\partial W_0}{\partial \tau'}\right)^2} \int_\infty^{\tau'} \frac{f_k(t,\tau'')\frac{\partial W_0}{\partial \tau''}}{VK(W_0)}\, d\tau'' d\tau'$$
$$= 1 - \Gamma \exp(-l\tau)\left[1 + \sum_{k=1}^{N} \varepsilon^k P_{2k}(\tau,t)\right] + o\big(\exp(-l\tau)\big), \tag{3.10}$$

holds; here $P_{2k}(\tau,t)$ are polynomials of degree $2k$ in τ with coefficients depending on t.

Proof. For $S = \beta^0(x+\varphi^0) + \varepsilon\beta^0\varphi^1$, the functions f_k, $k \geqslant 1$, on the left-hand side of (3.10) can be calculated as follows

$$\begin{aligned} f_k = \frac{1}{(\beta^0)^2\lambda(-\varphi^0,t)} \Bigg\{ & \frac{\partial W_{k-1}}{\partial t} + \frac{\tau - \beta^0\varphi^1}{\beta^0}\beta^0_t \frac{\partial W_{k-1}}{\partial \tau} \\ & - (\beta^0)^2 \sum_{j=1}^{k-1} \frac{\partial^j \lambda}{\partial x^j}\bigg|_{x=-\varphi^0} \frac{1}{j!}\left(\frac{\tau-\beta^0\varphi^1}{\beta^0}\right)^j \frac{\partial}{\partial \tau}\left(K(W_0)\frac{\partial W_{k-j}}{\partial \tau}\right) \\ & - \sum_{j=1}^{k} \frac{\partial^j \lambda}{\partial x^j}\bigg|_{x=-\varphi^0} \frac{1}{(j-1)!}\left(\frac{\tau-\beta^0\varphi^1}{\beta^0}\right)^{j-1} K(W_0)\frac{\partial W_{k-j}}{\partial \tau} \\ & - \sum_{j=1}^{k-1} \frac{\partial^j \gamma^2}{\partial x^j}\bigg|_{x=-\varphi^0} \left(\frac{\tau-\beta^0\varphi^1}{\beta^0}\right)^j \frac{1}{j!}F'(W_0)W_{k-j} \\ & - \frac{\partial^k \gamma^2}{\partial x^k}\bigg|_{x=-\varphi^0} \frac{1}{k!}\left(\frac{\tau-\beta^0\varphi^1}{\beta^0}\right)^k F(W_0) + \dots \Bigg\} \end{aligned} \tag{3.11}$$

(the dots stand for omitted nonlinear terms).

As $\tau \to \infty$, the functions $W_k(t,\tau)$, $k \geqslant 1$, given by the formula

$$W_k = \frac{\partial W_0}{\partial \tau} \int_0^\tau \frac{V}{\left(\frac{\partial W_0}{\partial \tau'}\right)^2} \int_\infty^{\tau'} \frac{f_k(t,\tau'')\frac{\partial W_0}{\partial \tau''}}{VK(W_0)}\, d\tau'' d\tau' \tag{3.12}$$

can be estimated as follows:

$$W_k = -\Gamma_k \exp(-l\tau)P_{2k}(\tau,t) + o\big(\exp(-l\tau)\big), \qquad \Gamma_k = \text{const},$$

where $P_{2k}(\tau,t) = \sum_{i=0}^{2k} A_{2k}^i(t)\tau^i$, $l = \frac{-b+\sqrt{b^2-4K(1)F'(1)}}{2K(1)}$. This follows from the estimate for W_0, namely,

$$W_0 = 1 - \Gamma_0 \exp(-l\tau) + o\big(\exp(-l\tau)\big),$$

and can be proved by standard induction: as $\tau \to \infty$, assume that W_{k-1} takes the form

$$W_{k-1} \sim -\Gamma_{k-1}\exp(-l\tau)P_{2k-2}(\tau,t),$$

then, by (3.12) and (3.11), the required statement holds.

Now we denote

$$G_N(\tau,t,\varepsilon) = \sum_{i=1}^{N} \varepsilon^i P_{2i}(\tau,t).$$

Lemma 3.2. *The relation holds:*

$$\sum_{k=0}^{N} \varepsilon^k P_{2k}(\tau,t) = \sum_{j=0}^{N} \frac{(G_N+R_N)^j}{j!} + O(\varepsilon^{N+1}\tau^{2N+2}),$$

where $R_N = \sum_{i=1}^{N} \varepsilon^i r_{2i}(\tau,t)$, $r_{2i} = \sum_{j=0}^{2i} r_{2i}^j(t)\tau^j$, $r_2 \equiv 0$, $P_0 \equiv 1$.

Proof. One can easily see that the polynomials $r_{2i}(\tau,t)$ are determined in succession by the formula

$$r_{2i} = -\frac{(n-1)!}{i!}\frac{\partial^i}{\partial\varepsilon^i}\sum_{n=0}^{i}\frac{(G_{i-1}+R_{i-1})^n}{n!}\bigg|_{\varepsilon=0} + \frac{n!-1}{(n-1)1}P_{2i}.$$

Indeed, we have

$$\sum_{k=0}^{i}\varepsilon^k P_{2k} = \sum_{n=0}^{i}\left[\frac{(G_{i-1}+R_{i-1})^n}{n!} + n\varepsilon^i(P_{2i}+r_{2i})\frac{(G_{i-1}+R_{i-1})^{n-1}}{n!} + O(\varepsilon^{2i})\right].$$

By differentiating the left- and right-hand sides with respect to ε i times and setting ε equal to zero, we obtain the required formula for r_{2i}, $i \geqslant 2$. The relation $G_1 = \varepsilon P_2(\tau,t)$ yields $r_2 \equiv 0$.

Collorary 3.1. *The following representation holds as* $\tau \to \infty$*:*

$$\begin{aligned}\sum_{k=0}^{N}\varepsilon^k W_k &= 1 - \Gamma\exp(-l\tau)\left(1+\sum_{i=1}^{N}P_{2i}\varepsilon^i\right) + o(\exp(-l\tau))\\ &= 1 - \Gamma\exp\{-l\tau + G_N + R_N\} + O\big(\varepsilon^{N+1}\exp(-l\tau)\tau^{2N+2}\big),\end{aligned} \tag{3.13}$$

and implies the estimate

$$\sum_{k=0}^{N}\varepsilon^k W_k\Big|_{\tau=S/\varepsilon} = 1-\exp\Big(-\frac{\Xi_N}{\varepsilon}\Big)+O\Big(\varepsilon^{(N+1)(\delta_2-1/2)}\exp\Big(-\frac{l(x+\varphi^0)}{\varepsilon}\Big)\Big)$$

in the domain $c_1\varepsilon^{\delta_1}\leqslant x+\varphi^0\leqslant c_2\varepsilon^{\delta_2}$, *where* $c_1, c_2>0$, $\frac{1}{2}<\delta_2<\delta_1<1$, *are arbitrary numbers,* $\Xi_N=\Xi_N(x+\varphi^0,t,\varepsilon)$, $\Xi_N(z,t,\varepsilon)$ *is a polynomial (with smooth coefficients) in* z *of the order not greater than* $2N$.

This statement is equivalent to the following one:

$$\frac{\partial^{n+m}}{\partial\tau^{n+m}}\frac{\partial^n}{\partial\varepsilon^n}(G_N+R_N)\Big|_{\substack{\tau=0\\ \varepsilon=0}}=0,\qquad n=2,\dots,N,\quad m=2,\dots,n. \tag{3.14}$$

The latter relations can be verified either by mathematical induction, or by using the characteristics of the initial equation.

Let us verify (3.14) for $N=2$ (in this case $n=2$, $m=2$). We obtain

$$\frac{\partial^4}{\partial\tau^4}\frac{\partial^2}{\partial\varepsilon^2}(G_2+R_2)\Big|_{\substack{\tau=0\\ \varepsilon=0}}=2(4!)\Big(A_4^4-\frac{1}{2}(A_2^2)^2\Big),$$

since

$$r_4=-\frac{1}{2}\frac{\partial^2}{\partial\varepsilon^2}\sum_{n=0}^{2}\frac{(\varepsilon P_2)^n}{n!}\Big|_{\varepsilon=0}=-\frac{(P_2)^2}{2}.$$

Here A_{2k}^i are the coefficients of the polynomials P_{2k}.

On the other hand, applying (3.11) and (3.12), we determine the coefficients A_2^2 and A_4^4:

$$\begin{aligned}A_2^2&=\frac{1}{2(\beta^0)^2}\Big\{K(1)\lambda(-\varphi^0,t)2l+\frac{1}{\beta^0}\frac{d\varphi^0}{dt}\Big\}^{-1}\\&\quad\times\Big\{\frac{d\beta^0}{dt}\frac{l}{\beta^0}+\frac{\partial\lambda}{\partial x}\Big|_{x=-\varphi^0}(l\beta^0)^2K(1)+F'(1)\frac{\partial\gamma^2}{\partial x}\Big|_{x=-\varphi^0}\frac{1}{\beta^0}\Big\},\\A_4^4&=A_2^2\frac{1}{(\beta^0)^2}\frac{1}{4}\Big\{K(1)\lambda(-\varphi^0,t)2l+\frac{1}{\beta^0}\frac{d\varphi^0}{dt}\Big\}^{-1}\\&\quad\times\Big\{\frac{d\beta^0}{dt}\frac{l}{\beta^0}+\frac{\partial\lambda}{\partial x}\Big|_{x=-\varphi^0}(l\beta^0)^2K(1)+F'(1)\frac{\partial\gamma^2}{\partial x}\Big|_{x=-\varphi^0}\frac{1}{\beta^0}\Big\}=\frac{1}{2}(A_2^2)^2.\end{aligned}$$

Consequently, (3.14) holds for $N=2$.

Let conditions (3.14) be satisfied for all $N\leqslant M$. We shall show that then they are also satisfied for $N=M+1$. By induction, for $n\leqslant M$,

$$\frac{\partial^{n+m}}{\partial\tau^{n+m}}\frac{\partial^n}{\partial\varepsilon^n}(G_{M+1}+R_{M+1})\Big|_{\substack{\tau=0\\ \varepsilon=0}}=\frac{\partial^{n+m}}{\partial\tau^{n+m}}\frac{\partial^n}{\partial\varepsilon^n}(G_M+R_M)\Big|_{\substack{\tau=0\\ \varepsilon=0}}=0.$$

It remains to show that

$$\frac{\partial^{M+1+m}}{\partial\tau^{M+1+m}}\frac{\partial^{M+1}}{\partial\varepsilon^{M+1}}(G_{M+1}+R_{M+1})\bigg|_{\substack{\tau=0\\ \varepsilon=0}}=0,\qquad m=2,\ldots,M+1. \tag{3.15}$$

By induction, we can transform the left-hand side as follows:

$$\frac{\partial^{M+1+m}}{\partial\tau^{M+1+m}}\frac{\partial^{M+1}}{\partial\varepsilon^{M+1}}(G_{M+1}+R_{M+1})\bigg|_{\substack{\tau=0\\ \varepsilon=0}}=(M+1)!\frac{\partial^{M+1+m}}{\partial\tau^{M+1+m}}(P_{2M+2}+r_{2M+2})\bigg|_{\tau=0}$$

$$=(M+1)!\,\frac{\partial^{M+1+m}}{\partial\tau^{M+1+m}}$$

$$\times\left[P_{2M+2}-\frac{1}{(M+1)!}\frac{\partial^M}{\partial\varepsilon^M}\Big((1+G_N)\frac{\partial}{\partial\varepsilon}(G_N+R_N)\Big)\bigg|_{\varepsilon=0}\right]\bigg|_{\tau=0}$$

$$=(M+1)!\,\frac{\partial^{M+1+m}}{\partial\tau^{M+1+m}}$$

$$\times\left[P_{2M+2}-\frac{1}{(M+1)}\sum_{i=1}^{M}(M-i+1)P_{2i}(P_{2(M-i+1)}+r_{2(M-i+1)})\right]\bigg|_{\tau=0}$$

$$=(M+1)!\left[\frac{\partial^{M+1+m}}{\partial\tau^{M+1+m}}P_{2M+2}-\frac{1}{(M+1)}\sum_{i=1}^{M}(M-i+1)\right.$$

$$\left.\times\sum_{j=M+1+m-2i\geqslant 0}^{M+1+m}C^j_{M+1+m}\frac{\partial^{M+1+m-j}}{\partial\tau^{M+1+m-j}}P_{2i}\frac{\partial^j}{\partial\tau^j}\big(P_{2(M-i+1)}+r_{2(M-i+1)}\big)\right]\bigg|_{\tau=0}.$$

Finally, we obtain (3.15) in the form

$$-\frac{1}{M+1}\sum_{i=1}^{M}(M-i+1)\sum_{j=M+1+m-2i\geqslant 0}^{M-i+2}A_{2i}^{M+1+m-j}(A^j_{2(M-i+1)}$$

$$+r^j_{2(M-i+1)})+A^{M+1+m}_{2M+2}=0,\qquad m=2,\ldots,M+1. \tag{3.15'}$$

These relations can be verified by substituting the coefficients. For example, for $m=M+1$, (3.15′) implies

$$-\frac{1}{M+1}A^{2M}_{2M}A^2_2+A^{2M+2}_{2M+2}=0.$$

This is satisfied identically, since, as $\tau\to\infty$, by (3.12), the coefficient at $\tau^{2M+2}e^{-l\tau}$ in the expansion of W_{M+1} takes the form

$$-\Gamma A^{2M+2}_{2M+2}=-\Gamma A^{2M}_{2M}\frac{1}{2(M+1)(\beta^0)^2}\Big(K(1)\lambda(-\varphi^0,t)2l+\frac{1}{\beta^0}\varphi^0_t\Big)^{-1}$$

$$\times\left(\beta^0_t\frac{l}{\beta^0}+\frac{\partial\lambda}{\partial x}\bigg|_{x=-\varphi^0}(l\beta^0)^2K(1)+F'(1)\frac{1}{\beta^0}\frac{\partial\gamma^2}{\partial x}\bigg|_{x=-\varphi^0}\right)=-\Gamma\frac{1}{M+1}A^{2M}_{2M}A^2_2.$$

The other relations in (3.15′) are proved similarly.

There is another proof of (3.13). By Corollary 3.1.,

$$\sum_{k=1}^{N} \varepsilon^k W_k = 1 - \Gamma \exp\{-l\tau + G_N + R_N\} + o\Big(\exp\Big(-\frac{\beta^0(x+\varphi^0)}{\varepsilon}\Big)\Big).$$

for $c_1\varepsilon^{\delta_1} \leqslant x + \varphi^0 \leqslant c_1\varepsilon^{\delta_2/2}$. In this domain, as $\varepsilon \to \infty$, we have $(x+\varphi^0) \ll 1$ and $(x+\varphi^0)/\varepsilon \sim \tau \gg 1$. Therefore

$$u_N(x,t,s) = \sum_{k=0}^{N} \varepsilon^k W_k(t,\tau)\Big|_{\tau=S/\varepsilon} = 1 - v_N(x,t,\varepsilon) + o\Big(\exp\Big(-\frac{\beta^0(x+\varphi^0)}{\varepsilon}\Big)\Big),$$

where

$$v_N(x,t,\varepsilon) = \Gamma \exp\{-l\tau + G_N + R_N\}\big|_{\tau=S/\varepsilon} \ll 1.$$

Consequently, the function v_N satisfies the linearized equation

$$\varepsilon\frac{\partial v_N}{\partial t} - \varepsilon^2 K(1)\frac{\partial}{\partial x}\Big(\lambda\frac{v_N}{\partial x}\Big) - \gamma^2 F'(1)v_N = O\Big(\varepsilon^{\gamma_N}\exp\Big(-\frac{\beta^0(x+\varphi^0)}{\varepsilon}\Big)\Big), \quad (3.16)$$

where

$$\gamma_N = (N+1)(\delta_2 - \tfrac{1}{2}).$$

The function v_N has the form

$$v_N(x,t,\varepsilon) = \Gamma \exp\{-\Phi(x,t,\varepsilon)/\varepsilon\},$$

where $\Phi(x,t,\varepsilon) = \varepsilon(l\tau - G_N - R_N)\big|_{\tau=S/\varepsilon}$ is a polynomial in $x+\varphi^0$ with coefficients smoothly depending on t and, generally speaking, possessing poles of a finite order at $\varepsilon = 0$. This follows from the fact that the functions G_N and R_N are polynomials in τ with coefficients depending on t.

Obviously, the function $\Phi = \Phi(x,t,\varepsilon)$ satisfies the following equation of the Hamilton–Jacobi type with low viscosity,

$$\frac{\partial\Phi}{\partial t} + K(1)\lambda(x,t)\Big(\frac{\partial\Phi}{\partial x}\Big)^2 + \gamma^2F'(1) - \varepsilon K(1)\frac{\partial}{\partial x}\Big(\lambda(x,t)\frac{\partial\Phi}{\partial x}\Big) = O(\varepsilon^{\gamma_N}). \quad (3.17)$$

By definition, the function $\Phi(x,t,\varepsilon)$ can be presented in the form

$$\Phi(x,t,\varepsilon) = \sum_{k=-N}^{N} \Phi_k(x,t)\varepsilon^k,$$

which, together with (3.17), implies

$$\Phi_k \equiv 0, \qquad k < 0.$$

On the other hand, it is known that the function $f(\Phi^0/\varepsilon, x, t, \varepsilon)$, such that $f(\tau, x, t, \varepsilon) \in C^\infty$, $\Phi^0 \in C^\infty$, $f \to 0$ as $\tau \to \infty$, satisfies the linear equation (3.16) and can be necessarily represented in the form

$$f\Big(\frac{\Phi^0}{\varepsilon}, x, t, \varepsilon\Big) = \exp\Big(-\frac{\Phi^0}{\varepsilon}\Big)\Big(\sum \varepsilon^k \nu_k(x,t) + O(\varepsilon^N)\Big), \qquad \nu_k \in C^\infty, \tag{3.18}$$

or, equivalently,

$$f\Big(\frac{\Phi^0}{\varepsilon}, x, t, \varepsilon\Big) = \exp\Big(-\frac{\Phi^0 + \varepsilon\Phi^1 + \dots + \varepsilon^N\Phi^N}{\varepsilon}\Big)\Big(1 + O(\varepsilon^N)\Big), \qquad \Phi^k \in C^\infty.$$

Here Φ^0 satisfies the Hamilton–Jacobi equation

$$\frac{\partial\Phi^0}{\partial t} + K(1)\lambda\Big(\frac{\partial\Phi^0}{\partial x}\Big)^2 + \gamma^2 F'(1) = 0, \tag{3.19}$$

while the functions Φ^i, $i > 0$ satisfy the transport equations. This follows from the fact that, by substituting $f(\Phi/\varepsilon, x, t, \varepsilon)$ into the equation and setting the coefficient at ε^0 equal to zero, we obtain an ordinary differential equation respect to τ for the function $f(\tau, x, t, 0)$, whose solution is a linear combination of exponential functions. We assume that the coefficients λ and γ are real analytic functions and, hence, we obtain that Φ_0 is a partial sum of the Taylor expansion of Φ^0. Since $\Phi_0\big|_{x=-\varphi^0} = 0$ by construction, we must supplement equation (3.19) with the boundary condition:

$$\Phi^0\big|_{x=-\varphi^0} = 0. \tag{3.20}$$

So, we arrive at the following statement.

Theorem 3.4. *In the domain $c_1\varepsilon^{\delta_1} \leqslant (x + \varepsilon^0) \leqslant c_2\varepsilon^{\delta_2}$ the following representation holds:*

$$u_N(x, t, \varepsilon) = 1 - \exp\Big(-\frac{\Phi^0 + \varepsilon\Phi^1 + \dots + \varepsilon^N\Phi^N}{\varepsilon}\Big)(1 + O(\varepsilon^{\gamma_N})),$$

where the function $\Phi^0 = \Phi^0(x,t)$ is the solution of the boundary-value problem (3.19), (3.20), *and the functions $\Phi^i = \Phi^i(x,t)$ satisfy the transport equations. Thus, the asymptotic solution has the following global structure: in a small neighborhood* ($O(\varepsilon^{1+\gamma})$) *of the weak discontinuity front the solution has the form* (3.9); *outside this neighborhood of the weak discontinuity front (when $x + \varphi > \varepsilon^{1-\gamma}$) the solution is close to the solution of the linearized problem.*

Theorem 3.4 also holds for the case considered in Theorem 2.2. In this case, to construct the solution in the form (3.9), we must replace the lower limit of the inner integral by 0.

Let us examine problem (3.19), (3.20) under the conditions of Theorem 2.2 for $k + q = 2$. Note that, in this case, $F'(1) < 0$. Therefore the Lagrangian of the Hamilton–Jacobi equation (3.19) is positive,

$$L = pH_p - H = K(1)\lambda p^2 + \gamma^2|F'(1)| > 0.$$

To investigate the solutions of (3.19), it is convenient to change variables: $y = x + \varphi^0(t)$. By setting $\Phi^0(x,t)\big|_{x=x(y,t)} = \Psi(y,t)$, we obtain the equation

$$\frac{\partial \Psi}{\partial t} + \varphi_t^0 \frac{\partial \Psi}{\partial y} + K(1)\widehat{\lambda}\left(\frac{\partial \Psi}{\partial y}\right)^2 + \widehat{\gamma}^2(y,t)F'(1) = 0, \tag{3.21}$$

where $\widehat{\gamma}(y,t) = \gamma(y-\varphi^0(t),t)$, $\widehat{\lambda} = \widehat{\lambda}(y,t) = \lambda(y-\varphi^0(t),t)$. The function $\varphi^0 = \varphi^0(t)$ is unknown, but, due to the results of the preceding section, it must satisfy the conditions

$$\varphi^0(t) \in C^\infty, \qquad \frac{\varphi^0(t)}{\gamma(-\varphi^0,t)\sqrt{\lambda(-\varphi^0,t)}} > b_{\min}.$$

By definition, the function $\Psi(y,t)$ satisfies the boundary condition

$$\Psi(y,t)\big|_{y=0} = 0. \tag{3.22}$$

The projections of the characteristics of equation (3.21) are shown in Figure 4.5

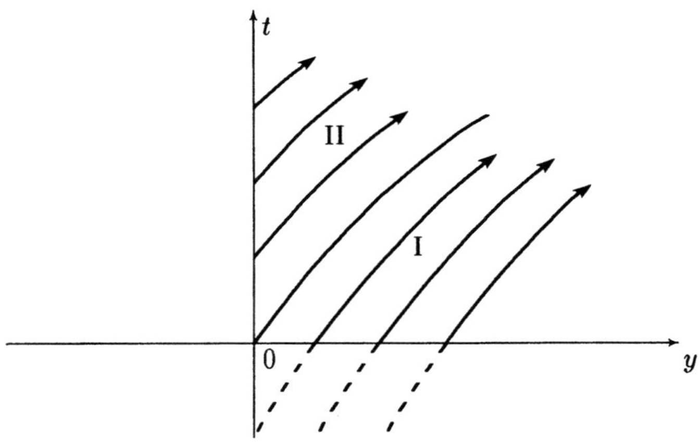

FIGURE 4.5

Indeed, the projections of trajectories on the (x,t)-plane are given by the equation

$$\dot{x} = 2\lambda K(1)p.$$

In a small neighborhood of the point $x = -\varphi^0$, we have $p \sim l\beta^0 > 0$. As is known, at the point (x,t), the solution of (3.21) is equal to the integral of the Lagrangian along the projection of the trajectory of the Hamilton system corresponding to (3.21), which comes to the point x at time t. Since the Lagrangian is positive, the solution $\Psi(y)$ is positive for $y > 0$ and, for any appropriate choice of the function φ^0, this solution is independent (see Figure 4.5) of the initial conditions (the conditions for $t = 0$) in domain II. This fact leads to an important conclusion. Namely, the law of motion of the weak discontinuity front is not determined by the initial boundary-value problem for equation (3.1) and must be defined in addition.

Now let us consider the case when the conditions of Theorem 2.2 are satisfied for the function W_0 having the following asymptotic expansion as $\tau \to \infty$ (recall that $b < 0$ in this case):

$$W_0(t,\tau) \sim 1 - \Gamma \exp\left(\frac{b + \sqrt{b^2 - 4|F'(1)|K(1)}}{2K(1)}\tau\right).$$

Here $b = b(t) \in C^\infty$, $|b| > |b_{\min}|$. In this case, Lemma 2.4 and Theorem 3.4 also hold, but the properties of the Hamilton–Jacobi equation (3.19) are essentially different. By changing the variables $t \to t$, $x \to y + \varphi^0(t)$, we reduce this equation to the form

$$\frac{\partial \Psi}{\partial t} + \varphi_t^0 \frac{\partial \Psi}{\partial y} + K(1)\widehat{\lambda}\left(\frac{\partial \Psi}{\partial y}\right)^2 + \widehat{\gamma}^2(y,t)F'(1) = 0, \tag{3.23}$$

where $\Psi(y,t) = \Phi^0(y + \varphi^0(t), t)$, $\widehat{\lambda} = \lambda(y + \varphi^0(t), t)$, $\widehat{\gamma} = \gamma(y + \varphi^0(t), t)$. The boundary condition remains:

$$\Psi(y,t)\big|_{y=0} = 0. \tag{3.24}$$

The Lagrangian L of equation (3.23) has the form

$$L = pH_p - H = K(1)\widehat{\lambda}p^2 - \gamma^2|F'(1)|$$

and, generally speaking, its sign is not fixed.

Let us examine the projections of trajectories of the Hamilton system corresponding to equation (3.23) near the boundary $y = 0$. These curves satisfy the equations

$$\dot{y} = \dot{\varphi}^0 + 2K(1)\widehat{\lambda}p. \tag{3.25}$$

By (3.25), for $y = 0$, we have

$$p\big|_{y=0} = \frac{\varphi_t^0 \pm \sqrt{(\varphi_t^0)^2 - 4\widehat{\lambda}\widehat{\gamma}^2 K(1)|F'(1)|}}{2K(1)\widehat{\lambda}}\Bigg|_{y=0}. \tag{3.26}$$

On the other hand, by Corollary 3.1, we get

$$p\big|_{y=0} = l\beta^0, \qquad l = \frac{-b - \sqrt{b^2 - 4K(1)|F'(1)|}}{2K(1)}, \qquad \beta^0 = \widehat{\gamma}/\sqrt{\widehat{\lambda}}. \tag{3.27}$$

Comparing (3.26) and (3.27), we find

$$p\big|_{y=0} = \frac{-\varphi_t^0 - \sqrt{(\varphi_t^0)^2 - 4\widehat{\lambda}\widehat{\gamma}^2 K(1)|F'(1)|}}{2K(1)\widehat{\lambda}}\Bigg|_{y=0}.$$

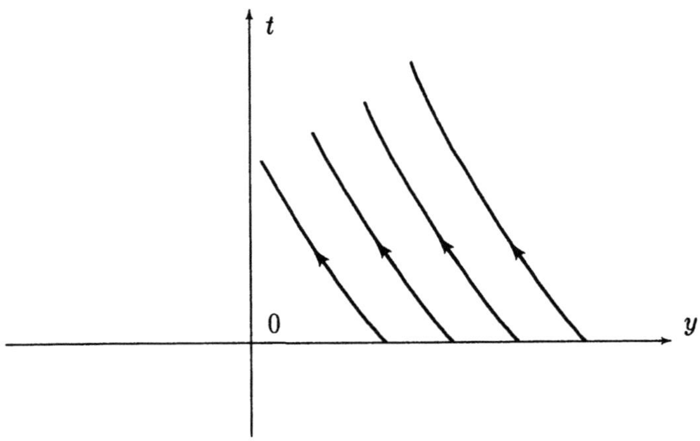

FIGURE 4.6

Hence, according to (3.25), we obtain for $|b| > |b_{\min}|$:

$$\dot{y}\big|_{y=0} = -\sqrt{(\varphi_t^0)^2 - 4\widehat{\lambda}\widehat{\gamma}^2 K(1)|F'(1)|}\,\Big|_{y=0} < 0.$$

The approximate behavior of trajectories near the boundary is shown in Figure 4.6, where the arrows indicate the direction of motion along the trajectory as t increases. In general, for $t = 0$, the trajectories intersecting the axis $t = 0$ start from the t-axis; thus, generally speaking, the boundary condition $\Psi\big|_{y=0} = 0$ is satisfied not for all initial conditions given at $t = 0$ and positive for $y > 0$.

Let us prove that, under the conditions of this theorem, for t small enough, the solution of problem (3.23), (3.24), positive for $y > 0$, exists in an (independent of ε) neighborhood of the axis $y = 0$. To prove this, it is sufficient (see Figure 4.6) to show that the Lagrangian of equation (3.23) is negative at $y = 0$. Taking into account the above expression for $p\big|_{y=0}$, we have:

$$\begin{aligned} L\big|_{y=0} &= \left(K(1)\widehat{\lambda}p^2 - \widehat{\gamma}^2|F'(1)|\right)\big|_{y=0} \\ &= \frac{\widehat{\gamma}}{2K(1)}\sqrt{b^2 - 4F'(1)K(1)}\left(\sqrt{b^2 - 4F'(1)K(1)} - |b|\right) < 0 \end{aligned} \tag{3.28}$$

(recall that here $b < 0$, $F'(1) > 0$). Since the Lagrangian is smooth, it is nonpositive in a neighborhood of the axis $y = 0$. This proves our statement. Thus, in our case the initial condition for the Hamilton–Jacobi equation completely determines the motion of the weak discontinuity front of the localized solution. Denote by $\Phi^0(x,t)$ the solution of the Cauchy problem for equation (3.19) satisfying the initial condition

$$\Phi^0\big|_{t=0} = \Phi_0^0(x), \qquad \Phi_0^0 \geqslant 0, \qquad \Phi_0^0(-\varphi_0^0) = 0, \qquad \frac{\partial \Phi_0^0}{\partial x}(x) > 0,$$

for $x \geqslant -\varphi_0^0$, then the function $\varphi^0(t)$ is determined by the implicit equation

$$\Phi^0(-\varphi^0, t) = 0, \tag{3.29}$$

which can be solved, at least, for sufficiently small t (independent of ε). Here the solution (3.9) turns out to consist of a family of simple waves $W_0(t,\tau)$ with parameter $b = b(t)$ given by the formula

$$b(t) = \frac{\varphi_t^0}{\gamma(-\varphi^0,t)\sqrt{\lambda(-\varphi^0,t)}}.$$

The condition $b(t) < b_{\min}$ will be satisfied, at least, for small t if $b(0) < b_{\min}$. For the latter inequality to hold, an additional condition must be imposed on $\Phi_0^0(x)$:

$$\frac{\partial \Phi_0^0}{\partial x}(-\varphi_0^0) = \frac{-b - \sqrt{b^2 - 4|F'(1)|K(1)}}{2K(1)}\frac{\gamma(-\varphi^0,0)}{\sqrt{\lambda(-\varphi_0^0,0)}}, \quad |b| > |b_{\min}|, \quad \varphi_0^0 = \varphi^0(0)$$

(this condition means that $p\big|_{x=-\varphi^0} = l\beta^0$). Hence we have the necessary and sufficient condition for the problem to have a solution (generally speaking, for small t):

$$|b_{\min}| < \left| \frac{K(1)\dfrac{\partial \Phi_0^0}{\partial x}(-\varphi_0^0)}{\beta_0^0} + \frac{\beta^0 F'(1)}{\dfrac{\partial \Phi_0^0}{\partial x}(-\varphi_0^0)} \right|,$$

where

$$\beta_0^0 = \frac{\gamma(-\varphi_0^0,t)}{\sqrt{\lambda(-\varphi_0^0,t)}}.$$

The correction $\varepsilon\varphi^1(t)$ to the coordinate of the weak discontinuity front φ^0 is given by the solution of the transport equation for Φ^1:

$$\frac{\partial \Phi^1}{\partial t} + 2K(1)\lambda(x,t)\frac{\partial \Phi^0}{\partial x}\frac{\partial \Phi^1}{\partial x} = K(1)\frac{\partial}{\partial x}\left(\lambda(x,t)\frac{\partial \Phi^0}{\partial x}\right).$$
$$\Phi^1\big|_{t=0} = \Phi_0^1(x).$$

The left-hand side of this equation is the derivative of the function Φ^1 along the projections on (x,t) of the trajectories of the Hamilton system corresponding to equation (3.19). Therefore the initial condition $\Phi_0^1(x)$ given in the domain $x \geqslant -\varphi_0^0$ completely determines the solution $\Phi^1(x,t)$ for $x = -\varphi^0(t)$. From the condition

$$\left(\Phi^0(x,t) + \varepsilon\Phi^1(x,t)\right)\big|_{x=-\varphi^0-\varepsilon\varphi^1} = -\ln\Gamma,$$

which is a consequence of (3.10), we find

$$\varphi^1 = \frac{\ln\Gamma + \Phi^1(-\varphi^0,t)}{\frac{\partial \Phi^0}{\partial x}(-\varphi^0,t)}.$$

Similarly, we can calculate the higher order corrections to the weak discontinuity front.

Note that the algorithm for constructing a global asymptotic solution cannot be applied for $|b| = |b_{\min}|$ because, as $\dot{y}\big|_{y=0} = 0$, the boundary $y = 0$ ($x = -\varphi^0(t)$) is the projection of the submanifold of singularities in the projection of the Lagrangian manifold corresponding to the Hamilton–Jacobi equation (3.25). Therefore the solution of equation (3.7) cannot be represented as $\exp(-\Phi/\varepsilon)$.

Thus, for equations satisfying the conditions of Theorem 2.2 for $|b| > |b_{\min}|$, the global asymptotic solution is well-posed. As for Theorem 2.1, $k + q = 2$, the problem cannot be regularized by matching a local asymptotic solution with the solution of the linearized equation. Therefore, one must pose some additional conditions to provide the uniqueness of the asymptotic solution of the initial-boundary-value problem. In order to formulate these conditions, we consider the Kolmogorov–Petrovskii–Piskunov equation as an example.

In Chapter I the boundary-value problems for semilinear equations of the Kolmogorov–Petrovskii–Piskunov type

$$\varepsilon\frac{\partial u}{\partial t} - \varepsilon^2\frac{\partial^2 u}{\partial x^2} - F(u) = 0, \qquad F(0) = F(1) = 0, \tag{3.30}$$
$$F'(0) \geqslant F'(u),$$
$$u(-\infty, t) = 0, \qquad u(+\infty, t) = 1.$$

were discussed (see [107]). To construct the unique solution of problem (3.30), we must define the asymptotic behavior of the initial condition:

$$u(x,0) \sim \exp\left(\frac{lx}{\varepsilon}\right), \qquad x < 0, \qquad l = \frac{b}{2} - \sqrt{\frac{b^2}{4} - 1}. \tag{3.31}$$

Thus, by fixing the constant l in (3.31),

$$l = \lim_{u\to 0} \varepsilon\frac{\partial u}{\partial x}\Big/u,$$

we uniquely choose the solution of problem (3.30). A characteristic property of this class of equations is that the solution of a boundary-value problem is not unique in the class of functions $u = W(S/\varepsilon, x, t, \varepsilon)$ (see the end of the Introduction t this book).

The solution of nonlinear problem (2.1)–(2.3) is also nonunique. Moreover, as proved in this section for $k+q=2$, $F(u) \geqslant 0$, $u \in [0,1]$, the position of the localized wave front is undetermined for both local and global asymptotic solutions.

The function $c_0(t)$ in

$$u(x,t,\varepsilon) = c_0(t)\left(\frac{x+\varphi(t,\varepsilon)}{\varepsilon}\right)^\alpha + o\left(\left(\frac{x+\varphi(t,\varepsilon)}{\varepsilon}\right)^\alpha\right),$$
$$x \to -\varphi(t,\varepsilon) + 0, \qquad \alpha = \frac{1}{k-1}$$

is an analog of the constant l in the localized solution of a semilinear parabolic equation. The function $c_0(t)$ and the constant $b(t)$ in the standard equation (2.5)

$$bW_0' - (K(W_0)W_0')' - F(W_0) = 0$$

are related as follows

$$b = l + \frac{1}{l}\frac{d}{dW_0}[K(W_0)F(W_0)]\bigg|_{W_0=0}, \qquad l = \alpha c_0^{-1}(K(W_0)/W_0^{k-1})\bigg|_{W_0=0}, \tag{3.32}$$

which actually coincides with (1.21) in Chapter I.

For $k+q>2$, $F(u)\leqslant 0$, the value of $b=b(t)$ was obtained in terms of $l=l(t)$ from the asymptotic solution in the "tail" region behind the wave front (see Theorem 3.4). For $k+q=2$, $F\geqslant 0$, there is no interaction between the "tail" region behind the wave front and a neighborhood of the front. Therefore, to determine the function $c_0(t)$, playing the same part as $l(t)$, it is necessary to have an additional boundary condition on the wave front (i.e., for $x=-\varphi^0(t)$). This condition allows to determine the function $c_0(t)$.

In the most simple case, the function $c_0(t)$ is determined by the expression

$$\frac{\partial u^{k-1}}{\partial x}\bigg|_{u=0} = f(t), \tag{3.33}$$

where f is a given smooth positive function satisfying the condition

$$f(t) > \left[\frac{\alpha(K(W_0)/W_0^{k-1})|_{W_0=0}}{2\sqrt{\frac{d}{dW_0}(K(K(W_0)F(W_0))|_{W_0=0}}}\right]^{k-1}$$

that is a consequence of the condition $b\geqslant b_{\min} = 2\sqrt{\frac{d}{dW_0}(K(K(W_0)F(W_0))|_{W_0=0}}$ and (3.32). Actually, by Theorem 1.2 from Chapter I, we have

$$l = (b - \sqrt{b^2 - b_{\min}^2})/2,$$

therefore $0 < l \leqslant l_{\min} = b_{\min}$. Since

$$u = c_0(t)\Big(\frac{x+\varphi(t,\varepsilon)}{\varepsilon}\Big)^{1/(k-1)} + o\Big(\Big(\frac{x+\varphi(t,\varepsilon)}{\varepsilon}\Big)^{1/(k-1)}\Big), \qquad k>1,$$

then, by (3.33), $c_0 = (f)^{\alpha}$, and, by (3.32), $c_0 > \alpha\big(K(W_0)W_0^{1-k}\big)|_{W_0=0}/l_{\min}$. This implies the required inequality.

The idea that the spatial localization of diffusion processes is a phenomenon caused by the difference in velocities of the diffusion and kinetic processes was developed by L. K. Martinson [46–48] who heuristically interpreted the nonuniqueness of solutions of boundary-value problems. He considered the wave mechanism as separate kinetic (related to the source $F(u)$) and diffusion (related to the divergent part of the equation) mechanisms. According to this theory, the kinematic waves with indeterminate velocity (this is a characteristic feature of these waves) dominate in the case under discussion ($k+q=2$).

To conclude this section we consider the conditions, under which formal asymptotic solutions (having the form $\exp(-S/h)\varphi$) of pseudodifferential equations with a small parameter $h \to +0$ are asymptotic solutions of these equations. It is assumed that the function S is not pure imaginary. In the case of real nonnegative functions, the way for constructing such asymptotic expansions was proposed in [49] for the problem about large deviations. Here we follow [33].

Let the function $H(p,x,t)$ analytically depend on $p \in C^n$ and have the derivatives of all orders with respect to $x \in \mathbf{R}^n$, $t \in [0,T]$; let $H(ip,x,t) \in S^\infty(\mathbf{R}^n \times \mathbf{R}^n \times [0,T])$, $p \in \mathbf{R}^n$, $x \in \mathbf{R}^n$, $t \in [0,T]$.

We consider the h-pseudodifferential operator,

$$L_h \stackrel{\text{def}}{=} -h\frac{\partial}{\partial t} + H\Big(-h\overset{1}{\frac{\partial}{\partial x}}, \overset{2}{x}, t\Big).$$

Recall that, by definition,

$$f\Big(-h\overset{1}{\frac{\partial}{\partial x}}, \overset{2}{x}\Big)u = (2\pi h)^{-n/2}\int_{\mathbf{R}^n} \exp\Big(\frac{ipx}{h}\Big) f(p,x)\widetilde{u}(p)\,dp,$$

$$\widetilde{u}(p) \stackrel{\text{def}}{=} (2\pi h)^{-n/2}\int_{\mathbf{R}^n} \exp\Big(-\frac{ipx}{h}\Big)u(x)\,dx.$$

Denote by $u_N(x,t,h)$ the formal asymptotic solution of the Cauchy problem

$$\begin{aligned} L_h u &= F(x,t,h),\\ u\big|_{t=0} &= u_0(x,h), \end{aligned} \tag{3.34}$$

satisfying the conditions

$$L_h u_N = F(x,t,h) + e^{-S/h}h^N g_N(x,t,h), \qquad u_N\big|_{t=0} = u_0(x,h). \tag{3.35}$$

Suppose that problem (3.34), (3.35) has solutions for each $h \in (0,1]$, and these solutions belong to $L_2(\mathbf{R}^n)$ for $t \in [0,T]$ (this condition defines the class of initial data and of the right-hand sides for problem (3.34)). In (3.35), by $S = S(x,t)$ we denote a function from $C^\infty(\mathbf{R}^n \times [0,T])$, and by $g_N(x,t,h)$ a function of $x \in \mathbf{R}^n_x$, $t \in [0,T]$, such that $g_N(x,t,h) \in L_2(\mathbf{R}^n_x) \cap C^\infty(\mathbf{R}^n_x)$ uniformly in h and t.

In the case

$$u_0 = \varphi_0(x)\exp(-S_0/h), \qquad F(x,t,h) = 0, \tag{3.36}$$

where $S_0(x) \in C^\infty(\mathbf{R}^n_x)$, $S_0 \geqslant 0$, we have $\varphi_0(x) \in C_0^\infty(\mathbf{R}^n)$. According to [49, 57], the formal asymptotic solution u_N must be sought in the form

$$u_N = \sum_{k=0}^{N} \exp(-S/h)\varphi_k h^k. \tag{3.37}$$

Here $S = S(x,t)$ is the solution of the Hamilton–Jacobi equation

$$\frac{\partial S}{\partial t} + H\Big(\frac{\partial S}{\partial x}, x, t\Big) = 0, \qquad S\big|_{t=0} = S_0(x), \tag{3.38}$$

while the functions φ_k are solutions of transport equations similar to those arising in the WKB method (see, for example, [57]). Thus, problem (3.36) has (at least for small t) a formal asymptotic solution u_N satisfying (3.35).

Theorem 3.5. *Let the solution of problem* (3.30) *exist in* $L_2(\mathbf{R}^n)$ *for* $t \in [0,T]$, *and let the following inequality hold for* $p \in \mathbf{R}^n$: $\operatorname{Re} H\big(\frac{\partial S}{\partial x} - ip, x, t\big) \leqslant -\operatorname{Re}\frac{\partial S}{\partial t}$. *Then, for* $t \in [0,T]$, $\|e^{S/h}(u - u_N)\|_{L_2(\mathbf{R}^n)} \leqslant ch^{N-1}$, *where* c *is independent of* h.

Corollary 3.2. *Let the solution of problem (3.30) exist and let the function $S(x,t)$ satisfy equation (3.36). Then, if the inequality*

$$\operatorname{Re} H\Big(\frac{\partial S}{\partial x} - ip, x, t\Big) \leqslant \operatorname{Re} H\Big(\frac{\partial S}{\partial x}, x, t\Big)$$

holds for $p \in \mathbf{R}^n$, we have $\|e^{-S/h}(u - u_N)\|_{L_2(\mathbf{R}^n)} \leqslant ch^{N-1}$ for $t \in [0,T]$ (c is independent of h).

Proof. The operator L_h can be represented in the form $L_h = -i\big[-ih\frac{\partial}{\partial t} + H_1\big(-ih\frac{\partial}{\partial x}, x, t\big)\big]$, where $H_1(p,x,t) = iH(-ip,x,t)$. We note that the difference $u_N - u$ satisfies the equation

$$L_h(u_N - u) = \exp(-S/h)h^N g_N(x,t,h)$$

or $\exp(S/h)L_h(u_N - u) = h^N g_N(x,t,h)$. Denoting

$$W_N = \exp(S/h)(u - u_N)$$

and applying the formulas of commutation between pseudodifferential operators and exponential functions [37, p.27; 54], we have

$$\Big[-ih\frac{\partial}{\partial t} + i\frac{\partial S}{\partial t} + H_1\Big(\overline{-ih\frac{\partial}{\partial x} - i\frac{\partial S}{\partial x}}^{1}, \overset{2}{x}, t\Big)\Big]W_N = -ih^N g_N(x,t,h),$$
$$W_N\big|_{t=0} = 0.$$

Hence, similarly to [54, p.111] and [37, p.31],

$$-ih\frac{\partial W_N}{\partial t} + i\frac{\partial S}{\partial t}W_N + H_1\Big(-ih\overset{1}{\frac{\partial}{\partial x}} - i\overset{2}{\frac{\partial S}{\partial x}}, \overset{2}{x}, t\Big)W_N + hQW_N = -ih^N g_N(x,t,h),$$

where

$$Q = \sum_{k\geqslant 0}^{M} Q_k\Big(-ih\overset{1}{\frac{\partial}{\partial x}}, \overset{2}{x}, t\Big)h^k + h^{M+1}R_M,$$

$Q_k(p,x,t) \in S^\infty(\mathbf{R}^n \times \mathbf{R}^n \times [0,T])$, $R_M\colon L_2 \to L_2$ is bounded.

Finally, we obtain the equation for W_N:

$$-ih\frac{\partial W_N}{\partial t} + \mathcal{H}\Big(-ih\frac{\partial}{\partial x}, x, t\Big)W_N + hQW_N = -ih^N g_N(x,t,h), \tag{3.39}$$

where

$$\mathcal{H}(p,x,t) = H_1\Big(p - i\frac{\partial S}{\partial x}, x, t\Big) - iH\Big(\frac{\partial S}{\partial x}, x, t\Big).$$

It is known that if $\operatorname{Im}\mathcal{H} \leqslant 0$, then, by the method of energy inequalities, one can estimate the solution of problem (3.39) in L_2. The condition imposed on $\operatorname{Im}\mathcal{H}$ coincides with that given in Theorem 3.5. By using the embedding theorems, one can easily obtain the following consequence.

Corollary 3.3. *Let the conditions of Theorem* 3.5 *be satisfied; then, for any* $\varepsilon > 0$, *we have* $\|e^{S/h}(u - u_N)\|_{C(\mathbf{R}^n)} \leqslant ch^{N-1-n/2-\varepsilon}$, *where c is independent of h.*

It is known that, if the condition $\operatorname{Im}\mathcal{H} \leqslant 0$ is not satisfied for equation (3.39), then there exists a solution satisfying the equation with bounded (as $h \to 0$) right-hand side and the initial data. This solution grows exponentially as $h \to 0$ (see [40]). Thus the conditions of Theorem 3.5 are necessary in the sense that, if they are not satisfied, there exists a formal asymptotic solution of (3.37), but it is not the asymptotic solution of problem (3.34) within the exponential precision.

4.4. Asymptotic behavior of localized solutions to equations with variable coefficients

In ecology, mathematical biology, in processes of diffusion with chemical reactions, the mathematical simulation often results in quasilinear parabolic equations, in which the function $F(x,t,u)$ (describing the distribution of sinks and sources) is such that the equation $F(x,t,u) = 0$ has the solution $u = A_0(x,t) \not\equiv \text{const}$. This situation occurs in models for systems with resources varying under the influence of environment.

In this section we present algorithms for the construction of asymptotic solutions for a class of problems including the example discussed in the Introduction to this book.

Consider the following parabolic equation with variable coefficients:

$$\begin{aligned} Lu \overset{\text{def}}{=} \varepsilon\frac{\partial u}{\partial t} - \varepsilon^2\frac{\partial}{\partial x}\left(\lambda(x,t)\frac{\partial u^k}{\partial x}\right) \\ - \varepsilon\delta(x,t)\frac{\partial u}{\partial x} - \gamma^2(x,t)u^q(A_0(x,t) - u^\mu)\nu = 0, x \in \mathbf{R}^n, \qquad t \in [0,T], \end{aligned} \tag{4.1}$$

where $k > 1$, $q > 0$, $\mu > 0$, $\lambda(x,t)$, $\gamma^2(x,t)$, $A_0(x,t)$ are given positive infinitely differentiable functions, $u \geqslant 0$, $A_0(x,t)$ is a function uniformly remote from zero $0 < c_1 < A_0$, $c_1 = \text{const}$, $\delta(x,t)$ is an infinitely differentiable function (possibly, with variable sign), $\nu = \pm 1$. The dimensionless variables are chosen so that

$$\sup_x u = 1, \qquad \sup_{x,t} A_0(x,t) = 1.$$

The physical meaning of the last summand $\gamma^2(x,t)u^q(A_0(x,t) - u^\mu)\nu$ in (4.1) is defined by the constant ν. If $\nu = +1$, this summand describes distributed sources and, if $\nu = -1$, distributed sinks.

There is an important difference between equation (4.1) and those considered above: the source-sink function has the variable root $u = A_0^{1/\mu}(x,t)$. Therefore the correction to the principal term of the asymptotic solution is calculated in a different way.

Let us construct an asymptotic solution of equation (4.1) satisfying the conditions:

$$u(x,t,\varepsilon)\big|_{x=-\varphi(t,\varepsilon)} = 0, \qquad \frac{\partial u^k}{\partial x}\bigg|_{x=-\varphi(t,\varepsilon)} = 0, \tag{4.2}$$

$$0 \leqslant u(x,t,\varepsilon) \leqslant A_0^{1/\mu}(x,t), \qquad \lim_{x\to\infty}(A_0^{1/\mu} - u) = 0. \tag{4.3}$$

Here $\varphi(t,\varepsilon)$ is an unknown function smooth in t, ε.

The function $u(x,t,\varepsilon)$ is one-sided localized. The conditions (4.2) mean that this function and its flow are continuous on the weak discontinuity front. Actually, the fact that problem (4.1)–(4.3) has a solution means that the wave remains stable under perturbations of the amplitude, velocity and shape of the wave caused by time and space variations in properties of the medium.

In this section the asymptotic solution is constructed by a method similar to the Poincare-Lighthill-Go method [18] for obtaining asymptotic solutions of ordinary differential equations.

Obviously, in the case $A_0(x,t) \equiv 1$, the asymptotic solution constructed by this method coincides with that obtained in the preceding sections. In the case $A_0(x,t) \not\equiv$ const, the solution constructed in this section can be reduced to the form that can be obtained by above methods, but sometimes it is more convenient to use a modification of the method given below.

Obviously, the method given below can be applied to equations with nonsingularities of general form, but for simplicity we shall consider certain specific equations.

Theorem 4.1. *Let $k>1$, $q>1$, $k+q>2$, $\mu>0$, $\nu=+1$. Then the asymptotic modulo $O(\varepsilon^2)$ solution of problem (4.1)–(4.3) has the form*

$$u(x,t,\varepsilon) = W\Big(\Big(\frac{S}{\varepsilon}+\varepsilon g\Big(\frac{S}{\varepsilon},t,\varepsilon\Big)+O(\varepsilon^2)\Big),t,x,\varepsilon\Big)$$
$$= A^{1/\mu}(x,t,\varepsilon)\chi\Big(\frac{S}{\varepsilon}+\varepsilon g\Big(\frac{S}{\varepsilon},t,\varepsilon\Big)+O(\varepsilon^2)\Big). \tag{4.4}$$

The function $\chi(\xi)$ satisfies the ordinary differential equation of the first order

$$\frac{d\chi^k}{d\xi} = \frac{d\theta}{d\zeta}(\zeta(\theta))\Big|_{\theta=\chi}, \tag{4.5}$$

where $\theta(\zeta)$ is the solution of the following problem for the Zeldovich equation:

$$b_0\frac{d\theta}{d\zeta} - \frac{d^2\theta}{d\zeta^2} - R(\theta) = 0, \qquad R(\theta) = \nu K(\theta)\theta^{k+q-1}(1-\theta^\mu),$$
$$\theta\big|_{\zeta\to\infty} = 1, \qquad \theta\big|_{\zeta\to-\infty} = 0.$$

where b_0 is the Zeldovich constant.

The function $\chi(\xi)$ can be estimated as follows:

$$\chi = c_0\xi^\alpha + c_1\xi^{\alpha+\alpha_1} + o(\xi^{\alpha+\alpha_1}), \qquad \xi\to 0, \tag{4.6}$$
$$c_0 = \Big[\frac{b_0(k-1)}{k}\Big]^{1/(k-1)}, \qquad \alpha = \frac{1}{k-1}, \quad \alpha_1 = \alpha q + 1 - \alpha = \frac{k+q-2}{k-1}$$
$$c_1 = \Big[\frac{b_0(k-1)}{k}\Big]^{q/(k-1)} \frac{k-1}{b_0(q+k-1)(3-2k-q)}$$
$$\chi = 1 - \Gamma_2\exp\Big(-\frac{l\xi}{k}\Big) + o\Big(\Big(-\frac{l\xi}{k}\Big)\Big), \qquad \xi\to\infty, \quad \Gamma_2 = \text{const} > 0,$$
$$l = -\frac{b_0}{2} + \sqrt{\frac{b_0^2}{4} + k\mu}.$$

The function $S(x,t,\varepsilon)$ *has the form*

$$S(x,t,\varepsilon)=\beta^0(t)(x+\varphi^0(t))-\varepsilon\beta^0\varphi^1(t),$$

where the functions $\beta^0(t)$, $\varphi^0(t)$ *satisfy the system of equations*

$$\begin{aligned}\beta^0(t)A_0^{(1-q)/\mu-1}(-\varphi^0,t)\left(\frac{d\varphi^0}{dt}-\delta(-\varphi^0,t)\right)&=b_0\gamma^2(-\varphi^0,t)\\(\beta^0(t))^2\lambda(-\varphi^0,t)A_0^{(k-q)/\mu-1}(-\varphi^0,t)&=\gamma^2(-\varphi^0,t).\end{aligned}\tag{4.7}$$

The function $A(x,t,\varepsilon)$ *has the form*

$$A(x,t,\varepsilon)=A_0(x,t)+\varepsilon A_1(x,t)+O(\varepsilon^2).$$

The functions φ', $g(\tau,t)$, $A_1(x,t)$ *are defined in Lemma* 4.1.

Remark. Note that here we have the condition

$$\frac{d}{du}[u^q(A(x,t)-u^\mu)]_{u=A^{1/\mu}}\leqslant\frac{d}{du}[u^q(A(x,t)-u^\mu)],$$

which is an analog of the condition $\frac{dF}{du}\big|_{u=1}<\frac{dF}{du}(u)$ introduced in Theorem 2.2.

Lemma 4.1. *The function* $g(\tau,t,0)$ *has the form*

$$\begin{aligned}g(\tau,t,0)&=-\int_0^\tau\exp\{I(\xi)\}\left[\int_\varkappa^\xi f(\eta,t)\exp\{-I(\eta)\}\,d\eta\right]d\xi,\\I(\xi)&=\int^\eta\left[\frac{b_0}{k}\chi^{1-k}-2\frac{\partial}{\partial\xi}\ln\frac{\partial\chi^k}{\partial\xi}\right]d\xi,\end{aligned}\tag{4.8}$$

where $\varkappa=0$ *and*

$$\begin{aligned}f(\xi,t)=&\left[A_0^{k/\mu}(-\varphi^0,t)(\beta^0)^2\lambda(-\varphi^0,t)\frac{d\chi^k}{d\xi}\right]^{-1}\\&\times\left[-\xi\frac{d\chi}{d\xi}A_0^{1/\mu}(-\varphi^0,t)\frac{1}{\beta^0}\frac{d\beta^0}{dt}+\xi\frac{d^2\chi^k}{d\xi^2}A_0^{k/\mu}(-\varphi^0,t)(\beta^0)^2\frac{\partial\lambda}{\partial x}\bigg|_{x=-\varphi^0}\right.\\&+\beta^0\frac{\partial\lambda}{\partial x}\bigg|_{x=-\varphi^0}A_0^{k/\mu}(-\varphi^0,t)\frac{d\chi^k}{d\xi}+2\lambda(-\varphi^0,t)\frac{\partial A_0^{k/\mu}}{\partial x}\bigg|_{x=-\varphi^0}\beta^0\frac{d\chi^k}{d\xi}\\&+\nu\frac{\partial\gamma^2}{\partial x}\bigg|_{x=-\varphi^0}\frac{\xi}{\beta^0}\chi^q(1-\chi^\mu)A_0^{1+q/\mu}(-\varphi^0,t)\\&+\frac{\partial\delta}{\partial x}(-\varphi^0,t)A_0^{1/\mu}(-\varphi^0,t)\frac{d\chi}{d\xi}+\frac{d\chi}{d\xi}A_0^{1/\mu}(-\varphi^0,t)\beta^0\frac{d\varphi^1}{dt}\\&+\frac{d^2\chi^k}{d\xi^2}\beta^0\frac{d\lambda}{dx}(-\varphi^0,t)\varphi^1A_0^{k/\mu}(-\varphi^0,t)\end{aligned}$$

$$
\begin{aligned}
&+\frac{\partial\gamma^2}{\partial x}(-\varphi^0,t)\varphi^1\nu\chi^q(1-\chi^\mu)A_0^{1+q/\mu}(-\varphi^0,t)\\
&-\beta^0\left(\frac{d\varphi^0}{dt}-\delta(-\varphi^0,t)\right)\frac{\partial A_0^{1/\mu}}{\partial x}\bigg|_{x=-\varphi^0}\left(\varphi^1+\frac{\xi}{\beta^0}\right)\frac{d\chi}{d\xi}\\
&+(\beta^0)^2\lambda(-\varphi^0,t)\frac{\partial A_0^{k/\mu}}{\partial x}\bigg|_{x=-\varphi^0}\left(\varphi^1+\frac{\xi}{\beta^0}\right)\frac{d^2\chi^k}{d\xi^2}\\
&+\gamma^2(-\varphi^0,t)\nu\frac{\partial A_0^{1+q/\mu}}{\partial x}\bigg|_{x=-\varphi^0}(1-\chi^\mu)\left(\varphi^1+\frac{\xi}{\beta^0}\right)\chi^q\\
&-\beta^0\left(\frac{d\varphi^0}{dt}-\delta(-\varphi^0,t)\right)\frac{1}{\mu}A_1(-\varphi^0,t)A_0^{1/\mu-1}(-\varphi^0,t)\frac{d\chi}{d\xi}\\
&+(\beta^0)^2\lambda(-\varphi^0,t)\frac{k}{\mu}A_1(-\varphi^0,t)A_0^{-1+k/\mu}\frac{d^2\chi}{d\xi^2}\\
&+\gamma^2(-\varphi^0,t)\nu\frac{q}{\mu}A_1(-\varphi^0,t)A_0^{q/\mu}(-\varphi^0,t)\chi^q(1-\chi^\mu)\Big]
\end{aligned}
\tag{4.9}
$$

Under the condition

$$
\int_0^\infty f(\eta,t)e^{-I(\eta)}\,d\eta=0 \tag{4.10}
$$

the estimates hold:

$$
\begin{aligned}
g&=O(\tau^1), \qquad \tau\to 0,\\
g&=O(\tau^2), \qquad \tau\to\infty.
\end{aligned}
$$

The function $A_1=A_1(x,t)$ *has the form*

$$
A_1=\frac{\delta(x,t)\frac{\partial A_0^{1/\mu}}{\partial x}-\frac{\partial A_0^{1/\mu}}{\partial t}}{\gamma^2(x,t)A_0^{q/\mu}}.
$$

The function φ^1 *satisfies equation (4.10) that can be written as follows:*

$$
\begin{aligned}
&\beta^0A_0^{1/\mu}(-\varphi^0,t)\frac{d\varphi^1}{dt}I_5+\varphi^1(\beta^0)^2\frac{\partial\lambda}{\partial x}(-\varphi^0,t)A_0^{1/\mu}(-\varphi^0,t)I_4\\
&\quad+\frac{1}{\beta^0}\frac{d\beta^0}{dt}A_0^{1/\mu}(-\varphi^0,t)I_1+\frac{\partial\gamma^2}{\partial x}(-\varphi^0,t)\varphi^1A_0^{1/\mu}(-\varphi^0,t)I_6\\
&\quad-\beta^0\frac{\partial\lambda}{\partial x}(-\varphi^0,t)A_0^{k/\mu}(-\varphi^0,t)I_2-\beta^0\frac{\partial\lambda}{\partial x}(-\varphi^0,t)A_0^{k/\mu}(-\varphi^0,t)I_0\\
&\quad-\nu\frac{\partial\gamma^2}{\partial x}(-\varphi^0,t)\frac{1}{\beta^0}A_0^{q/\mu+1}(-\varphi^0,t)I_3-2\lambda(-\varphi^0,t)\frac{\partial A_0^{k/\mu}}{\partial x}\bigg|_{x=-\varphi^0}\beta^0I_2\\
&\quad-\frac{\partial\delta}{\partial x}\bigg|_{x=-\varphi^0}A_0^{1/\mu}(-\varphi^0,t)I_1-A_0^{1/\mu}(-\varphi^0,t)\beta^0\frac{\partial\delta}{\partial x}\bigg|_{x=-\varphi^0}I_1\\
&\quad+A_1(-\varphi^0,t)\Big[\beta^0\left(\frac{d\varphi^0}{dt}-\delta(-\varphi^0,t)\right)\frac{1}{\mu}A_0^{1/\mu-1}I_5
\end{aligned}
$$

$$- (\beta^0)^2\lambda(-\varphi^0,t)\frac{k}{\mu}A_0^{k/\mu-1}(-\varphi^0,t)I_4 - \gamma^2(-\varphi^0,t)\frac{\nu q}{\mu}A_0^{q/\mu}(-\varphi^0,t)I_6\Big]$$

$$+ \beta^0\left(\frac{d\varphi^0}{dt} - \delta(-\varphi^0,t)\right)\frac{\partial A_0^{1/\mu}}{\partial x}\bigg|_{x=-\varphi^0}\left(\varphi^1 I_5 + \frac{I_1}{\beta^0}\right)$$

$$- (\beta^0)^2\lambda(-\varphi^0,t)\frac{\partial A_0^{k/\mu}}{\partial x}\bigg|_{x=-\varphi^0}\left(\varphi^1 I_4 + \frac{I_0}{\beta^0}\right)$$

$$- \gamma^2(-\varphi^0,t)\nu\frac{\partial A_0^{q/\mu+1}}{\partial x}\left(\varphi^1 I_6 + \frac{I_3}{\beta^0}\right) = 0, \tag{4.11}$$

where

$$I_0 = \int_0^\infty \frac{\exp\big(-I(\eta)\big)\eta\frac{d^2\chi^k}{d\eta^2}}{\frac{d\chi^k}{d\eta}}\,d\eta,$$

$$I_1 = \int_0^\infty \frac{\exp\big(-I(\eta)\big)\eta}{k\chi^{k-1}(\eta)}\,d\eta,$$

$$I_2 = \int_0^\infty \exp\big(-I(\eta)\big)\,d\eta,$$

$$I_3 = \int_0^\infty \frac{\eta\chi^q(1-\chi^\mu)\exp\big(-I(\eta)\big)}{\frac{d\chi^k}{d\eta}}\,d\eta,$$

$$I_4 = \int_0^\infty \frac{\frac{d^2\chi^k}{d\eta^2}\exp\big(-I(\eta)\big)}{\frac{d\chi^k}{d\eta}}\,d\eta,$$

$$I_5 = \int_0^\infty \frac{\exp\big(-I(\eta)\big)}{k\chi^{k-1}(\eta)}\,d\eta,$$

$$I_6 = \int_0^\infty \frac{\chi^q(1-\chi^\mu)}{\frac{d\chi^k}{d\eta}}\exp\big(-I(\eta)\big)\,d\eta.$$

Let us prove these statements. By substituting the solution (4.4) into (4.1), we obtain

$$\frac{\partial W}{\partial \xi}\left(1+\varepsilon\frac{\partial g}{\partial \tau}\right)\left[\beta^0\frac{d\varphi^0}{dt} - \varepsilon\left(\beta^1\frac{d\varphi^1}{dt} + \beta^0\frac{d\varphi^1}{dt}\right) + \varepsilon\frac{\xi+\beta^0\varphi^1}{\beta^0}\frac{d\beta^0}{dt}\right]$$

$$+ \varepsilon\frac{\partial W}{\partial t} + \varepsilon^2\frac{\partial W}{\partial \xi}\frac{\partial g}{\partial t} - \left[\left(1+2\varepsilon\frac{\partial g}{\partial \tau} + O(\varepsilon^2)\right)\frac{\partial^2 W^l}{\partial \xi^2} + \varepsilon\frac{\partial W^k}{\partial \xi}\frac{\partial^2 g}{\partial \tau^2}\right]$$

$$\times\left[\lambda(-\varphi^0,t) + \frac{\partial\lambda}{\partial x}\bigg|_{x=-\varphi^0}\varepsilon\varphi^1 + \varepsilon\frac{\tau}{\beta^0}\frac{\partial\lambda}{\partial x}\bigg|_{x=-\varphi^0} + O(\varepsilon^2)\right]$$

$$- \beta^0\left[\varepsilon\frac{\partial\lambda}{\partial x}\bigg|_{x=-\varphi^0} + O(\varepsilon^2)\right]\frac{\partial W^k}{\partial \xi}$$

$$- 2\varepsilon\left[\lambda(-\varphi^0,t) + \varepsilon\frac{\partial\lambda}{\partial x}\bigg|_{x=-\varphi^0} + O(\varepsilon^2)\right]\frac{\partial^2 W^k}{\partial \xi\partial x}\left(1+\varepsilon\frac{\partial g}{\partial \tau}\right)$$

$$
\begin{aligned}
&-\left[\delta(-\varphi^0,t)+\frac{\partial\delta}{\partial x}\bigg|_{x=-\varphi^0}\varepsilon\varphi^1+\frac{\varepsilon\tau}{\beta^0}\frac{\partial\delta}{\partial x}\bigg|_{x=-\varphi^0}+O(\varepsilon^2)\right]\frac{\partial W}{\partial\xi}\left(1+\varepsilon\frac{\partial g}{\partial\tau}\right)\\
&-\varepsilon\delta(x,t)\frac{\partial W}{\partial x}\\
&-\left[\gamma^2(-\varphi^0,t)+\frac{\partial\gamma^2}{\partial x}\bigg|_{x=-\varphi^0}\varepsilon\varphi^1+\frac{\varepsilon\tau}{\beta^0}\frac{\partial\gamma^2}{\partial x}\bigg|_{x=-\varphi^0}+O(\varepsilon^2)\right]W^q(A-W^\mu)=0.
\end{aligned}
\tag{4.12}
$$

Here the following expressions for the derivatives of $S(x,t,\varepsilon)$ were used:

$$
\begin{aligned}
\frac{\partial S}{\partial t}&=\beta^0\frac{d\varphi^0}{dt}+\varepsilon\frac{\tau+\beta^0\varphi^1}{\beta^0}\frac{d\beta^0}{dt}-\frac{d}{dt}(\beta^0\varphi^1)\bigg|_{\tau=S/\varepsilon},\\
\frac{\partial S}{\partial x}&=\beta^0,\qquad (x+\varphi)=\frac{\tau\varepsilon}{\beta^0}+\varepsilon\varphi^1\bigg|_{\tau=S/\varepsilon}.
\end{aligned}
$$

In (4.12) we set the summands with equal powers of the parameters ε equal to each other and, as $\varepsilon\to 0$, get

$$
\begin{aligned}
\beta^0\left(\frac{d\varphi^0}{dt}-\delta(-\varphi^0,t)\right)\frac{\partial W}{\partial\xi}&-(\beta^0)^2\lambda(-\varphi^0,t)\frac{\partial^2W^k}{\partial\xi^2}\\
&-\gamma^2(-\varphi^0,t)\nu W^q(A_0(x,t)-W^\mu)=0
\end{aligned}
\tag{4.13}
$$

We put

$$
W=A^{1/\mu}(x,t,\varepsilon)\chi(\xi)=\big(A_0^{1/\mu}(x,t)+\frac{\varepsilon}{\mu}A_1(x,t)A_0^{1/\mu-1}(x,t)+\dots\big)\chi(\xi)
$$

and in (4.13) collect the summands of the orders $\varepsilon^0,\varepsilon^1$:

$$
\begin{aligned}
\varepsilon^0:\ &\beta^0\left(\frac{d\varphi^0}{dt}-\delta(-\varphi^0,t)\right)A_0^{1/\mu}(x,t)\frac{d\chi}{d\xi}-(\beta^0)^2\lambda(-\varphi^0,t)A_0^{k/\mu}(x,t)\frac{d^2\chi^k}{d\xi^2}\\
&-\gamma^2(-\varphi^0,t)\nu A_0^{q/\mu+1}(x,t)\chi^q(1-\chi^\mu), \qquad (4.14)\\
\varepsilon^1:\ &\beta^0\left(\frac{d\varphi^0}{dt}-\delta(-\varphi^0,t)\right)\frac{1}{\mu}A_1(x,t)A_0^{1/\mu}(x,t)\frac{d\chi}{d\xi}\\
&-(\beta^0)^2\lambda(-\varphi^0,t)\frac{k}{\mu}A_1(-\varphi^0,t)A_0^{k/\mu-1}(-\varphi^0,t)\frac{d^2\chi^k}{d\xi^2}\\
&-\gamma^2(-\varphi^0,t)\nu\frac{q}{\mu}A_1(x,t)A_0^{q/\mu}(x,t)\chi^q(1-\chi^\mu)\\
&+\underline{\gamma^2(-\varphi^0,t)\nu A_0^{q/\mu}(x,t)A_1(x,t)\chi^{q+\mu}}, \qquad (4.15)
\end{aligned}
$$

By above considerations, all functions of the argument x multiplied by functions, decreasing as $\xi\to\infty$, can be expanded into the Taylor series at the point $x=-\varphi^0$ except the last (underlined) summand in (4.15). For example, for the function $A_0^{1/\mu}(x,t)$, we have

$$
A_0^{1/\mu}(x,t)=A_0^{1/\mu}(-\varphi^0,t)+\frac{\partial A_0^{1/\mu}}{\partial x}\bigg|_{x=-\varphi^0}\left(\frac{\varepsilon\tau}{\beta^0}+\varepsilon\varphi^1\right)+O(\varepsilon^2).
$$

By (4.14), setting the sum of summands of the order ε^0 equal to zero, we get the equation

$$\beta^0\left(\frac{d\varphi^0}{dt} - \delta(-\varphi^0,t)\right)A_0^{1/\mu}(-\varphi^0,t)\frac{d\chi}{d\xi} - (\beta^0)^2\lambda(-\varphi^0,t)A_0^{k/\mu}(-\varphi^0,t)\frac{d^2\chi^k}{d\xi^2} - \gamma^2(-\varphi^0,t)A_0^{q/\mu+1}\chi^q(1-\chi^\mu) = 0$$

that can be reduced to the form:

$$b\frac{d\chi}{d\xi} - \frac{d^2}{d\xi}\chi^k - \nu\chi^q(1-\chi^\mu) = 0$$

if

$$\beta^0(t)A_0^{(1-q)/\mu-1}(-\varphi^0,t)\left(\frac{d\varphi^0}{dt} - \delta(-\varphi^0,t)\right) = b_0\gamma^2(-\varphi^0,t),$$
$$(\beta^0)^2\lambda(-\varphi^0,t)A_0^{(k-q)/\mu-1}(-\varphi^0,t) = \gamma^2(-\varphi^0,t).$$

These relations coincide with (4.6). Thus we get the following boundary-value problem for the ordinary differential equation

$$\begin{aligned} b_0\frac{d\chi}{d\xi} - \frac{d^2\chi^k}{d\xi^2} - \chi^q(1-\chi^\mu) = 0,\\ \chi(0) = 0, \qquad \chi\big|_{\xi\to\infty} = 1. \end{aligned} \tag{4.16}$$

The summands of the order ε in (4.14), all summands in (4.17) and the summands of the order ε in (4.12) appear in the equation for the function $g(\tau,t)$ written below.

Equations of form (4.16) were studied in Chapter I. Thus there exists a one-sided localized solution of problem (4.16) related to a running wave.

Recall that, if $k+q>2$, the desired solution exists only for one value of b, $b=b_0$, called the Zeldovich constant. For example, for $k+q=3$, $\mu=1$, using (4.5), we have

$$b_0 = \sqrt{k/2}, \qquad \theta(\zeta) = (1+\exp\{-\zeta\sqrt{k/2}\})^{-1},$$

and the function $\chi(\xi)$ is defined as follows

$$\int\frac{\chi^{k-2}\,d\chi}{1-\chi} = \frac{\xi}{\sqrt{2k}}.$$

Let us calculate the asymptotic expansion of the function $\chi(\xi)$ for $\xi\to 0$. As stated in Theorem 4.1 the desired asymptotic expansion has the form

$$\chi = c_0\xi^\alpha + c_1\xi^{\alpha_1+\alpha} + \dots, \qquad \alpha_1 > 0. \tag{4.17}$$

As an example, consider the case $q>1$, $\nu=1$ (then $k+q>2$). By substituting (4.17) into (4.16), we get

$$\begin{aligned} &b_0[\alpha c_0\xi^{\alpha-1} + c_1(\alpha+\alpha_1)\xi^{\alpha+\alpha_1-1} + \dots]\\ &\quad - [\alpha k c_0^k\xi^{\alpha k-2}(\alpha k-1) + kc_1c_0^{k-1}\xi^{\alpha k+\alpha_1-2}(\alpha k+\alpha_1)(\alpha k+\alpha_1-1)]\\ &\quad - c_0^q\xi^{\alpha q} - qc_1c_0^{q-1}\xi^{\alpha q+\alpha_1} - \dots = 0. \end{aligned} \tag{4.18}$$

Setting the sum of coefficients at the main singularity of $\xi^{\alpha-1}$, $\alpha = 1/(k-1)$, equal to zero, we obtain

$$c_0 = \left[\frac{b_0(k-1)}{k}\right]^{1/(k-1)}.$$

To calculate the exponent in the second term in (4.19), we must compare the powers of ξ with exponents $\alpha+\alpha_1-1$, $\alpha k+\alpha_1-2$, αq. Note that if $\alpha = 1/(k-1)$, the first two exponents coincide: $\alpha k - 1 = \alpha$, and if $\alpha_1 = \alpha(q-1)+1$, all three exponents coincide. Thus we obtain the equation for c_1:

$$c_1(\alpha_1+\alpha)b_0 - kc_1c_0^{k-1}(\alpha k+\alpha_1)(\alpha k+\alpha_1-1) - c_0^q = 0,$$

and consequently

$$c_1 = \left(\frac{b_0(k-1)}{k}\right)^{q/(k-1)} \frac{k-1}{b_0(q+k-1)(3-2k-q)}.$$

Actually, to calculate the function g is equivalent to calculate the next term of the expansion. The properties of g essentially depend on the combination of values of the parameters and coefficients in equation (4.1). The solution presented in Theorem 4.1 can be written (with respect to the variables τ, x, t) in the form:

$$W(\tau+\varepsilon g, t, \varepsilon) = A^{1/\mu}(x,t,\varepsilon)\left[\chi(\tau) + \varepsilon\frac{d\chi}{d\tau}(\tau)g(\tau,t,\varepsilon) + O(\varepsilon^2)\right].$$

If the estimate $O(\varepsilon^2)$ is uniform in τ, t, then this solution can be constructed by the method considered in Section 4.1 of this chapter.

Note that $g(0,t,\varepsilon) = 0$. The function $g(\tau,t,0)$ satisfies the equation obtained from (4.12):

$$\frac{\partial^2 g(\tau,t,0)}{\partial\tau^2} - \frac{\partial g(\tau,t,0)}{\partial\tau}\left(b\frac{d\chi}{d\xi} - 2\frac{d^2\chi^k}{d\xi^2}\right)\left(\frac{d\chi^k}{d\xi}\right)^{-1} = -\big(f(\xi,t) + f_0(\xi,t,x)\big). \quad (4.19)$$

By (4.12), (4.14), (4.15), we get fhe function f

$$\begin{aligned} f(\xi,t) = {} & \left[A_0^{k/\mu}(-\varphi^0,t)(\beta^0)^2\lambda(-\varphi^0,t)\frac{d\chi^k}{d\xi}\right]^{-1} \\ & \times\left[-\left(\frac{\xi}{\beta^0}+\varphi^1\right)\frac{d\chi}{d\xi}A_0^{1/\mu}(-\varphi^0,t)\frac{d\beta^0}{dt} + \xi\frac{d^2\chi^k}{d\xi^2}A_0^{k/\mu}(-\varphi^0,t)\beta^0\frac{\partial\lambda}{\partial x}\Big|_{x=-\varphi^0}\right. \\ & + A_0^{k/\mu}(-\varphi^0,t)\beta^0\frac{\partial\lambda}{\partial x}\Big|_{x=-\varphi^0}\frac{d\chi^k}{d\xi} + 2\lambda(-\varphi^0,t)\frac{\partial A_0^{k/\mu}}{\partial x}\Big|_{x=-\varphi^0}\beta^0\frac{d\chi^k}{d\xi} \\ & + \nu\frac{\partial\gamma^2}{\partial x}\Big|_{x=-\varphi^0}\frac{\xi}{\beta^0}\chi^q(1-\chi^\mu)A_0^{1+q/\mu}(-\varphi^0,t) \\ & + \frac{\partial\delta}{\partial x}\Big|_{x=-\varphi^0}\frac{\xi}{\beta^0}A_0^{1+q/\mu}(-\varphi^0,t)\frac{d\chi}{d\xi} \end{aligned}$$

$$
\begin{aligned}
&- \frac{\partial \chi}{\partial \xi} A_0^{1/\mu}(-\varphi^0,t)\beta^0 \frac{d\varphi^1}{dt} + \frac{d^2\chi^k}{d\xi^2}(\beta^0)^2 \frac{\partial \lambda}{\partial x}(-\varphi^0,t)\varphi^1 A_0^{1/\mu}(-\varphi^0,t) \\
&+ \left.\frac{\partial \gamma^2}{\partial x}\right|_{x=-\varphi^0} \varphi^1 \nu \chi^q (1-\chi^\mu) A_0^{1/\mu}(-\varphi^0,t) \\
&- \beta^0 \left(\frac{d\varphi^0}{dt} - \delta(-\varphi^0,t)\right) \frac{1}{\mu} A_1(-\varphi^0,t) A_0^{1/\mu-1}(-\varphi^0,t) \frac{d\chi}{d\xi} \\
&+ (\beta^0)^2 \lambda(-\varphi^0,t) \frac{k}{\mu} A_1(-\varphi^0,t) A_0^{k/\mu-1}(-\varphi^0,t) \frac{d^2\chi^k}{d\xi^2} \\
&+ \gamma^2(-\varphi^0,t)\nu \frac{q}{\mu} A_0^{q/\mu} A_1(-\varphi^0,t) \chi^q (1-\chi^\mu) \\
&- \beta^0 \left(\frac{d\varphi^0}{dt} - \delta(-\varphi^0,t)\right) \left.\frac{\partial A_0^{1/\mu}}{\partial x}\right|_{x=-\varphi^0} \left(\varphi^1 + \frac{\tau}{\beta^0}\right) \frac{d\chi}{d\xi} \\
&+ (\beta^0)^2 \lambda(-\varphi^0,t) \left.\frac{\partial A_0^{k/\mu}}{\partial x}\right|_{x=-\varphi^0} \left(\varphi^1 + \frac{\tau}{\beta^0}\right) \frac{d^2\chi^k}{d\xi^2} \\
&+ \gamma^2(-\varphi^0,t)\nu \left.\frac{\partial A_0^{1+q/\mu}}{\partial x}\right|_{x=-\varphi^0} \left(\varphi^1 + \frac{\tau}{\beta^0}\right) \chi^q (1-\chi^\mu) \Bigg], \qquad (4.20)
\end{aligned}
$$

$$
\begin{aligned}
f_0 = &\left\{ \chi(\xi) \left[\delta(x,t) \frac{\partial A_0^{1/\mu+1}}{\partial x}(x,t) - \frac{\partial A_0^{1/\mu}}{\partial t}(x,t) \right] \right. \\
&\left. - \gamma^2(x,t)\nu \frac{q}{\mu} A_0^{q/\mu}(x,t) A_1(x,t) \chi^{q+\mu}(\xi) \right\} \\
&\times \left[A_0^{k/\mu}(-\varphi^0,t)(\beta^0)^2 \lambda(-\varphi^0,t) \frac{d\chi^k}{d\xi} \right]^{-1}.
\end{aligned}
$$

On the right-hand side of (4.19) all functions in the variables x, t, except those grouped in f_0, are expanded into the Taylor series at the point $x = -\varphi^0(t)$. The function $\chi(\xi)$ becomes constant as $\xi \to \infty$ (see estimate (4.7)) and the function f_0 is bounded. In equation (4.19) we replace the function $g(\tau, t, 0)$ by $g(\xi, t, 0)$. Note that

$$
\begin{aligned}
g(\tau,t,0) &= g(\tau + \varepsilon g(\tau,t,) - \varepsilon g(\tau,t,0), t, 0) \\
&= g(\xi,t,0) - \frac{\partial g}{\partial \xi}(\xi,t,0)\varepsilon g(\tau,t,0) + O(\varepsilon^2).
\end{aligned}
$$

The singularity in the next terms of this expansion decreases as $\tau \to 0$, i.e., the remainder becomes smoother. This follows from the estimates:

$$
\frac{\partial g}{\partial \tau} = O(1), \qquad g^j(\tau,t,\varepsilon)\frac{\partial^j g}{\partial \tau^j}(\tau,t,\varepsilon) = O(\tau^1), \qquad j \in N.
$$

The latter expression is the estimate of the remainder in the Taylor formula in the Lagrange form.

The function $\partial g(\tau,t,0)/\partial\tau$ was calculated by the formula:

$$\begin{aligned}\frac{\partial g}{\partial\tau}(\tau,t,0) &= \frac{\partial g}{\partial\tau}(\tau+\varepsilon g(\tau,t,0)-\varepsilon g(\tau,t,0),t,0) = \frac{\partial g}{\partial\tau}(\xi-\varepsilon g(\tau,t,0),t,0)\\ &= \frac{\partial g}{\partial\xi}(\xi,t,0)-\varepsilon g\frac{\partial^2 g}{\partial\tau}(\tau,t,0)+O(\varepsilon^2).\end{aligned}$$

Obviously, we have the necessary condition for the function $g(\xi,t,0)$ to increase (as $\xi\to\infty$) not faster than at a polynomial rate:

$$|f_0(\xi,t,x)| = O(\xi^L), \qquad \xi\to\infty,$$

for a certain $L>0$.

This yields $f_0(\xi,t,x)=0$ or

$$A_1(x,t) = \frac{\delta(x,t)\dfrac{\partial A_0^{1/\mu}}{\partial x}(x,t)-\dfrac{\partial A_0^{1/\mu}}{\partial t}(x,t)}{\gamma^2(x,t)A_0^{q/\mu}(x,t)}.$$

Thus, the function $A_1(x,t)$ is determined. The function $g(\xi,t,0)$ satisfies the equation

$$\frac{\partial^2 g}{\partial\xi^2}(\xi,t,0)-\frac{\partial g(\xi,t,0)}{\partial\xi}\left[\frac{b}{k}\chi^{1-k}-2\frac{\partial}{\partial\xi}\ln\frac{d\chi^k}{d\xi}\right] = -f(\xi,t), \tag{4.21}$$

where f is given by (4.10).

Equation (4.21) has the solution

$$\begin{aligned} g(\xi,t,0) &= -\int_0^{\xi} e^{I(\xi_1)}\left[\int_{\varkappa}^{\xi_1} f(\eta,t)e^{-I(\eta)}\,d\eta\right]d\xi_1,\\ I(\eta) &= \int^{\eta}\left[\frac{b\chi^{1-k}}{k}-2\frac{d}{d\xi}\ln\left(\frac{d\chi^k}{d\xi}\right)\right]d\xi. \end{aligned} \tag{4.22}$$

The function f is defined by (4.20).

As $\xi\to 0$, we have the estimate $f=O(\xi^{\alpha(1-k)})$ since

$$\frac{d\chi}{d\xi}\left(\frac{d\chi^k}{d\xi}\right)^{-1} = O(\xi^{\alpha(1-k)}), \qquad \xi\to 0.$$

As $\xi\to +0$, we have

$$e^{I(\xi)} = \exp\left\{\int^{\xi}\frac{b_0\chi^{1-k}}{k}\,d\xi\right\}\left(\frac{d\chi^k}{\xi}\right)^{-2} \sim \tau^{b_0/kC_0^{k-1}-2\alpha}.$$

Since (4.20) implies that b_0/kC_0^{k-1}, we get

$$g(\xi,t,0)\sim\int_0^{\xi}\xi_1^{-\alpha}\int_{\varkappa}^{\xi}\eta^{\alpha-1}\,d\alpha\sim\xi, \qquad \xi\to 0.$$

Incidentally, we see that the inner integral in (4.8) converges as $\xi_1 \to 0$, hence we can put $\varkappa = 0$ in (4.8).

Let us estimate the function $g(\xi, t, 0)$ at infinity (as $\xi \to \infty$). As $\xi \to \infty$, we have

$$e^{I(\xi)} = \exp\left\{\int^{\xi} \frac{b_0 \chi^{1-k}}{k}\, d\xi\right\}\left(\frac{d\chi^k}{\xi}\right)^{-2} \sim \exp\left\{-\frac{\xi}{k}\left(\frac{b_0^2}{4} + \left|\frac{\partial R}{\partial \theta}\right|_{\theta=1}\right)^{1/2}\right\}.$$

Therefore, since $f \sim \xi$, $\xi \to \infty$, the inner integral in (4.19) converges within the limits 0 and infinity.

In order the function $\frac{d\chi}{d\xi} g(\xi, t, 0)$ decrease as $\tau \to \infty$, it is necessary and sufficient that

$$\int_0^{\infty} f(\eta, t) e^{-I(\eta)}\, d\eta = 0. \tag{4.23}$$

Let us calculate (4.24) by using the explicit formula for the function f. The detailed equation (4.23) for the function φ^1 has the form (4.12).

Theorem 4.1 is proved.

Let us consider the case $k + q = 2$, $q < 1$, $\nu = +1$.

Theorem 4.2. *Let $k > 1$, $q > 0$, $k + q = 2$, $\mu > 0$, $\nu = +1$. Then the asymptotic $O(\varepsilon^2)$ solution of problem (4.1)–(4.3) exist and has the form*

$$\begin{aligned} u(x, t, \varepsilon) &= W\left(\left(\frac{S}{\varepsilon} + \varepsilon g\left(\frac{S}{\varepsilon}, t, \varepsilon\right) + O(\varepsilon^2)\right), t, x, \varepsilon\right) \\ &= A^{1/\mu}(x, t, \varepsilon)\chi\left(\frac{S}{\varepsilon} + \varepsilon g\left(\frac{S}{\varepsilon}, t, \varepsilon\right) + O(\varepsilon^2)\right). \end{aligned} \tag{4.24}$$

The function $\chi(\xi)$ satisfies the ordinary differential equation of the first order

$$\frac{d\chi^k}{d\xi} = \frac{d\theta}{d\zeta}(\zeta(\theta))\bigg|_{\theta=\chi},$$

where the function $\theta(\zeta)$ is the solution of the following problem for equation of the Kolmogorov–Petrovskii–Piskunov type:

$$\begin{gathered} b\frac{d\theta}{d\zeta} - \frac{d^2\theta}{d\zeta^2} - k\theta(1 - \theta^{\mu}) = 0, \\ \theta\big|_{\zeta\to\infty} = 1, \qquad \theta\big|_{\zeta\to-\infty} = 0. \end{gathered}$$

The function $S(x, t, \varepsilon)$ has the form

$$S(x, t, \varepsilon) = \beta^0(t)(x + \varphi^0(t)) - \varepsilon\beta^0\varphi_1(t),$$

where the functions $\beta^0(t)$, $\varphi^0(t)$ satisfy the following relations:

$$\begin{aligned} &\beta^0(t) A_0^{(1-q)/\mu-1}(-\varphi^0, t)\left(\frac{d\varphi^0}{dt} - \delta(-\varphi^0, t)\right) = b\gamma^2(-\varphi^0, t), \\ &(\beta^0(t))^2 \lambda(-\varphi^0, t) A_0^{(k-q)/\mu-1}(-\varphi^0, t) = \gamma^2(-\varphi^0, t), \qquad b \geqslant b_{\min} = 2\sqrt{k}. \end{aligned} \tag{4.25}$$

The function $\chi(\xi)$ satisfies the estimate

$$\chi(\xi) = C_0\xi^\alpha + o(\xi^\alpha), \qquad \xi \to 0,$$
$$\alpha = \frac{1}{k-1} = \frac{1}{1-q}, \qquad C_0 = \left(\frac{b-\sqrt{b^2-4k}}{2\alpha k}\right)^\alpha, \qquad b \geqslant 2\sqrt{k}.$$

As $\xi \to \infty$,

$$\chi = 1 - \exp(-l\xi/k) + o\big(\exp(-l\xi/k)\big), \qquad \xi \to \infty,$$
$$l = -b/2 + \sqrt{b^2/4 + k\mu}.$$

The function $A(x,t,\varepsilon)$ has the form

$$A(x,t,\varepsilon) = A_0(x,t) + \varepsilon A_1(x,t) + O(\varepsilon^2),$$

The functions $A_1(x,t)$, $g(\tau,t,0)$ are defined in the following lemma.

Lemma 4.2. *The function $g(\tau,t,0)$ has the form (4.8), where $\varkappa = 0$, the function f is the same as in Lemma (4.1).*

For arbitrary $\varphi^1(t)$, the function $g(t,\tau,\varepsilon)$ satisfies the estimates:

$$g = O(\tau\ln\tau), \qquad b = b_{\min} = 2\sqrt{k}, \qquad \tau \to 0,$$
$$g \sim O(\tau^1), \qquad b > 2\sqrt{k}, \qquad \tau \to 0,$$
$$g \sim O(\tau^2), \qquad \tau \to \infty,$$

The function $A_1(x,t)$ is defined by the equality:

$$A_1(x,t) = \frac{\delta(x,t)\dfrac{\partial A_0^{1/\mu}}{\partial x}(x,t) - \dfrac{\partial A_0^{1/\mu}}{\partial t}(x,t)}{\gamma^2(x,t)A_0^{q/\mu}(x,t)}.$$

Now let us prove these statements. The principal equations are obtained by a similar method as in the proof of Theorem 4.1. By (4.12), as above, we get equation (4.16) and the following relation

$$\beta^0(t)\left(\frac{d\varphi^0}{dt} - \delta(-\varphi^0,t)\right)A_0^{1/\mu)}(-\varphi^0,t)\frac{d\chi}{d\xi} - (\beta^0(t))^2\lambda(-\varphi^0,t)A_0^{k/\mu}(-\varphi^0,t)\frac{d^2\chi^k}{d\xi^2}$$
$$-\gamma^2(-\varphi^0,t)A_0^{q/\mu+1}\chi^q(1-\chi^\mu) = 0.$$

Thus, we have the boundary-value problem for the ordinary differential equation

$$b\frac{d\chi}{d\xi} - \frac{d^2\chi}{d\xi^2} - \chi^q(1-\chi^\mu) = 0, \qquad \chi(0) = 0, \qquad \chi\big|_{\xi\to\infty} = 1. \tag{4.26}$$

Let us calculate the asymptotic expansion of the function $\chi(\xi)$ as $\xi \to 0$. By Theorem 4.1 of Chapter III, we have the asymptotic expansion

$$\chi = C_0\xi^{\alpha} + C_1\xi^{\alpha_1+\alpha} + \ldots, \qquad \alpha_1 > 0. \tag{4.27}$$

Substituting (4.27) into (4.26), we get (4.7) with b instead of b_0.

In the case under consideration we have $q < 1$, $k + q = 2$, $\nu = 1$. Then $\alpha k - 2 = \alpha - 1$, $\alpha = 1/(1-q) = 1/(k-1)$. The coefficient C_0 is defined as follows

$$b\alpha - \alpha^2 k C_{0\pm}^{k-1} - C_{0\pm}^{1-k} = 0, \qquad C_{0\pm}^{k-1} = \frac{b \pm \sqrt{b^2 - 4k}}{2\alpha k}.$$

The Kolmogorov–Petrovskii–Piskunov inequality $b \geqslant 2\sqrt{k}$ implies that the coefficient C_0 is real. The subsequent term in the expansion was calculated in Chapter III (see also the end of Section 4.2).

Now let us prove Lemma 4.2. The proof is similar to that of Lemma 4.1.

Recall that we consider the case $k + q = 2$. In this case the solution χ of the standard equation exists for $b \geqslant b_{\min}$. According to Lemma 1.10 of Chapter III, we have

$$\frac{b_{\min}}{kC_{0\pm}^{k-1}} = 2\alpha,$$

since

$$b_{\min} = 2\sqrt{k}, \qquad C_{0\pm}^{k-1}\Big|_{b=b_{\min}} = \frac{1}{\alpha\sqrt{k}}.$$

Hence

$$\frac{b}{kC_{0-}^{k-1}} \geqslant 2\alpha, \qquad C_{0\pm} = \left(\frac{b \pm \sqrt{b^2 - 4k}}{2\alpha k}\right)^{\alpha}.$$

Let us calculate the asymptotic expansion of the integrand in the inner integral in (4.8):

$$f \sim \xi^{\alpha(1-k)}, \quad e^{-I(\xi)} \sim \xi^{2\alpha - b/(kC_{0-}^{k-1})}, \quad fe^{-I(\xi)} \sim \xi^{2\alpha - 1 - b/(kC_{0-}^{k-1})} \quad \xi \to 0, .$$

Since $2\alpha - 1 - b/(kC_{0-}^{k-1}) > 1$, the inner integral in (4.8) diverges near the lower limit. Then one must put $\varkappa = \infty$ in (4.23). The latter estimate implies that, for arbitrary φ^1, the inner integral in (4.8) has the following asymptotics as $\xi \to 0$:

$$\int_{\infty}^{\xi} f(\eta, t)e^{-I(\eta)}\,d\eta \sim \begin{cases} \ln\xi, & b = b_{\min}, \\ \xi^{2\alpha - b/(kC_{0-}^{k-1})}, & b > b_{\min}. \end{cases}$$

Actually, for $b = b_{\min}$, the estimate holds

$$e^{I(\xi)} \sim \xi^{2\alpha k - 2 - b/(kC_{0-}^{k-1})} \sim \xi^0, \qquad \xi \to 0.$$

Thus

$$\int_0^{\xi} e^{I(\xi')}\left[\int_{\infty}^{\xi'} f(\eta, t)e^{-I(\eta)}\,d\eta\right] d\xi' \sim \int_0^{\xi} \ln\xi'\,d\xi'.$$

By using the L'Hopital rule, we get

$$\lim_{\xi\to 0} L\frac{\xi\ln\xi}{\int_0^\xi \ln\xi'\,d\xi'} \sim O(1).$$

Thus the estimates hold

$$g(\xi,t,0)=\begin{cases}\xi^1\ln\xi, & b=b_{\min},\\ \xi, & b>b_{\min},\end{cases}\qquad \xi\to 0.$$

For $b=b_{\min}$, as $\tau\to 0$, the "correction" $\varepsilon g(\tau,t,0)$ to τ in the argument of function χ in (4.4) is less smooth, i.e.,

$$\lim_{\tau\to 0}\frac{\tau}{g(\tau,t,0)}=\lim_{\tau\to 0}\frac{\tau}{\tau\ln\tau}=0.$$

As $\xi\to\infty$, the function $g(\xi,t,0)$ is estimated in the same way as in Lemma 4.1. Lemma 4.2 is proved.

Theorem 4.3. *Let $k>1$, $0<q<1$, $k+q>2$, $\mu>0$, $\nu=-1$. Then the asymptotic modulo $O(\varepsilon^2)$ solution of problem (4.1)–(4.3) has the form*

$$\begin{aligned}u(x,t,\varepsilon)&=W\big(S/\varepsilon+\varepsilon g(S/\varepsilon,t,\varepsilon)+O(\varepsilon^2),t,x,\varepsilon\big)\\ &=A^{1/\mu}(x,t,\varepsilon)\chi\big(S/\varepsilon+\varepsilon g(S/\varepsilon,t,\varepsilon)+O(\varepsilon^2)\big).\end{aligned}$$

The function $\chi(\xi)$ satisfies the ordinary differential equation of the first order

$$\frac{d\chi^k}{d\xi}=\frac{d\theta}{d\zeta}(\zeta(\theta))\bigg|_{\theta=\chi},$$

where the function $\theta(\zeta)$ is the solution of the following problem:

$$b\frac{d\theta}{d\zeta}-\frac{d^2\theta}{d\zeta}+k\theta^{k+q-1}(1-\theta^\mu)=0,\qquad b\leqslant -2\sqrt{k\mu},$$
$$\theta\big|_{\zeta\to\infty}=1,\qquad \theta\big|_{\zeta\to-\infty}=0.$$

The function θ satisfies the estimates

$$\theta=O\bigg(\frac{1}{|\xi|^{1/(k+q-2)}}\bigg),\qquad \xi\to-\infty,$$
$$\theta=1-e^{\widehat{l}\xi}+o(e^{\widehat{l}\xi}),\qquad \xi\to\infty$$
$$l=\frac{b}{2}+\sqrt{\frac{b^2}{4}-k\mu}<0.$$

The function $S(x,t,\varepsilon)$ has the form

$$S(x,t,\varepsilon)=\beta^0(x+\varphi^0(t))-\varepsilon\beta^0\varphi^1(t)+\dots,$$

where the functions β^0, φ^0 satisfy the following relations

$$\begin{gathered}\beta^0(t)A_0^{(1-q)/\mu-1}(-\varphi^0,t)\left(\frac{d\varphi^0}{dt}-\delta(-\varphi^0,t)\right)=b\gamma^2(-\varphi^0,t),\\ (\beta^0(t))^2\lambda(-\varphi^0,t)A_0^{(k-q)/\mu-1}(-\varphi^0,t)=\gamma^2(-\varphi^0,t),\qquad b\leqslant-2\sqrt{k\mu}.\end{gathered}\tag{4.28}$$

The function $\chi(\xi)$ satisfies the estimate

$$\chi(\xi)=C_0\xi^\alpha+o(\xi^{\alpha+\alpha_1}),\qquad \xi\to0,$$
$$C_0=\left(\frac{1-q}{-b}\right)^{1/(1-q)},\qquad b<0.\qquad \alpha=\frac{1}{1-q}.$$

The estimates hold

$$\chi(\xi)\sim1-\Gamma_2\exp\left\{\frac{l\xi}{k}\right\}+o\left(\exp\left\{\frac{l\xi}{k}\right\}\right),\qquad \Gamma_2=\text{const},\quad \xi\to\infty.$$

The function $A(x,t,\varepsilon)$ has the form

$$A(x,t,\varepsilon)=A_0(x,t)+\varepsilon A_1(x,t)+O(\varepsilon^2).$$

The functions A_1 and g are defined in Lemma 4.3.

Lemma 4.3. *The function $g(t,\tau,0)$ has the form (4.8) where $\varkappa=0$, the function f is defined by (4.10). For arbitrary $\varphi^1\in C^\infty$, $b<-2\sqrt{\mu k}$, the following estimates hold:*

$$g=O(\tau),\qquad \tau\to0,$$
$$g=O(\tau^2),\qquad \tau\to\infty.$$

The function $A_1(x,t)$ has the form

$$A_1(x,t)=\frac{\delta(x,t)\dfrac{\partial A_0^{1/\mu}}{\partial x}(x,t)-\dfrac{\partial A_0^{1/\mu}}{\partial t}(x,t)}{\gamma^2(x,t)\dfrac{q}{\mu}A_0^{q/\mu}(x,t)}.\tag{4.29}$$

The function φ^1 is defined by (4.12).

Proof. Obviously, Theorem 4.3 and Lemma 4.3 are proved by the same arguments and calculations as the preceding lemmas and theorems. Relations (4.26) yield the equation for the function $\chi(\xi)$:

$$\begin{gathered}b\frac{d\chi}{d\xi}-\frac{d^2\chi^k}{d\xi^2}+\chi^q(1-\chi^\mu)=0\\ \chi(0)=0,\qquad \chi\big|_{\xi\to\infty}=1,\end{gathered}\tag{4.30}$$

and the system of equations for β^0, φ^0 (4.28).

As shown in Theorem 4.1 in Chapter III, the desired asymptotic expansion has the form

$$\chi(\xi) = C_0\xi^\alpha + o(\xi^\alpha),$$

where

$$C_0 = \left(\frac{1-q}{-b}\right)^{1/(1-q)}, \qquad b < 0, \qquad \alpha = \frac{1}{1-q}.$$

The function $g(\xi, t, 0)$ satisfies equation (4.19) which can be analyzed as usually. Relation $f_0 \equiv 0$ implies (4.29) for the function $A_1(x, t)$. Thus we get the same equation for $g(\xi, t, 0)$ as above:

$$\frac{\partial^2 g}{\partial \xi^2}(\xi, t, 0) - \frac{\partial g}{\partial \xi}(\xi, t, 0)\left[\frac{b}{k}\chi^{1-k} - 2\frac{d}{d\xi}\ln\left(\frac{d\chi^k}{d\xi}\right)\right] = f(\xi, t).$$

Obviously, the solution of this equation has the form (4.8). However, the asymptotic behavior of the integrand is different. Precisely as above, $f \sim \xi^{\alpha(1-k)}$, $\xi \to 0$, but since $\alpha = 1/(1-q)$, we have

$$f \sim \xi^{(1-k)/(1-q)}, \qquad \xi \to 0.$$

Further,

$$e^{I(\xi)} = \xi^{-2(\alpha k-2)} \exp\left\{-\frac{b\xi^{1-\alpha k+\alpha}}{(\alpha k - \alpha - 1)C_0^{k-1}}\right\}, \qquad \xi \to 0,$$

$$e^{-I(\xi)} = \xi^{2(\alpha k-2)} \exp\left\{\frac{b\xi^{1-\alpha k+\alpha}}{(\alpha k - \alpha - 1)C_0^{k-1}}\right\}, \qquad \xi \to 0.$$

Note that in the case considered $\alpha k - \alpha - 1 = (k+q-2)/(1-q) > 0$, $b < 0$, therefore the inner integral in (4.8) converges as $\xi \to 0$ and we can put $\varkappa = 0$. As above, this implies the estimate

$$g \sim \xi, \qquad \xi \to 0.$$

As $\xi \to \infty$, it is easy to calculate that, according to the asymptotics

$$1 - \chi(\xi) \sim \exp\left\{\left(\frac{b + \sqrt{b^2 - 4\mu k}}{2k}\right)\xi\right\}$$

and

$$e^{-I(\xi)} \sim (\chi')^2 \exp\left\{-\frac{b}{k}\xi\right\} \sim \exp\left\{\frac{\sqrt{b^2-4\mu k}}{k}\xi\right\},$$

the inner integral in (4.8) diverges as $\xi_1 \to \infty$ for $|b| > 2\sqrt{\mu k}$. Therefore, there are no additional conditions on the function $\varphi^1(t)$ and it is easy to obtain the estimate:

$$g(\xi, t, 0) \sim \xi^2, \qquad \xi \to \infty.$$

As above, this estimate follows from the asymptotics of $e^{-I(\xi)}$ and the fact that $f \sim \xi$, $\xi \to \infty$. Theorem 4.3 is proved.

Clearly, the contents of this section is similar to the results of Section 4.2, and hence there must be an analog of Section 4.3. We do not present the corresponding constructions, since they literally repeat the constructions given in Section 4.3. In particular, the nonuniqueness of the asymptotic solution in the situations of Theorems 4.2 and 4.3 allows to state that the Cauchy problem is asymptotically ill-posed (like in Theorem 3.1). In the situation of Theorem 4.3, precisely as in Theorem 3.4, the regularization is performed by matching a local asymptotic solution with the solution of the linearized equation.

But in the case considered in Theorem 4.2, no regularization is obtained by matching the solution with the solution of the linearized equation. To construct a unique asymptotic solution, we must impose additional conditions, for example, (3.33).

4.5. Heat wave propagation in nonlinear media. Asymptotic solutions to hyperbolic heat (diffusion) equation

Hyperbolic equations describing the heat conduction are used in mathematical models for temperature (diffusion) waves propagating on a nonzero background [45]. Equations of this type can be considered as equations describing waves propagating in dispersive media [55] (see also [73]).

Nonlinear mathematical models for the transfer processes, constructed on the basis of quasilinear parabolic equations considered above, are often criticized. The reason is that it is not always possible to introduce a dimensionless function $u(x,t)$ such that it varies from 0 to a constant. In some models, it is necessary to neglect the magnitude of the background in order to be able to use the advantages of nonlinear models, namely, the finite velocity of perturbation propagation and the localization of these perturbations in space.

When deriving a hyperbolic equation, the second derivative with respect to time is usually neglected, i.e., in fact, only slow variations of the transferred quantity are considered. Otherwise, we obtain a hyperbolic equation describing the heat and mass transfer, for which it is well-known that the velocity of perturbation propagation is finite (including the propagation on a nonsmooth background).

In this section an asymptotic solution for the hyperbolic equation of heat transfer is constructed under the assumption that the thermal conductivity (diffusivity) varies sharply for a certain value of the transferred quantity.

Such equations describe the problems about propagation of heat waves in nonlinear media (e.g., plasma, condensing media) widely used in the literature during the last thirty years.

In [1, 2] the heat waves radiating energy from their fronts are investigated by means of approximate methods. The constructed solution consists of two parts: the self-similar part of the solution to the heat equation and a certain constant. The graphical results obtained by this approach are quite satisfactory from the physical viewpoint. However, from the mathematical viewpoint, the heat wave, constructed in these work, does not satisfy the partial differential equation in the whole, since the derivatives of this solution are not continuous. Below (see Figures 4.7, 4.8), such solutions are constructed for a *hyperbolic* heat equation, and the obtained solution satisfies all necessary conditions.

In many papers the problem about a heavy explosion in a gas is solved in the gas dynamical setting (see [10, 42]). The same setting of the problem is considered in papers dealing with the combustion theory (see [6, 7]).

In the first part of this section we consider models related to the propagation of heat waves in heterogeneous media with phase transition. In the second part of this section we present some results related to the propagation of heat waves in plasma [19] under the following assumptions. For high temperatures, the processes of radiant heat conduction are very important. In these processes the thermal conductivity is usually assumed to depend on temperature exponentially, $\sim T^{k-1}$, $k > 1$. This thermal conductivity rapidly increases in the domain, where the dimensionless temperature varies from one characteristic value a_0 (e.g., the ionization temperature) to another characteristic value a_1.

When a heat wave begins to propagate, the medium is stable, but, because of hydrodynamic effects, the medium is also set in motion and its particles begin to accelerate. This phenomenon can be modeled by the convective term $\delta(x,t)\partial u/\partial x$, where $\delta(x,t)$ is a given function of the medium velocity, describing the medium acceleration.

Besides of this, we assume that the medium properties vary slowly, and introduce a function $\lambda(x,t)$. It is always possible to take it equal to 1 ($\lambda \equiv 1$). The algorithm for constructing asymptotic solutions, applied in this section, was first published in [86].

Thus we have the equation

$$\varepsilon\frac{\partial u}{\partial t} + \varepsilon^2\frac{\partial^2 u}{\partial t^2} - \varepsilon^2\frac{\partial}{\partial x}\Big(\lambda(x,t)\frac{\partial u}{\partial x}\Big) + \varepsilon\delta(x,t)\frac{\partial u}{\partial x} - F(u) = 0, \tag{5.1}$$

where ε is a small parameter related to the Peclet number.

Consider the problem about the construction of an asymptotic solution to equation (5.1) satisfying the conditions

$$\begin{gathered} a_0 \leqslant u(x,t,\varepsilon) \leqslant a_1, \\ u\big|_{x\to-\infty} \to a_0 + 0, \qquad u\big|_{x\to+\infty} \to a_1 - 0. \end{gathered} \tag{5.2}$$

As we see below, the boundary-value problem (5.1), (5.2) is closely related to the following boundary-value problem for the ordinary differential equation:

$$\begin{gathered} b\frac{dW}{d\xi} + \frac{d}{d\xi}\left[(K(W) - b^2)\frac{dW}{d\xi}\right] - F(W) = 0, \\ W\big|_{\xi=0} = a_0, \qquad W\big|_{\xi\to\infty} \to a_1. \end{gathered} \tag{5.3}$$

In its turn, similarly to Section 1.1 of Chapter I, the solution of the boundary-value problem (5.3) can be expressed in terms of the monotone solution of the problem:

$$\begin{gathered} b\frac{d\theta}{d\zeta} - \frac{d^2\theta}{d\zeta^2} - (K(\theta) - b^2)F(\theta) = 0, \\ \theta\big|_{\zeta\to-\infty} = a_0, \qquad \theta\big|_{\zeta\to\infty} = a_1. \end{gathered} \tag{5.4}$$

Here the following formula is used:

$$\big(K(W)-b^2\big)\frac{dW}{d\xi}=\frac{d\theta}{d\zeta},\qquad W(\xi)=\theta(\zeta). \tag{5.5}$$

By assumption, $F(\theta)=0$ for $\theta=a_0$ and $\theta=a_1$. =If $K(\theta)-b^2\neq 0$ for $\theta\in[a_0,a_1]$, then (5.4) is a semilinear equation of the Kolmogorov–Petrovskii–Piskunov or Zeldovich type, see Chapter I. By (5.5), it is easy to show that, in this case, there is no solution $W(\xi)$ satisfying the boundary condition $W(0)=a_0$ (this value can be attained only in the limit as $\xi\to-\infty$). Thus, for $K(W)-b^2\neq 0$ and $W\in[a_0,a_1]$, there is no localized (on the background a_0) solution W_0 satisfying the boundary-value problem (5.3).

If $K(\widehat{W})-b^2=0$ for $\widehat{W}\in[a_0,a_1]$, then, as is easy to see, the boundary-value problem (5.3) also has no solution. This situation is similar to that we have in equations with cubic nonlinearity examined in Chapter II, namely, a structurally stable solution consists of two waves with values in the strip regions $[a_0,\widehat{W}]$ and $[\widehat{W},a_1]$.

If $\widehat{W}=a_0$ or $\widehat{W}=a_1$, then, if the functions $F(W)$ and $K(W)-b^2$ behave like polynomials near the roots, a localized solution of the boundary-value problem (5.3) may exist.

Therefore, we can formulate the following assertion:

Lemma 5.1. *Let $\big(K(W)-b^2\big)F(W)\neq 0$ for $W\in(a_0,a_1)$. Then $[K(a_0)]^{1/2}=b$ is the necessary condition for the solution of the boundary-value problem (5.3) to exist.*

Example 5.1. In equation (5.1) let

$$K(u)=16+\sqrt{u-1},\qquad F(u)=\sqrt{u-1}(2\sqrt{u-1}-6u+10).$$

Then $a_0=1$, $a_1=2$, $q=1/2$, $k=3/2$. For $b=4$, the solution, satisfying the standard equation (5.4), exists and has the form

$$W=\begin{cases}1+(1-e^{-\xi})^2, & \xi>0,\\ 1, & \xi\leqslant 0.\end{cases}$$

This solution is plotted in Figure 4.7.

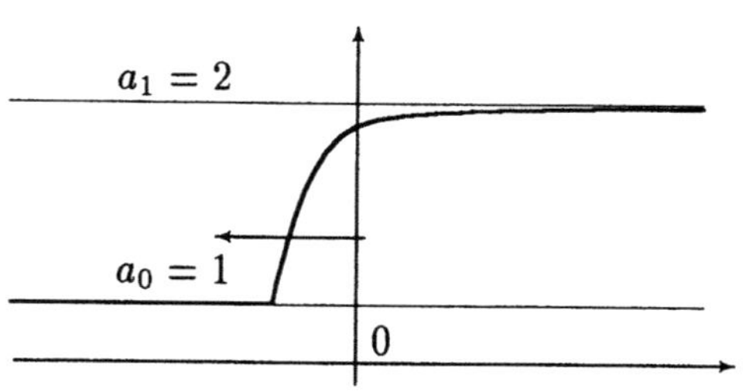

FIGURE 4.7

In accordance with Example 5.1, we assume that

$$K(u) = K(a_0) + \mu(u - a_0)^{k-1} + \dots, \qquad u \to a_0, \quad k > 1, \tag{5.6}$$

i.e., that the dimensionless thermal conductivity behaves like a power near the zero root of the function $K(u) - b^2$. The function $F(u)$ describes the heat release in medium and energy losses for ionization and recombination.

Let

$$\begin{gathered} F(u) = \nu(u - a_0)^q + (u - a_0)^{q+1} + \dots, \qquad q > 0, \quad u \to a_0 + 0, \quad \nu > 0, \\ F(a_1) = 0, \qquad \left.\frac{dF}{du}\right|_{u=a_1} < 0. \end{gathered} \tag{5.7}$$

Thus we assume that the source-sink function vanishes at the ends of the temperature interval.

Indeed, similarly to the preceding cases, one must assume that the inequality $k + q \geqslant 2$ holds.

First, we consider the case

$$\big(K(W) - b^2\big)F(W) > 0, \qquad W \in (a_0, a_1). \tag{5.8}$$

Here, for $k+q > 2$, equation (5.4) belongs to the class of Zeldovich equations and, for a certain $b = b_0$, there exists an exponentially decreasing (as $\xi \to -\infty$) solution of the boundary-value problem. The assumptions of Lemma 5.1 imply severe conditions of coincidence, namely, $[K(a_0)]^{1/2} = b_0$.

For $k+q = 2$, equation (5.4) belongs to the class of equations of the Kolmogorov–Petrovskii–Piskunov type and the boundary-value problem (5.4) has the solution for

$$b \geqslant 2\sqrt{\left.\frac{dR}{d\theta}\right|_{\theta=a_0}}, \qquad R = (K(\theta) - b^2)F(\theta)$$

(obviously, we shall assume that $\mu\nu = \frac{dR}{d\theta}(0) > \frac{dR}{d\theta}(\theta)$, $\theta \in (a_0, a_1)$, see (5.5), (5.7)). Then there exists a localized solution of the boundary-value problem (5.3): for

$$K(a_0) \geqslant 4\mu\nu. \tag{5.9}$$

Clearly, in this case, a localized solution of the boundary-value problem (5.3) always exists, and $b = [K(a_0)]^{1/2}$. Note that, by putting the value of a_0 that satisfies (5.9), by Lemma 5.1, we define the velocity of the simple wave in (5.3) and (5.4), and thus eliminate the nonuniqueness peculiar to equations of the Kolmogorov–Petrovskii–Piskunov type.

In this case the asymptotic solution to problem (5.1), (5.2) will be sought in the form

$$u(x, t, \varepsilon) = W\Big(\frac{S}{\varepsilon} + \varepsilon g\Big(\frac{S}{\varepsilon}, t\Big) + O(\varepsilon^2)\Big), \tag{5.10}$$

where

$$S(x, t, \varepsilon) = \beta^0(t)(x + \varphi^0(t)) - \varepsilon\beta^0(t)\varphi^1(t). \tag{5.11}$$

Theorem 5.1. *Let the conditions $k > 1$, $k + q = 2$,*

$$\sqrt{K(a_0)} + \max_{x,t} \frac{\delta}{\sqrt{\lambda}} > 2\sqrt{\mu\nu}$$

hold. Then the asymptotic localized solution of the boundary-value problem (5.1), (5.2) *has the form* (5.10). *The function $W(\xi, t)$ is the solution of the boundary-value problem* (5.3) *for*

$$b = b(t) = \sqrt{K(a_0)} + \frac{\delta(-\varphi^0, t)}{\sqrt{\lambda(-\varphi^0, t)}},$$

where the function $\varphi^0(t)$ satisfies the ordinary differential equation

$$\varphi_t^0 = \big(K(a_0)\lambda(-\varphi^0, t)\big)^{1/2}. \tag{5.12}$$

The function β^0 in (5.11) *has the form $\beta^0(t) = (\lambda(-\varphi^0, t))^{-1/2}$. As $\xi \to 0$, the asymptotic expansion of the function W has the form*

$$W = a_1 - c_1\xi^\alpha + o(\xi^\alpha), \qquad \alpha = 1/(k-1). \tag{5.13}$$

As $\xi \to \infty$, we have the estimate

$$a_1 - W \sim \exp\Big\{ - \Big|\frac{l}{K(a_0) - K(a_1)}\Big|\xi\Big\}, \tag{5.14}$$

where the constant l is given by the formula

$$l = -\frac{b}{2} + \sqrt{\frac{b^2}{4} + \frac{dR}{d\theta}\Big|_{\theta=a_1}}, \qquad R = F(\theta)(K(\theta) - K(a_1)).$$

The function $g(\tau, t, 0)$ is defined in Lemma 5.2.

Lemma 5.2. *The function $g(\tau, t, 0)$ has the form*

$$g(\tau, t, 0) = -\int_0^\tau e^{I(\xi)}\Big[\int_{\varkappa}^{\xi} f_0(\eta, t)e^{-I(\eta)}\,d\eta\Big]\,d\xi, \tag{5.15}$$

where

$$\varkappa = 0,$$

$$I(\xi) = 2\ln\Big|K(W)\frac{\partial W}{\partial \xi} - K(a_0)\frac{\partial W}{\partial \xi}\Big| - \int^{\xi} \frac{b_0}{K(W) - K(a_0)}\,d\xi,$$

$$f_0 = \Big\{\frac{\partial W}{\partial \xi}\Big(\beta^0\frac{d\varphi^1}{dt} + \frac{\xi + \varphi^1\beta^0}{\beta^0}\frac{d\beta^0}{dt}\Big) - (\beta^0)^2\frac{\partial\delta}{\partial x}\Big|_{x=-\varphi^0}\Big(\varphi^1 + \frac{\xi}{\beta^0}\Big)\frac{\partial W}{\partial \xi}$$
$$+ \frac{d^2W}{d\xi^2}2\beta^0\frac{d\varphi^0}{dt}\Big(\beta^0\frac{d\varphi^1}{dt} + \frac{\xi + \varphi^1\beta^0}{\beta^0}\frac{d\beta^0}{dt}\Big) + \frac{dW}{d\xi}\Big[\frac{d}{dt}\Big(\beta^0\frac{d\varphi^0}{dt}\Big) + \frac{d\varphi^0}{dt}\frac{d\beta^0}{dt}\Big]$$

$$
- \frac{d}{d\xi}\Big(K(W)\frac{dW}{d\xi}\Big)\beta^0 \frac{\partial \lambda}{\partial x}\Big|_{x=-\varphi^0}\Big(\varphi^1 + \frac{\xi}{\beta^0}\Big)
$$
$$
- \frac{\partial \lambda}{\partial x}\Big|_{x=-\varphi^0}\beta^0 K(W)\frac{dW}{d\xi} - F(W)\Big\}\Big\{(K(W)-K(a_0))\frac{dW}{d\xi}\Big\}^{-1}.
$$

For arbitrary $\varphi^1 \in C^\infty$ *the estimates hold*:

$$
\begin{aligned}
g &= O(\tau^1), & \tau &\to 0,\\
g &= O(\tau^2), & \tau &\to \infty.
\end{aligned}
$$

Remark. In the theorem, for $\lambda \equiv 1$, the identity $\varphi_t = \sqrt{K(a_1)}$ means that the heat wave front moves with the speed of sound. This assumption, based on physical considerations, was first introduced by A.S Kompaneits [35]. This heat wave is shown in Figure 4.7.

Proof. Substituting the solution (5.10) into (5.1), we obtain

$$
\begin{aligned}
&\frac{\partial W}{\partial \xi}\Big(1+\varepsilon\frac{\partial g}{\partial \tau}\Big)\Big[\beta^0\frac{d\varphi^0}{dt} + \varepsilon\beta^0\frac{d\varphi^1}{dt} + \frac{\varepsilon\xi}{\beta^0}\frac{d\beta^0}{dt}\Big]\\
&\quad + \varepsilon\frac{\partial W}{\partial t} + \varepsilon^2\frac{\partial W}{\partial \xi}\frac{\partial g}{\partial t} + \frac{\partial^2 W}{\partial \xi^2}\Big(1 + 2\varepsilon\frac{\partial g}{\partial \tau} + O(\varepsilon^2)\Big)\\
&\quad \times\Big[\Big(\beta^0\frac{d\varphi^0}{dt}\Big)^2 + 2\varepsilon\beta^0\frac{d\varphi^0}{dt}\Big(\beta^0\frac{d\varphi^1}{dt} + \frac{\xi+\varphi^1\beta^0}{\beta^0}\frac{d\beta^0}{dt}\Big)\Big]\\
&\quad + \varepsilon\frac{\partial W}{\partial \xi}\Big[\frac{d}{dt}\Big(\beta^0\frac{d\varphi^0}{dt}\Big) + O(\varepsilon)\Big] + \varepsilon\frac{\partial W}{\partial \xi}\frac{\partial^2 g}{\partial \tau^2}\Big[\Big(\beta^0\frac{d\varphi^0}{dt}\Big)^2 + O(\varepsilon)\Big] + O(\varepsilon^2)\\
&\quad - \Big[\Big(1 + 2\varepsilon\frac{\partial g}{\partial \tau} + O(\varepsilon^2)\Big)\frac{\partial}{\partial \xi}\Big(K(W)\frac{\partial W}{\partial \xi}\Big) + \varepsilon K(W)\frac{\partial W}{\partial \xi}\frac{\partial^2 g}{\partial \tau^2}\Big]\\
&\quad \times(\beta^0)^2\Big[\lambda(-\varphi^0,t) + \frac{\partial \lambda}{\partial x}\Big|_{x=-\varphi^0}\varepsilon\Big(\varphi^1 + \frac{\tau}{\beta^0}\Big) + O(\varepsilon^2)\Big]\\
&\quad - \Big[\varepsilon\frac{\partial \lambda}{\partial x}\Big|_{x=-\varphi^0} + O(\varepsilon^2)\Big]\beta^0 K(W)\frac{\partial W}{\partial \xi} + \beta^0\delta(-\varphi^0,t)\frac{\partial W}{\partial \xi}\Big(1+\varepsilon\frac{\partial g}{\partial \tau}\Big)\\
&\quad + \frac{\partial \delta}{\partial x}\Big(\varphi^1 + \frac{\xi}{\beta}\Big)\frac{\partial W}{\partial \xi} + O(\varepsilon^2) - F(W) = 0.
\end{aligned}
\tag{5.16}
$$

Here

$$
\frac{\partial S}{\partial t} = \beta^0\frac{d\varphi^0}{dt} + (x+\varphi^0)\frac{d\beta^0}{dt} - \varepsilon\frac{d}{dt}(\beta^0\varphi^1),
$$
$$
\frac{\partial^2 S}{\partial t^2} = \frac{d}{dt}\Big(\beta^0\frac{d\varphi^0}{dt}\Big) + \varepsilon\frac{\tau + \beta^0(t)\varphi^1(t)}{\beta^0(t)}\frac{d^2\beta^0}{dt^2} + \frac{d\varphi^0}{dt}\frac{d\beta^0}{dt} - \varepsilon\frac{d^2}{dt^2}(\beta^0(t)\varphi^1(t)),
$$
$$
x + \varphi^0(t) = \varepsilon\frac{\tau + \beta^0(t)\varphi^1(t)}{\beta^0(t)},
$$

$$\frac{\partial S}{\partial x} = \beta^0(t), \qquad \frac{\partial^2 S}{\partial t^2} = 0,$$

$$\lambda(x,t) = \lambda(-\varphi^0,t) + \frac{\partial \lambda}{\partial x}\bigg|_{x=-\varphi^0} \varepsilon\Big(\varphi^1 + \frac{\tau}{\beta^0}\Big) + O(\varepsilon^2\tau^2)\bigg|_{\tau=S/\varepsilon}.$$

By setting in (5.13) the sum of terms (that do not tend to zero as $\varepsilon \to 0$) equal to zero, we obtain the equation for W:

$$\beta^0 \frac{d\varphi^0}{dt}\frac{\partial W}{\partial \xi} + \Big(\beta^0 \frac{d\varphi^0}{dt}\Big)^2 \frac{d^2 W}{d\xi^2} + \beta^0 \delta(-\varphi^0,t)\frac{\partial W}{\partial \xi}$$
$$- (\beta^0)^2 \lambda(-\varphi^0,t)\frac{d}{d\xi}\Big(K(W)\frac{dW}{d\xi}\Big) - F(W) = 0. \quad (5.17)$$

By setting the sum of terms of the order ε^1 in (5.16) equal to zero, we get

$$\frac{\partial W}{\partial \xi}\frac{\partial g}{\partial \tau}\beta^0\frac{d\varphi^0}{dt} + \frac{\partial^2 W}{\partial \xi^2} 2\frac{\partial g}{\partial \tau}\Big(\beta^0\frac{d\varphi^0}{dt}\Big)^2$$
$$+ \frac{\partial W}{\partial \xi}\frac{\partial^2 g}{\partial \tau^2}\Big(\beta^0\frac{d\varphi^0}{dt}\Big)^2 - 2\frac{\partial g}{\partial \tau}\frac{\partial}{\partial \xi}\Big(K(W)\frac{\partial W}{\partial \xi}\Big)(\beta^0)^2\lambda(-\varphi^0,t)$$
$$+ \beta^0\delta(-\varphi^0,t)\frac{\partial g}{\partial \tau}\frac{\partial W}{\partial \xi} - \frac{\partial^2 g}{\partial \tau^2}K(W)\frac{\partial W}{\partial \xi}(\beta^0)^2\lambda(-\varphi^0,t) = f, \quad (5.18)$$

where

$$f = \frac{\partial W}{\partial \xi}\Big(\beta^0\frac{d\varphi^1}{dt} + \frac{\xi + \varphi^1\beta^0}{\beta^0}\frac{d\beta^0}{dt}\Big) + \frac{\partial W}{\partial t}$$
$$+ \frac{\partial^2 W}{\partial \xi^2} 2\beta^0\frac{d\varphi^0}{dt}\Big(\beta^0\frac{d\varphi^1}{dt} + \frac{\xi + \varphi^1\beta^0}{\beta^0}\frac{d\beta^0}{dt}\Big) + \frac{\partial W}{\partial \xi}\Big[\frac{d}{dt}\Big(\beta^0\frac{d\varphi^0}{dt}\Big) + \frac{d\varphi^0}{dt}\frac{d\beta^0}{dt}\Big]$$
$$- \frac{\partial}{\partial \xi}\Big(K(W)\frac{\partial W}{\partial \xi}\Big)\beta^0\frac{\partial \lambda}{\partial x}\bigg|_{x=-\varphi^0}\Big(\varphi^1 + \frac{\tau}{\beta^0}\Big)$$
$$- \frac{\partial \lambda}{\partial x}\bigg|_{x=-\varphi^0}\beta^0 K(W)\frac{\partial W}{\partial \xi} + \frac{\partial \delta}{\partial x}(-\varphi^0,t)(\xi + \varphi^1\beta^0)\frac{\partial W}{\partial \xi}$$
$$- \lambda(-\varphi^0,t)\Big[\frac{\partial}{\partial \xi}\Big(K(W)\frac{\partial W}{\partial x}\Big) + \frac{\partial}{\partial x}\Big(K(W)\frac{\partial W}{\partial \xi}\Big)\Big]\beta^0 = 0.$$

Let us consider equation (5.17). We rewrite it as follows

$$\beta^0\Big(\frac{d\varphi^0}{dt} + \delta(-\varphi^0,t)\Big)\frac{\partial W}{\partial \xi} - (\beta^0)^2\lambda\frac{\partial}{\partial \xi}\Big(\Big(K(W) - \Big(\frac{d\varphi^0}{dt}\Big/\sqrt{\lambda}\Big)^2\Big)\frac{\partial W}{\partial \xi}\Big) - F(W) = 0.$$

By Lemma 5.1, the relation (5.12) is the necessary condition for a localized solution to exist. By putting

$$(\beta^0)^2\lambda = 1, \qquad \beta^0\Big(\frac{d\varphi^0}{dt} + \delta(-\varphi^0,t)\Big) = b(t),$$

we get equation (5.3) and the formula for $b(t)$ given in Theorem 5.1.

The function g is determined by equation (5.18) and the condition $g(0,t,\varepsilon)=0$. After easy transformations, we get

$$\frac{\partial^2 g}{\partial\xi^2}+\frac{\partial g}{\partial\xi}\left[b\frac{dW}{d\xi}+2K(a_0)\frac{d^2W}{d\xi^2}-2\frac{d}{d\xi}\left(K(W)\frac{dW}{d\xi}\right)\right]\left[K(a_0)\frac{dW}{d\xi}-K(W)\frac{dW}{d\xi}\right]^{-1}=f, \tag{5.19}$$

where

$$\begin{aligned} f=\Bigg\{&\frac{dW}{d\xi}\left(\beta^0\frac{d\varphi^1}{dt}+\frac{\xi+\varphi^1\beta^0}{\beta^0}\frac{d\beta^0}{dt}\right)+\frac{d^2W}{d\xi^2}\left(\beta^0\frac{d\varphi^1}{dt}+\frac{\xi+\varphi^1\beta^0}{\beta^0}\frac{d\beta^0}{dt}\right)2\beta^0\frac{d\varphi^0}{dt}\\ &+\frac{\partial W}{\partial\xi}\left[\frac{d}{dt}\left(\beta^0\frac{d\varphi^0}{dt}\right)+\frac{d\varphi^0}{dt}\frac{d\beta^0}{dt}\right]-\frac{\partial}{\partial\xi}\left(K(W)\frac{\partial W}{\partial\xi}\right)\beta^0\left(\varphi^1+\frac{\xi}{\beta^0}\right)\frac{\partial\lambda}{\partial x}\bigg|_{x=-\varphi^0}\\ &-\frac{\partial\lambda}{\partial x}\bigg|_{x=-\varphi^0}\beta^0K(W)\frac{dW}{d\xi}-F(W)\Bigg\}\left\{\left(K(W)-K(a_0)\right)\frac{dW}{d\xi}\right\}^{-1}. \end{aligned}$$

The formulas

$$\tau=\tau+\varepsilon g(\tau,t,0)-\varepsilon g(\tau,t,0)=\xi-\varepsilon g(\tau,t,0)$$

imply that the functions $g(\xi,t)$, $f(\xi,t)$ in (5.18) can be expressed in terms of the functions $g(\tau,t)$, $f(\tau,t)$.

Obviously, the solution of equation (5.19) is given by (5.15). As $\xi\to 0$, we have

$$e^{I(\xi)}\sim\xi^{-b/\mu C_0^{k-1}}\left[\frac{\partial W}{\partial\xi}(K(W)-K(a_0))\right]^2\sim\xi^{-b/\mu C_0^{k-1}+2\alpha},$$

where $C_0=\lim_{\xi\to 0}(W-a_0)\tau^{1-k}$. Note that the formula, determining the function $f=f(\xi,t)$ as $\xi\to 0$, implies the estimate

$$f\sim\tau^{-1-\alpha},$$

and the results of Section 3.3 of Chapter III give

$$2\alpha<\frac{b}{\mu C_0^{k-1}}$$

for $b>b_{\min}$. The latter inequality follows from the assumptions of Theorem 5.1. Therefore the inner integral in (5.15) converges as $\xi\to 0$.

Let us calculate the asymptotics of $e^{I(\xi)}$ as $\xi\to\infty$. We have

$$\begin{aligned} e^{I(\xi)}&\sim\exp\left\{\int\frac{b\,d\xi}{K(a_0)-K(a_1)}\right\}\left(\frac{\partial W}{\partial\xi}\right)^2\\ &\sim\exp\left\{-\frac{b\xi}{|K(a_0)-K(a_1)|}\right\}\exp\left\{-\frac{2l\xi}{|K(a_0)-K(a_1)|}\right\}\Bigg\}. \end{aligned}$$

Further

$$-b-2l=-\frac{\sqrt{b^2+4(K(a_0)-K(a_1))|F'(a_1)|}}{|K(a_0)-K(a_1)|}.$$

The formula for the right-hand side of equation (5.19) implies the estimate

$$f \sim \xi, \qquad \xi \to \infty.$$

Therefore

$$fe^{-I(\xi)} \sim \xi \exp\left\{ \frac{\sqrt{b^2 + 4(K(a_0) - K(a_1))|F'(a_1)|}}{|K(a_0) - K(a_1)|}\xi \right\}.$$

Hence, like in Lemma 2.2, the inner integral in (5.15) converges on the domain $[0, \infty)$ and, just as in Section 4.2, we have $\varkappa = 0$ and there is no equation for the function φ^1.

Theorem 5.1 and Lemma 5.1 are proved.

So, precisely as in Theorems 2.1 and 2.2, the formula for the function $\varphi^1(t)$ is not obtained. This fact allows to formulate an analog of Theorem 3.1 about the ill-posedness.

To determine the function $\varphi^1(t)$, one must complete the boundary-value problem (5.1), (5.2), following Section 4.2.

Obviously, using the same methods as in Section 4.2, we can consider problem (5.1)–(5.4) for $q < 1$, $k + q > 2$, $\mu F(u) \geqslant 0$ or $\mu F(u) \leqslant 0$. However, these formulas are not given here.

If the rate of a chemical reaction varies slowly with time and the position in space, the distribution of sinks and sources in equation (5.1) can be modeled by a function of the form $\gamma^2(x,t)F(u)$. (The function $\gamma^2(x,t)$ describes slow variations of the rate of a chemical reaction under the action of external effects, the case $\gamma^2 \equiv 1$ is considered above.)

Because of the function $\gamma^2(x,t) \neq \text{const}$, the equation (5.17), obtained by the substitution of solution (5.10) into the equation (5.1), can be written in the form

$$\beta^0 \frac{d\varphi^0}{dt}\frac{dW}{d\xi} + \left(\beta^2 \frac{d\varphi^0}{dt}\right)^2 \frac{d^2W}{d\xi^2} + \beta^0\delta(-\varphi^0,t)\frac{\partial W}{\partial \xi}$$
$$- (\beta^0)^2\lambda(-\varphi^0,t)\frac{d}{d\xi}\left(K(W)\frac{dW}{d\xi}\right) - \gamma^2(-\varphi^0,t)F(W) = 0.$$

In order to reduce this equation to the form (5.3), we put

$$\beta^0 = \gamma(-\varphi^0,t)\big(\lambda(-\varphi^0,t)\big)^{-1/2}$$
$$\frac{d\varphi^0}{dt} = \big(K(a_0)\lambda(-\varphi^0,t)\big)^{1/2}/\gamma(-\varphi^0,t),$$
$$b(t) = \frac{(K(a_0))^{1/2}}{\gamma(-\varphi^0,t)} + \frac{\delta(-\varphi^0,t)\gamma(-\varphi^0,t)}{\sqrt{\lambda(-\varphi^0,t)}}.$$

Actually, the difference between these formulas and the corresponding formulas of Theorem 5.1 is precisely the difference between the cases $\gamma = 1$ and $\gamma = \gamma(x,t)$.

Example 5.2. In equation (5.1) let

$$K(u) = 6 + \sqrt{2-u}, \qquad F(u) = \sqrt{2-u}(8 - 2\sqrt{2-u} - 6u).$$

Then $a_0 = 1$, $a_1 = 2$, $q = 1/2$, $k = 3/2$. For $b = -4$, the standard equation (5.3) has the solution of the form

$$W = \begin{cases} 2, & \xi \geqslant 0, \\ 2 - (1 - e^{\xi})^2, & \xi < 0. \end{cases}$$

This solution is plotted in Figure 4.8.

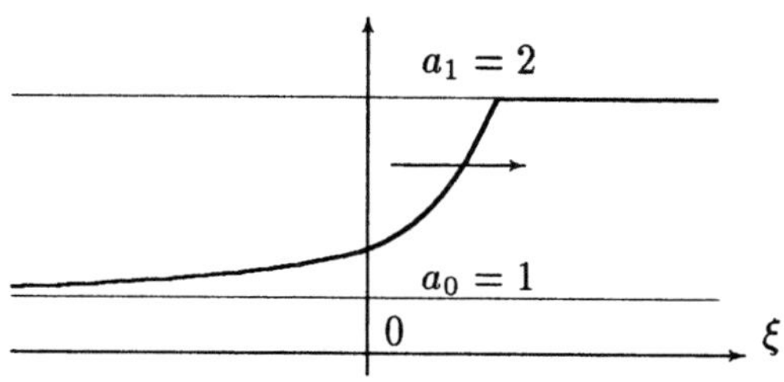

FIGURE 4.8

Assume that $F(u), K(u) \in C^{\infty}$ for $u \neq a_1$, $K(u) > 0$ for $u \in]a_0, a_1[$ and for $u \to a_{1-0}$. We have the representation

$$\begin{aligned} K(u) &= K(a_1) + \mu(a_1 - u)^{k-1} + \dots, \\ F(u) &= \nu(u - a_1)^q + \dots, \qquad k > 1, \quad q > 0, \quad \mu\nu = \text{const}. \end{aligned} \tag{5.20}$$

Example 5.2 demonstrates another situation, in which an asymptotic solution can be constructed. An important point is that, in the class of equations considered, the number of situations, when localized asymptotic solutions can be constructed, is much greater than in the class of parabolic equations, since here we can also consider the case $K(W) - b^2 < 0$ for $W \in (a_0, a_1)$. According to (5.3), this means that we, so to say, consider a parabolic equation with negative coefficient of diffusion. First we consider the situation of Example 5.2. We have

$$\nu < 0, \qquad \mu > 0. \tag{5.21}$$

Then the boundary-value problem (5.3) takes the form

$$\begin{gathered} b\frac{\partial W}{\partial \xi} + \frac{\partial}{\partial \xi}\left[(K(W) - b^2)\frac{\partial W}{pa\xi}\right] - F(W) = 0, \\ W\big|_{\xi=0} = a_1, \qquad W\big|_{\xi\to-\infty} \to a_0. \end{gathered} \tag{5.22}$$

We assume that the inequalities hold

$$F(W) < 0, \qquad K(W) - K(a_1) > 0, \qquad W \in (a_0, a_1), \qquad \left.\frac{dF}{dW}\right|_{W=a_0} < 0. \tag{5.23}$$

Theorem 5.2. *Let the coefficients of equation* (5.1) *satisfy the conditions* (5.20), (5.21), (5.23), $k+q=2$, *and*

$$\max_{x,t} \frac{\delta}{\sqrt{\lambda}} < \sqrt{K(a_1)} - 2\sqrt{|\nu|\mu}.$$

Then the localized asymptotic solution of the boundary-value problem (5.1), (5.2) *has the form* (5.10).

The function $W = W(\xi, t)$ *is the solution of the boundary-value problem* (5.22) *for*

$$b(t) = -\sqrt{K(a_1)} + \frac{\delta(-\varphi^0, t)}{\sqrt{\lambda(-\varphi^0, t)}},$$

where $\varphi^0(t)$ *satisfies the ordinary differential equation*

$$\frac{d\varphi^0}{dt} = -\big(K(a_1)\lambda(-\varphi^0, t)\big)^{1/2}.$$

The function β^0 *in* (5.11) *has the form* $\beta^0(t) = \big(\lambda(-\varphi^0, t)\big)^{-1/2}$. *The following relations hold*

$$\begin{aligned} a_1 - W &= C_0|\xi|^\alpha + o(|\xi|^\alpha), && \xi \to 0, \\ W - a_0 &\sim \exp\left\{\frac{l\xi}{K(a_0) - K(a_1)}\right\}, && \xi \to -\infty, \\ l &= \frac{b + \sqrt{b^2 + 4(K(a_0) - K(a_1))|F'(a_0)|}}{2}, && b < 0. \end{aligned}$$

The function $g(\tau, t, 0)$ *is defined in Lemma* 5.2, *where the constant* $K(a_0)$ *is replaced by the constant* $K(a_1)$, *and* $\varkappa = 0$.

Proof of this theorem differs from that of Theorem 5.1 only in details. The main distinction is the boundary-value problem (5.22). First, we verify that the inequality

$$b < 0$$

is the necessary condition for the solution to exist. Actually, multiplying the equation (5.22) by $\partial W/\partial \xi$, integrating from $-\infty$ to 0, and taking the boundary conditions into account, we get

$$b \int_{-\infty}^{0} \left(\frac{\partial W}{\partial \xi}\right)^2 d\xi = \int_{a_0}^{a_1} F(W)\, dW < 0. \tag{5.24}$$

This gives the required inequality and the signs for $b(t)$ and $d\varphi^0/dt$ in the statement of our theorem. The first inequality in Theorem 5.2 follows from the fact that the semilinear equation (an analog of (5.4)) corresponding to (5.22) is an equation of the Kolmogorov–Petrovskii–Piskunov type.

In this case the formulas of Lemma 5.2 are proved similarly.

A more interesting situation occurs if we assume that

$$K(W) - K(a_1) < 0, \qquad F(W) < 0, \qquad \frac{dF}{dW}(a_0) < 0, \qquad W \in (a_0, a_1), \quad (5.25)$$

and the equality (5.20) holds for

$$k + q = 2, \qquad \mu < 0, \qquad \nu < 0. \quad (5.26)$$

Obviously, in this case, the boundary-value problem for the function W_0, used when we construct an asymptotic solution, coincides with (5.22). However its semi-linear analog is constructed in a different way. Assume that, according to the formula

$$\big(K(W) - K(a_1)\big)\frac{\partial W}{\partial \xi} = \frac{\partial \widehat{\theta}}{\partial \zeta}, \qquad W(\xi) = \widehat{\theta}(\zeta), \quad (5.27)$$

the solution of the boundary-value problem (5.22) can be expressed in terms of the solution of the following boundary-value problem

$$\widehat{b}\frac{\partial \widehat{\theta}}{\partial \zeta} - \frac{\partial^2 \widehat{\theta}}{\partial \zeta^2} - \big(K(\widehat{\theta}) - K(a_1)\big)F(\widehat{\theta}) = 0, \quad (5.28)$$

$$\widehat{\theta}\big|_{\zeta \to -\infty} \to a_0, \qquad \widehat{\theta}\big|_{\zeta \to \infty} \to a_1. \quad (5.29)$$

Obviously, $\frac{d\widehat{\theta}}{d\zeta} > 0$. On the other hand, by (5.27), $\frac{\partial W}{\partial \xi}\frac{d\widehat{\theta}}{d\zeta} < 0$. Hence, (5.27) cannot relate the monotone solutions of problems (5.28), (5.29) and (5.22).

Lemma 5.3. *Let the conditions* (5.25), (5.26) *hold. Then the monotone solution of the boundary-value problem* (5.22) *can be expressed in terms of the monotone solution of the problem*

$$\begin{gathered} b\frac{\partial \theta}{\partial \xi} - \frac{\partial^2 \theta}{\partial \xi^2} - R(\theta) = 0, \\ R(\theta) = \big(K(\theta) - K(a_1)\big)F(\theta), \qquad \theta\big|_{\zeta \to -\infty} \to a_1, \quad \theta\big|_{\zeta \to \infty} \to a_0, \end{gathered} \quad (5.30)$$

by means of the formula

$$\big(K(W) - K(a_1)\big)\frac{\partial W}{\partial \xi} = \frac{\partial \theta}{\partial \zeta}, \qquad W(\xi) = \theta(\zeta).$$

As $\xi \to 0_{-0}$,

$$a_1 - W \sim |\xi|^{1/(k-1)}.$$

Under the condition $\frac{dR}{d\theta}(a_0) > \frac{dR}{d\theta}$, $\theta \in (a_0, a_1)$, *as* $\xi \to -\infty$,

$$W - a_0 \sim \exp\Big\{ - \frac{l^- |\xi|}{|K(a_0) - K(a_1)|}\Big\}, \qquad l^- = \frac{|b| - \sqrt{b^2 - 4R'(a_0)}}{2}.$$

Proof. First note that, by assumption, $R(\theta) > 0$, $\theta \in (a_0, a_1)$. Therefore, under the conditions

$$\frac{dR}{d\theta}(a_0) > 0, \qquad \frac{dR}{d\theta}(a_1) < 0,$$

that follow from (5.25), (5.26), the equation (5.30) belongs to the class of the Kolmogorov–Petrovskii–Piskunov type equations. The fact that the solutions of equations (5.22) and (5.28) are related by the formula given in this lemma can be verifies just as in Theorem 1.1 of Chapter I. It is easy to see that the necessary condition for the monotone solution of problem (5.30) to exist is the inequality

$$b < -2\sqrt{R'(a_0)}.$$

Actually, similarly to Theorem 5.2, it is proved that $b < 0$. Then, calculating the asymptotics of θ as $\zeta \to \pm\infty$, we get

$$\begin{aligned} \theta - a_0 &\sim e^{-l^{\pm}\zeta}, \qquad l^{\pm} = \frac{|b| \pm \sqrt{b^2 - 4R'(a_0)}}{2}, \qquad \xi \to \infty, \\ a_1 - \theta &\sim e^{l\zeta}, \qquad l = \frac{-|b| + \sqrt{b^2 + 4R'(a_1)}}{2}, \qquad \xi \to -\infty. \end{aligned}$$

This yields the required Kolmogorov–Petrovskii–Piskunov inequality.

As $\xi \to 0$, we get

$$\mu\big((a_1 - W)^{k-1} + O(a_1 - W)^k\big)\frac{\partial W}{\partial \xi} = -l\big((a_1 - W) + O\big((a_1 - W)^2\big)\big)$$

or

$$a_1 - W \sim |\xi|^{\alpha}, \qquad \alpha = 1/(k-1).$$

As $\xi \to -\infty$, we have

$$\big(K(a_0) - K(a_1)\big)\frac{\partial W}{\partial \xi} = l^{-}\big(W + O(W^2)\big),$$

and thus we obtain the representation given in this lemma. The lemma is proved.

The results of Lemma 5.3 allow to formulate the following statement.

Theorem 5.3. *Let the conditions of Lemma* 5.3 *be satisfied.*

Then the localized asymptotic solution of the boundary-value problem (5.1), (5.2) *has the form* (5.10). *The function* $W = W(\xi, t)$ *is the solution of the boundary-value problem* (5.12) *for*

$$b(t) = -\sqrt{K(a_1)} + \frac{\delta(-\varphi^0, t)}{\sqrt{\lambda(-\varphi^0, t)}},$$

where $\varphi^0(t)$ *satisfies the ordinary differential equation*

$$\frac{d\varphi^0}{dt} = -\big(K(a_1)\lambda(-\varphi^0, t)\big)^{1/2}.$$

The function $\beta^0(t)$ is the same as in Theorem 5.2, the function $g(t,\tau,0)$ is defined in Lemma 5.2 for $\varkappa = -\infty$ with constant $K(a_0)$ replaced by $K(a_1)$.

Proof. Obviously, it is sufficient to analyze the integrals in formula given in Lemma 5.2, i.e., it is sufficient to calculate the function $\exp\{I(\xi)\}$ as $\xi \to 0$ and $\xi \to -\infty$. By using the formulas of Lemma 5.3, as $\xi \to -\infty$, we have

$$e^{I(\xi)} \sim \exp\left\{\frac{b\xi}{K(a_1)-K(a_0)} + l^-\xi\right\} = \exp\left\{-\frac{\sqrt{b^2-4R'(a_0)}}{K(a_1)-K(a_0)}\xi\right\}.$$

The latter relation and the estimate

$$f \sim \xi, \qquad \xi \to -\infty,$$

imply that the inner integral in (5.15) converges as $\xi \to -\infty$. Therefore, one can take $\varkappa = 0$ if the inner integral in (5.5) converges as $\xi \to 0$ and $\varkappa = \text{const} < 0$ if this integral diverges as $\xi \to 0$.

As $\xi \to 0$, we have

$$e^{I(\xi)} \sim \exp\left\{-\int \frac{b}{\mu(a_1-W)^{k-1}}\,d\xi\right\}(-\xi)^{2\alpha} \sim (-\xi)^{b/\mu C_0^{k-1}+2\alpha}.$$

Obviously, $e^{I(\xi)} \to 0$ as $\xi \to 0$. Further, the estimate

$$f \sim \xi^{-1-\alpha}, \qquad \xi \to 0,$$

implies that the inner integral in (5.15) diverges as $\xi \to 0$. Therefore, we obtain formula (5.15) for $\varkappa = -\infty$ and the estimates given in Lemma 5.2.

4.6. Localized solutions in the multidimensional case

In this section we generalize the results, obtained in Sections 4.2, to the multidimensional case. The waves of combustion and frost penetration present structures, whose evolution is described by solutions of the following semilinear parabolic equations:

$$\varepsilon\frac{\partial u}{\partial t} - \varepsilon^2\langle\nabla, \lambda(x,t)K(u)\nabla u\rangle - \gamma^2(x,t)F(u) = 0, \tag{6.1}$$
$$x \in \mathbf{R}^n, \qquad t \in [0,T],$$

where $u(x,t,\varepsilon)$ is an unknown function with values in $[0,1]$.

The functions $K(u)$, $F(u)$ satisfy the conditions $K(0) = 0$, $K(u) \geqslant 0$, $F(0) = F(1) = 0$. The functions $F(u)$, $K(u)$, $\lambda(x,t)$, $\gamma^2(x,t)$ are positive and infinitely differentiable (the first two of them for $u > 0$). The functions $\gamma(x,t)$, $\lambda(x,t)$ describe slowly varying properties of the medium. Precisely as in Section 4.1, we assume that

$$K(u) = u^{k-1}\rho(u), \qquad k > 1, \qquad F(u) = u^q G(u), \qquad q > 0,$$

where the functions $\rho(u)$ and $G(u)$ are smooth for $u \geqslant 0$ and $\rho(0) > 0$, $G(0) \neq 0$.

If $G(u) > 0$ for $u \in [0,1[$, then the solution of equation (6.1) describes the propagation of waves of combustion wave type.

If $G(u) < 0$ for $u \in [0,1[$, then the solution of equation (6.1) simulates the frost penetration waves in nonlinear media.

Definition 6.1. A continuous nonnegative solution $u(x,t,\varepsilon)$ of equation (6.1) such that

1) $0 \leqslant u(x,t,\varepsilon) \leqslant 1$, $u(x,t,\varepsilon) = 1 - \delta(\varepsilon)$, $\delta(\varepsilon) > 0$, $\delta(\varepsilon) = O(\varepsilon^m)$, $m > 0$, $x \in \Omega_1$;

2) $u(x,t,\varepsilon) \equiv 0$ for $x \in \Omega_2$, where Ω_1, Ω_2 are domains in $\mathbf{R}^n$;

3) $\lim_{u\to+0} K(u)\nabla u = 0$, $\Omega_1 \cup \omega_2 \to \mathbf{R}^n$ as $\varepsilon \to 0$. (6.2)

will be called a solution of *formed wave* type.

Precisely as in Section 4.1, we shall seek the asymptotic solution of formed wave type, satisfying equation (6.1), in the form

$$u = \left(\sum_{i=0}^{N} W_i(\tau, x)\varepsilon^i\right)\Bigg|_{\tau=S/\varepsilon}, \qquad W_i(\tau,x) \in B^\alpha. \tag{6.3}$$

In contrast to the one-dimensional case considered above, in the multidimensional case, the function S has the form

$$S(x,t,\varepsilon) = \psi(x)(t + \varphi(x)) - \varepsilon\psi(x)\varphi^1(x), \tag{6.4}$$

where the functions $\psi(x)$, $\varphi(x)$ are infinitely differentiable. By definition of functions from the set B^α (see Chapter III, Section 3.1), the set $\{(x,t)\,|\,S(x,t,\varepsilon) = 0\}$ is the boundary of the support of an asymptotic solution (i.e., this set is the weak discontinuity front).

By definition of the function S, we have that, for $\varepsilon = 0$, the zero set of this function coincides with the set $\{(x,t)\,|\,\varphi(x) = -t\}$. Obviously, for a given function $\varphi(x)$, the position of the weak discontinuity front $\{\varphi(x) = -t\}$ is determined by the problem, before the solution is constructed. This leads to a contradiction. Thus it is natural to determine only the zero level surface $\Gamma_0 \subset \mathbf{R}^n$ of the function $\varphi(x)$, $\Gamma_0 = \{x\,|\,\varphi(x) = 0\}$, which corresponds to the definition of the weak discontinuity front of the asymptotic solution at the initial time $t = 0$.

At the initial instant of time it is assumed that the boundary of the solution support is a smooth submanifold $\Gamma_0 \subset \mathbf{R}^n$, $\dim \Gamma_0 = n - 1$.

The algorithm for constructing the asymptotic solution is given in the following

Theorem 6.1. *Let $k > 1$, $q > 1$, $F(u) > 0$ for $u \in (0,1)$. The problem (6.1), (6.2) has an asymptotic solution of the form (6.3), where the function W_0 satisfies the equation for simple waves:*

$$\begin{gathered} b\frac{dW_0}{d\tau} - \frac{d}{d\tau}\left(K(W_0)\frac{dW_0}{d\tau}\right) - F(W_0) = 0, \\ W_0(\tau)\Big|_{\tau=0} = 0, \qquad W_0(\tau)\big|_{\tau\to\infty} = 1 - 0. \end{gathered} \tag{6.5}$$

The function $S(x,t,\varepsilon)$ has the form

$$S(x,t,\varepsilon) = \psi(x)(t + \varphi(x)) - \varepsilon\psi(x)\varphi^1(x). \tag{6.6}$$

The function $\psi(x)$ *has the form*

$$\psi(x) = b_0\gamma^2(x, -\varphi(x)),$$

where b_0 *is the Zeldovich constant and the function* $\varphi(x)$ *is the solution of the problem*

$$|\nabla\varphi|^2 = \Phi(x, \varphi), \qquad \varphi\big|_{\Gamma_0} = 0, \tag{6.7}$$

where

$$\Phi(x, \varphi(x)) = \{b_0^2\gamma^2(x, -\varphi(x))\lambda(x, -\varphi(x))\}^{-1}.$$

The weak discontinuity front of the wave at time t is the level surface $\Gamma_t = \{x \mid \varphi(x) = -t\}$, *then* $\Omega_1 \subset \{x \mid \varphi(x) > t\}$, $\Omega_2 \subset \{x \mid \varphi(x) \leqslant -t\}$. *For the function* W_0 *the estimates hold:*

$$W_0(\tau) = O(\tau^\alpha), \qquad \tau \to 0, \qquad \alpha = \frac{1}{k-1},$$

$$W_0(\tau) = 1 - \Gamma\exp\Big\{-\frac{l\tau}{K(1)}\Big\} + o\Big(\exp\Big\{-\frac{l\tau}{K(1)}\Big\}\Big), \qquad \tau \to \infty,$$

$$l = -\frac{b_0}{2} + \sqrt{\frac{b_0^2}{4} + \left|\frac{dR}{d\theta}\right|_{\theta=1}}.$$

The functions W_1, $\varphi^1(x)$ *are defined in Lemma* 6.1.

Lemma 6.1. *The function* $W_1(x, \tau)$ *has the form*

$$W_1(\tau, x) = \frac{dW_0}{d\tau}\int_0^\tau V\Big(\frac{dW_0}{d\tau}\Big)^{-2}\int_0^\tau\Big(\frac{f}{VK(W_0)}\frac{dW_0}{d\xi}\,d\xi\Big)\,d\tau. \tag{6.8}$$

The functions V *and* f *are given by the relations*

$$V(\tau) = \frac{1}{K^2(W_0)}\exp\Big\{\int_0^\tau \frac{b_0}{K(W_0)}\,d\tau\Big\}, \tag{6.9}$$

$$\begin{aligned}
f = {} & \frac{1}{\psi(x)|\nabla\varphi|^2\lambda(x, -\varphi(x))}\Big\{\langle\nabla\lambda(x, -\varphi(x)), \psi\nabla\varphi\rangle K(W_0)\frac{dW_0}{d\tau} \\
& + \frac{\partial\lambda}{\partial t}\Big|_{t=-\varphi(x)}(\tau + \psi\varphi^1)\frac{d}{d\tau}\Big(K(W_0)\frac{dW_0}{d\tau}\Big)\psi|\nabla\varphi|^2 \\
& + 2\lambda(x, -\varphi(x))\frac{d}{d\tau}\Big(K(W_0)\frac{dW_0}{d\tau}\Big) \\
& \times \big\langle\nabla\varphi, \big[-\psi\nabla(\psi\varphi^1) + 2(\tau + \psi\varphi^1)\nabla\psi + \nabla(\psi\varphi^1)\big]\big\rangle \\
& - \lambda(x, -\varphi(x))\big(\langle\nabla, \psi\nabla\varphi\rangle + \langle\nabla\psi, \nabla\varphi\rangle\big)K(W_0)\frac{dW_0}{d\tau} \\
& - \frac{\partial\gamma^2}{\partial t}\Big|_{t=-\varphi(x)}(\tau + \psi\varphi^1)\frac{1}{\psi}F(W_0)\Big\}.
\end{aligned} \tag{6.10}$$

The following estimates hold:

$$W_1/W_0 = O(1), \qquad \tau \to 0,$$
$$W_1 = O\Big(\tau^2 \exp\Big(-\frac{|l_1|\tau}{K(1)}\Big)\Big), \qquad \tau \to \infty.$$

The function $\varphi^1(x)$ is defined by the equation:

$$\begin{aligned}
&\langle \nabla\lambda(x, -\varphi(x)), \psi\nabla\varphi\rangle I_2 + \psi|\nabla\varphi|^2 \frac{\partial\lambda}{\partial t}\Big|_{t=-\varphi(x)} (I_0 + I_1\psi\varphi^1) \\
&\quad + 2\lambda(x, -\varphi(x))[\langle\nabla\varphi, \nabla\psi\rangle(I_0 + I_1\psi\varphi^1) + \psi\langle\nabla\varphi, \nabla(\psi\varphi^1)\rangle I_1] \\
&\quad + \lambda(x, -\varphi(x))\langle\nabla, \psi\nabla\varphi\rangle I_2 + \frac{1}{\psi}\frac{\partial\gamma^2}{\partial t}\Big|_{t=-\varphi(x)} (I_3 + I_4\psi\varphi^1) = 0,
\end{aligned} \tag{6.11}$$

$$I_0 = \int_0^\infty \frac{\tau}{VK(W_0)} \frac{dW_0}{d\tau}\frac{d}{d\tau}\Big(K(W_0)\frac{dW_0}{d\tau}\Big)\,d\tau, \tag{6.12}$$
$$I_1 = \int_0^\infty \frac{1}{VK(W_0)} \frac{dW_0}{d\tau}\frac{d}{d\tau}\Big(K(W_0)\frac{dW_0}{d\tau}\Big)\,d\tau,$$
$$I_2 = \int_0^\infty \frac{1}{V}\Big(\frac{dW_0}{d\tau}\Big)^2 d\tau,$$
$$I_3 = \int_0^\infty \frac{\tau}{VK(W_0)} F(W_0)\,d\tau, \qquad I_4 = \int_0^\infty \frac{1}{VK(W_0)} F(W_0)\,d\tau.$$

Let us briefly prove Theorem 6.1.

Substituting the function u defined by (6.2) into equation (6.1) and setting the coefficients at ε^0 equal to zero, we get

$$\frac{\partial S}{\partial t}\frac{\partial W_0}{\partial \tau} - |\nabla S|^2\lambda(x,t)\frac{\partial}{\partial\tau}\Big(K(W_0)\frac{\partial W_0}{\partial\tau}\Big) - \gamma^2(x,t)F(W_0) = 0. \tag{6.13}$$

We expand the coefficients in (6.13) by the Taylor formula in a neighborhood of the point $t = -\varphi(x)$. For example,

$$\lambda(x,t) = \Big[\lambda(x, -\varphi(x)) + \varepsilon\frac{\tau + \psi\varphi^1}{\psi}\frac{\partial\lambda}{\partial t}\Big|_{\tau=-\varphi(x)} + O((\varepsilon\tau)^2)\Big]\Big|_{\tau=S(x,t,\varepsilon)/\varepsilon}.$$

Calculate the derivative of the function $S(x,t,\varepsilon)$:

$$\begin{aligned}
\frac{\partial S}{\partial t} &= \psi(x), \\
\nabla S &= \varepsilon(\tau/\psi + \varphi^1)\nabla\psi + \psi\nabla\varphi - \varepsilon\nabla(\psi\varphi^1) + O(\varepsilon^2), \\
|\nabla S|^2 &= \psi^2|\nabla\varphi|^2 + 2\varepsilon\psi\big[(\tau/\psi + \varphi^1)\langle\nabla\varphi, \nabla\psi\rangle - \langle\nabla\varphi, \nabla(\psi\varphi^1)\rangle\big] + O(\varepsilon^2), \\
\Delta S &= \langle\nabla, \psi\nabla\varphi\rangle + \langle\nabla\psi, \nabla\varphi\rangle + O(\varepsilon^2).
\end{aligned}$$

We substitute the obtained expressions into equation (6.13). By setting the sum of summands of the order ε^0 equal to zero, we obtain the following equation (the summands of the order ε^1 will be taken into account in the next approximation):

$$\psi(x)\frac{\partial W_0}{\partial \tau} - |\nabla\varphi|^2\psi^2\lambda(x, -\varphi(x))\frac{\partial}{\partial \tau}\left(K(W_0)\frac{\partial W_0}{\partial \tau}\right) - \gamma^2(x, -\varphi(x))F(W_0) = 0. \tag{6.14}$$

We assume that the relations

$$\frac{\psi(x)}{\gamma^2(x, -\varphi(x))} = b, \qquad \psi^2|\nabla\varphi|^2\psi^2\lambda(x, -\varphi(x)) = \gamma^2(x, -\varphi)$$

hold, then we get an ordinary differential equation for the function W_0:

$$b_0\frac{\partial W}{\partial \tau} - \frac{\partial}{\partial \tau}\left(K(W_0)\frac{\partial W_0}{\partial \tau}\right) - F(W_0) = 0. \tag{6.15}$$

Denote by $W_0(\tau)$ the monotone nonnegative solution of equation (6.15), such that $W_0 = 0$ for $\tau \leqslant 0$, $W_0 \in [0, 1]$,

$$\begin{aligned} &W_0(\tau) \in C^\infty(]0, \infty[), \\ &W_0 = 1 - O\left(\exp\left(-\frac{l\tau}{K(1)}\right)\right), \qquad \tau \to \infty. \end{aligned}$$

Thus, for the function $\varphi(x)$, we obtain the equation

$$|\nabla\varphi|^2 = \Phi(x, \varphi),$$

whose right-hand side $\Phi(x, \varphi)$ is known. The general setting of the problem about the weak discontinuity position implies that, for $t - 0$, the front of weak discontinuity of Γ_0 is known and determined as the zero set of the function $\varphi(x)$: $\Gamma_0 = \{x \mid \varphi(x) = 0\}$. One must determine the level surfaces $\Gamma_t = \{x \mid \varphi(x) = t\}$ of the function $\varphi(x)$, satisfying (6.7), when the initial surface Γ_0 is fixed.

Assume that a smooth surface Γ_0 is given by parametrical equations:

$$\Gamma_0 = \{x = x(\zeta), \zeta \in \Omega \subset \mathbf{R}^{n-1}\}.$$

The condition $\dim \Gamma_0 = n - 1$ implies that $\operatorname{rank} \partial x/\partial\zeta = n - 1$.

Recall that, to solve equation (6.7), one must solve the system of equations

$$\begin{aligned} &\frac{dx_i}{ds} = 2p_i, \\ &\frac{dp_i}{ds} = \frac{\partial \Phi}{\partial \varphi}p_i + \frac{\partial \Phi}{\partial x_i}, \qquad i = 1, \ldots, n, \\ &\frac{d\widetilde{\varphi}}{ds} = 2|p|^2, \end{aligned} \tag{6.16}$$

where

$$p = p(s,\zeta), \qquad x = x(s,\zeta), \qquad \widetilde{\varphi} = \widetilde{\varphi}(s,\zeta), \qquad \Phi = \Phi(x(s,\zeta), \varphi(s,\zeta)). \quad (6.17)$$

We have the following initial conditions for system (6.16):

$$x_i\big|_{s=0} = x_i(\zeta), \qquad p_i\big|_{s=0} = p_i(\zeta), \qquad \varphi\big|_{s=0} = 0. \quad (6.18)$$

The solution of problem (6.7) has the form

$$\varphi(x) = \widetilde{\varphi}(s(x), \zeta(x)), \quad (6.19)$$

where $(s(x), \zeta(x))$ is the solution of the system:

$$x(s,\zeta) = x.$$

This solution exists if the Jacobian $\partial x/\partial \zeta$ does not vanish. In our case this conditions is:

$$\begin{pmatrix} p_1 & \dots & p_n \\ \partial x_1/\partial \zeta_1 & \dots & \partial x_n/\partial \zeta_1 \\ \dots & \dots & \dots \\ \partial x_1/\partial \zeta_{n-1} & \dots & \partial x_n/\partial \zeta_{n-1} \end{pmatrix} \neq 0. \quad (6.20)$$

From the geometrical viewpoint, this means that the projections of trajectories of (6.16) on $\mathbf{R}^n_x$ must not intersect in a neighborhood of the surface Γ_0, in which the solution to problem (6.7) exists. Let us verify that the condition (6.19) holds for $s = 0$. Actually, the left null-vector $Y = (y_1, \dots, y_n)$ of the matrix on the left-hand side of (6.20) (if this vector exists) must satisfy the system of equation:

$$\sum_{i=1}^{n} p_i y_i = 0, \qquad Y^t \frac{\partial x}{\partial \zeta} = 0. \quad (6.21)$$

However, by the latter equation in (6.21) and the condition $\operatorname{rank}(\partial x/\partial \zeta) = n - 1$, we have $Y = cP$, where $c \neq 0$. Then, by (6.7) and $\Phi(x, \varphi) \neq 0$, the first relation in (6.21) cannot be satisfied.

Thus the solution to problem (6.7) exists in a certain neighborhood of the surface Γ_0.

Note that even in the case of constant coefficients in (6.1), the solution to problem about a weak discontinuity propagation can be nontrivial, since the surface Γ_0 can be of a complicated form, for example, not axially (spherically) symmetric. (In contrast to the one-dimensional case, the solution can be not self-similar because of the curvature of the weak discontinuity front.)

Denote by W_1 the solution of the boundary-value problem:

$$\psi(x)\frac{\partial W_1}{\partial \tau} - \psi^2(x)|\nabla\varphi|^2\lambda(x,-\varphi)\frac{\partial^2}{\partial \tau^2}(K(W_0)W_1) - \gamma^2(x,-\varphi(x))\frac{\partial F}{\partial W_0}W_1 = f,$$

$$W_1\big|_{\tau=0} = 0, \qquad W_1\big|_{\tau\to+\infty} \to 0,$$

where f is defined by formula (6.10). (This equation is obtained by substituting (6.3) into (6.1) and setting the coefficient at ε equal to zero.) The solution has the form:

$$W_1(\tau,x)=\frac{-\partial W_0/\partial\tau}{\psi(x)|\nabla\varphi(x)|^2\lambda(x,-\varphi)}\times\int_0^{\tau}\left(\int_0^{\tau}\frac{f}{VK(W_0}\frac{\partial W_0}{\partial\tau}\,d\tau\right)V\left(\frac{\partial W_0}{\partial\tau}\right)^{-2}d\tau. \quad (6.22)$$

The function (6.22) is examined in the same way as in Section 4.1. The inner integral in (6.22) converges and this is the necessary condition for the decreasing solution W_1 to exist:

$$\int_0^{\infty}\frac{f}{VK(W_0)}\frac{\partial W_0}{\partial\tau}\,d\tau=0.$$

This yields equation (6.11) for the function φ^1. Theorem 6.1 and Lemma 6.1 are proved.

Now let us formulate the Huygens principle for formed waves (known in the physical literature [104]).

The projection on $\mathbf{R}^n_x$ of the trajectory $(x(t),p(t),\tilde{\varphi}(t))$ of the system of ordinary differential equations (6.16), such that $x(0)=x_0$, is called a *ray* starting from the point $x_0\in\mathbf{R}^n$.

The set of endpoints of rays starting from x_0, where the function φ takes a fixed value β_0, is called the *wave-front* $\Gamma_{x_0\beta_0}$.

Huygens' principle (for localized formed waves) [14]. The wave-front $\Gamma_{t+\Delta}$ is the envelope of wave-fronts $\Gamma_{x,-\Delta}$ for x running along the front Γ_t.

For the solution constructed, the width of the region, where the solution varies from 0 to 1, is of the order ε. As $\varepsilon\to 0$, our solution converges pointwise to the θ-function (the Heaviside function) with jump on the surface of the weak discontinuity front of the asymptotic solution, i.e., the limiting (as $\varepsilon\to 0$) solution is a shock wave. However, in our case, in contrast to the equation of fluid or gas dynamics with small viscosity ε, there is no differential equation for the limiting function for $\varepsilon=0$.

Let us discuss different types of solutions. Consider the simplest case

$$\Phi(x,\varphi)\equiv 1,\qquad \Gamma_0=\{|x|=R_1\},\qquad n=2.$$

Then the solution to problem (6.7) has the form $\varphi_+=|x|-R_1$ or $\varphi_-=R_1-|x|$, and the solution to problem (6.1) has the form:

$$u_-=W_0\Big(\frac{R_1-|x|+t}{\varepsilon}\Big),\qquad u_+=W_0\Big(\frac{t+|x|-R_1}{\varepsilon}\Big),\qquad R_1=\text{const},$$

where W_0 is the solution to problem (6.5) (for $\Phi(x,\varphi)\equiv 1$). The solution u_- is equal to zero for $|x|>R_1+t$, and tends exponentially to 1 for $|x|<R_1+t+O(\varepsilon^{1-\delta})$, $\delta>0$. The solution u_+ is equal to zero in the ball $|x|<R_1-t$. Note that, for $x=0$, the solution u_- is not smooth. However (see above), for $O(\varepsilon^{1-\delta})<|x|<R_1+t+\gamma\varepsilon^{1-\delta}$ $(\gamma,\delta=\text{const}>0)$, the solution u_- can be estimated as $u_-=1-O(\varepsilon^N)$, $\forall N\in\mathbb{Z}$.

Therefore, in a small neighborhood of the point $x = 0$, the function u_- can be replaced by a constant $u \equiv 1$, using the partition of unity as in Section 3.2 of Chapter III.

Another solution has the form

$$u = W_0\Big(\frac{|x| - R + t}{\varepsilon}\Big)W_0\Big(\frac{-|x| + R + t - t_0}{\varepsilon}\Big).$$

Similar solutions exist in the general case when the function $\Phi(x, \varphi)$ is defined by (6.7). Then the motion of the wave-front is determined by the medium properties and the fact that such solutions exist means that the class of such solutions is stable with respect to perturbations of the external medium.

Let us expose the way for constructing solutions generalizing the solutions u_+ and u_-. Let the surface $\Gamma_0 = \{x \mid \varphi(x) = 0\}$, which is the weak discontinuity front at the initial instant of time, be given. Recall that the surface Γ_0 is supposed to be smooth, of dimension $n-1$ ($x \in \mathbf{R}^2$). Assume that the surface Γ_0 is closed. Let

$$\Gamma_0 = \{x = x(\zeta),\ \zeta \in \Omega \subset \mathbf{R}^{n-1}\}.$$

As mentioned above, to construct any asymptotic solution satisfying equation (6.1) in a neighborhood of the surface Γ_0, one must construct a function $\varphi(x)$, such that the zero set of this function coincides with Γ_0 and

$$|\nabla\varphi|^2 = \Phi(x, \varphi).$$

An algorithm for solving this problem was presented above. Note that the choice of the initial conditions for the vector p in system (6.21) is not unique. More precisely, if the vector $p(\zeta)$ satisfies equations (6.21), then the vector $-p(\zeta)$ also satisfies these conditions.

By construction, the vector $p(\zeta)$ does not belong to the plane tangent to Γ_0 at the point $x(\zeta)$. Hence, one of the vectors p or $-p$ is directed inside the surface Γ_0. Denote this vector by $P_0(\zeta)$. The function $u = W_0\big(\frac{t+\varphi(x)}{\varepsilon}\big)$, where $\varphi(x)$ is constructed according to algorithm (6.16)–(6.19) (the initial condition in (6.18) has the form $p_i\big|_{s=0} = (P_0(\zeta))_i$, $i = 1, \dots, n$), is the generalization of the solution $u_- = W_0\big(\frac{R_1 - |x| + t}{\varepsilon}\big)$ to the case of arbitrary variable coefficients.

Similarly, taking the appropriate components of the vector $-P_0(\zeta)$ as the initial conditions for p_i, we obtain an analog of the solution u_+. The evolution of the weak discontinuity front in other situations was studied in [91].

CHAPTER V

FINITE ASYMPTOTIC SOLUTIONS OF DEGENERATE EQUATIONS

5.1. An example of constructing an asymptotic solution

The asymptotic solutions to semilinear parabolic equations constructed in the preceding chapter are similar to the asymptotic solutions describing shock waves in low-viscous media or soliton-like waves in weakly dispersive media. Such solutions have been already discussed in physical and mathematical literature. In this chapter we construct asymptotic solutions with compact support that, at interior points of the support, increase infinitely as a small parameter ε tends to zero. Such solutions have no analogs in the theory of wave processes. An algorithm for constructing such solutions was first published in [87] (see also [51, 23].

An example of structures, whose evolution is described by solutions with such properties, is the temperature distribution of a heated gas, established as the result of the heat explosion (the heating of some gas by an intense source). At certain stages of this process the convective motion of gas may be neglected [10, 42, 75]. In this case the temperature distribution is determined by the radiative heat transfer. The mathematical simulation of such processes, under the assumption that the internal energy is proportional to the temperature [10, 42, 75], leads to semilinear heat transfer equations. Another example of structures, whose evolution can be described by the method presented in this chapter, is a liquid drop on a hydrophylic surface.

First let us consider an equation, whose solution possesses such properties, namely, the equation

$$\varepsilon\frac{\partial u}{\partial t} - \varepsilon^2\frac{\partial}{\partial x}\Big(K(u,t)\frac{\partial u}{\partial x}\Big) = 0, \qquad u \geq 0, \quad x \in \mathbf{R}, \quad t \in \mathbf{R}_+, \tag{1.1}$$

$$K(u,t) = \frac{u^{k-1}}{1+\rho_\infty(t)u^{k-1}}, \qquad \rho_\infty(t) = 4\Big[\frac{k+1}{2}(t+1)\Big]^{(k-1)/(k+1)},$$

where $0 < \varepsilon \ll 1$ is a small parameter and $k > 1$. The exact solution of (1.1) has the form (for other examples, see [13]):

$$u(x,t,\varepsilon) = W_0(t,\tau)\big|_{\tau=S_0(x,t)/\varepsilon},$$

$$W_0(t,\tau) = \begin{cases} \Big[\dfrac{e^{\tau/\alpha}-1}{\rho_\infty(t)}\Big]^\alpha, & \tau \geq 0, \\ 0, & \tau < 0, \end{cases} \qquad \alpha = \frac{1}{k-1}, \tag{1.2}$$

$$S_0 = c^2 - \frac{x^2}{\varphi(t)}, \qquad \varphi(t) = \Big[\frac{1+k}{2}(t+1)\Big]^{2/(1+k)}, \qquad c = \text{const} > 0.$$

This solution is shown in Figure 5.1. Curves 1, 2, 3 correspond to times $t = 1, 2, 3$ respectively; $\varepsilon = 1$, $k = 2$, $c = 1$. The coordinates of the front $x_f(t)$ are the roots of the equation $S_0(x_f, t) = 0$.

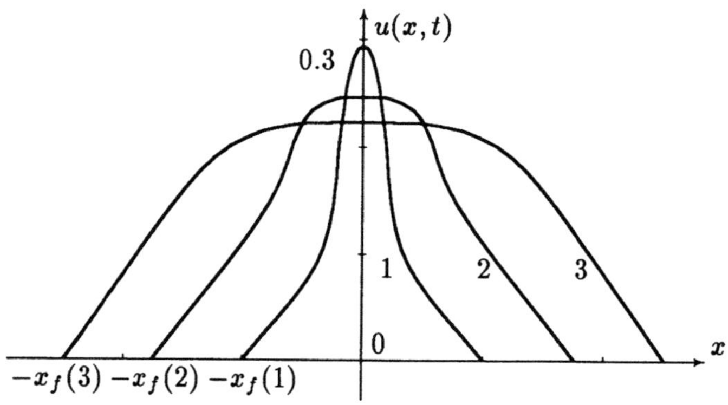

FIGURE 5.1

In this example, unlike the problems considered in the preceding chapter, the equation $S_0(x_f, t) = 0$ has two solutions, while in the multidimensional case considered below ($x \in \mathbf{R}^n$) the position of the front is determined by the zero set of the function S_0 (a closed surface of dimension $n - 1$). At interior points of the support, the function S_0 is strictly positive. Therefore, the function S_0 is bounded on the support of the solution, and there exist points, at which this function takes its maximum.

Here we wish to point out the major contribution of A. A. Samarsky and S. P. Kurdyumov, and their colleagues [43, 63–68] to the investigation of such solutions by numerical and qualitative methods.

In models of actual thermal and diffusion processes the parameter ε, arising when we pass to dimensionless variables, is small, but fixed. Therefore, in models with actual parameters, the asymptotic solutions of the form (1.2) turn out to be bounded in absolute value. However, when asymptotics are constructed within mathematical models, the values of ε are taken arbitrarily small while, consequently, the values of the unknown function u become arbitrarily large. In the above example, the transport coefficient $K(u, x, t)$ is uniformly bounded for any $u \geq 0$ and, as $u \to \infty$, tends to the limit in C^∞. There is the question: to what extent do these properties of the function $K(u, x, t)$ satisfy the problem of heat and mass transfer? Experimental data [74, 75] show that, for a medium in a single aggregate state (liquid, gas, plasma), starting from a certain sufficiently large value, the transport coefficient is independent of the transported quantity (temperature, concentration). The behavior of actual transport coefficients is shown in Figure 5.2.

As $u \to \infty$, there are two possibilities for the source-sink function $F(u, x, t)$: 1) $F(u, x, t) \to \gamma(x, t)u$, 2) $F(u, x, t) \to \gamma(x, t)$, (the function $\gamma(x, t)$ can, for example, vanish identically) (see Figure 5.3).

In this chapter the main attention is paid to the construction of asymptotic solutions for equations, in which the transport coefficient $K(u, x, t)$ and the source-sink

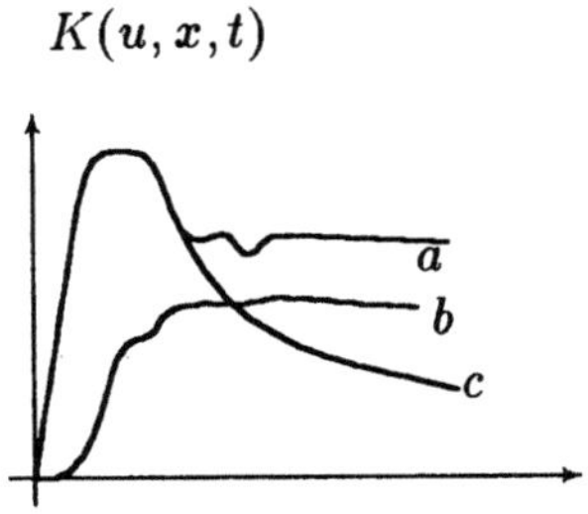

FIGURE 5.2

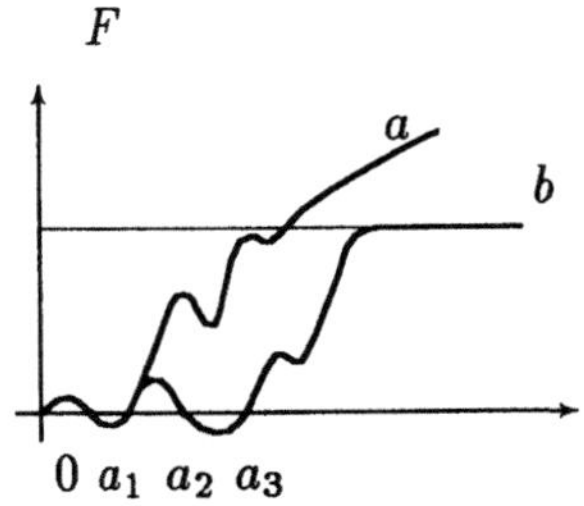

FIGURE 5.3

function $F(u,x,t)$ possess properties described above. In this case, the chain of necessary conditions, which follow from the existence condition for a solution belonging to $\mathfrak{U}^{\alpha}_{\beta}$, is a recurrent system of equations. By solving these equations in succession, one can construct an asymptotic solution belonging to $\mathfrak{U}^{\alpha}_{\beta}$ and satisfying the equation up to any precision with respect to ε. In particular, it turns out that the function $S(x,t,0)$ satisfies the Hamilton–Jacobi equation, i.e., the characteristic equation for the corresponding limiting (for large values of u) linear parabolic equation with a small parameter at the derivatives.

In this chapter equation (1.1) is taken as an example to explain the idea of how asymptotic solutions are constructed. We set $k=2$, then $\alpha=1$ and the solution $W_0(t,\tau)$ takes the form

$$W_0(t,\tau)=\begin{cases}(e^{\tau}-1)/\rho_\infty(t), & \tau\geq 0,\\ 0, & \tau<0.\end{cases}$$

Note that the function $W_0(t,\tau)$ can be written as follows

$$W_0(t,\tau)=\frac{e^{\tau}}{\rho_\infty(t)}(1-e^{-\tau})=\frac{e^{\tau}}{\rho_\infty(t)}M(\tau). \tag{1.3}$$

Substituting (1.3), where $\tau=S/\varepsilon$, into (1.1), we obtain the equation for the function $M(S/\varepsilon)$:

$$\begin{aligned}\frac{1}{\rho_\infty}\frac{\partial S}{\partial t}M+\frac{\varepsilon}{\rho_\infty}\frac{\partial M}{\partial t}-\frac{1}{\rho_\infty^2}\left(\frac{\partial S}{\partial x}\right)^2\widehat{K}(M)M-\frac{\varepsilon}{\rho_\infty^2}\frac{\partial S}{\partial x}K(M)\frac{\partial M}{\partial x}\\ -\frac{\varepsilon}{\rho_\infty^2}\frac{\partial S}{\partial x}\frac{\partial}{\partial x}(\widehat{K}(M)M)-\frac{\varepsilon^2}{\rho_\infty^2}\frac{\partial}{\partial x}\left(\widehat{K}(M)\frac{\partial M}{\partial x}\right)\\ +\varepsilon M\frac{\partial\rho_\infty^{-1}}{\partial t}-\frac{\varepsilon}{\rho_\infty^2}\widehat{K}(M)M\left(\frac{\partial^2 S}{\partial x^2}\right)=0,\quad \widehat{K}(M)=M.\end{aligned} \tag{1.4}$$

One can see that in this example the function $S(x,t)$ satisfies the equation

$$\frac{\partial S}{\partial t}-\frac{1}{\rho_\infty}\left(\frac{\partial S}{\partial x}\right)^2=0, \tag{1.5}$$

while the function ρ_∞ is taken so that

$$\frac{\partial \rho_\infty^{-1}}{\partial t} = \frac{S_{xx}}{\rho_\infty^2}. \tag{1.6}$$

Note that, by (1.5), (1.6), equation (1.4) has the solution $M \equiv 1$ and can be rewritten in the form

$$\varepsilon\frac{\partial M}{\partial t} - \frac{\varepsilon^2}{\rho_\infty}\frac{\partial}{\partial x}\left(\widehat{K}(M)\frac{\partial M}{\partial x}\right) - \frac{\varepsilon}{\rho_\infty}\left(\frac{\partial S}{\partial x}\right)\widehat{K}(M)\frac{\partial M}{\partial x} - \frac{\varepsilon}{\rho_\infty}\frac{\partial}{\partial x}(\widehat{K}(M))$$
$$+ \frac{\partial S}{\partial t}(M - \widehat{K}(M)M) + \frac{\varepsilon}{\rho_\infty}\left(\frac{\partial^2 S}{\partial x^2}\right)(M - \widehat{K}(M)M) = 0. \tag{1.7}$$

The solution of this equation is known. However, for further considerations, it is important to note that the substitution of a solution of the form (1.3) into (1.1) leads to equation (1.7) for the "pre-exponential" function M. This equation is similar to those, for which the method of constructing asymptotic solutions was developed in Chapter IV. We shall consider the solution of equation (1.7) in detail. By (1.3), it has the form $M(S/\varepsilon)$, where

$$M(\tau) = \begin{cases} 1 - e^{-\tau}, & \tau \geq 0, \\ 0, & \tau < 0, \end{cases} \qquad S(x,t,) = c - \frac{x^2}{\varphi(t)}. \tag{1.8}$$

We note that this solution is similar to asymptotic solutions with compact supports, for which the construction algorithm was developed in Section 4.2. We recall that an asymptotic solution with compact support is constructed as the product of two localized bounded asymptotic solutions with semibounded supports (see Example 2.2 and Figure 3.2 in Chapter III). In our example, the function $S(x,t)$ is a quadratic polynomial in x, therefore the solution $M(S/\varepsilon)$ cannot be presented as a product. But this is not an obstacle to formulate the algorithm for constructing asymptotic solutions to equation (1.4). In our case, this algorithm gives the exact solution, but below it is applied to equations, for which the formulas of exact solutions are unknown.

Since the solution of equation (1.4) belongs to B^1, the function $S(x,t)$ can be represented as follows

$$S(x,t) = \beta(x + \psi(t)) + \beta_1(x + \psi(t))^2 \tag{1.9}$$

(the explanations given in the Introduction and the results obtained in Chapter IV imply that to construct solutions with precision up to $\bmod O(\varepsilon^2)$, it is sufficient to take only two terms of the Taylor expansion of the function $S(x,t)$).

By (1.5), we obtain the relations for ψ, β, β_1:

$$\frac{d\psi}{dt} = \frac{\beta}{\rho_\infty}, \qquad \frac{d\beta}{dt} = \frac{2\beta\beta_1}{\rho_\infty}. \tag{1.10}$$

Following Chapter IV, we shall look for the function $M = M(\tau,t,\varepsilon)\big|_{\tau=S/\varepsilon}$ (for the solution of (1.4)) in the form

$$M = M_0 + \varepsilon M_1 + \varepsilon^2 M_2. \tag{1.11}$$

Here equation (1.7) is different. Namely, by substituting the solution in the form (1.3) with M defined by (1.11) into (1.1), we obtain equation (1.7), where

$$\widehat{K}(M) = \frac{M_0 + \varepsilon M_1}{1 + \varepsilon M_1} = M_0 + \varepsilon M_1(1 - M_0) + \varepsilon^2 M_2(1 - M_0) + \varepsilon^2 M_1^2 + O(\varepsilon^3). \quad (1.12)$$

In other words, there are new summands of the order ε and ε^2 in (1.7). Like in Chapter IV, the following relations hold:

$$\begin{gathered} x + \psi(t) = \frac{\varepsilon\tau}{\beta} + \frac{\varepsilon^2\tau^2\beta_1}{\beta^3} + O(\varepsilon^3)\Big|_{\tau=S/\varepsilon}, \\ \frac{\partial S}{\partial t} = \beta\frac{d\psi}{dt} + \left(\frac{d\beta}{dt} + 2\beta\frac{d\psi}{dt}\right)\frac{\varepsilon\tau}{\beta} + O(\varepsilon^2)\Big|_{\tau=S/\varepsilon}, \\ \left(\frac{\partial S}{\partial x}\right)^2 = \beta^2 + 4\beta_1\varepsilon\tau + O(\varepsilon^2)\Big|_{\tau=S/\varepsilon}, \qquad \frac{\partial^2 S}{\partial x^2} = 2\beta_1 + O(\varepsilon)\Big|_{\tau=S/\varepsilon}. \end{gathered} \quad (1.13)$$

By substituting the expansion (1.11) into the new equation (1.7), precisely as in Chapter IV, we obtain at ε^0:

$$\beta\frac{d\varphi}{dt}\frac{dM_0}{d\tau} - \frac{\beta^2}{\rho_\infty}\frac{d}{dt}\left(M_0\frac{dM_0}{d\tau}\right) - \frac{\beta^2}{\rho_\infty}M_0\frac{dM_0}{d\tau} - \frac{\beta^2}{\rho_\infty}\frac{d}{d\tau}M_0^2 + \beta\frac{d\psi}{dt}M_0(1 - M_0) = 0$$

or, taking into account (1.10),

$$\frac{dM_0}{dt} - \frac{d}{d\tau}\left(M_0\frac{dM_0}{d\tau}\right) - 3M_0\frac{dM_0}{d\tau} + M_0(1 - M_0) = 0. \quad (1.14)$$

The solution of equation (1.14) has the form

$$M_0 = \begin{cases} 1 - e^{-\tau}, & \tau \geq 0, \\ 0, & \tau < 0. \end{cases} \quad (1.15)$$

By setting the terms of the order ε equal to zero, we get the equation for the function M_1:

$$\begin{aligned} LM_1 = {} & \frac{\partial M_1}{\partial\tau} - 2\frac{\partial}{\partial\tau}(M_0M_1) - \frac{\partial}{\partial\tau}\left(M_0\frac{\partial M_1}{\partial\tau}\right) - \frac{\partial}{\partial\tau}\left(M_1(1 - M_0)\frac{dM_0}{d\tau}\right) \\ & - M_1(1 - M_0)\frac{\partial M_0}{\partial\tau} - M_0\frac{\partial M_1}{\partial\tau} + \frac{\partial}{\partial\tau}(M_1M_0^2) + M(1 - M_0)^2 \\ = {} & -\frac{\rho_\infty}{\beta}F, \end{aligned} \quad (1.16)$$

where

$$\begin{aligned} F = {} & M_0^\alpha\left[\frac{\tau}{\rho_\infty\beta}\left(\frac{d\beta}{dt} + 2\beta_1\frac{d\psi}{dt}\right) + \frac{\partial\rho_\infty^{-1}}{\partial t}\right] \\ & - 4\frac{\beta_1\tau}{\rho_\infty^2}\left[3M_0\frac{\partial M_0}{\partial\tau} + \frac{\partial}{\partial\tau}\left(M_0\frac{\partial M_0}{\partial\tau}\right) + M_0^2\right] - \frac{2\beta_1}{\rho_\infty^2}M_0\frac{\partial M_0}{\partial\tau} \\ & + \frac{\tau}{\rho_\infty\beta}\left(\frac{d\beta}{dt} + 2\beta_1\frac{d\psi}{dt}\right)\frac{\partial M_0}{\partial\tau} - \left(\frac{\partial\rho_\infty^{-1}}{\partial t} - \frac{2\beta_1}{\rho_\infty^2}\right)\left(M_0 + \frac{\partial M_0}{\partial\tau}\right). \end{aligned}$$

Substituting the derivative $d\beta/dt$ from (1.10) into the formula for F, we obtain

$$F \equiv 0, \qquad \text{if} \quad \frac{\partial \rho_\infty^{-1}}{\partial t} = \frac{2\beta_1}{\rho_\infty^2}.$$

It can be shown that nontrivial solutions of (1.16) with zero right-hand side do not satisfy the boundary conditions

$$M_1\big|_{\tau=0} = 0, \qquad M_1 \to 0 \quad \text{as} \quad \tau \to \infty.$$

Therefore, here, $M_1 \equiv 0$.

Consequently, in this example, the solvability conditions for M_i $(i = 0, 1)$ do not imply any additional condition that would permit to close the system (1.10).

This condition follows from (1.6) and has the form $\beta_t = 2\beta\beta_1/\rho_\infty$. By solving this system of conditions, we determine the functions ψ, β and β_1. Comparing (1.9) with formula (1.2) for the function S, we can obtain formulas for these functions, but we do not write them here.

Let us describe the algorithm, considered in our example, in the general situation. In the case when the coefficients of the equation become stable as $u \to \infty$, we represent the solution $u \in \mathfrak{U}_{\varepsilon^r}^{\alpha}$ in the form

$$u(x,t,\varepsilon) = \exp\left\{\frac{S(x,t,\varepsilon)}{\varepsilon}\right\} M(x,t,\varepsilon), \tag{1.17}$$

where $S \geq 0$, $0 \leq M \leq 1$. On the intersection Ω of the domain $S > 0$ with the support of M, the function $u(x,t,\varepsilon)$ increases infinitely as $\varepsilon \to 0$. Therefore the function $S(x,t,\varepsilon)$ is defined by a linear equation so that, for $(x,t) \in \Omega$, the discrepancy (arising after the substitution of (1.17) into the equation) has the form $O(\varepsilon^2)e^{S/\varepsilon}$ and the function $M(x,t,\varepsilon) \equiv 1$ satisfies the equation for M, obtained by eliminating the exponential function from the initial equation.

Then the equation for M_i turns out to be similar to that considered in Section 4.1, and its asymptotic solution can be constructed by applying the methods of Chapter IV.

Note that here some difficulties arise. The equation for $S(x,t,\varepsilon)$ (the Hamilton–Jacobi equation) was obtained independently of the asymptotic expansion for M. Consequently, the surface $\Gamma_\varepsilon = \{(x,t) \mid S(x,t,\varepsilon) = 0\}$ is determined uniquely by the initial conditions for $S(x,0,\varepsilon)$. But, constructing the asymptotic solution of equation for M, we can also determine the front of the weak discontinuity. Obviously, this front must coincide with the surface Γ_ε. Thus, there is a problem of matching. To explain how it can be solved, we turn, once again, to the example considered at the beginning of this section.

Recall that $S = S(x,t) = c - x^2/\varphi(t)$. Curve 1 in Figure 5.4 shows the surface (point) $\Gamma_\varepsilon = \{(x,t) \mid S(x,t,\varepsilon) = 0\}$. The family of curves 2 shows the projections of trajectories of the Hamilton system corresponding to the Hamilton–Jacobi equation (1.5) on the (x,t)-plane. The projections of trajectories of the Hamilton system,

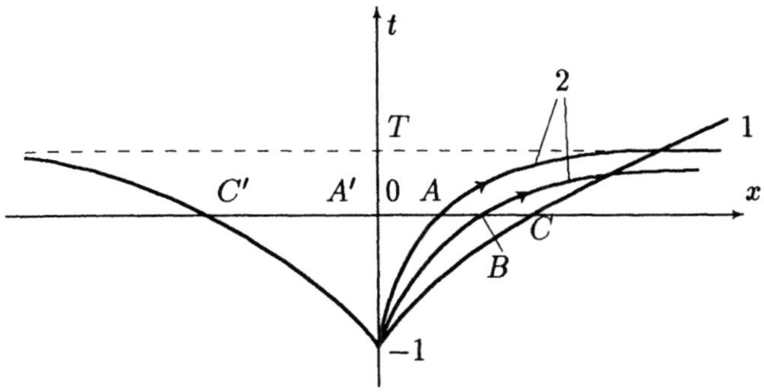

FIGURE 5.4

crossing the x-axis to the left of the point A, intersect curve 1 for $t > T$ (we shall consider only the right-hand half of the figure, symmetric with respect to the t-axis).

If the solution of the problem were a priori unknown, the function $S(x,t)$ could be obtained as follows. Having constructed the solution of equation (1.7), we determine the curves Γ. Setting $S = 0$ on Γ, we solve the boundary-value problem for the Hamilton–Jacobi equation (1.5). Fixing the value of $T > 0$, we determine the domain in the x-space (i.e., for $t = 0$), where the function $S(x,0)$ is determined by the solution of the boundary-value problem (the segments $[A, C]$ and $[C', A']$ in Figure 5.4). Next, we extend the function $S(x,0)$ over the interval $[C', C]$ arbitrarily, but in a smooth way, so as to obtain a smooth positive function. For example, for $x \geq A - \delta$ ($\delta > 0$ is a small number), this can be done by calculating the values of $S(x,0)$ from the solution of the boundary-value problem and using of the partition of unity. Obviously, to establish this extension, it is necessary for the function $S(x,0)$, defined by the boundary-value problem, to be positive inside the intervals $[A, C]$ and $[C', A']$.

We recall that the solution $S(x,t)$ of the boundary-value problem is given by the formula

$$S(x,t) = \int_{\gamma}^{(x,t)} L\,dt, \tag{1.18}$$

where the integral is taken along the projection of the trajectory starting from the point $\gamma \in \Gamma$ and coming to the point (x,t), $L = pH_p - H$ is the Lagrangian calculated in terms of the Hamiltonian $H(p,x,t)$. If $H = p_0 - A(x,t)p^2$, $A > 0$, then

$$L = -Ap^2 < 0 \tag{1.19}$$

(here $p_0 \leftrightarrow \partial S/\partial t$).

Curve 1 is given by the equation $x = \psi(t)$ and, by (1.10), $d\psi/dt > 0$. The Hamilton system, corresponding to the function $H = p_0 - Ap^2$, has the form

$$\begin{aligned} &\dot{t} = \frac{\partial H}{\partial p_0} = 1, && \dot{x} = \frac{\partial H}{\partial p} = -2Ap \\ &\dot{p}_0 = -\frac{\partial H}{\partial t} = \frac{\partial A}{\partial t}p^2, && \dot{p} = -\frac{\partial H}{\partial x} = \frac{\partial A}{\partial x}p^2, \qquad x\Big|_{\xi_j=0} = x_0. \end{aligned} \tag{1.20}$$

Here ξ is a parameter on the trajectory, while the initial data for the momenta p_0^0, p^0 are defined by the system of equations

$$p_0^0 - A(x_0, t_0)\left(p^0\right)^2 = 0, \qquad \frac{d\Psi}{dt}p^0 + p_0^0 = 0. \tag{1.21}$$

The last equation is equivalent to the relation $dS(\psi(t), t)/dt = 0$. By (1.21), the condition $d\psi/dt > 0$ implies

$$p^0 = \frac{1}{A}\frac{d\Psi}{dt} < 0.$$

Consequently, on Γ, we have

$$\dot{x} > 0.$$

Therefore, since the point $\xi = 0$ on the trajectory corresponds to the point γ on Γ (for example, let the point γ coincide with point D, see Figure 5.4), there is a certain value $\xi_1 < 0$ corresponding to the point of intersection of this trajectory with the x-axis (point B in Figure 5.4).

By (1.19), we calculate the integral in (1.18) along the segment DB of the trajectory within the limits $\xi_1 \leq \xi \leq 0$ and obtain $S(B, 0) > 0$. Similarly, one can consider the function $S(x, 0)$ for $x \in [C', A']$, taking into account that the curve symmetric to curve 1 is given by the equation $x = -\psi(t)$.

Note that we have obtained an important property of the function $S(x, 0)$ satisfying the boundary-value problem. Namely, $S(x, 0)$ increases, as x varies from C to A. Indeed, the system (1.20) implies

$$P = -\int_0^{\xi} A\, d\xi' + p^0,$$

and, consequently,

$$S = -\int_0^{\xi}\left(p^0 - \int_0^{\xi'} A\, d\xi''\right)^2 d\xi'.$$

Above we saw that negative values of ξ correspond to points on the x-axis. Therefore, since $A > 0$ and the motion along the x-axis from point C to point A means the increase of $|\xi|$, the required statement follows from the last formula. Thus, the smooth extension of $S(x, 0)$ onto the whole interval $[C', C]$ has an extremum inside the interval $[A', A]$.

Note that, if $H = p_0 - Ap^2$, the solution $S(x, t)$ of the Cauchy problem for equation (1.5) in the domain $\{x \mid S \geq 0\}$ depends only on the values of $S(x, 0)$ in the domain $\{(x, t) \mid S(x, 0) \geq 0\}$. Indeed, in Lagrangian coordinates, the general solution of the Cauchy problem for the Hamilton–Jacobi equation has the form

$$S(x_0, t) = S(x_0, t) + \int_0^t L\, dt',$$

where L is the Lagrangian. In our case $L < 0$, therefore the trajectory, on which the inequality $S(x_0, t) \geq 0$ holds for $t \leq t_0$, must start from a point x_0 such that $S(x_0, 0) > 0$. Thus, by fixing a smooth nonnegative function as the initial condition for equation (1.5) on the interval $[C', C]$, we can construct the solution for all $t > 0$. However, in general, (see Figure 5.4), for $t > T$, the solution S of the Cauchy problem may not satisfy the boundary condition $S|_\Gamma = 0$. Clearly, for the Hamiltonian $-Ap^2$, it is possible to choose the initial data so that the solution of the Cauchy problem satisfy the boundary condition for $t \le O(\varepsilon^{-1})$. (Here we mean that the curves Γ are taken precisely as in the preceding example, while the initial condition does not coincide with the condition that follows from the exact solution, given at the beginning of this section).

In the next section the algorithm considered here is used for the construction of asymptotic solutions to equations, whose exact solutions are unknown.

5.2. Asymptotic solutions in the one-dimensional case

Consider a semilinear parabolic equation with a small parameter at the derivatives

$$\frac{\partial u}{\partial t} - \varepsilon \frac{\partial}{\partial x}\left(K(x,t,u)\frac{\partial u}{\partial x}\right) - F(x,t,u) = 0. \tag{2.1}$$

Assume that the functions $K(x,t,u)$ and $F(x,t,u)$ satisfy the conditions

$$\begin{gathered} K(x,t,u) = u^{k-1}\rho(x,t,u), \qquad \rho(x,t,u) \in C^\infty, \\ \rho(x,t,u) > 0, \qquad F(x,t,u) \in C^\infty. \end{gathered} \tag{2.2}$$

Further, assume that, as $u \to \infty$, we have

$$\begin{aligned} K(x,t,u) &= r(x,t) + r_1(x,t)u^{-\gamma} + o(u^{-\gamma}), \\ F(x,t,u) &= \varkappa(x,t)u + \varkappa_1(x,t)u^{-\gamma} + o(u^{-\gamma}), \qquad \gamma > 0. \end{aligned} \tag{2.3}$$

As an example of the function $K(x,t,u)$ satisfying conditions (1.2), (1.3), we can take the function

$$K(x,t,u) = \frac{u^{k-1}\rho_0(x,t)}{1 + u^{k-1}\rho_\infty(x,t)}$$

similar to the transport coefficient from equation (1.1).

A physically meaningful solution $u(x,t,\varepsilon)$ to equation (2.1) must satisfy the flow continuity condition on the weak discontinuity front:

$$K(x,t,u)\frac{\partial u}{\partial x}\bigg|_{u=0} = 0.$$

The exact solution given in the preceding section is not a traveling-wave solution and belongs, so to say, to an intermediate class between the class of solutions to semilinear parabolic equations describing localized traveling waves (see Chapter IV) and the class of solutions describing the blow-up regimes. Exact solutions belonging to these classes show that equations with constant coefficients may have a solution of traveling wave type if the "principal part" of this solution is invariant under a group of translations. The existence of solutions of the form given above depends on a group of scale transformations.

In particular, if a source-sink function in equation (2.3) is replaced by a function of the form $F(u,t) = u(1-u)/(u+1)$, then the obtained equation has both an asymptotic solution of the form similar to (2.4) and the exact solution of localized traveling wave type $u = \chi\big((x\varepsilon^{1/2} + \varepsilon bt)/\varepsilon\big)$, where the function $\chi = \chi(\tau)$ is the monotone solution of the ordinary differential equation

$$b\chi' - \left(\frac{k\chi^{k-1}\chi'}{1+\chi^{k-1}}\right)' - \frac{\chi(1-\chi)}{1+\chi} = 0,$$
$$\chi(0) = 0, \qquad \chi \to 1, \qquad \tau \to \infty.$$

The existence of such a solution for a certain value of b can be proved by the methods given in Chapter I.

We shall point out another specific feature of this class of solutions. Though it is impossible to obtain explicit formulas even in the case when the principal term of the solution $f(\tau, x, t, 0)$ depends on the fast variable $\tau = S/\varepsilon$, one can write the equation describing the motion of the boundary of the solution support. Precisely as in [21] (see also Chapter IV), the asymptotic solution can be constructed in two steps. First, an asymptotic solution is constructed in a neighborhood of the boundary of the support (this corresponds to construction of an asymptotic solution near the wave front). Second, an asymptotic solution is constructed for the "linearized" equation (the limiting equation as $u \to \infty$).

In our case, an exponentially increasing solution will be constructed for the linearized equation (formula (2.4) implies that, for $S/\varepsilon \gg \nu$, the solution has the form $u \sim \exp(S(x,t)/\varepsilon)$). Such (and more general) asymptotic solutions for linear problems are justified in [22] and at the end of Chapter IV.

As in the preceding section, we shall construct an asymptotic solution in the form (compare with (1.3)):

$$u(x,t,\varepsilon) = W_0\Big(\frac{S}{\varepsilon}, x, t\Big) + W_1\Big(\frac{S}{\varepsilon}, x, t\Big) + \dots \tag{2.4}$$

and take only the sum of the first two terms. Similarly to (1.3), we assume that the functions $W_i(\tau, x, t)$ can be written as

$$W_i(\tau, x, t) = e^{\tau} m_i(\tau, x, t),$$

where $m_i = m_i(\tau, x, t)$ is a self-stabilizing function, i.e.,

$$\frac{\partial^\alpha}{\partial\tau^\alpha} m_i(\tau, x, t) = O(\tau^{-N}), \qquad \tau \to \infty, \tag{2.5}$$

for any $N \in \mathbb{Z}_+$, $\alpha \geq 1$, and, if $i > 0$, relation (2.5) holds for $\alpha \geq 0$. Besides of this, it is assumed that $m_i(0, x, t) = 0$.

Recall that for self-stabilizing functions, not the complete phase function, but only its expansion by the Taylor formula can be used, namely, if $x = -\psi_0(t)$ satisfies the equation $S_0(-\psi_0, t) = 0$, then

$$\begin{aligned} S(x,t,\varphi) &= S_0(x,t) + \varepsilon S_1(x,t) \\ &= \beta(t)(x + \psi_0(t)) + \varepsilon\psi_1(t) + O(\varepsilon^2 + (x+\psi_0(t))^2) \\ &\overset{\text{def}}{=} \breve{S} + O(\varepsilon^2 + (x_0 + \psi_0(t))^2) \end{aligned}$$

and (compare with the Introduction)

$$m_i(\tau, x, t)\Big|_{\tau=S/\varepsilon} = m_i\left(\frac{\check{S}}{\varepsilon}, x, t\right) + O(\varepsilon). \tag{2.6}$$

Thus, by the above argument, formula (2.4) can be rewritten as follows

$$u(x,t,\varepsilon) = \exp(S/\varepsilon)\left[m_0\left(\frac{\check{S}}{\varepsilon}, x, t\right) + \varepsilon m_1\left(\frac{\check{S}}{\varepsilon}, x, t\right) + \dots\right].$$

But if the derivatives with respect to x satisfy (2.5), then

$$u(x,t,\varepsilon) = \exp(S/\varepsilon)\left[m_0\left(\frac{\check{S}}{\varepsilon}, -\psi_0(t), t\right) + \varepsilon m_1\left(\frac{\check{S}}{\varepsilon}, -\psi_0(t), t\right) + \dots\right]. \tag{2.7}$$

Obviously, by (2.6), in the last two formulas the functions m_i, $i > 0$, differ from the functions in (2.4), but possess the same properties. Therefore, since formula (2.7) is most important for us, we preserve the notation.

Formula (2.7) plays the central role in the proposed algorithm for the construction of asymptotic solutions. This formula implies that the solution of a nonlinear equation with self-stabilizing nonlinearities (see (2.2)) has the form similar to that of WKB-solutions to nonlinear equations. The distinction is that there is a pre-exponential function (in brackets) that plays the part of the boundary layer and allows to match the solution of the linear equation in the domain $S > 0$ (where it is essential that the coefficients are self-stabilizing) with the solution in the domain $S \approx 0$ (where the solution is close to zero).

Obviously, one can expect that the function $S_0(x,t)$ satisfies the Hamilton–Jacobi equation

$$\frac{\partial S_0}{\partial t} - r(x,t)\left(\frac{\partial S_0}{\partial x}\right)^2 = 0, \tag{2.8}$$

and the function $S_1(x,t)$ satisfies the transport equation

$$\frac{\partial S_1}{\partial t} - 2r(x,t)\frac{\partial S_0}{\partial x}\frac{\partial S_1}{\partial x} - \frac{\partial}{\partial x}\left(r(x,t)\frac{\partial S_0}{\partial x}\right) - \varkappa = 0. \tag{2.9}$$

Clearly, the boundary Γ_t of the support of $u(x,t,\varepsilon)$ is defined as the surface, on which the function $S(x,t,\varepsilon)$ is equal to zero. In turn, this implies that Γ_t (with precision up to $O(\varepsilon)$) is the surface, on which the solution of (2.8) vanishes. Therefore, to construct the "principal" term of the surface Γ_t (we denote it by Γ_t^0), it is sufficient to solve the Cauchy problem for equation (2.8).

In the next section (see also Section 5.1) we show that the solution satisfying (2.8) in the domain $S_0 \geq 0$ is defined by the values of the initial conditions $S_0\big|_{t=0} = S_0^0$ only in the domain $S_0^0 \geq 0$. Thus the function S_0 is defined by the initial conditions of the form (2.4).

The violation of this property for wave solutions of semilinear parabolic equations implies the asymptotic ill-posedness (see [20]). For (2.9) (and for other equations obtained for S_i, $i > 1$, when the solution is constructed with greater precision), the boundary condition on Γ_t^0 is unknown and can be obtained from the condition that the solution in a neighborhood of the boundary Γ_t must be matched with the solution inside the support.

Denote by $m_0(\tau, x, t)$ the solution of the equation

$$r(x,t)\frac{\partial m_0}{\partial \tau} - K(x,t,e^\tau m_0)\Big(m_0 + \frac{\partial m_0}{\partial \tau}\Big) = 0, \qquad m_0\Big|_{\tau=0} = 0,$$

and assume that

$$\frac{\partial}{\partial x}\Big[\big(r(x,t) - K(x,t,e^\tau m_0)\big)M_0 - K(x,t,e^\tau m_0)\frac{\partial M_0}{\partial \tau}\Big]\Big|_{S_0=0} \neq 0,$$

where $M_0 = M_0(\tau, t) = m_0\big|_{S_0(x,t)=0}$. Under this assumption we have the following theorem.

Theorem 2.1. *For $t \in [0,T]$ let equations (2.8) and (2.9), (2.28) have smooth solutions such that the domain $\Omega_t = \{x, S_0(x,t) \geq 0\}$ is an interval, $\Omega_t = \{-\psi_{01}^0(t) \leq x \leq -\psi_{02}^0(t)\}$. Then the asymptotic solution of equation (2.1) has the form*

$$u(x,t,\varepsilon) = \exp\Big\{\frac{S(x,t,\varepsilon)}{\varepsilon}\Big\}\prod_{i=1}^{2}\big(M_{0i}(t,\tau) + \varepsilon M_{1i}(t,\tau)\big)\Big|_{\tau=\beta_i^0(x+\psi_{0i}(t))/\varepsilon+\beta_i^0\psi_{1i}},$$

where the function $M_{0i}(t,\tau)$ satisfies the equation

$$\big[r(-\psi_{0i}(t),t) - K(-\psi_{0i},t,e^\tau M_{0i})\big]M_{0i} - K(-\psi_{0i},t,e^\tau M_{0i})\frac{\partial M_{0i}}{\partial \tau} = 0,$$
$$M_{0i}\big|_{\tau=0} = 0, \qquad M_{0i}\big|_{\tau\to\infty} \to 1,$$

the functions M_{1i} have the form (2.24), where $C_{ki} = 0$, the functions ψ_{1i} are defined in (2.28), $\beta_i^0 = \partial S_0/\partial x\big|_{x=-\psi_{0i}}$.

Proof. Following to [23], we substitute the function

$$u(x,t,\varepsilon) = \big[W_0(x,t,\tau) + \varepsilon W_1(x,t,\tau)\big]_{\tau=(S_0+\varepsilon S_1)/\varepsilon}$$

into (2.1). By setting the sum of coefficients at ε^{-1} equal to zero, we obtain the equation

$$\frac{\partial S_0}{\partial t}\frac{\partial W_0}{\partial \tau} - \Big(\frac{\partial S}{\partial x}\Big)^2\frac{\partial}{\partial \tau}\Big(K(x,t,W_0)\frac{\partial W_0}{\partial \tau}\Big) = 0. \tag{2.10}$$

Precisely as in the preceding section, we put $W_0 = e^\tau m_0(x,t,\tau)$ and obtain the equation for the function m_0:

$$\frac{\partial S}{\partial t}m_0 - \Big(\frac{\partial S}{\partial x}\Big)^2 K(x,t,e^\tau m_0)\Big(m_0 + \frac{\partial m_0}{\partial \tau}\Big) = 0. \tag{2.11}$$

Here we have used the condition that the flow must be continuous on the weak discontinuity front. This condition in terms of the function m_0 has the form

$$K(x,t,e^{\tau}m_0)\frac{\partial m_0}{\partial \tau}\bigg|_{m_0=0} = 0. \tag{2.12}$$

It is required that the solution of equation (2.11) satisfy the condition: $m_0 \to 1$ as $\tau \to \infty$.

Hence, by (2.11), (2.3), we obtain the equation for the function $S_0(x,t)$ (compare with (1.5)):

$$\frac{\partial S_0}{\partial t} - r(x,t)\left(\frac{\partial S_0}{\partial x}\right)^2 = 0. \tag{2.13}$$

By (2.13), equation (2.11) can be rewritten either as

$$\mathcal{L}_0 m_0 \overset{\text{def}}{=} \left(\frac{\partial S_0}{\partial x}\right)^2 \left[(r - K(x,t,e^{\tau}m_0))m_0 - K(x,t,e^{\tau}m_0)\frac{\partial m_0}{\partial \tau}\right] = 0$$

or

$$(r - K(x,t,e^{\tau}m_0))m_0 - K(x,t,e^{\tau}m_0)\frac{\partial m_0}{\partial \tau} = 0.$$

The latter equation implies that the function $m_0 = m_0(x,t,\tau)$

(1) is continuous, and its derivatives can have a singularity as $\tau \to +0$

$$m_0 = O(\tau^{\alpha}), \qquad \alpha = 1/(k-1),$$

(2) is self-stabilizing as $\tau \to \infty$, i.e.,

$$1 - m_0 = O(e^{-\gamma\tau}), \qquad \tau \to \infty.$$

Denote by $\psi_{0i} = \psi_{0i}(t)$, $i = 1, 2$ the solutions of equation $S_0(-\psi_0, t) = 0$, such that $S_0 > 0$ for $-\psi_{01} < x < -\psi_{02}$. By setting

$$M_{0i}(t,\tau) = m_0(-\psi_{0i}, t, \tau),$$

we get the equation for $M_{0i}(t,\tau)$:

$$(r(-\psi_{0i},t) - K(-\psi_{0i},t,e^{\tau}M_{0i}))M_{0i} - K(-\psi_{0i},t,e^{\tau}M_{0i})\frac{\partial M_{0i}}{\partial \tau} = 0, \tag{2.14}$$

and the expression for the transport coefficient $K(x,t,e^{\tau}M_{0i})$, important for further considerations:

$$K(-\psi_{0i},t,e^{\tau}M_{0i}) = r(-\psi_{0i},t)M_{0i}\left(M_{0i} + \frac{\partial M_{0i}}{\partial \tau}\right)^{-1}. \tag{2.15}$$

Substituting the function $u = W_0 + \varepsilon W_1$ into (2.1), collecting the terms of the order ε^0 and setting their sum equal to zero, we get the equation

$$\begin{aligned}
&\frac{\partial W_1}{\partial \tau}\frac{\partial S_0}{\partial \tau} + \frac{\partial W_0}{\partial \tau}\frac{\partial S_1}{\partial t} + \frac{\partial W_0}{\partial t} \\
&\quad - \frac{\partial S_0}{\partial x}\frac{\partial}{\partial \tau}\left[K(x,t,W_0)\left(\frac{\partial W_1}{\partial x}\frac{\partial S_0}{\partial x} + \frac{\partial W_0}{\partial \tau}\frac{\partial S_1}{\partial x}\right) + \frac{\partial K}{\partial \tau}W_1\frac{\partial S_0}{\partial x}\right] \\
&\quad - \left(\frac{\partial S_1}{\partial x}\frac{\partial}{\partial \tau} + \frac{\partial}{\partial x}\right)\left(K(x,t,W_0)\frac{\partial W_0}{\partial \tau}\frac{\partial S_0}{\partial x}\right) - F(x,t,W_0) = 0.
\end{aligned} \tag{2.16}$$

Hence, by putting $W_1 = e^\tau m_1(x,t,\tau)$, we get the equation for m_1:

$$\begin{aligned}
\mathcal{L}_1 m_1 &\stackrel{\text{def}}{=} e^\tau \frac{\partial S_0}{\partial t}\left(m_1 + \frac{\partial m_1}{\partial \tau}\right) - \left(\frac{\partial S_0}{\partial x}\right)^2 \frac{\partial^2}{\partial \tau^2}[K(x,t,e^\tau m_0)e^\tau m_1] \\
&= -\left\{\frac{\partial S_1}{\partial t}e^\tau\left(m_0 + \frac{\partial m_0}{\partial \tau}\right) + e^\tau\frac{\partial m_0}{\partial \tau}\right. \\
&\quad - 2\frac{\partial S_0}{\partial x}\frac{\partial S_1}{\partial x}\frac{\partial}{\partial \tau}\left[K(x,t,e^\tau m_0)e^\tau\left(m_0 + \frac{\partial m_0}{\partial \tau}\right)\right] \\
&\quad \left. - e^\tau\left(m_0 + \frac{\partial M_0}{\partial \tau}\right)\frac{\partial}{\partial x}K(x,t,e^\tau m_0)\frac{\partial S_0}{\partial x} - F(x,t,e^\tau m_0)\right\} = 0.
\end{aligned} \tag{2.17}$$

The function m_1 must satisfy the boundary condition: $m_1(x,t,\tau) \to 0$ as $\tau \to \infty$. Then, if the boundary condition for the function m_0 is taken into account, the boundary condition for m_1 implies the relation

$$\frac{\partial S_1}{\partial t} - 2\frac{\partial S_0}{\partial x}\frac{\partial S_1}{\partial x} - \frac{\partial}{\partial x}\left(r\frac{\partial S_0}{\partial x}\right) - \varkappa = 0. \tag{2.18}$$

By substituting the solution $u(x,t,\varepsilon) = W_0 + \varepsilon W_1$ into the equation, we obtain the sum

$$\left(\varepsilon^{-1}e^\tau\mathcal{L}_0 m_0 + \varepsilon^0 e^\tau \mathcal{L}_1 m_1 + e^\tau O(\varepsilon^2)\right)\big|_{\tau=S/\varepsilon} = \left(\varepsilon^{-1}e^\tau[\mathcal{L}_0 m_0 + \varepsilon\mathcal{L}_1 m_1 + O(\varepsilon^2)]\right)\big|_{\tau=S/\varepsilon},$$

where the O-estimate is uniform with respect to x, t, τ.

We shall consider the sum of the first two terms in brackets on the right-hand side. Note that the boundary conditions imposed on the functions m_0 and m_1 (determined by the Hamilton–Jacobi and transport equations) imply that these summands are self-stabilizing functions as $\tau \to \infty$. Therefore, denoting

$$M_{ji}(t,\tau) = m_j(x,t,\tau)\big|_{x=-\psi_{0i}}, \qquad L_{ji} = \mathcal{L}_j\big|_{x=-\psi_{0i}},$$

we get

$$\mathcal{L}_0 m_0 + \varepsilon\mathcal{L}_1 m_1 = \begin{cases} L_{01}M_{01} + \varepsilon L_{11}M_{11} + O(\varepsilon^2), \\ L_{02}M_{02} + \varepsilon L_{12}M_1 + O(\varepsilon^2), \end{cases} \tag{2.19}$$

with (compare with (2.14))

$$L_{0i}M_{0i} = \big(r(-\psi_{0i},t) - K(-\psi_{0i},t,e^{\tau}M_{0i})\big)M_{0i} - K(-\psi_{0i},t,e^{\tau}M_{0i})\frac{\partial M_{0i}}{\partial\tau} = 0$$

and

$$\begin{aligned}
L_{1i}M_{1i} = \beta_i^0\frac{d\psi_{0i}}{dt}\bigg[&\frac{\partial^2 M_{1i}}{\partial\tau^2}M_{0i}\left(\frac{\partial M_{0i}}{\partial\tau}+M_{0i}\right)^{-1}\\
&-\frac{\partial M_{1i}}{\partial\tau}\left\{1-2\frac{\partial}{\partial\tau}\left(M_{0i}\left(M_{0i}+\frac{\partial M_{0i}}{\partial\tau}\right)^{-1}\right)-2M_{0i}\left(\frac{\partial M_{0i}}{\partial\tau}+M_{0i}\right)^{-1}\right\}\\
&-M_{1i}\left\{1-\frac{\partial^2}{\partial\tau^2}\left(M_{0i}\left(M_{0i}+\frac{\partial M_{0i}}{\partial\tau}\right)^{-1}\right)\right.\\
&\left.-2\frac{\partial}{\partial\tau}\left(M_{0i}\left(M_{0i}+\frac{\partial M_{0i}}{\partial\tau}\right)^{-1}\right)-M_{0i}\left(M_{0i}+\frac{\partial M_{0i}}{\partial\tau}\right)^{-1}\right\}\bigg]\\
&+\left\{\left(M_{0i}+\frac{\partial M_{0i}}{\partial\tau}\right)^{-1}\left(2\beta_{1i}^0 r+\beta^0\frac{\partial r}{\partial x}+\varkappa-\beta_i^0\frac{\partial K}{\partial x}(x,t,e^{\tau}M_{0i})\right)\right.\\
&+\frac{\partial M_{0i}}{\partial t}-\left(2\beta_{1i}^0 rM_{0i}+e^{\tau}F(x,t,e^{\tau}M_{0i})+\beta_1^0(\tau-\beta_1^0\psi_{1i})\right)\left(1+\frac{\partial}{\partial\tau}\right)\\
&\times\left.\frac{\partial}{\partial x}\left(rM_{0i}-K(x,t,e^{\tau}M_{0i})\left(M_{0i}+\frac{\partial M_{0i}}{\partial\tau}\right)^{-1}\right)\right\}_{x=-\psi_{0i}},
\end{aligned}\tag{2.20}$$

where ψ and β with indices are the coefficients of the expansion of the function $S(x,t,\varepsilon) = S_0(x,t)+\varepsilon S_1(x,t)$ according to the Taylor formula. Denoting by $\psi_i(t,\varepsilon)$ the solution to the equation $S(-\psi,t,\varepsilon)=0$,

$$\psi_i = \psi_i(t,\varepsilon) = \psi_{0i}(t)+\varepsilon\psi_{1i}(t)+O(\varepsilon^2),$$

we obtain

$$\begin{aligned}
S &= \beta_i^0(x+\psi_i)+\beta_i^1(x+\psi_i)^2+O((x+\psi_i)^3)\\
&= (\beta_i^0+\varepsilon\beta_{1i}^0+\dots)(x+\psi_{0i}+\varepsilon\psi_{1i})+(\beta_i^1+\dots)(x+\psi_i)^2+O((x+\psi_i)^3)\\
&= \beta_i^0(x+\psi_i)+\beta_i^1(x+\psi_i)^2+\dots\\
&\quad+\varepsilon\{\beta_i^0\psi_{1i}+(\beta_{1i}^0+2\beta_i\psi_{1i})(x+\psi_{0i})+\dots\}.
\end{aligned}\tag{2.21}$$

Hence

$$\beta_i^0(x+\psi_{0i}) = \left[\varepsilon(\tau-\beta_i^0\psi_{1i})+O(\varepsilon^2)\right]_{\tau=S/\varepsilon},$$

$$\frac{\partial S_0}{\partial x} = \beta_i^0+2\beta_i^1(x+\psi_{0i})+O((x+\psi_{0i})2),$$

$$\left.\frac{\partial^2 S_0}{\partial x^2}\right|_{x=-\psi_{0i}} = 2\beta_i^1,\qquad \left.\frac{\partial S_1}{\partial x}\right|_{x=-\psi_{0i}} = \beta_{1i}^0+2\beta_i^1\psi_{1i},$$

$$\frac{\partial S_0}{\partial t} = \beta_i^0\frac{d\psi_{0i}}{dt}\left(\frac{d\beta_i^0}{dt}+2\beta_i^1\frac{d\psi_{0i}}{dt}\right)(x+\psi_{0i})+O((x+\psi_{0i})^2).$$

These relations and equations (2.13), (2.18) were used to derive formula (2.20).

By setting the summands of the order ε^0 in (2.19) equal to zero, we obtain equation (2.14) for the function M_{0i}, and by setting the summand $L_{1i}M_{1i}$ equal to zero, we obtain the equation for the function M_{1i}

$$\frac{\partial^2}{\partial\tau^2}\left(M_{1i}M_{0i}\left(\frac{\partial M_{0i}}{\partial\tau}+M_{0i}\right)^{-1}\right)$$
$$-\frac{\partial M_{1i}}{\partial\tau}\left\{1-2\frac{\partial}{\partial\tau}\left(M_{0i}\left(M_{0i}+\frac{\partial M_{0i}}{\partial\tau}\right)^{-1}\right)-2M_{0i}\left(\frac{\partial M_{0i}}{\partial\tau}+M_{0i}\right)^{-1}\right\}$$
$$-M_{1i}\left\{1-\frac{\partial^2}{\partial\tau^2}\left(M_{0i}\left(M_{0i}+\frac{\partial M_{0i}}{\partial\tau}\right)^{-1}\right)\right.$$
$$\left.-2\frac{\partial}{\partial\tau}\left(M_{0i}\left(\frac{\partial M_{0i}}{\partial\tau}+M_{0i}\right)^{-1}\right)-M_{0i}\left(\frac{\partial M_{0i}}{\partial\tau}+M_{0i}\right)^{-1}\right\}=f, \qquad (2.22)$$

where

$$f=\left(\beta_i\frac{d\psi_{0i}}{\partial t}\right)^{-1}\left[\left(M_{0i}+\frac{\partial M_{0i}}{\partial\tau}\right)\left(2\beta^0_{1i}r+\beta^0_i\frac{\partial r}{\partial x}+\varkappa-\beta^0_i\frac{\partial K}{\partial x}(x,t,e^\tau M_{0i})\right)\right.$$
$$+\frac{\partial M_{0i}}{\partial t}-2r(-\psi_{0i},t)\beta^0_{1i}M_{0i}-e^{-\tau}F(x,t,e^\tau M_{0i})+\beta^0_i(\tau-\beta^0_{1i}\psi_{1i})\left(1+\frac{\partial}{\partial\tau}\right)$$
$$\left.\times\frac{\partial}{\partial x}\left(rM_{0i}-K(x,t,e^\tau M_{0i})\left(M_{0i}+\frac{\partial M_{0i}}{\partial\tau}\right)\right)\right]_{x=-\psi_{0i}} \qquad (2.23)$$

Obviously, equation (2.11) is the first variation equation with right-hand side corresponding to equation (2.14). To solve (2.11), one must know the solution of the homogeneous equation, which can be solved as follows [23, 87]. The function $\partial W_0/\partial\tau$ satisfies the homogeneous variation equation (2.16), and our construction implies that (2.22) is the equation for the pre-exponential function m_1 in terms of W_1, i.e, $m_1 = e^{-\tau}W_1$. Thus, the solution M^0_{1i} of the homogeneous equation is the "pre-exponential" function for the function $\partial W_0/\partial\tau$, or

$$M^0_{1i}=e^{-\tau}\dot W_0\Big|_{x=-\psi_{0i}}=M_{0i}+\frac{\partial M_{0i}}{\partial\tau}.$$

One can directly verify that the function M^0_{1i}, calculated in this way, satisfies the homogeneous equation (2.22).

Denoting $K(M_{0i})=M_{0i}(\partial M_{0i}/\partial\tau+M_{0i})^{-1}$, we obtain the Wronskian of equation (2.22):

$$V=e^{-2\tau}K^2(M_{0i})\exp\left\{\int K(M_{0i})^{-1}\,d\tau\right\}=e^{-\tau}M_{0i}^{-1}(M^0_{1i})^2.$$

The general solution of equation (2.22) has the form

$$M_{1i}=C_{1i}M^0_{1i}+C_{2i}M^0_{1i}\int V(M^0_{1i})^{-2}\,d\tau$$
$$+M^0_{1i}\int_0^\tau V(M_{1i})^{-2}\int_0^{\tau'}fM^0_{1i}(VK(M_{0i}))^{-1}\,d\tau''d\tau'. \qquad (2.24)$$

Let us study the asymptotic behavior of integrands. By (2.23) and the expression for the Jacobian, we have

$$\frac{fM_{1i}^{0}}{VK(M_{0i}} \sim \begin{cases} e^{(1-\gamma)\tau}, & \tau \to \infty, \\ \tau^{2\alpha-1/m-1}, & \tau \to 0, \end{cases}$$

where $m = \lim_{\tau \to 0} M_{0i}/(\tau \partial M_{0i}/\partial \tau)$. A priori two situations are possible:

$$2\alpha > 1/m \qquad \text{and} \qquad 2\alpha < 1/m.$$

However, since the function M_{0i} has the exponential asymptotics as $\tau \to 0$, (2.14) implies $m = 1/\alpha$. Thus, if

$$\int_0^\infty V(M_{0i})^{-2} \int_0^\infty fM_{1i}^0 (VK(M_{0i}))^{-1} \, d\tau' d\tau = 0, \tag{2.25}$$

the solution M_{11} satisfies the boundary conditions

$$\begin{aligned} M_{1i} &\sim \tau^\alpha, & \tau &\to 0, \\ M_{1i} &\to 0, & \tau &\to \infty, \end{aligned}$$

for $c_1 = c_2 = 0$. Relation (2.25) implies

$$M_{1i} \sim \begin{cases} \tau^2 e^{-\gamma\tau}, & \gamma \geq 1, \\ e^{-\tau}, & 0 < \gamma < 1, \quad \tau \to \infty. \end{cases} \tag{2.26}$$

In our case, relation (2.25) is an analog of the orthogonality condition providing that the variation equation has a decreasing solution.

Further we must consider two different situations:

$$\begin{aligned} &\text{I:} \quad \frac{\partial}{\partial x}\left[(r - K(x,t,e^\tau M_{0i}))M_{0i} - K(x,t,e^\tau M_{01})\frac{\partial M_{0i}}{\partial \tau}\right]\Bigg|_{x=-\psi_{0i}} \neq 0, \\ &\text{II:} \quad \frac{\partial}{\partial x}\left[(r - K(x,t,e^\tau M_{01}))M_{0i} - K(x,t,e^\tau M_{0i})\frac{\partial M_{0i}}{\partial \tau}\right]\Bigg|_{x=-\psi_{0i}} = 0. \end{aligned} \tag{2.27}$$

For example, below we have the second situation if the transport coefficient depends only on the transferred variable, $K \equiv K(u)$ (in general, the case $K = K(t,u)$ can lead to the first situation, since, instead of the expansion with respect to the monomial $x + \psi_{0i}(t)$, we can use the expansion with respect to $t + \hat{\psi}_{0i}(x)$ (similarly to the multidimensional case, see the next section).

Assume that condition I holds. Then (2.25) implies the relation for $\psi_{1i}(t) =$

$S_1\big|_{x=-\psi_{0i}(t)}$:

$$\begin{aligned}\psi_{1i} = \Bigg(\int_0^\infty V(M_{1i}^0)^{-1}\int_0^\tau (VK(M_{0i}))^{-1} \\ \times\Bigg[\Big(M_{0i}+\frac{\partial M_{0i}}{\partial\tau}\Big)\Big(\beta_i^0\frac{\partial r}{\partial x}\Big|_{x=-\psi_{0i}^0} + \varkappa(-\psi_i^0,t) - \beta_i^0\frac{\partial K}{\partial x}\Big|_{x=-\psi_i^0}\Big) \\ + \frac{\partial M_{0i}}{\partial t} + 2r(-\psi_i^0,t)\beta_{1i}^0\frac{\partial M_{0i}}{\partial\tau} - e^{-\tau}F(-\psi_i^0,t,e^\tau M_{0i}) \\ + \tau\beta_i^0\Big(1+\frac{\partial}{\partial\tau}\Big)\frac{\partial}{\partial x}\Big(rM_{0i} - K\Big(M_{0i}+\frac{\partial M_{0i}}{\partial\tau}\Big)\Big)_{x=-\psi_i^0}\Bigg]\,d\tau'd\tau\Bigg) \\ \times\Bigg((\beta_i^0)^2\int_0^\infty V(M_{1i}^0)^{-1}\int_0^\tau (VK(M_{0i}))^{-1}\Bigg[\Big(1+\frac{\partial}{\partial\tau}\Big) \\ \times\frac{\partial}{\partial x}\Big(rM_{0i} - K\Big(M_{0i}+\frac{\partial M_{0i}}{\partial\tau}\Big)\Big)_{x=-\psi_i^0}\Bigg]\,d\tau'd\tau\Bigg)^{-1}. \qquad (2.28)\end{aligned}$$

Thus, in our case, the principal coordinate of the weak discontinuity front is determined by the solution of the Cauchy problem for equation (2.8):

$$S_0(-\psi_i^0,t)=0,$$

and the correction ψ_{1i}^0 to this coordinate is given by (2.28), which was obtained independently of the solution of the Cauchy problem for the transport equation (2.9). Thus, for the order ε^1, there is a problem of matching the solution of the Cauchy problem for equation (2.9) with the solution of this equation satisfying the boundary condition (2.28). To solve this problem, one must choose the initial condition as it was done for the Hamilton–Jacobi equation in the preceding section.

Theorem 2.1 is proved.

For (2.27) II, the relation between ψ_i^0 and β_{1i}^0 can be obtained from (2.25):

$$\begin{aligned}\beta_{1i}^0 = -\Bigg(\frac{1}{2r}\int_0^\infty V(M_{1i}^0)^{-2} \\ \times\int_0^\tau M_{1i}^0\big(\varkappa(-\psi_i^0,t)M_{1i}^0 - e^{-\tau'}F\big)(VK(M_{0i}))^{-1}\,d\tau'd\tau\Bigg) \\ \times\Bigg(\int_0^\infty V(M_{1i}^0)^{-2}\int_0^\tau M_{1i}^0\frac{\partial M_{0i}}{\partial\tau'}(VK(M_{0i}))^{-1}\,d\tau'd\tau\Bigg)^{-1}. \qquad (2.29)\end{aligned}$$

On the other hand, substituting the expansion (2.21) into the Hamilton–Jacobi equation (2.8), we find the relation between ψ_i^0 and β_{1i}^0. Namely,

$$2\beta_{1i}^0\frac{d\psi_i^0}{dt} + \frac{d}{dt}\Big(\frac{d\psi_i^0}{dt}\Big/r\Big) = 4\frac{d\psi_i^0}{dt}\beta_{1i}^0. \qquad (2.29')$$

Note that, by assumptions made after condition II was taken, $r=\text{const}$ and hence (2.29) implies that $\beta_{11}^0=\text{const}$ if $\varkappa=\text{const}$.

Therefore equation (2.29) can be written in the form

$$2r\beta^0_{1i}\frac{d\psi^0_i}{dt} = \frac{d^2\psi^0_i}{dt^2}. \tag{2.30}$$

Integrating this equation, we get

$$\psi^0_i(t) = \psi^0_i(0) + C_i r\int_0^t \exp\{2r\beta^0_{1i}t'\}\,dt', \qquad C_i = \beta^0_i\big|_{t=0}.$$

Then we obtain a family (2.30) of laws of motion of the boundary. As is easily seen, in this case we cannot take arbitrary initial conditions for equation (2.8), since the law of motion of the boundary is given beforehand. We choose the constant $C_1 > 0$ and thus define the law of motion of the weak discontinuity for $t > 0$. Then, as in the preceding section, we can define the initial condition for equation (2.8) in the domain $x \geq -\psi_{01}(0)$ in a certain neighborhood of the point $-\psi_{01}(0) \leq x \leq d_1$, so that the solution of the Cauchy problem for (2.8) for $0 < t < T$ satisfy the zero boundary condition for $x = -\psi_{01}(t)$. (Obviously, the size d_1 of this neighborhood determines the time of existence T for our asymptotic solution.)

Similarly, if we choose the initial position for the right-hand boundary of the support of the solution $x = -\psi_{02}(0)$, so that $\psi^0_{01}(0) > \psi_2(0)$, and put $C_2 < 0$, then equation (2.30) describes the law of motion of the right-hand boundary of the support and the neighborhood $d_2 \leq x \leq -\psi^0_{02}(0)$ of the point $x = -\psi^0_{02}(0)$. (At this point, the initial condition implies that, for $0 \leq t \leq T$, the Cauchy problem for (2.8) has the solution satisfying the zero boundary condition for $x = -\psi^0_{02}(t)$.)

Then, to determine the initial condition, one must smoothly match the initial conditions in a neighborhood of the points $x = -\psi_{01}(0)$ and $x = -\psi_{02}(0)$, by filling the interval $d_1 \leq x \leq d_2$, so that to obtain a positive smooth function in the whole interval $-\psi_{01}(0) < x < -\psi_{02}(0)$.

Further, when the next terms of the asymptotic expansion are constructed, one must solve a similar problem of choosing the initial conditions (like in case I (2.27)) for the transport equation (2.9).

5.3. Asymptotic finite solutions of degenerate quasilinear parabolic equations with small diffusion

In this section we briefly discuss an algorithm for constructing an asymptotic solution to the equation

$$\frac{\partial u}{\partial t} - \varepsilon\langle\nabla, K(x,t,u)\nabla u\rangle - F(x,t,u) = 0 \tag{3.1}$$

in the multidimensional case following [23]. Precisely as in the preceding section, we assume that the stabilization condition (3.3) holds and look for the solution $u = u(x,t,\varepsilon)$ in the form

$$u = W(S/\varepsilon, x, \varepsilon),$$

where

$$W(\tau,x,t,\varepsilon) \stackrel{\text{def}}{=} W_0(\tau,x,t) + \varepsilon W_1(\tau,x,t) + \dots, \qquad W_i = e^{\tau}M_i. \tag{3.2}$$

It is assumed that the functions M_0 and M_1 satisfy the conditions

$$\begin{aligned} M_0: &\quad \tau \leq 0; \quad M_0 \to 1, \quad \tau \to \infty, \\ M_1: &\quad \tau \leq 0; \quad M_1 \to 0, \quad \tau \to \infty. \end{aligned} \tag{3.3}$$

The function $S(x,t,\varepsilon)$ has the form

$$S(x,t,\varepsilon) = S_0(x,t) + \varepsilon S_2(x,t). \tag{3.4}$$

Similarly to (2.4), these formulas imply that in any interior subdomain of the support the solution has the form $u \approx e^{S/\varepsilon}$, $S > 0$. By assumption, for large u, equation (3.1) becomes linear parabolic:

$$\frac{\partial u}{\partial t} - \varepsilon\langle\nabla, r(x,t,)\nabla u\rangle - \varkappa(x,t)u = 0. \tag{3.5}$$

Therefore, as above, it is natural to assume that the function $S_0(x,t)$ satisfies the Hamilton–Jacobi equation

$$\frac{\partial S_0}{\partial t} - r(x,t)|\nabla S_0|^2 = 0, \tag{3.6}$$

and the function $S_1(x,t)$ satisfies the transport equation

$$\frac{\partial S_1}{\partial t} - 2r(x,t)\langle\nabla S_0, \nabla S_1\rangle - \langle\nabla, r(x,t)\nabla S_0\rangle - \varkappa = 0. \tag{3.7}$$

Obviously, the boundary Γ_t of the support of $u(x,t,\varepsilon)$ is determined as the zero surface of the function $S(x,t,\varepsilon)$. In it turn, this implies that, with precision up to $O(\varepsilon)$, Γ_t is the surface where the solution to equation (3.6) vanishes. Therefore to construct the surface Γ_t in the principal approximation (we denote it by Γ_t^0), it is sufficient to solve the Cauchy problem for equation (3.6).

At the end of this section we shall show that the solution to equation (3.6) in the domain $S_0 \geq 0$ is determined by the values of the initial conditions $S_0\big|_{t=0} = S_0^0$ only in the domain $S_0^0 \geq 0$. Thus, as above, the function S_0 is determined by initial data of the form (3.8).

The boundary condition for equation (3.7) on Γ_t^0 (as well as for the successive equations for S_i, $i > 1$, that appear if the solution is constructed with greater precision) is unknown and can be obtained from the condition of matching the solution in a neighborhood of the boundary Γ_t with the solution of equation (3.5). As shown below, this condition has the form

$$S_1\big|_{\Gamma_t^0} = \nu(x), \tag{3.8}$$

where the function $\nu(x)$ is defined by (3.24) if the denominator in (3.24) does not vanish. Note that, since S_0 is nonnegative on the solution support, we cannot construct the solution of the linearized equation similarly to the Taylor expansions of S_i, $i \geq 0$, which were used instead of these functions in construction of asymptotic solutions in the previous section. On the other hand, the above methods can be also used in the case "$u \to \infty$", when we look for the functions M_i that play the role of nonlinear boundary layers matching the expansions near the boundary. The matching condition can be expressed in terms of M_i:

$$M_i\big|_{\tau=0} = 0, \qquad M_0 \to 1, \qquad M_i \to 0, \qquad i \geq 1, \qquad \tau \to \infty.$$

Theorem 3.1.. *Let equations (3.6) and (3.7), (3.8) have smooth solutions and* $\Omega_t = \{x \mid S_0(x,t) \geq 0\}$ *be a compact simply connected domain with smooth boundary.*

Then the asymptotic solution of (3.1) has the form (3.2), the function $M_0(\tau, x)$ *satisfies the ordinary differential equation*[1] *

$$[M_0' - K(x,t,e^\tau M_0)(M_0 + M_0')]_{S_0=0} = 0, \qquad M_0\big|_{\tau=0} = 0,$$

the function M_1 *is defined by formula (3.19), and the following estimates hold:*

$$M_0 = O(\tau^\alpha), \quad M_1 = O(\tau^\alpha), \quad \tau \to +0, \quad \alpha = (k-1)^{-1},$$

$$1 - M_0 = O(e^{-\gamma\tau}), \quad M_1 \sim \begin{cases} O(e^{-\tau}), & \gamma > 1, \\ O(\tau e^{-\tau}), & \gamma = 1, \end{cases} \quad \tau \to \infty.$$

Proof. By substituting expression (3.2) into equation (3.1) and setting the obtained coefficients at ε^0 equal to zero, we get

$$\frac{\partial S_0}{\partial t}\frac{\partial W_0}{\partial \tau} - |\nabla S_0|^2 \frac{\partial}{\partial \tau}\left(K(x,t,W_o)\frac{\partial W_0}{\partial \tau}\right) = 0. \tag{3.9}$$

Integrating this equation and taking into account the flow continuity condition, we obtain

$$\frac{\partial S_0}{\partial t} W_0 - |\nabla S_0|^2 K(x,t,W_o)\frac{\partial W_0}{\partial \tau} = 0. \tag{3.10}$$

We take the function W_0 in the form $W_0 = e^\tau M_0(\tau, x)$ and assume that $M_0(\tau,x) \to 1$ as $\tau \to \infty$. Then, by (3.3), (3.10), we obtain the Hamilton–Jacobi equation for the function S_0

$$\frac{\partial S_0}{\partial t} W_0 - r(x,t)|\nabla S_0|^2 = 0,$$

and equation (3.10) takes the form

$$r(x,t,)W_0 - K(x,t,W_0)\frac{\partial W_0}{\partial \tau} = 0.$$

In a neighborhood of Γ_t^0, the functions $S_0(x,t)$ and $S_1(x,t)$ can be represented in the form

$$\begin{aligned} S_0(x,t) &= \beta(x)(t + \psi(x)) + \beta_1(t + \psi(x))^2 + \dots, \\ S_1(x,t) &= \nu(x) + \alpha(t + \psi(x)) + \dots. \end{aligned} \tag{3.11}$$

Then we have

$$r(x,t)W_0 - K(x,t,W_0)\frac{\partial W_0}{\partial \tau} = \left[r(x,-\psi)W_0 - K(x,-\psi,W_0)\frac{\partial W_0}{\partial \tau}\right]$$
$$+ \varepsilon\beta^{-1}(r - \nu)\frac{\partial}{\partial t}\left[rW_0 - K\frac{\partial W_0}{\partial \tau}\right] + \dots,$$

[1] $M_0' \equiv \partial M/\partial \tau$.

where dots stand for the next terms of the Taylor formula. Now, putting $W_0 = e^\tau M_0(\tau, x)$, we get

$$r(x,t)W_0 - K(x,t,W_0)\frac{\partial W_0}{\partial \tau} = e^\tau \Big\{ \Big[r(x,-\psi)M_0 - K(x,-\psi,e^\tau)\Big(M_0 + \frac{\partial M_0}{\partial \tau}\Big)\Big]$$
$$+ \varepsilon\beta^{-1}(\tau - \nu)\Big[M_0\Big(\frac{\partial r}{\partial t} - \frac{\partial K}{\partial t}(x,-\psi,e^\tau)\Big) - \frac{\partial M_0}{\partial \tau}\frac{\partial K}{\partial t}(x,-\psi,e^\tau)\Big] + O(\varepsilon^2)\Big\}.$$

This equation implies that the expression in the first brackets on the right hand side is equal to zero, and the expression in the second brackets will be taken into account in the next (with respect to ε) approximation.

Finally, we obtain the equation for the function M_0:

$$r(x,-\psi)M_0 - K(x,-\psi,e^\tau)\Big(M_0 + \frac{M_0}{\partial \tau}\Big) = 0 \tag{3.12}$$

with boundary condition $M_0\big|_{\tau=0} = 0$.

Let us construct the asymptotic expansion of the function M_0 as $\tau \to 0$ and as $\tau \to \infty$. By (3.10) with (2.2), (2.3), we have

$$W_0 \sim \tau^\alpha, \qquad \alpha = 1/(k-1), \qquad \tau \to 0.$$

Therefore

$$M_0 \sim \tau^\alpha, \qquad \tau \to 0. \tag{3.13}$$

To obtain the asymptotic expansion of the function M_0 as $\tau \to \infty$, we first construct the asymptotic expansion of the function W_0 taking into account (3.10). By (2.3), we get

$$W_0 - \Big(1 + \frac{r_1}{rW_0^\gamma} + \dots\Big)\frac{\partial W_0}{\partial \tau} = 0.$$

Therefore, as $\tau \to \infty$, we have

$$W_0 = e^\tau + \frac{r_1}{r}e^{(1-\gamma)\tau} + o(e^{(1-\gamma)\tau}).$$

Hence, for the function M_0, we have the asymptotic expansion as $\tau \to \infty$:

$$M_0 = 1 + \Big(\frac{r_1}{r}\Big)e^{-\gamma\tau} + o(e^{-\gamma\tau}).$$

By substituting the function $W(S/\varepsilon, x, \varepsilon)$ into equation (3.1), setting the coefficient at ε^0 equal to zero, and taking into account (3.12), we obtain

$$M_1 + \frac{\partial M_1}{\partial \tau} - re^{-\tau}\frac{\partial^2}{\partial \tau^2}\Big(M_0M_1e^\tau \Big/ \Big(M_0 + \frac{\partial M_0}{\partial \tau}\Big)\Big)$$
$$= -\frac{\partial S_0}{\partial t}\Big\{\Big(M_0 + \frac{\partial M_0}{\partial \tau}\Big)\frac{\partial S_1}{\partial t} - 2r\langle \nabla S_0, \nabla S_1\rangle - \langle \nabla, M_0 r \nabla S_0\rangle$$
$$- r\Big\langle \nabla S_0, \Big(1 + \frac{\partial}{\partial \tau}\Big)\Big(M_0\Big(M_0 + \frac{\partial M_0}{\partial \tau}\Big)^{-1}\Big)\nabla M_0\Big\rangle$$
$$+ \beta^{-1}(\tau - \nu)\Big[M_0\Big(\frac{\partial r}{\partial t} - \frac{\partial K}{\partial t}\Big)\Big|_{t=-\psi} - \frac{\partial M_0}{\partial \tau}\frac{\partial K}{\partial t}\Big|_{t=-\psi}\Big]$$
$$+ e^{-\tau}F(x,t,e^\tau M_0)\Big\} \stackrel{\text{def}}{=} f(\tau, x).$$

Here $M_i = M_i(\tau, x)$, $i = 0, 1$, and, as in the preceding section, we take into account the relation

$$K(x, t, e^{\tau} M_0) = rM_0\left(M_0 + \frac{\partial M_0}{\partial \tau}\right)^{-1},$$

which is a consequence of (3.12). The above equation can be rewritten as follows

$$\begin{aligned} &M_0\left(M_0 + \frac{\partial M_0}{\partial \tau}\right)^{-1} \frac{\partial^2 M_1}{\partial \tau^2} \\ &\quad - \left[r^{-1} - 2\left(\frac{\partial}{\partial \tau}\left(M_0\left(M_0 + \frac{\partial M_0}{\partial \tau}\right)^{-1}\right) - M_0\left(M_0 + \frac{\partial M_0}{\partial \tau}\right)^{-1}\right]\frac{\partial M_1}{\partial \tau} \\ &\quad - \left[r^{-1} - M_0\left(M_0 + \frac{\partial M_0}{\partial \tau}\right)^{-1} - 2\frac{\partial}{\partial \tau}\left(M_0\left(M_0 + \frac{\partial M_0}{\partial \tau}\right)^{-1}\right)\right. \\ &\quad \left. - \frac{\partial^2}{\partial \tau^2}\left(M_0\left(M_0 + \frac{\partial M_0}{\partial \tau}\right)^{-1}\right)\right] M_1 = -f(r)^{-1}. \end{aligned} \tag{3.14}$$

Note that for $M_0 \equiv 1$ the homogeneous equation (3.14) has the solution M_1^0 satisfying the condition $\partial M_1^0/\partial \tau = 0$. We shall require that the solution of equation (3.14) with right-hand side satisfy the condition $M_1(\tau, x) = 0$ for $M_0 = 1$ (i.e., as $\tau \to \infty$). This yields a condition to be imposed on the right-hand side f of (3.14), namely,

$$\frac{\partial S_1}{\partial t} - 2r\langle \nabla S_0, \nabla S_1\rangle - \langle \nabla, r(x,t)\nabla S_0\rangle = 0. \tag{3.15}$$

By (3.15), the right-hand side of (3.14) can be transformed as follows

$$\begin{aligned} f(\tau, x) = &-\left(\frac{\partial S_0}{\partial t}\right)^{-1}\left\{\left(M_0 + \frac{\partial M_0}{\partial \tau}\right)(\langle \nabla, r\nabla S_0\rangle - \varkappa)\right. \\ &+ \langle \nabla, M_0 r \nabla S_0\rangle + \beta^{-1}(\tau - \nu)\left[M_0\left(\frac{\partial r}{\partial t} - \frac{\partial K}{\partial t}\right) - \frac{\partial M_0}{\partial \tau}\frac{\partial K}{\partial t}\right] \\ &\left. - M_0 r\left\langle \nabla S_0, \left(1 + \frac{\partial}{\partial \tau}\right)\left(M_0\left(M_0 + \frac{\partial M_0}{\partial \tau}\right)^{-1}\right)\nabla M_0\right\rangle + e^{\tau} F(x, -\psi(x), e^{\tau} m_0)\right\}. \end{aligned}$$

First of all, to obtain the final equation for M_1, we must eliminate the contradiction in (3.14) (the right-hand side depends on t, the left-hand side does not). This can be easily done if we note that, as $\tau \to \infty$, the right-hand side f tends to zero faster than any power of τ^{-1}. Therefore, precisely as in the preceding section, we use the Taylor expansion (3.11) and put

$$\frac{\partial S_0}{\partial t} = \left[\beta + 2\varepsilon\beta_1(\tau - \nu)/\beta + O(\varepsilon^2\tau^2)\right]_{\tau = (\beta(t+\psi) + \varepsilon\nu + \dots)/\varepsilon},$$

$$\nabla S_0 = \left[\beta \nabla \psi + +O(\varepsilon\tau)\right]_{\tau = (\beta(t+\psi) + \varepsilon\nu + \dots)/\varepsilon}.$$

Finally, putting $f = f\big|_{t=-\psi}$, we obtain

$$\check{f} = -\beta^{-1}\Bigg[\Big(M_0 + \frac{\partial M_0}{\partial \tau}\Big)(\nabla, r\beta\nabla\psi)2r\beta_1|\nabla\psi|^2 - \varkappa)
- \langle\nabla, M_0 r\beta\nabla\psi\rangle - 2M_0 r\beta_1|\nabla\psi|^2
- M_0 r\Big\langle\nabla\psi, \Big(1 + \frac{\partial}{\partial\tau}\Big)\Big(M_0\Big(M_0 + \frac{\partial M_0}{\partial\tau}\Big)^{-1}\Big)\nabla M_0\Big\rangle
+ e^{\tau}F(x, -\psi, e^{\tau}M_0) + \beta^{-1}(\tau - \nu)\Big(M_0\frac{\partial}{\partial t}(r - K) - \frac{\partial M_0}{\partial\tau}\frac{\partial K}{\partial t}\Big)\Bigg]\Bigg|_{t=-\psi}. \tag{3.16}$$

Note that the function $M_{10} = M_0 + \partial M_0/\partial\tau$ satisfies the homogeneous equation (3.14). Therefore, we can write the solution of (3.14) in the form

$$M_1 = c_1 M_{10} + c_2 M_{10}\int V(M_{10})^{-2}\,d\tau + M_{10}\int V(M_{10})^{-2}\int \frac{\check{f}M_{10}}{VK(M_0)}\,d\tau' d\tau, \tag{3.17}$$

where $K(M_0) = M_0/(M_0 + \partial M_0/\partial\tau)$, and V is the Wronskian

$$V = \exp(-2\tau)M_0^2\Big(M_0 + \frac{\partial M_0}{\partial\tau}\Big)^{-2}\exp\Big\{\int K(M_0)^{-1}\,d\tau\Big\} = \exp(-\tau)M_0^{-1}M_{10}^2.$$

Let us calculate the asymptotic expansion of the integrand in the inner integral in (3.17). By (3.12), (2.3), as $\tau \to \infty$, we have[2]*

$$\frac{\check{f}M_{10}}{VK(M_0)} \sim e^{(1-\gamma)\tau}, \tag{3.18}$$

and, as $\tau \to 0$,

$$\frac{\check{f}M_{10}}{VK(M_0)} \sim \tau^{2\alpha - 1/m - 1}, \qquad m = \lim_{\tau\to 0} M_0\Big/\Big(\tau\frac{\partial M_0}{\partial\tau}\Big).$$

A priori, two situations are possible:

$$2\alpha > 1/m \qquad \text{and} \qquad 2\alpha < 1/m.$$

As is easy to verify, since the numerator and denominator in the formula for m are linear in M_0 and the function M_0 behaves like a polynomial as $\tau \to 0$, we have $m = 1/\alpha$. Therefore $2\alpha > 1/m$, the function M_1 is determined by the relation

$$M_1 = M_{10}\int_0^{\tau} V(M_{10})^{-2}\int_0^{\tau'} \frac{fM_{10}}{VK(M_0)}\,d\tau'' d\tau' \tag{3.19}$$

[2]Recall that $v \sim w$ as $\tau \to a \Longleftrightarrow 0 < |\lim_{\tau\to a} v/w| < \infty$.

and exhibits the following asymptotic behavior as $\tau \to 0$:

$$M_1 \sim \tau^\alpha. \tag{3.20}$$

As $\tau \to \infty$, we have for the integrand in the inner integral:

$$\frac{f M_{10}}{VK(M_0)} \sim f e^\tau. \tag{3.21}$$

Since formula (3.16) implies

$$f \sim e^{-\gamma\tau},$$

the integrand (as $\tau \to \infty$) has the form

$$V(M_{10})^{-2} \int_0^\tau \frac{\check{f} M_{10}}{VK(M_0)}\, d\tau' \sim \begin{cases} e^{-\tau}, & \gamma > 1, \\ \tau e^{-\tau}, & \gamma = 1. \end{cases} \tag{3.22}$$

Therefore, if

$$\int_0^\infty V(M_{10})^{-2} \int_0^\tau \frac{\check{f} M_{10}}{VK(M_0)}\, d\tau' d\tau = 0, \tag{3.23}$$

we have the estimate for M_1 given in the statement of the theorem. Condition (3.23) determines the function $\nu(x)$:

$$\begin{aligned} \nu(x) = \Bigg\{ \int_0^\infty V(M_{10})^{-2} \int_0^\tau (VK(M_0))^{-1} \Bigg[\left(M_0 + \frac{\partial M_0}{\partial \tau} \right) ((\nabla, r\beta\nabla\psi) - \varkappa) \\ - \langle \nabla, M_0 r\beta\nabla\psi \rangle - M_0 r\beta \left\langle \nabla\psi, \left(1 + \frac{\partial}{\partial\tau}\right) \left(M_0 + \frac{\partial M_0}{\partial\tau} \right)^{-1} \nabla M_0 \right\rangle \\ + \tau' \left(M_0 \frac{\partial}{\partial t}(r - K) - \frac{\partial M_0}{\partial\tau}\frac{\partial K}{\partial t} \right) \Bigg] d\tau' d\tau \Bigg\} \\ \times \left\{ \int_0^\infty V(M_{10})^{-2} \int_0^\tau VK(M_0)^{-1} \left[M_0 \frac{\partial}{\partial t}(r - K) - \frac{\partial M_0}{\partial\tau}\frac{\partial K}{\partial t} \right] d\tau' d\tau \right\}^{-1} \Bigg|_{t=-\psi(x)} \end{aligned} \tag{3.24}$$

provided that the denominator does not vanish. The theorem is proved.

Precisely as in the preceding section, for $K \equiv K(u)$, the denominator in (3.24) vanishes, moreover, the integrand in (3.23) is independent of $\nu(x)$ (or the function S_1). Therefore, formula (3.23) relates the coefficients of the Taylor expansion of S_0 for $t = -\psi(x)$, namely, $\beta_1 = G(\nabla\psi, \nabla\beta, \beta)$, where $G = G(z, y, w)$ is a smooth function. Moreover, by substituting the expansion of $S_0(x,t)$ from (3.11) into (3.6) with $r \equiv 1$, we get

$$\langle \nabla\psi, \nabla|\nabla\psi|^{-2} \rangle = 2\beta_1 |\nabla\psi|^2, \qquad \beta = |\nabla\psi|^{-2}.$$

Thus, as in the preceding section, here the position of the boundary (the function $\psi(x)$) is determined independently of the solution of the Cauchy problem for equation (3.6), and hence, the constructed solution exists only for a special choice of initial conditions for equation (3.6).

Let us study the equation for characteristics (3.6) and the transport equation (3.7). Consider the Cauchy problem for (3.6) and assume that the set $\Omega_0 = \{x \mid S_0(x,0) \geq 0\}$ is a connected compact domain with smooth boundary.

Lemma 3.1. *In the domain $\Omega_t = \{x \mid S_0(x,t) \geq 0\}$ the solution of the Cauchy problem for equation (3.6) is determined only by the values of the initial conditions in the domain Ω_0.*

Proof. In coordinates on trajectories of the Hamilton system for equation (3.6), we have

$$S_0(x_0,t) = S_0(x_0,o) - \int_0^t r(x(x_0,t'),t')p^2(x_0,t')\,dt'$$

The second summand in the last formula is nonpositive, hence, if the left-hand side is nonnegative, then the first summand on the right-hand side is nonnegative.

The lemma is proved.

Note that the extremum points of the function $S_0^0(x) = S_0(x_0)$ are the projections of the fixed points of the Hamilton system

$$\begin{aligned} \dot{x}_i &= -2r(x,t)p_i, \qquad & x_i\big|_{t=0} &= x_0, \\ \dot{p}_i &= \frac{\partial r}{\partial x_i}|p|^2, \qquad & p_i\big|_{t=0} &= p_0 = \frac{\partial S_0^0}{\partial x_i}(x_0), \qquad i = 1,\dots,n. \end{aligned} \tag{3.25}$$

Obviously, if t increases, the trajectories, starting from narrowing neighborhoods of extremum points of the function $S_0^0(x)$ (they are also extremum points of the function $S^0(x,t)$), come to the boundary of the domain Ω_t. Further, for simplicity, we assume that the function $S_0^0(x)$ has only one extremum (the maximum) at the point $\overline{x} \in \Omega_0$, $r = r(t)$, and the denominator in (3.24) does not vanish. Then, as is easy to see, the solution of system (3.25) has the form

$$x = x_0 - 2p(x_0)\int_0^t r(t')\,dt', \qquad p(x_0,t) = p_0(x_0),$$

and the Jacobian

$$\frac{\partial x}{\partial x_o} \geq 1.$$

Thus, in this case, the solution exists for all $t > 0$, and the closer the initial point x_0 is to a point $\bar{x}$, the greater is the time $(\sim 1/p(x_0))$ that the trajectory, starting from x_0, takes to come to a point outside a fixed neighborhood of the point $\bar{x}$.

Recall another important property of equations with variable coefficients, namely, the fact that the wave front accelerates infinitely. For combustion waves (localized and nonlocalized) in the one-dimensional case, this follows from the formulas given in Chapter IV. Actually (see Theorem 2.1 in Chapter IV), the coordinate of the wave front $\varphi(t)$ satisfies the equation

$$\frac{d\varphi}{dt} = b\big(\gamma(-\varphi,t)\lambda(-\varphi,t)\big)^{1/2}, \tag{3.26}$$

where b is the velocity of the combustion wave in homogeneous media, $\lambda(x,t)$ is the thermal conductivity (diffusion), $\gamma(x,t)$ is the intensity of nonlinearity (the rate of chemical reactions). Obviously, if the inequality $|\gamma(x,t)\lambda(x,t)| \geq O(|x|^{2+\delta})$, $\delta > 0$, holds, then the solution of equation (3.26) becomes infinite during a finite interval of time. A similar property also takes place for equations considered here. Indeed, if $|r(x,t)| \geq O(|x|^{2+\delta})$, $|x| \to \infty$, then a localized asymptotic solution exists only for a finite time.

We restrict our consideration to the following example. Let

$$r(x,t) = (|x|^2 + 1)^2, \qquad S_0\big|_{t=0} = 1 - |x|^2,$$

provided that the denominator in (3.24) does not vanish. Then equation (3.6) can be solved in polar coordinates. Preserving the notation S_0 for the unknown function in variables $r = |x|$, t, we obtain

$$\frac{\partial S_0}{\partial t} - (r^2+1)^2\left(\frac{\partial S_0}{\partial r}\right)^2 = 0.$$

The radius $r = r(t)$ of the ball, inside of which the solution is positive, is determined by the system of equations:

$$\dot{r} = 2r_0(r_0^2+1)(r^2+1), \qquad 4r_0^2(r_0^2+1)^2 t = 1 - r_0^2. \tag{3.27}$$

This system has been derived from formulas (3.24), (3.25) under the condition that the Hamiltonian $-(r^2+1)^2p^2$ is constant along the trajectories of system (3.25). The system (3.27) can be written in the form

$$\sqrt{t}\dot{r} = (1 - r_0^2)^{1/2}(r^2+1).$$

The above arguments imply that there exists $t_2 > 0$, such that $1 - r_0^2 \geq r_1 > 0$ for $t \geq t_1$, $r_1 = \text{const}$. This can be proved by differentiating the second equation in (3.27) with respect to t and verifying that $\dot{r}_0 < 0$.

Hence, for $t \geq t_1$, we have the equation

$$\frac{dr}{dt'} = (1 - r_0^2)^{1/2}(r^2+1), \qquad t' = \frac{\sqrt{t}}{2},$$

whose right-hand side increases (as $r \to \infty$) faster than a linear function. Thus the solution to the last equation (as well as the solution to equation (3.27)) becomes infinite during a finite interval of time.

The algorithm for constructing asymptotic solutions given above can be used for constructing asymptotic solutions to the following quasilinear equations

$$\varepsilon\frac{\partial u}{\partial t} - \varepsilon^2\frac{\partial}{\partial x}K(x,t,u)\frac{\partial u}{\partial x} - \gamma(x,t)u = 0, \tag{3.28}$$

where, for $u > 0$, the transport coefficients $K(x,t,u)$ are smooth functions, satisfying the conditions

$$K(x,t,u) > 0, \quad u > 0, \qquad K(x,t,u) \sim u^{k-1}, \quad u \to 0,$$
$$K(x,t,u) = r(x,t) + \frac{r_1(x,t)}{u^{\gamma_0}} + \frac{r_2(x,t)}{u^{\gamma_1}} + \dots, \qquad u \to \infty,$$

where r, r_1, r_2 are smooth functions, $\gamma_0, \gamma_1 = \text{const} > 0$, $\gamma_1 > \gamma_0$.

We have already seen that the Lagrangian plays a large part in investigation of the Hamilton–Jacobi equation arising when asymptotic solutions are constructed. In our case the Lagrangian has the form

$$L = -Ap^2 - \gamma(x,t).$$

Obviously, the inequality $\gamma(x,t) \geq 0$ implies

$$L < 0. \tag{3.29}$$

Therefore, in this case, the algorithm for constructing the function $S(x,t,\varepsilon)$ is completely preserved. Generally speaking, for $\gamma < 0$, the inequality (3.29) does not hold. Therefore, in the domain $S(x,t) \geq 0$, the solution $S(x,t)$ of the corresponding Hamilton–Jacobi equation cannot be determined only by the initial conditions posed in this domain for $S(x,0)$.

Let us briefly review the results obtained by the above algorithm applied to equations of the form

$$\frac{\partial u}{\partial t} - \varepsilon\frac{\partial}{\partial x}K(x,t,u)\frac{\partial u}{\partial x} = 0$$

with coefficient $K(x,t,u)$ increasing as $u \to \infty$.

As an example consider the equation

$$\frac{\partial u}{\partial t} - \varepsilon\frac{\partial^2 u^2}{\partial x^2} = 0. \tag{3.30}$$

By substituting the solution of the form

$$u(x,t,\varepsilon) = W(S/\varepsilon, x, t) + \varepsilon W(S/\varepsilon, x, t),$$

we obtain the equation for $W_0(\tau, x, t)$:

$$\frac{\partial S}{\partial t}\frac{\partial W_0}{\partial \tau} - \left(\frac{\partial S}{\partial x}\right)^2 \frac{\partial^2 W}{\partial \tau^2} = 0.$$

Thus $W_0 = A(x,t)\tau$. Hence, according to our algorithm, the solution must be constructed in the form

$$u(x,t,\varepsilon) = (S/\varepsilon)(A_0 + \varepsilon A_1 + \ldots). \tag{3.31}$$

Note that, in this case, by the theorem about the structure of singularities (see Chapter III), a localized solution satisfying the flow continuity condition must satisfy the condition

$$u(x,t,\varepsilon) = O(S) \qquad \text{for} \quad S \to 0.$$

Hence, $A_0\big|_{S=0} \neq 0$, $A_1\big|_{S=0} \neq 0$. This is the difference between equations with growing coefficients and equations with already stabilized coefficients. The substitution of the solution in the form (3.31) gives a system of equations for the functions A and S. The system turns out to be exactly the same as that obtained by the substitution of the solution of the form $u = AS/\varepsilon$ into equation (3.30) (i.e., we can put $A \equiv 1$). This fact can be easily explained: the change of variables $u = \varepsilon v$ eliminates the parameter ε from (3.30) and, hence, any solution of the form (3.31) is a regular perturbation to the exact solution of the initial equation.

In the following section we present a certain algorithm for constructing asymptotic solutions of equations of the form (3.30) (and of more general equations) based on generalization of the localizing transformation introduced in Chapter I.

5.4. Relation between approximate solutions of quasilinear parabolic and parabolic equations

In this section, based on the results of [11], we present formulas for asymptotic solutions of the quasilinear equation

$$\frac{\partial z}{\partial \delta} - \varepsilon \frac{\partial}{\partial \xi}\left(K(z)\frac{\partial z}{\partial \xi}\right) = 0 \tag{4.1}$$

in terms of solutions of the nonlinear equation with variable coefficients and nonlinearity

$$\varepsilon \frac{\partial u}{\partial t} - \varepsilon^2 \frac{u_x}{(u\mu)_x}\frac{\partial^2 u}{\partial x^2} = \varepsilon \mathcal{F}(u, \varepsilon u_x, \varepsilon u_t, \varepsilon^2 u_{xx}, \varepsilon^2 u_{tx}), \tag{4.2}$$

where $\mu = \mu(x,t,\varepsilon) = \mu_0(x,t) + \varepsilon\mu_1(x,t) + \dots$ are unknown functions, $\varepsilon \le 1$ is a small parameter. In particular, the formulas obtained in this section relate simple wave solutions (i.e., functions of $x + bt$, $b =$ const) of quasilinear and semilinear parabolic equations (see [15], [51, p.67], [92]).

Here and below, we use the notation $\partial(\cdot)/\partial\varkappa = (\cdot)_\varkappa$, where $\varkappa$ is a variable from the set $(x, t, \delta, \xi, \dots)$. Similarly, $\partial^2(\cdot)/\partial\varkappa^2 = (\cdot)_{\varkappa\varkappa}$, etc. In the leading (with respect to ε) terms, the functions $\mathcal{F}$ are homogeneous of 0-degree with respect to the function u. Therefore, though the initial equation is nonlinear, we can construct the solution in the form of WKB-asymptotics

$$u = \exp\frac{S(x,t,r)}{\varepsilon}.$$

To find the function $z = z(\delta, \xi, \varepsilon)$, satisfying equation (4.1), we must solve the following system of equations, which is an analog of the transformation introduced in [15]:

$$z(\xi(x,t,\varepsilon), \delta(x,t,\varepsilon)) = \mu(x,t,\varepsilon)u(x,t,\varepsilon), \tag{4.3}$$

$$K(z)z_\xi = u_x, \quad K(z)z_\delta = u_t, \tag{4.4}$$

$$u(x,t,\varepsilon) = \frac{1}{\varepsilon^\alpha}\exp\frac{S(x,t,\varepsilon)}{\varepsilon}, \tag{4.5}$$

where $S(x,t,\varepsilon) = S_0(x,t) + \varepsilon S_1(x,t) + \dots$, $\alpha = (k-1)^{-2}$, $K(z) = kz^{k-1}$. The functions $\xi(x,t,\varepsilon)$, $\delta(x,t,\varepsilon)$ are determined by the system of equations

$$\begin{aligned} \mu u_x \det M &= K(\mu u)u((-K(\mu u) + \delta_t/\mu)\mu_x - \delta_x\mu_t/\mu), \\ \mu u_t \det M &= K(\mu u)u((-K(\mu u) + \xi_x/\mu)\mu_t - \xi_t\mu_x/\mu), \end{aligned} \tag{4.6}$$

where $\det M = K(\mu u)(K(\mu u) - \xi_x/\mu - \delta_t/\mu) + \det J/\mu^2$. The matrices M and J are determined below. To obtain formulas suitable for calculations, we perform some useful transformations. By (4.3), (4.4), we get

$$\begin{aligned} Kz_\xi &= u_x = (z_\xi\xi_x + z_\delta\delta_x)/\mu - z\mu_x/\mu^2, \\ Kz_\delta &= u_t = (z_\xi\xi_t + z_\delta\delta_t)/\mu - z\mu_t/\mu \end{aligned} \tag{4.7}$$

By using the relations

$$\frac{\partial x}{\partial \xi}=\frac{\partial \delta}{\partial t}\frac{1}{\det J},\qquad \frac{\partial x}{\partial \delta}=-\frac{\partial \xi}{\partial t}\frac{1}{\det J},\qquad \frac{\partial t}{\partial \xi}=-\frac{\partial \delta}{\partial x}\frac{1}{\det J},\qquad \frac{\partial t}{\partial \delta}=\frac{\partial \xi}{\partial x}\frac{1}{\det J},$$

we obtain two systems from (4.7):

$$M\widetilde{\nabla} z=\frac{u\nabla\mu}{\mu}, \tag{4.8}$$

where

$$M=\begin{pmatrix} -K(\mu u)+\mu^{-1}\xi_x & \mu^{-1}\delta_x \\ \mu^{-1}\xi_t & -K(\mu u)+\mu^{-1}\delta_t \end{pmatrix},\qquad \left(\widetilde{\nabla}=\frac{\partial}{\partial\xi},\frac{\partial}{\partial\delta}\right),$$

and

$$J\widetilde{\nabla} z=\mu\nabla u, \tag{4.9}$$

where

$$\left(\nabla=\frac{\partial}{\partial x}\frac{\partial}{\partial t}\right),\qquad J=\begin{pmatrix} \xi_x & \delta_x \\ \xi_t & \delta_t \end{pmatrix}.$$

Assume that $\det M\neq 0$. Then, by (4.8),

$$\widetilde{\nabla} z=\frac{u}{\mu}M^{-1}\nabla\mu, \tag{4.10}$$

and hence we obtain system (4.6).

Definition. The transformation given by formulas (4.3)–(4.6) will be called an *lr-transformation.*

Note that, for any functions $\xi=f_1(S/\varepsilon,x,t,\varepsilon)$ and $\delta=f_2(S/\varepsilon,x,t,\varepsilon)$, the coefficient at ε^{-2} vanishes in the expression $\det J=\xi_x\delta_t-\delta_x\xi_t$, i.e., the power of ε in the contribution of the determinant J is always less by one than the maximum possible. Therefore, we have

$$\det J=\frac{\mu}{u_x}\left[-K^2(\mu u)\cdot(\mu u)_x+K(\mu u)\cdot\frac{(\mu u)_x}{\mu}+\frac{K(\mu u)\cdot(-u\mu_t\delta_x+\xi_x\mu u_x)}{\mu}\right], \tag{4.11}$$

where

$$\begin{aligned} K(z)L(z) &\equiv K(z)\left[\varepsilon z_\delta-\varepsilon^2(K(z)z_\xi)_\xi\right] \\ &=\varepsilon u_t-\varepsilon^2K(\mu u)\left(\frac{\partial\delta}{\partial t}\frac{\partial^2 u}{\partial x^2}-\frac{\partial\delta}{\partial x}\frac{\partial^2 u}{\partial t\partial x}\right)\frac{1}{\det J} \\ &=\varepsilon u_t-\frac{\varepsilon^2 u_x}{(u\mu)_x}u_{xx}+\varepsilon^2\frac{K\mu}{\det J}\left[\left(\frac{(-u\mu_t\delta_x+\xi_x\mu u_x)}{(u\mu)_x}-K\right)u_{xx}+\frac{\delta_x u_{tx}}{\mu}\right]. \end{aligned} \tag{4.12}$$

Let us show that the first two summands in the latter expression play the leading part as $\varepsilon\to 0$. Let $f_i(\tau,x,t,\varepsilon)\sim\varepsilon^{-\beta}$, $i=1,2$. Then,

$$\det J=O(\varepsilon^{-(2\beta+1)}).$$

Similarly, $\det M = O(\varepsilon^{-(2\beta+1)})$. By setting $\beta = \alpha(k-1)$, we can verify that the first two summands on the right-hand side of (4.11) are of the order $\varepsilon^{-\alpha}$, and the other satisfy the estimate $O(\varepsilon^{-\alpha+1})$. If we set the right-hand side of (4.12) equal to zero, we get equation (4.2).

For each specific type of nonlinearity in (4.1), the existence of a small parameter allows to construct approximate formulas for functions $\xi(x,t,\varepsilon)$, $\delta(x,t,\varepsilon)$ (see Examples).

The functions $\xi(x,t,\varepsilon)$, $\delta(x,t,\varepsilon)$ are sought in the form:

$$\begin{aligned} \xi(x,t,\varepsilon) &= \frac{B(x,t,\varepsilon)}{\varepsilon^{\beta}} \exp\frac{S}{\varepsilon}, & B &= B_0(x,t) + \varepsilon B_1(x,t) + \dots, \\ \delta(x,t,\varepsilon) &= \frac{A(x,t,\varepsilon)}{\varepsilon^{\beta}} \exp\frac{S}{\varepsilon}, & A &= A_0(x,t) + \varepsilon A_1(x,t) + \dots. \end{aligned} \tag{4.13}$$

Obviously, for system (4.4) to be solvable, it is necessary and sufficient that the gradients are equal to each other:

$$\widetilde{\nabla}\Big(\int K(z)\,dz\Big)\Big|_{\xi=\xi(x,t,\varepsilon),\delta=\delta(x,t,\varepsilon)} = \nabla u. \tag{4.14}$$

This equation for the functions ξ and δ can be rewritten as the condition for the derivatives to be equal to each other:

$$(K(z)z_\xi)_\delta = (K(z)z_\delta)_\xi. \tag{4.15}$$

Clearly, this equality can be examined from the viewpoint of asymptotic expansions of the functions z, ξ, δ with respect to a small parameter. Namely, if this equality holds $\mathrm{mod}\,O(\varepsilon^N)$, then the function z exists $\mathrm{mod}\,O(\varepsilon^M)$, and the function $M = M(N)$ increases infinitely as $N \to \infty$.

By (4.4), (4.5), for the function z, we obtain the following equation of the first (!) order

$$z_\xi + g(\xi,\delta,\varepsilon)z_\delta = 0, \tag{4.16}$$

where $g(\xi,\delta) = (S_x/S_t)\big(x(\xi,\delta,\varepsilon), t(\xi,\delta,\varepsilon)\big)$.

Thus, we have already proved the main result of this section:

Theorem 4.1. *Let the functions u, ξ, δ satisfy system* (4.3), (4.4), (4.5) *with* $O(\varepsilon^N)$, $N > 1$. *Then the function z satisfies equation* (4.1) *with* $O(\varepsilon^N)$.

Further, we want to obtain a closed noncontradictory system of equations for the functions S_0, S_1, μ_0, A_0, B_0, A_1, B_1 and to show that, using this method, we can construct asymptotic solutions of (4.1).

As already pointed out, equation (4.2) can be written as follows

$$\varepsilon\frac{\partial u}{\partial t} - \varepsilon^2\frac{u_x}{(u\mu)_x}u_{xx} + \varepsilon^2\frac{K(\mu u)\mu}{\det J}\left[\left(\frac{(-u\mu_t\delta_x + \xi_x\mu u_x)}{(u\mu)_x} - K\right)u_{xx} + \frac{\delta_x u_{tx}}{\mu}\right] = 0. \tag{4.17}$$

In the leading term, this equation is approximately homogeneous with respect to the parameter ε and precisely homogeneous for $k = 2$. We illustrate this statement in Example 4.

Substituting expressions for u from (4.5) and for δ from (4.13) into (4.17) and setting the coefficients at equal powers of ε equal to zero, we obtain

$$\varepsilon^{-\alpha}: S_{0t} - \frac{1}{\mu_0} S_{0x}^2 = 0, \tag{4.18}$$

$$\varepsilon^{-\alpha+1}: S_{1t} - S_{0xx}/\mu_0 - 2S_{0x}S_{1x}/\mu_0 - 2\mu_1(S_{0x})^2/\mu_0 + S_{0x}\mu_{0x}/\mu_0^2 + \frac{[S_{0x}^2(B_0 S_{0x} + A_0 S_{0t}/\mu_0]}{(A_0 + B_0)} = 0. \tag{4.19}$$

Example 4.1. Let us illustrate our method for $K(z) = 2\lambda z$. The functions $\xi(x,t,\varepsilon)$, $\delta(x,t,\varepsilon)$ are determined by system (4.6) and have the form

$$\begin{aligned} \xi(x,t,\varepsilon) &= [B_0(x,t) + \varepsilon B_1(x,t,\varepsilon)]\frac{1}{\varepsilon}\exp\frac{S(x,t,\varepsilon)}{\varepsilon}, \\ \delta(x,t,\varepsilon) &= [A_0(x,t) + \varepsilon A_1(x,t,\varepsilon)]\frac{1}{\varepsilon}\exp\frac{S(x,t,\varepsilon)}{\varepsilon} + \widehat{C}, \quad \widehat{C} = \text{const}, \\ \mu(x,t,\varepsilon) &= \mu_0(x,t) + \varepsilon\mu_1(x,t). \end{aligned} \tag{4.20}$$

By (4.6), we get

$$\widetilde{A}\rho = 2\lambda\mu^2\rho, \qquad \widetilde{A} = \begin{pmatrix} S_t & S_x \\ S_t & S_x \end{pmatrix}, \tag{4.21}$$

where $\rho = (A_0, B_0)$ is a column vector. For the next approximation, we have

$$\widetilde{A}(A_1, B_1) = F_1. \tag{4.22}$$

The function B_0 can be found from the equation

$$\langle l^*, F_1 \rangle = 0. \tag{4.23}$$

This is the orthogonality condition for the right-hand side of (4.21) and the vector $l^* = (1, -1)$, which is the zero eigenvector of the matrix $\widetilde{A}^*$. Here, though $\det \widetilde{A} = 0$, the system (4.22) is solvable.

The condition (4.15) has the form

$$u_{xt}\xi_x - u_{xx}\xi_t - u_{xt}\delta_t + u_{tt}\delta_x = 0. \tag{4.24}$$

Thus, we obtain a chain of relations at the powers of the small parameter ε (we write the first two equations):

$$\frac{1}{\varepsilon^4}: -S_x^2 S_t B + S_x S_t(BS_x - S_t A) + S_t^2 A S_x = 0 \tag{4.25}$$

$$\frac{1}{\varepsilon^3}: -S_{0x}^2 B_{0t} - S_{0xx}S_{0t}B_0 + S_{0x}S_{0t}(B_{0x} - A_{0t}) + S_{0xt}(B_0 S_{0x} - A_0 S_{0t}) + S_{0tt}A_0 S_{0xt} + S_{0t}^2 A_{0x} = 0, \tag{4.26}$$

As is easy to see, (4.25) holds for any functions S, A, B. Thus, we have seven equations (4.18), (4.19), (4.21), (4.22), (4.23), (4.26) for seven unknown functions S_0, S, μ_0, A_0, B_0, A_1, B_1, which we must solve under certain initial conditions.

Example 4.2. For $\mu = \text{const}$, by (4.6), the matrix M is degenerate and equalities (4.6) take the form

$$\begin{aligned} \delta_x &= \mu(K(\mu u) - \xi_x/\mu)u_x/u_t, \\ \xi_t &= \mu(K(\mu u) - \delta_t/\mu)u_t/u_x. \end{aligned} \tag{4.27}$$

Actually, in this case, the matrix of system (4.8) has the zero eigenvalue and the corresponding eigenvector l, i.e., the gradient of z is proportional to the *recalculation function*:

$$\widetilde{\nabla} z = l \cdot \operatorname{lr}(x,t,\varepsilon), \tag{4.28}$$

where

$$l = \begin{pmatrix} -K(\mu u) + \mu^{-1}\delta_t \\ -\mu^{-1}\xi_t \end{pmatrix} = \begin{pmatrix} \mu^{-1}\delta_x \\ K(\mu u) - \mu^{-1}\xi_t \end{pmatrix}$$

is the eigenvector of (4.8). The recalculation formula has the form

$$\operatorname{lr}(x,t,\varepsilon) = \frac{\langle \widetilde{\nabla} z, l\rangle}{|l|^2} = \frac{\langle \nabla u, l\rangle}{|l|^2 K(\mu u)}. \tag{4.29}$$

Then, (4.28), (4.29) imply (4.27). We obtain

$$\begin{aligned} K(z)L(z) &\equiv K(z)\big[\varepsilon z_\delta - \varepsilon^2 (K(z) z_\xi)_\xi\big] \\ &= \varepsilon u_t - \frac{\varepsilon^2}{\mu} u_{xx} + \varepsilon^2 \frac{K}{\det J}\delta_x u_x \left(\frac{u_t}{u_x}\right)_x, \end{aligned} \tag{4.30}$$

where

$$\det J = \xi_x \delta_t - \xi_t \delta_x = K(\mu u)\mu^2(\xi/\mu + \delta_t/\mu - K(\mu u)).$$

So, we have a system of equations for the coefficients in (4.13):

$$\begin{aligned} A_0 S_{0x} + B_0 S_{0x}^2/S_{0t} &= 2\mu^2 \lambda S_{0x}/S_{0t}, \\ B_0 S_{0t} + A_0 S_{0t}^2/S_{0x} &= 2\mu^2 \lambda S_{0t}/S_{0x}. \end{aligned} \tag{4.31}$$

The general solution of system (4.31) has the form

$$\begin{pmatrix} A_0 \\ B_0 \end{pmatrix} = \begin{pmatrix} C(x,t)S_{0x} + 2\mu^2\lambda S_{0x}/[S_{0t}^2 + S_{0t}S_{0x}] \\ -C(x,t)S_{0t} + 2\mu^2\lambda S_{0t}/[S_{0x}^2 + S_{0t}S_{0x}] \end{pmatrix}. \tag{4.32}$$

For the next approximation (A_1, B_1), we have the system

$$\widehat{A}(A_1, B_1) = \widetilde{F}_1, \tag{4.33}$$

where $\widehat{A}$ is the matrix of system (4.31), and the right-hand side $\widetilde{F}_1$ can be easily written out. The function $C(x,t)$ in (4.32) is determined by the equation

$$\langle l^*, \widetilde{F}_1\rangle = 0. \tag{4.34}$$

This is the condition that the right-hand side of (4.33) and the zero eigenvector $l^* = (S_{0t}^2, -S_{0x}^2)$ of matrix $\widehat{A}^*$ are orthogonal to each other.

The last summand in the transport equation (4.19) has the form

$$f = A_0 S_{0x}^2 \frac{\partial}{\partial x}\left(\frac{S_{0t}}{S_{0x}}\right)(\mu_0 B_0 S_{0x} + \mu_0 A_0 S_{0t} - 2\mu_0^3 \lambda)^{-1}. \tag{4.35}$$

Example 4.3. We consider the case $K = 2z$. In particular, if $S_0 = \mu_0 x^2/(4(c-t))$, $c = \text{const}$, we get $\mu \equiv \mu_0 = 3/2$, $S \equiv S_0$, and equation (4.15) takes the form

$$z_\xi + \frac{2\xi}{\widehat{C} - \delta} z_\delta = 0, \quad \widehat{C} = \text{const}.$$

Further,

$$\xi = \frac{18(c-t)}{x\varepsilon} \exp \frac{S}{\varepsilon}, \qquad \delta = \widehat{C} - \frac{18(c-t)^2}{x^2\varepsilon} \exp \frac{S}{\varepsilon}, \qquad z = \frac{\xi^2}{12(\widehat{C} - \delta)\varepsilon}.$$

Remark. By (4.24), we obtain the second version of the transformation $\xi_x = \mu K$, $\delta_t = \mu K$. The analysis of this situation is beyond the frames of our book.

Note the following property of (4.12).

Consider $\det J$ (4.11). The estimates hold

$$\begin{aligned} K^2(\mu u) \cdot (\mu u)_x &= O(\varepsilon^{-(2\beta+\alpha+1)}), \\ K\delta_t \frac{(\mu u)_x}{\mu} &= O(\varepsilon^{-(2\beta+\alpha+2)}), \\ K \cdot \xi_x u_x &= O(\varepsilon^{-(2\beta+\alpha+2)}). \end{aligned} \tag{4.36}$$

Thus the summand in (4.11), quadratic with respect to K, is the lowest in ε. The leading summands in equation (4.12) are homogeneous in ε. Consider equation (4.1), where $K(z) = z^2$. This nonlinearity corresponds to $k = 3$, $\alpha = 1/4$, $\beta = 1/2$. For this equation, we can write the system (4.18), (4.19), (4.22), (4.25), (4.26). On the other hand, if $K(z) = 3z^2 + 4z^3 > 0$, then we can choose either $k = 3$, $\alpha = 1/4$, $\beta = 1/2$, or $k = 4$, $\alpha = 1/9$, $\beta = 1/3$. In both cases, we can again write the system of equations (4.18), (4.19), (4.22), (4.25), (4.26). It turns out that, by (4.36), equation (4.19) is the same in all three cases.

Example 4.4.

Consider the quasilinear parabolic equation with linear source function

$$\varepsilon \frac{\partial w}{\partial \varkappa} - \varepsilon^2 \frac{\partial^2 w^k}{\partial \xi^2} - \gamma w = 0, \qquad w = w(\varkappa, \xi, \varepsilon). \tag{4.37}$$

As shown in Chapter I (Theorem 1.1), if we consider the simple wave solutions $w = W(b\varkappa + \xi)$ and use the localizing transformation, we get the semilinear equation

$$\varepsilon \frac{du}{d\tau} - \varepsilon^2 \frac{d^2 u}{d\tau^2} - \gamma u^{k-2} = 0.$$

If in (4.37) we change the variables:

$$w = \exp(\gamma\varkappa) z(\delta(\varkappa), \xi), \quad \delta(\varkappa) = \frac{\exp(\gamma(k-1)\varkappa) - 1}{\gamma(k-1)},$$

we obtain the quasilinear parabolic equation without source:

$$\frac{\partial z}{\partial \delta} - \frac{\partial^2 z^k}{\partial \xi^2} = 0.$$

By using the lr-transformation, we reduce the latter equation to a linear parabolic equation (see Example 4.3.).

CHAPTER VI

MODELS FOR MASS TRANSFER PROCESSES

In this chapter we discuss mathematical models for processes of precipitation and coprecipitation of ions. From the viewpoint of differential equations these models are boundary-value problems for systems of quasilinear parabolic equations. In some cases considered in this chapter the system consists of two equations: a quasilinear (or linear) parabolic equation and an ordinary (or algebraic) nonlinear equation. For example, the processes described by these models can be used to extract, concentrate and separate radionuclides. Not only qualitative, but also quantitative estimates of parameters of these processes are of great practical and scientific importance. In particular, it is necessary to study the laws of precipitation and coprecipitation of radionuclides in order to be able to predict how they spread in ecological chains, which is important for environment protection and public health services.

The dynamic precipitation and coprecipitation of ions occur when electrolyte solutions are filtered through dispersive media. The parameters of this process can vary in time and space with temperature gradient, inhomogeneities of filtering media, etc.

In this chapter we construct asymptotic solutions satisfying mathematical models of physical and chemical processes of heterogeneous ion distribution and the distribution of ions between the phases. Different versions of the isotherm equation describe the precipitation of hydroxides of metals both with and without aquahydrocomplex formation. The obtained results can be used for calculation of the precipitation chromatography [32] and in other fields of physical chemistry and chemical technology.

6.1. Nonstationary models of mass transfer

We shall consider the heterogeneous process of filtering a solution of an electrolyte through dispersive media. The rate of filtration may be variable because of changes in external conditions. The electrolyte solution contains ions of a precipitated substance (cations) and ions of a precipitator (anions). Between these ions of two types a chemical reaction takes place. The precipitate contains ions of the substance precipitated, and a layer of sorbate grows on the surface of the dispersive medium through which the solution is filtered. It is assumed that the diffusion of the precipitated substance takes place.

Let us consider the dimensionless equation describing the law of mass conservation:

$$\frac{\partial v}{\partial t} + u(y,t)\frac{\partial v}{\partial y} + \frac{1}{m}\frac{\partial \varphi}{\partial t} = \varepsilon\frac{\partial}{\partial y}\Big(D(y,t)\frac{\partial v}{\partial y}\Big), \qquad (1.1)$$

$$\varepsilon = \mathrm{Pe}^{-1}, \qquad m = \mathrm{const}.$$

Here v is the dimensionless concentration of the sorbate in the solution (the moving phase), φ is the dimensionless concentration of the same substance in the stationary phase, $u(y,t)$ is the linear dimensionless transfer velocity of the moving phase, $D(y,t) > 0$ is the dimensionless coefficient of quasidiffusion (depending on the spatial coordinate and time in the case of a slowly perturbed medium), Pe is the Péclet number, y is the dimensionless coordinate along the flow, t is the dimensionless time.

To this equation we add a simple dimensionless relation describing the kinetics of sorption:

$$\frac{\partial\varphi}{\partial t} = F(v,\varphi). \tag{1.2}$$

In the case of dynamic equilibrium, the kinetic equation can be replaced by the isotherm equation

$$\varphi = f(v). \tag{1.3}$$

Then the obtained system describes the equilibrium sorption. In particular, the isotherm can be of the form $\varphi = a - b/v^{1/\lambda}$, $a, b, \lambda = \text{const} > 0$ and the relation $a - b = 1$ always holds. This form of isotherm results form the mass balance in the chemical reaction and the law of mass action, λ is a constant equal to the ratio between the stoichiometric coefficients of the reaction.

The system of equations describing the processes of precipitation is derived in [32, 81].

We shall construct an asymptotic solution of the problem modeling the equilibrium dynamics of molecule transport:

$$\begin{gathered}
\frac{\partial v^-}{\partial t} + u\frac{\partial v^-}{\partial y} + \frac{1}{m}\frac{\partial\varphi}{\partial t} = \varepsilon D\frac{\partial^2 v^-}{\partial y^2}, \\
v^- = \left(\frac{b}{a-\varphi}\right)^\lambda, \qquad \text{for} \quad \varphi > 0, \\
\frac{\partial v^+}{\partial t} + u\frac{\partial v^+}{\partial y} = \varepsilon D\frac{\partial^2 v^+}{\partial y^2}, \qquad \text{for} \quad \varphi \equiv 0, \quad -\infty < y < +\infty, \\
v^-\Big|_{y\to 0} \to 1_{-0}, \\
v^\pm\Big|_{y=-ct} = \left(\frac{b}{a}\right)^\lambda, \qquad \frac{\partial v^-}{\partial y}\Big|_{y=-ct} = \frac{\partial v^+}{\partial y}\Big|_{y=-ct}, \\
\varphi\Big|_{y\geqslant -ct} \equiv 0, \qquad v^+\Big|_{y\to\infty} \to 0.
\end{gathered} \tag{1.4}$$

This boundary conditions correspond to the situation when the precipitation column input is at a far distance in the negative area, and inside the column the ion front (discontinuous for $D \equiv 0$) propagates along the y-axis. The front velocity c is an unknown constant we are looking for. Thus the filtration problem can be formulated as a problem with free boundary.

Assume that the flow velocity and the diffusion coefficient are constant. By setting

$$v^- = v^-(\tau), \qquad \varphi = \varphi(\tau), \qquad \tau = \frac{y+ct}{\varepsilon},$$

we obtain the following ordinary differential equation for the function $v^-(\tau)$:

$$c\frac{dv^-}{d\tau} + u\frac{dv^-}{d\tau} + \frac{c}{m}\frac{d}{d\tau}\left(a - \frac{b}{v^{1/\lambda}}\right) = D\frac{d^2v}{d\tau^2}. \tag{1.5}$$

The boundary conditions for $v^-(\tau)$ have the form:

$$v^-(-\infty) = 1, \qquad \frac{dv^-}{d\tau}(-\infty) = 0, \qquad v^-(0) = \left(\frac{b}{a}\right)^\lambda.$$

By integrating (1.5), we obtain the equation of the first order

$$(c+u)v^- + \frac{c}{m}\left(a - \frac{b}{v^{1/\lambda}}\right) = D\frac{dv^-}{d\tau} + E, \tag{1.6}$$

where E is a constant of integration.

By setting (due to the boundary conditions) $v^- = 1$, $\frac{dv^-}{d\tau} \to 0$ as $\tau \to -\infty$ in (1.6) and taking into account that $a - b = 1$, we get

$$E = c + u + \frac{c}{m}.$$

The solution of equation (1.6) is given by the formula:

$$\tau = D\int_{(a/b)^\lambda}^{v^-} \frac{v^{1/\lambda}\,dv}{(u+c)v^{1+1/\lambda} - \left(\frac{c}{m}(1-a) + u + c\right)v^{1/\lambda} - \frac{cb}{m}}. \tag{1.7}$$

The denominator in the integrand vanishes at the point $v^- = 1$ since $1 - a = b$. We denote the improper integral on the right-hand side by $I_1(v^-)$, then $\tau = DI_1(v^-)$. Generally speaking, the integral $I_1(v^-)$ diverges as $v^- \to 1 - 0$ (i.e., $\tau \to -\infty$, $v^- \to 1 - 0$), but the inverse function exists because the assumptions of the inverse function theorem are satisfied.

By (1.7), the function v^- tends exponentially to 1 as $\tau \to -\infty$ since

$$-c < \frac{u}{1 + \frac{b}{m\lambda}}. \tag{1.8}$$

Obviously, the solution determined by (1.7) is not unique, since it depends on an unknown constant c. However, in our case, in the domain $\varphi \equiv 0$ the solution to the linear problem is not of the exponential boundary layer character, as it was in Chapter IV. Therefore, here the matching conditions are the continuity conditions for the functions $v^+\big|_{y=-ct} = v^-\big|_{y=-ct}$ and the flows $\frac{\partial v^+}{\partial y}\big|_{y=-ct} = \frac{\partial v^-}{\partial y}\big|_{y=-ct}$. We restrict ourselves to matching only the principal terms of the solutions v^- and v^+.

For $\tau = 0$ ($y = -ct$), equation (1.6) implies

$$\left.\frac{\partial v^-}{\partial y}\right|_{y=-ct} = (D\varepsilon)^{-1}\left[(u+c)\left(\frac{b}{a}\right)^\lambda - \left(u + c + \frac{c}{m}\right)\right]. \tag{1.9}$$

We shall calculate the solution in the domain $\varphi \equiv 0$. We change the variables so that the left boundary of this domain remains stationary: $y \to z - ct$. Preserving the notation for the unknown function, we obtain the following problem

$$\frac{\partial v^+}{\partial t} + (u+c)\frac{\partial v^+}{\partial z} = \varepsilon D \frac{\partial^2 v^+}{\partial t^2},$$
$$v^+\Big|_{z=0} = +\left(\frac{b}{a}\right)^{\lambda}, \qquad v^+\Big|_{z\to\infty} \to 0. \tag{1.10}$$

Further, we show that, for $t > 0$ and any smooth initial conditions $\Psi(z)$, the contribution of this smooth initial condition to the derivative

$$\frac{\partial v^+}{\partial y}\Big|_{z=0}$$

is of order less than ε^{-1}, more precisely, the contribution is of the order $\varepsilon^{-1/2}$ or less.

Actually, the problem

$$\frac{\partial v^+}{\partial t} = (u+c)\frac{\partial v^+}{\partial z} + \varepsilon D \frac{\partial^2 v^+}{\partial z^2},$$
$$v^+\big|_{z=0} = 0, \qquad v^+\big|_{t=0} = \psi(z,\varepsilon), \qquad z \geqslant 0,$$

has the solution of the form [28]:

$$v^+(z,t) = \frac{1}{2\sqrt{\pi t\varepsilon}} \int_0^\infty \left(\exp\left\{ -\frac{\left(z-\xi-(u+c)t\right)^2}{4t\varepsilon} \right\} \right.$$
$$\left. \times \exp\left\{ -\frac{(z+\xi)^2 + (2\xi - 2z + (u+c)t)(u+c)t}{4t\varepsilon} \right\} \right) \psi(\xi,\varepsilon)\, d\xi.$$

Hence, for $t \geqslant \delta > 0$,

$$\frac{\partial v^+}{\partial z}\Big|_{z=\infty} = \frac{1}{2\sqrt{\pi t\varepsilon}} \int_0^\infty \frac{\xi + (u+c)t}{t\varepsilon} \exp\left\{ -\frac{\left(\xi+(u+c)t\right)^2}{4t\varepsilon} \right\} \psi(\xi,\varepsilon)\, d\xi$$
$$= O\left(\|\psi\|_{C(\mathbf{R}^1)}\right).$$

Here we took into account the inequality $u + c > 0$ that is a consequence of (1.8).

Thus, to obtain the required derivative, it is sufficient to consider the case $C \equiv 0$. Here it is convenient to use the formula for asymptotic solution of the boundary-value problem

$$u = -\int_0^t [2\varepsilon G'_x(x,t,\tau)]\nu(\tau)\, d\tau,$$

where

$$G = 2(\pi\varepsilon(t-\tau))^{-1/2} \exp\left\{ -\frac{(z+(u+c)(t-\tau))^2}{4\varepsilon(t-\tau)} \right\},$$

and $\nu(t)$ is the boundary condition, namely, $v^+\big|_{z=0} = \nu(t)$.

Analysing this formula, we get the following natural result. For $u+c<0$, $t>0$, in a neighborhood of the boundary $z=0$, an exponential boundary layer is formed, and in the case $u+c>0$, $t>0$, in the domain $0<z<(u+c)t-\gamma$ (γ is an arbitrary constant, $0<\gamma<(u+c)t$), the solution coincides in the main (mod$O(\varepsilon^\infty)$) with the solution of the boundary-value problem

$$\widehat{v}_t + (u+c)\widehat{v}_t = 0,$$
$$\widehat{v}\big|_{z=0} = \mu(t).$$

In our case $u+c>0$ due to (1.8), $\mu(t) = (b/a)^\lambda = \text{const}$, therefore we have

$$\widehat{v}(z,t) = \left(\frac{b}{a}\right)^\lambda, \qquad 0<z<(u+c)t-\gamma,$$

and

$$\frac{\partial v^+}{\partial z}\bigg|_{z=0} = \frac{\partial v^+}{\partial y}\bigg|_{y=-ct} = \frac{\partial \widehat{v}}{\partial z}\Big|_{z=0} = 0. \tag{1.11}$$

Thus, the boundary conditions (1.11) and (1.9) imply

$$(u+c)\left(\frac{b}{a}\right)^\lambda - \left(u+c+\frac{c}{m}\right) = 0.$$

Or, finally,

$$c = -\frac{[1-(\frac{b}{a})^\lambda]u}{1+\frac{1}{m}-(\frac{b}{a})^\lambda} < 0. \tag{1.12}$$

By (1.12), the inequality $b>m\lambda$ implies (1.8). The solution is approximately plotted in Figure 6.1.

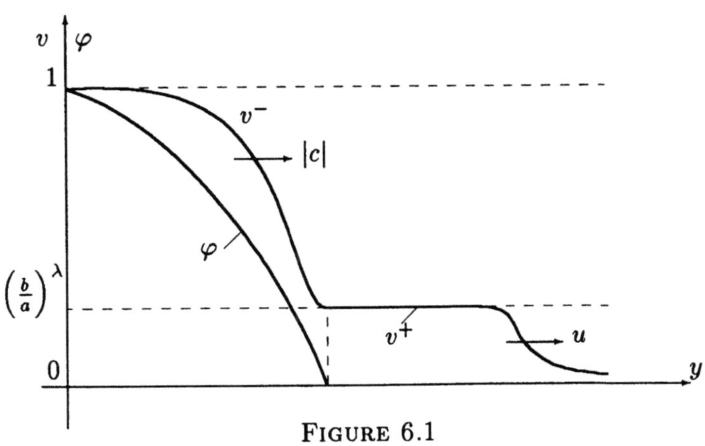

FIGURE 6.1

6.2. Asymptotic solution to the kinetics equation of nonequilibrium molecular processes with external diffusion effects

In this section we shall consider a mathematical model for nonequilibrium process of precipitation. These processes are described by the system of equations

$$\begin{aligned}&\frac{\partial v^-}{\partial t}+u\frac{\partial v^-}{\partial y}+\frac{\varepsilon}{m}\frac{\partial \varphi}{\partial t}-\varepsilon\frac{\partial}{\partial y}\left(D\frac{\partial v^-}{\partial y}\right)=0,\\&\frac{\partial \varphi}{\partial t}=\mu_1\left[v^- -(f(\varphi))\right],\qquad \mu_1=\text{const}>0,\end{aligned}\tag{2.1}$$

in the domain $\varphi > 0$; and by the equation

$$\frac{\partial v^+}{\partial t}+u\frac{\partial v^+}{\partial y}-\varepsilon\frac{\partial}{\partial y}\left(D\frac{\partial v^+}{\partial y}\right)=0$$

in the domain $\varphi \equiv 0$. The boundary conditions have the form:

$$\begin{aligned}&\varphi\big|_{y\to-\infty}\to 1,\qquad \varphi\big|_{y\geqslant-\Psi(t,\varepsilon)}\equiv 0,\qquad v^-\big|_{y\to-\infty}\to 1,\\&v^\pm\big|_{y=-\Psi(t,\varepsilon)}=f(0)+\frac{1}{\mu_1}\frac{\partial\varphi}{\partial t}\bigg|_{y=-\Psi(t,\varepsilon)},\\&\frac{\partial v^-}{\partial y}\bigg|_{y=-\Psi(t,\varepsilon)}=\frac{\partial v^+}{\partial y}\bigg|_{y=-\Psi(t,\varepsilon)},\qquad v^+\big|_{y\to+\infty}\to 0.\end{aligned}\tag{2.2}$$

Here $0 \leqslant v^\pm \leqslant 1$, $0 \leqslant \varepsilon \leqslant 1$, $f(\varphi)$ is a smooth monotone function satisfying the conditions $0 < f(0) < 1$, $f(1) = 1$, $\Psi = \Psi(t,\varepsilon)$ is the unknown coordinate of the front.

We assume that the inequality $u > 0$ holds and $\mu_1 = \varepsilon^{-1}\mu$, where $\mu = \text{const}$. This means that the rate of the external diffusion of the substance into the sorbent particles is sufficiently high.

We shall seek an asymptotic solution of problem (2.1), (2.2) in the form:

$$\begin{aligned}\varphi&=\left[\Phi_0(y,t,\tau)+\varepsilon\Phi_1(y,t,\tau)+\dots\right]\big|_{\tau=S/\varepsilon},\\v^-&=\left[W_0(y,t,\tau)+\varepsilon W_1(y,t,\tau)+\varepsilon^2W_2(y,t,\tau)\dots\right]\big|_{\tau=S/\varepsilon},\end{aligned}\tag{2.3}$$

where $S(y,t,\varepsilon) = (y + \Psi(t,\varepsilon))$, $\Psi(t,\varepsilon) = \Psi_0(t) + \varepsilon\Psi_1(t) + \dots$, $\Phi_i, W_i \in B^1$. Here we have the following analogs to conditions (1.4):

$$\begin{aligned}&\Phi_0(y,t,0)=0,\qquad \lim_{\tau\to-\infty}\Phi_0(y,t,\tau)=1,\\&\Phi_1(y,t,0)=0,\qquad \lim_{\tau\to-\infty}\Phi_1(y,t,\tau)=0,\end{aligned}\tag{2.4}$$

and

$$\begin{aligned}&W_0(y,t,0)=f(0)+\frac{1}{\mu}\frac{d\Psi_0}{dt}\frac{\partial\Phi_0}{\partial\tau}\bigg|_{\tau=0},\qquad \lim_{\tau\to-\infty}W_0(y,t,\tau)=1,\\&W_1(y,t,0)=\frac{1}{\mu}\left(\frac{d\Psi_0}{dt}\frac{\partial\Phi_1}{\partial\tau}\bigg|_{\tau=0}+\frac{\partial\Phi_0}{\partial t}\bigg|_{\tau=0}+\frac{d\Psi_1}{dt}\frac{\partial\Phi_0}{\partial\tau}\bigg|_{\tau=0}\right),\\&\lim_{\tau\to-\infty}W_i(y,t,\tau)=0,\qquad i\geqslant 1.\end{aligned}\tag{2.5}$$

For $\tau = 0$, the conditions on the functions W_i follow from the expansion

$$v^-\Big|_{y=-\Psi(t,\varepsilon)} = f(0) + \frac{\varepsilon}{\mu}\frac{\partial\varphi}{\partial t}\Big|_{y=-\Psi(t,\varepsilon)}$$
$$= f(0) + \frac{\varepsilon}{\mu}\frac{\partial(\Phi_0+\varepsilon\Phi_1)}{\partial\tau}\Big|_{\tau=0}\frac{d}{dt}(\Psi_0+\varepsilon\Psi_1) + \frac{\varepsilon^1}{\mu}\frac{\partial\Phi_0}{\partial\tau}\Big|_{\tau=0} + O(\varepsilon^3).$$

Theorem 2.1. *For any smooth function $\Psi(t,\varepsilon)$ such that*

$$\frac{\partial\Psi_0}{\partial t} < 0, \qquad \frac{\partial\Psi}{\partial t} + u_0(-\Psi,t) > 0,$$

*there exists an asymptotic solution to system (2.1) such that conditions (2.2) are satisfied. This solution has the form (2.3), where the function $\Phi_0 = \Phi_0(t,\tau)$ solves the equation**

$$\Psi_0'\frac{\partial^2\Phi_0}{\partial\tau^2} + \left\{\mu f(\Phi_0) - \frac{1}{D}\Psi_0'(\Psi_0'+u)\Phi_0'\right\}_\tau'$$
$$- \frac{\mu}{D}\left[f(\Phi_0)(\Psi_0'+u) + \frac{1}{m}\Phi_0\Psi_0' - \left(\Psi_0'+u+\frac{1}{m}\Psi_0'\right)\right] = 0 \quad (2.6)$$

with boundary conditions

$$\Phi_0\Big|_{\tau=0} = 0, \qquad \Phi_0\Big|_{\tau\to-\infty} \to 1, \qquad \frac{\partial\Phi_0}{\partial\tau}\Big|_{\tau=0} < 0. \quad (2.7)$$

The function $W_0 = W_0(t,\varepsilon)$ has the form

$$W_0 = \frac{1}{\mu}\Psi_{0t}'\frac{\partial\Phi_0}{\partial\tau} + f(\Phi_0).$$

Proof. Substituting the solutions of the form (2.3) into (2.1) and setting the coefficients at the highest powers of ε^{-1} equal to zero, we obtain the system

$$\begin{gathered}\frac{\partial W_0}{\partial\tau}(\Psi_{0t}'+u) + \frac{1}{m}\frac{\partial\Phi_0}{\partial\tau}\Psi_0' - D\frac{\partial^2 W_0}{\partial\tau^2} = 0,\\ \Psi_0'\frac{\partial\Phi_0}{\partial\tau} = \mu W_0 - \mu f(\Phi_0).\end{gathered} \quad (2.8)$$

Expressing the function W_0 from the second equation and substituting it into the first equation of system (2.8), we obtain an equation for the function Φ_0. By integrating this equation and taking into account the boundary conditions as $\tau\to\infty$, we obtain equation (2.6).

*Here and below $u = u(-\Psi_0,t)$.

Let us consider the asymptotic behavior of the function Φ_0 as $\tau \to -\infty$. By setting $1-\Phi_0 \sim e^{\gamma\tau}$, we get a squared equation for the exponent γ. The solution to this equation has the form

$$\gamma_{+,-} = \frac{1}{2\Psi_{0t}}\Big\{ -\mu f'(1) + \frac{1}{d}\Psi'_{0t}(\Psi'_{0t}+u)$$

$$\pm\Big[\Big(-\mu f'(1) + \frac{\Psi'_{0t}}{D}(\Psi'_0+u)\Big)^2 + \frac{4\mu}{D}(f'(1)(\Psi'_{0t}+u) + \frac{1}{m}\Psi'_{0t})\Psi'_{0t}\Big]^{1/2}\Big\}.$$

In order the function $1-\Phi_0$ decrease as $\tau \to -\infty$, it is necessary that one of the numbers $\gamma_{+,-}$ be positive. This is always possible since, by the theorem, the sum of the first two terms in curly brackets is negative and the factor before the curly brackets is also negative. Hence, the point $\Phi_0 = 1$ is stationary.

Let us calculate the asymptotics of the function $\Phi_0(t,\tau)$ as $\tau \to 0$. Assuming $\Phi_0 = C_1\tau^\alpha + O(\tau^{\alpha+\beta})$, $\beta > 0$, and substituting this expansion into the equation, it is easy to verify that there are two possibilities: $C_1 = 0$, $\Phi_0 \sim \tau^2$ or $C_1 \neq 0$, $\alpha = 1$. Thus the boundary conditions do not uniquely define the function Φ_0 (the solution given in the statement of theorem). The last boundary conditions in (2.7) together with the condition for Φ_0 follow from the matching condition precisely as in the preceding section. The theorem is proved.

The solution to equation (2.6) with boundary conditions (2.7) can be constructed by the method of shooting.

To derive the second boundary condition in (2.7) and the law of motion for the concentration front, we pass to the second stage of constructing an asymptotic solution, i.e., we match the solution constructed in Theorem 2.1 with the solution of the linear equation in the domain $\varphi \equiv 0$. First, as in the preceding section, we calculate the derivative of the solution v^- for $y = -\Psi(t,\varepsilon)$. We get

$$\frac{\partial v^-}{\partial y}\Big|_{y=-\Psi(t,\varepsilon)} = \Big(\frac{\partial W_1}{\partial\tau} + O(\varepsilon)\Big)\Big|_{\tau=0}$$

$$= -D^{-1}\Big(\Big(\frac{d\Psi_0}{dt}+u\Big) + m^{-1}\frac{d\Psi_0}{dt}\Big) + O(\varepsilon). \tag{2.9}$$

On the other hand, in the domain $\varphi \equiv 0$ the equation has the form (comp. with (1.10))

$$\frac{\partial v^+}{\partial t} + \Big(u + \frac{d\Psi_0}{dt}\Big)\frac{\partial v^+}{\partial z} = \varepsilon D\frac{\partial^2 v^+}{\partial z^2}, \quad u = u(-\Psi_0, t). \tag{2.10}$$

According to (2.5), the boundary conditions have the form

$$v^+\Big|_{z=0} = f(0) + \frac{1}{\mu}\frac{\partial\Phi_0}{\partial\tau}\Big|_{\tau=0}\Psi'_{0t}, \qquad v^+\Big|_{z\to\infty} \to 0.$$

The asymptotic solution to equation (2.10), satisfying the boundary conditions and zero initial conditions, has the form

$$v^+(z,t) = \int_0^t \Big[2\varepsilon\frac{\partial\Gamma}{\partial\xi}(z,0,t,\tau) - \Big(u + \frac{d\Psi_0}{dt}\Big)\Gamma(z,0,t,\tau)\Big]$$

$$\times\Big(f(0) + \frac{1}{\mu}\frac{\partial\Phi_0}{\partial\tau}\Big)\Psi'_{0t}(1 + O(h))\,d\tau,$$

where

$$\Gamma = 2(\pi(t-\tau))^{-1/2}\exp\left\{-\frac{\left(z-\xi+Z(t)-Z(\tau)\right)^2}{4\varepsilon(t-\tau)}\right\},$$

$$Z(t) = \int_0^t \left[u(-\Psi_0(t'),t') + \frac{d\Psi_0}{dt}(t')\right] dt'.$$

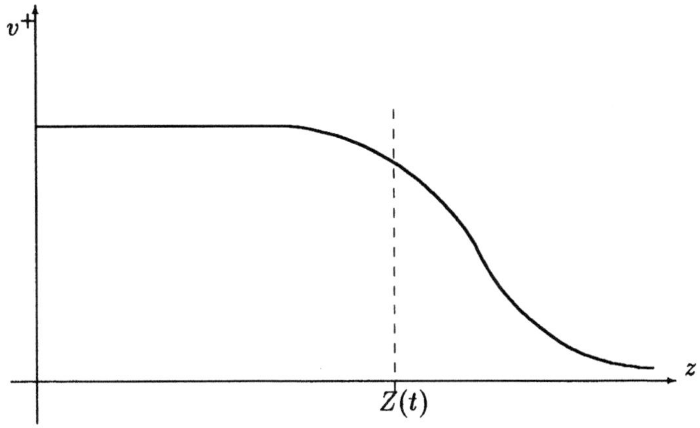

FIGURE 6.2

The function $v^+(z,t)$ is, in fact, the error integral written in a regular system of coordinates, see Figure 6.2.

Similarly to the preceding section, the fact that the solution is continuous does not imply any additional conditions, but the requirement that the derivatives in y be continuous for $y = -\Psi_0(t,\varepsilon)$ necessarily implies:

$$\left.\frac{\partial W_0}{\partial \tau}\right|_{\tau=0} = 0. \tag{2.11}$$

On the other hand, by integrating the first equation in (2.8) and, similarly to the preceding section, taking into account the boundary conditions, we obtain

$$W_0\left(\Psi'_{0t}+u\right) + \frac{1}{m}\Phi_0\Psi'_{0t} - D\frac{\partial W_0}{\partial \tau} = \left(\Psi'_{0t}+u\right) + \frac{1}{m}\Psi'_{0t}. \tag{2.12}$$

Hence, by setting $\tau = 0$ and having in mind relation (2.11) and the second equation in (2.8), we get an analog of the law of motion of the concentration front, derived in the preceding section:

$$\begin{aligned}\frac{d\Psi_0^{\pm}}{dt} = \frac{\mu}{2}\left(\frac{\partial\Phi_0}{\partial\tau}\right)^{-1}\Bigg[&-\frac{u(-\Psi_0,t)}{\mu}\left.\frac{\partial\Phi_0}{\partial\tau}\right|_{\tau=0} + 1 - f(0) + \frac{1}{m}\\ &\pm\Bigg[\left(-\frac{u(-\Psi_0,t)}{\mu}\frac{\partial\Phi_0}{\partial\tau} + 1 - f(0) + \frac{1}{m}\right)^2\\ &+\frac{4u(-\Psi_0,t)}{\mu}\left.\frac{\partial\Phi_0}{\partial\tau}\right|_{\tau=0}(1-f(0))\Bigg]^{1/2}\Bigg].\end{aligned} \tag{2.13}$$

In this case, by (2.13), two different laws of motion of the concentration front are possible. Note that, by condition

$$\left.\frac{\partial \Phi_0}{\partial \tau}\right|_{\tau=0} < 0,$$

both functions $\Psi_0^{\pm}$ satisfy the inequalities given in the statement of Theorem 2.1.

However, in the limiting (as $\mu \to \infty$) kinetics (i.e., in the stationary kinetics), only one function, namely Ψ_0^{-}, turns into the function describing the law of motion of the concentration front obtained in the preceding section.

For the functions $\Psi_0^{\pm}(t)$ the distributions of concentration are plotted in Figures 6.3 and 6.4 respectively.

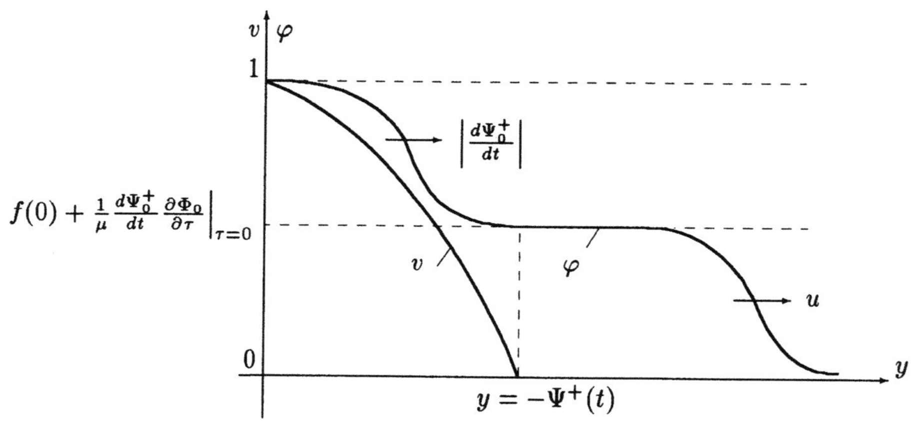

FIGURE 6.3

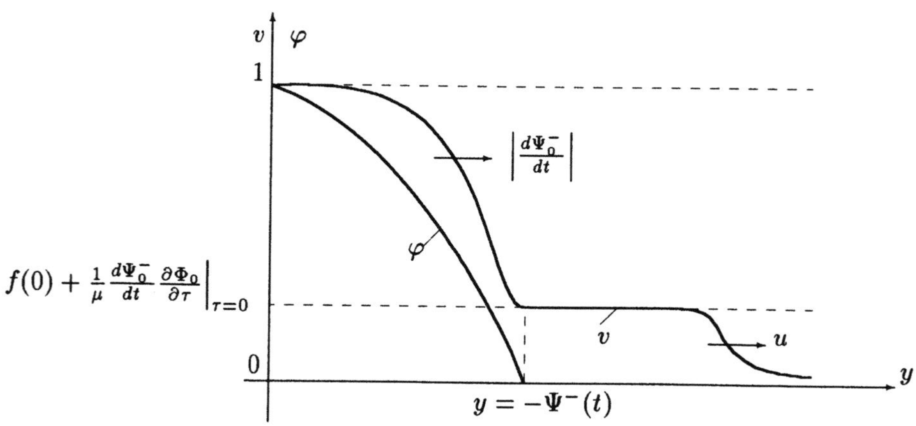

FIGURE 6.4

Obviously,

$$\frac{d\Psi_0^+}{dt} + u < \frac{d\Psi_0^-}{dt} + u,$$

and

$$\left|\frac{d\Psi_0^+}{dt}\right| > \left|\frac{d\Psi_0^-}{dt}\right|.$$

Relation (2.13) implies that even in the case of constant external velocity (the input velocity), to obtain the law of motion (i.e., to obtain the velocity that is constant in this case), one must solve equations (2.6), (2.7), taking into account (2.13). Obviously, the solution is independent of $\Psi_0(0)$, i.e., of the initial position of the concentration front.

In the case of a variable velocity, by defining the position of the concentration front for $t = t_0$, one can determine the derivative $\frac{d\Psi_0^\pm}{dt}(t_0)$ precisely as for the constant velocity, then define $\Psi_0^\pm(t_0 + \Delta t)$, by setting, for example,

$$\Psi_0^\pm(t_0 + \Delta t) \cong \Delta t \frac{d\Psi_0^\pm}{dt}(t_0) + \Psi_0^\pm(t_0), \qquad \text{etc.}$$

The fact that only the function Ψ_0^- allows to pass to the limiting (stationary) kinetics suggests that the regime corresponding to Ψ_0^- is stable, and the regime corresponding to Ψ_0^+ is unstable.

Thus the following adsorption process becomes possible. At the beginning, the output concentration of the substance that must be adsorbed (of impurities) is large (this corresponds to the regime "+"), but later on, after this regime is destroyed, one can observe that the output concentration of the substance that is being adsorbed decreases (!) (this effect corresponds to the regime "+" and this regime exists for a long time).

It must be emphasized that our result is conditional, since we do not pose conditions for the existence of the regime "+" (the explicit formulas of the preceding section imply that the regime "−" always exists for a sufficiently large t).

However, one can propose an algorithm for calculating the derivatives $d\Psi_0^\pm/dt$, for example, for $u = \text{const}$. Note that, according to (2.13), for $u = \text{const}$ (or for a fixed t) the value of the required derivative is determined by the derivative $A = \partial\Phi_0/\partial\tau\big|_{\tau=0}$, depending, in turn, on the derivatives $d\Psi_0^\pm/dt$. Therefore one can propose the following method for calculations. We define $\partial\Phi_0/\partial\tau\big|_{\tau=0}$ and choose, for example, the sign "+". Then we can calculate $d\Psi_0^+/dt$, and hence we know the coefficients of equation (2.6). Thus, for this equation, we can construct the solution satisfying the boundary conditions

$$\Phi_0\big|_{\tau\to-\infty} \to 1, \qquad \Phi_0(0) = 0.$$

Further, by calculating the derivative of this solution for $\tau = 0$, we obtain the transformation

$$\left.\frac{\partial\Phi_0}{\partial\tau}\right|_{\tau=0} = G^+(A).$$

Obviously, to solve this problem, it is necessary to find the fixed point of this transformation. The existence of such a fixed point is equivalent to the existence of the regime "+".

Similarly, taking the sign "−" in (2.13) and repeating the calculations, we construct the transformation G^-, for which the existence of a fixed point is equivalent to the existence of the regime "−".

As an example, we consider the function $f(\Phi) = (2 - \Phi)^{-1}$ satisfying the conditions of Theorem 2.1.

The results of calculations for $u = 1$, $m = 2$ are shown in Figures 6.5 (regime "+") and 6.6 (regime "−"), $p(\Phi) = \partial\Phi/\partial\tau^*$.

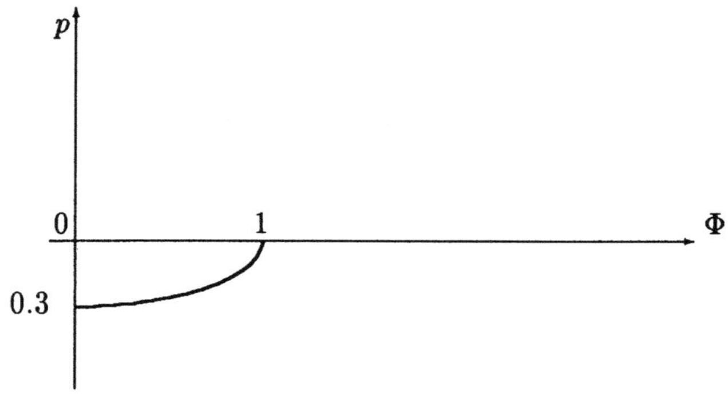

FIGURE 6.5

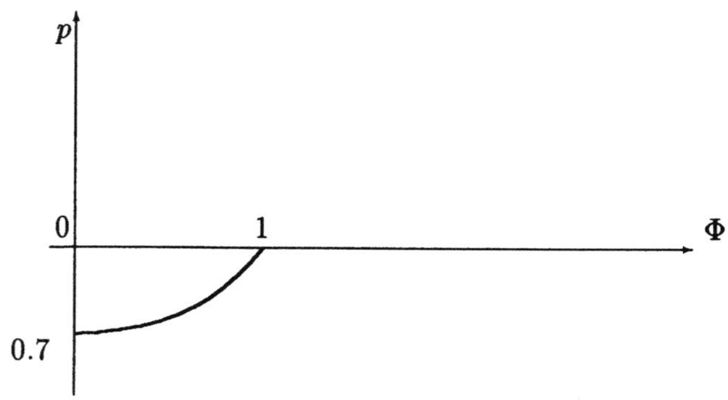

FIGURE 6.6

Note that previously E. A. Lukashov observed the effect of sorbent "adaptation" in sorption experiments carried out for carbon sorbents [70, 71]. This effect is quite similar to the passage from the unstable regime "+" to the stable regime "−".

Some peculiar effects were observed in adsoption of pesticides by carbon sorbents. Namely, for certain pesticides, the sorbent adsorptivity increases as the total time of filtration goes on.

---

*Calculations were performed by M. V. Gavrilova.

So, after water containing metaphos (200 dm^{-1}) was filtered (the input concentration was 60–80 mkg dm^{-1}), this pesticide was not found in the output liquid. The water samples, taken in the process of filtration, show that the concentration of this pesticide decreases steadily (the "inverse front"). A similar phenomenon was observed in adsorption of DDT (the input concentration was 80–140 mkg dm^{-1}) and of 0,0-dimethyl-0 (2,2-dichlorovinyl) posphate (the input concentration was 80 mkg dm^{-1}). However, this effect takes place not for all pesticides and not for all regimes of filtration. The common law of filtration, namely, the increase of the pollution concentration with time of filtration (the "direct front") was observed for such pesticides as rogor and hexachlorocyclohexane (as well as for metaphos and DDT under certain parameters of filtration).

This effect (observed at the initial stage of filtration) can be explained by the special molecular mechanism of adsorption. Further investigations of this effect may lead to effective final purification of drinkable water from toxic impurities.

Similar effects can be observed when phenols and chlororganic compounds [71] are being adsorbed from drinkable water in statics, when the increase of adsorptivity is not proportional to the increase of equilibrium concentration of impurities. In the case of phenol adsorption, the mechanism of purification is rather complicated. First, a monolayer of phenol molecules is formed on the sorbent surface. Then the concentration increases, and the dimerization of phenol molecules begins, which explains why the adsorption increase is not proportional. In [71], as a working hypothesis, the following explanation is proposed for the concave character of the isomerism of chloroform adsorption for concentrations greater than 600 mkg dm^{-1}. In other words, it is explained why the adsorptivity increase is not proportional to the increase of chloroform concentration in the equilibrium liquid phase (in water).

The sorption of chlororganic compounds by carbon sorbents (they have the properties of polyelectrolytes) decreases (as compared with electrolytes) the adsorption of the liquid phase in the porous structure of sorbent (considered as a mixed water-organic solute). This leads to a decrease of effective concentration of a strong electrolyte since pairs of ions are being formed. In turn, this phenomenon decreases the capacity of polyelectrolytes (in a salt form) to salt out the chlororganic compounds. Further, this implies that the adsorption of chlororganic compounds decreases.

Comparing the capacity of different sorbents (of activated carbon type) to adsorb chlororganic compounds, the authors abandoned the commonly accepted hypothesis [71]: isotherms of S-type indicate that the intermolecular interaction in the sorbate substance is stronger than the interaction between the sorbate molecules and the adsorbent surface.

As a consequence, they point out that not a monolayer, but a polymolecular layer is being formed on the adsorbent surface.

Moreover, a similar behavior, namely, an increase of the pollution adsorption with time, was observed in the process of water clarification by high-rate filters after the preliminary treatment by coagulates and the stage of sedimentation.

This effect is best shown when the height of the filtering material is small. In models describing the work of high-rate filters, this fact is either not described or taken into account formally. One can assume that, as the particles of a granular filtering material are being covered by the sorbate, this process is intensified due to as the particles of a granular filtering material are being covered by the sorbate, this process is intensified due to changes in physical and chemical properties of the surface.

6.3. The simplest one-dimensional model

A physico-mathematical model for thermal oxidation of silicon is described in detail in [51, p. 280, p.336]. This book also contains the bibliography concerning different models of thermal oxidation. The process considered in this section is one of the most important fundamental processes in modern microelectronic technology and is widely used in production of high-quality insulator-semiconductor systems.

Figure 6.7 shows the results of computer simulation of the process of silicon dioxide formation on the surface of a silicon crystal (for details, see [51]). Here $y_0(t)$ is the moving silicon dioxide–silicon interface inside a solid, y_0 is the interface between the solid and a gaseous oxidant. The internal Si – Si O_2 interface (see Figure 6.7), on which a solid-state chemical reaction takes place, moves with velocity $\frac{dy_0}{dt}$. Curve 1 shows the distribution of the oxidant O_2^- in Si O_2 at different instants of time ($v = c/c_0$ is the dimensionless concentration). Curve 2 presents the distribution of electric potential φ at the same instants of time. These plots give grounds to make a hypothesis that there is a boundary layer of thickness $\sim \varepsilon$ (in dimensionless variables $0 < \varepsilon \ll 1$) in which the gradients of unknown functions are large.

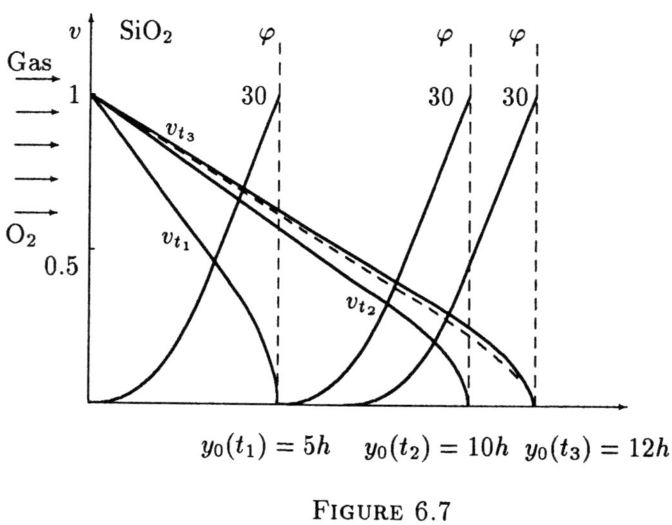

FIGURE 6.7

The oxidation process considered in this chapter is characterized by the following quantities [51]:

$$x_0 = 2 \cdot 10^{-7}\,\text{m}, \qquad t \geqslant 10^2\,\text{sec},$$
$$\varepsilon^2 = x_0^2 (D t_0)^{-1} \sim 10^{-2} \div 10^{-3}, \qquad \varepsilon^{3/2} = 0.1,$$
$$D = D_{O_2^- \text{ in SiO}_2} \sim 10^{-13}\,\text{m}^2/\text{sec}$$

(the temperature is supposed to be constant, $T \sim 1100°$C), where x_0 is the thickness of the silicon dioxide film, t_0 is the characteristic time, during which the film grows to the thickness such that the space charge is formed, D is the coefficient of diffusion of the oxidant O_2^- in silicon dioxide.

In dimensionless form, the diffusion equation has the form

$$\frac{\partial v}{\partial t} - \frac{1}{\mathrm{Pe}}\frac{\partial^2 v}{\partial y^2} + \frac{1}{\mathrm{Pe}'}\frac{\partial}{\partial y}\left(v\frac{\partial \Phi}{\partial y}\right) = 0, \tag{3.1}$$

where

$$v = \frac{c(x,t)}{c_0}, \qquad y = \frac{x}{x_0}, \qquad t = \frac{\widetilde{t}}{t_0},$$
$$\Phi = \frac{e\varphi}{kT}, \qquad \mathrm{Pe} = \frac{x_0^2}{Dt_0}, \qquad \mathrm{Pe}' \sim \mathrm{Pe},$$

are, respectively, the dimensionless concentration of oxygen ions, the dimensionless coordinate and time, the unknown dimensionless potential, and the diffusion Péclet number.

In Appendix to [51], it is shown that the oxidant concentration on the silicon–dioxide silicon interface $y = y_0(t,\varepsilon)$ is greater than the oxidant concentration on the gas–solid interface ($y = 0$), i.e,

$$\frac{v(y_0(t,\varepsilon),t,\varepsilon)}{v(0,t,\varepsilon)} = 10^{-4} \sim \varepsilon^4.$$

Therefore, the boundary conditions for equation (3.1) have the form

$$\begin{aligned} &v(0,t,\varepsilon) = n(0,t,\varepsilon) = 1,\\ &v(y_0(t,\varepsilon),t,\varepsilon) = \widehat{c}_0\varepsilon^4,\\ &n(y_0(t,\varepsilon),t,\varepsilon) = \widehat{c}_0\varepsilon^4, \qquad \widehat{c}_0 = \mathrm{const}, \end{aligned} \tag{3.2}$$

where n is the dimensionless concentration of oxygen vacancies. Here $\widetilde{y}_0(t,\varepsilon)$ is the unknown dimensionless coordinate of the silicon–silicon dioxide interface, which we are looking for.

To calculate the concentration of oxygen vacancies, we rewrite the diffusion equation as follows

$$\frac{\partial n}{\partial t} - \frac{1}{\mathrm{Pe}_1}\frac{\partial^2 n}{\partial y^2} - \frac{1}{\mathrm{Pe}_1'}\frac{\partial}{\partial y}\left(n\frac{\partial \Phi}{\partial y}\right) = 0. \tag{3.3}$$

The Péclet numbers Pe_1 and Pe_1' are similar to those in (3.1). Here D is the coefficient of diffusion of oxygen vacancies in the silicon oxide.

To calculate the distribution of potential in the oxide, we must solve the following one-dimensional Poisson equation

$$\varepsilon_1^2\frac{\partial^2 \varphi}{\partial y^2} = v - n, \tag{3.4}$$

where $\varepsilon_1 = L_D/x_0 a$, $a = L_D\sqrt{Dt_0}/x_0^2$; $L_D^2 = \bar{\varepsilon}\varepsilon_0 kT/1|e|^2 c_0$ is the Debye shielding length, $a = O(1)$ (see [51]). Here $\bar{\varepsilon}\varepsilon_0$ is the product of the relative and absolute dielectric constants, k is the Boltzmann constant, T is the temperature, e is the electron charge.

Assume that the small parameters satisfy the relation $\varepsilon_1^2 = \varepsilon^3$. This means that the region occupied by the space charge has been formed. The boundary conditions for equation (3.4) have the form

$$\begin{aligned}&\varphi(0,t,\varepsilon) = O(\varepsilon^2),\\&\varphi\big(y_0(t,\varepsilon),t,\varepsilon\big) = \Phi_0 < 0.\end{aligned} \tag{3.5}$$

The magnitude of the electric potential on the interface between the phases can be measured experimentally. Thus, summing up, we obtain the following system of equations

$$\begin{aligned}&\varepsilon^2\frac{\partial v}{\partial t} - \varepsilon\frac{\partial^2 v}{\partial y^2} + \varepsilon\frac{\partial}{\partial y}\left(v\frac{\partial\varphi}{\partial y}\right) = 0,\\&\varepsilon 2\frac{\partial n}{\partial t} - \varepsilon\frac{\partial^2 n}{\partial y^2} - \varepsilon\frac{\partial}{\partial y}\left(n\frac{\partial\varphi}{\partial y}\right) = 0,\\&\varepsilon^3\frac{\partial^2\varphi}{\partial y^2} = v - n.\end{aligned} \tag{3.6}$$

The motion of the silicon–silicon dioxide interface is described by the flux balance equation

$$\frac{\partial\widehat{y}_0(t,\varepsilon)}{\partial t} = c^*\left(-\frac{\partial v}{\partial y} + v\frac{\partial\varphi}{\partial y}\right)\bigg|_{y=\widetilde{y}_0(t,\varepsilon)}, \tag{3.7}$$

where c^* is a given constant.

The system (3.6) must be solved together with conditions (3.2), (3.5), (3.7).

Thus the system of equations (3.6) together with boundary conditions (3.2), (3.5), (3.7) gives a mathematical model for the process of thermal oxidation of silicon in the case when the kinetics of thermal oxidation is close to the standard kinetics.

The asymptotic solution of this problem will be sought in the form

$$\begin{aligned}&v(y,t,\varepsilon) = \big[\varepsilon W(\tau,t,\varepsilon) + \widehat{V}(t,y,\varepsilon) + O(\varepsilon^2)\big]\bigg|_{\tau=\frac{S(y,t,\varepsilon)}{\varepsilon}},\\&n(y,t,\varepsilon) = \big[\varepsilon N(\tau,t,\varepsilon) + V(t,y,\varepsilon) + O(\varepsilon^2)\big]\bigg|_{\tau=\frac{S(y,t,\varepsilon)}{\varepsilon}},\\&\varphi(y,t,\varepsilon) = \big[F(\tau,t,\varepsilon)\big]\bigg|_{\tau=\frac{S(y,t,\varepsilon)}{\varepsilon}},\end{aligned} \tag{3.8}$$

where each function from the set $\{W, N, F\}$ is a function of boundary layer type in a neighborhood of the point $y = \widetilde{y}_0(t,\varepsilon)$ (i.e., $\tau = 0$), such that

$$\lim_{\tau\to\infty} K = 0, \qquad \lim_{\tau\to\infty}\frac{\partial^j K}{\partial\tau^j} = 0, \qquad \forall j > 0 \qquad K \subset \{W, N, F\}. \tag{3.9}$$

The function $S(y,t,\varepsilon)$ will be sought in the form

$$S(y,t,\varepsilon) = y_0(t) + \varepsilon y_{01}(t) - y,$$

where y_0, y_{01} are smooth functions.

We have the expansions:

$$\begin{aligned} W(\tau,t,\varepsilon) &= W_0(\tau,t) + \varepsilon W_1(\tau,t) + O(\varepsilon^2), \\ N(\tau,t,\varepsilon) &= N_0(\tau,t) + \varepsilon N_1(\tau,t) + O(\varepsilon^2), \\ \varphi &= F_0(\tau,t,\varepsilon) + \varepsilon F_1(\tau,t,\varepsilon) + O(\varepsilon^2), \\ \widetilde{V}(t,y,\varepsilon) &= \widetilde{V}_0(t)(y_0(t) - y) + \varepsilon \widetilde{V}_1(t) y + O(\varepsilon^2), \\ \widetilde{y}_0(t,\varepsilon) &= y_0(t) + \varepsilon y_{01}(t) + O(\varepsilon^2), \\ V(t,y,\varepsilon) &= V_0(t)(y_0(t) - y) + \varepsilon V_1(t) y + O(\varepsilon^2). \end{aligned} \tag{3.10}$$

The functions W_0, W_1, N_0, N_1, F_0, F_1 are infinitely differentiable with respect to t, τ.

To describe the structure of solutions, we shall formulate the following theorem.

Theorem 3.1. *An asymptotic solution to problem* (3.6), (3.7), (3.2), (3.5)? *has the form* (3.10), *where*

$$V_0 = \left(\sqrt{2c^*t + \bar{y}_0^2}\right)^{-1}, \qquad y_0(0) = \bar{y}_0,$$

the functions $W_0(\tau,t)$, $N_0(\tau,t)$ have the form

$$\begin{aligned} W_0(\tau,t) &= e^{F_0(\tau,t)} \int_\infty^\tau e^{-F_0(\xi,t)} \left[V_0(t)(\xi - y_{01}(t)) \frac{\partial F_0}{\partial \xi} + \widetilde{V}_1(y_0) \frac{\partial F_0}{\partial \xi} \right] d\xi, \\ N_0(\tau,t) &= -e^{-F_0(\tau,t)} \int_\infty^\tau e^{F_0(\xi,t)} \left[V_0(t)(\xi - y_{01}(t)) \frac{\partial F_0}{\partial \xi} + V_1(y_0) \frac{\partial F_0}{\partial \xi} \right] d\xi. \end{aligned} \tag{3.11}$$

The function F_0 is determined by the problem

$$\begin{aligned} &\frac{\partial^2 F_0}{\partial \tau^2} = W_0 - N_0, \\ &F_0\big|_{\tau=0} = \Phi_0, \qquad F_0\big|_{\tau\to\infty} \to 0. \end{aligned} \tag{3.12}$$

The function S has the form $S = y_0(t) + \varepsilon y_{01}(t) - y$, and

$$\begin{aligned} V_1(t) &= -\frac{1}{y_0^2(t)} \int_\infty^0 e^{F_0} \tau \frac{dF_0}{d\tau} + \frac{y_{01}}{y_0^2}, \\ \widetilde{V}_1(t) &= \frac{1}{y_0^2(t)} \int_\infty^0 e^{-F_0} \tau \frac{dF_0}{d\tau} d\tau - \frac{y_1}{y_0^2}. \end{aligned}$$

The function $y_{01}(t)$ is determined by the equation

$$\frac{dy_{01}}{dt} = c^* \widetilde{V}_1(t). \tag{3.12'}$$

The functions W_1, F_1, N_1 *are determined by equations* (3.15), (3.17), (3.19) *with boundary conditions.*

Proof. Substituting the solution (3.10) into the first equation in (3.6), we obtain

$$\varepsilon^4\left[\frac{\partial W}{\partial t}+\frac{\partial V}{\partial t}+\frac{\partial W}{\partial \tau}\frac{\partial S}{\partial t}\frac{1}{\varepsilon}\right]-\varepsilon^2\left[\frac{\partial^2 W}{\partial \tau^2}\left(\frac{\partial S}{\partial y}\right)^2\frac{1}{\varepsilon^2}+\frac{\partial w}{\partial \tau}\frac{\partial^2 S}{\partial y^2}\frac{1}{\varepsilon}+\frac{\partial^2 V}{\partial y^2}\right]$$
$$+\varepsilon^2\frac{\partial}{\partial y}\left[(W+V)\frac{\partial F}{\partial y}\frac{\partial S}{\partial y}\frac{1}{\varepsilon}\right]=0.$$

The substitution of (3.10) into the other two equations in (3.6) gives similar expressions.

By using the properties of F, W for $y=0$, $\tau\to\infty$ and setting the sum of coefficients at ε^2 equal to zero, we get

$$\frac{\partial^2\widetilde{V}}{\partial y^2}=0,\qquad \frac{\partial^2 \dot{V}}{\partial y^2}=0. \tag{3.13}$$

Then, integrating with respect to τ, we obtain equations for the functions W_0, W_1, N_0, N_1, F_0, F_1:

$$-\frac{\partial W_0}{\partial \tau}+W_0\frac{\partial F_0}{\partial \tau}+\widetilde{V}_0(t)\big(\tau-y_{01}(t)\big)\frac{\partial F_0}{\partial \tau}+\widetilde{V}_1(t)y_0(t)\frac{\partial F_0}{\partial \tau}=0, \tag{3.14}$$

$$-\frac{\partial W_1}{\partial \tau}+W_1\frac{\partial F_0}{\partial \tau}+W_0\frac{\partial F_1}{\partial \tau}=f_1, \tag{3.15}$$

$$f_1=-\left[\widetilde{V}_0\big(\tau-y_{01}(t)\big)\frac{\partial F_1}{\partial \tau}+\widetilde{V}_1(\tau-y_{01})\frac{\partial F_1}{\partial \tau}+\widetilde{V}_2 y_0\frac{\partial F_0}{\partial \tau}\right],$$

$$\frac{\partial N_0}{\partial \tau}+N_0\frac{\partial F_0}{\partial \tau}+V_0\big(\tau-y_{01}(t)\big)\frac{\partial F_0}{\partial \tau}+V_1 y_0\frac{\partial F_0}{\partial \tau}=0, \tag{3.16}$$

$$\frac{\partial N_1}{\partial \tau}+N_1\frac{\partial F_0}{\partial \tau}+N_0\frac{\partial F_1}{\partial \tau}7=f_2, \tag{3.17}$$

$$f_2=-\left[V_0\big(\tau-y_{01}(t)\big)\frac{\partial F_1}{\partial \tau}+V_1(\tau-y_{01})\frac{\partial F_1}{\partial \tau}-V_2 y_0\frac{\partial F_0}{\partial \tau}\right],$$

$$\frac{\partial^2 F_0}{\partial \tau^2}=W_0-N_0, \tag{3.18}$$

$$\frac{\partial^2 F_1}{\partial \tau^2}=W_1-N_3. \tag{3.19}$$

The following relations were used to obtain (3.14)–(3.19):

$$\frac{\partial S}{\partial y}=-\alpha(t),\qquad \left(\frac{\partial S}{\partial y}\right)^2=\alpha^2,$$
$$y_0(t)-y=\varepsilon\big(\tau-y_{01}(t)\big)+O(\varepsilon^2),$$
$$y_1-y=\big(y_1(t)-y_0(t)\big)+\varepsilon\big(\tau-y_{01}(t)\big)+O(\varepsilon^2),$$
$$y_3-y=\big(y_3(t)-y_0(t)\big)+\varepsilon\big(\tau-y_{01}(t)\big)+O(\varepsilon^2),\qquad \tau=S/\varepsilon.$$

The system of coordinates is chosen so that the point $S = 0$ be the point $y = \widetilde{y}_0(t,\varepsilon)$ corresponding to the point $\tau = 0$. Hence, due to properties of the function W_0, the boundary conditions for system (3.11)–(3.16) have the form

$$\widetilde{V}_0 y - 0 = 1, \qquad W_1(0,t) - \widetilde{V}_0 y_{01} + \widetilde{V}_1 y_0 = 0, \qquad W_0\big|_{\tau\to\infty} = 0, \tag{3.20}$$

$$V_0 y_0 = 1, \qquad N_1 - V_0 y_{01} + V_1 y_0 = 0, \qquad N_0\big|_{\tau\to\infty} = 0, \tag{3.21}$$

$$F_0(0,t) = \Phi_0, \qquad F_0\big|_{\tau\to\infty} = 0. \tag{3.22}$$

Finally, (3.14) gives

$$W_0(\tau,t) = e^{F_0(\tau,t)} \int_\infty^\tau e^{-F_0(\xi,t)} \left(V_0(t)(\xi - y_{01}(t)) \frac{\partial F_0}{\partial \xi} + \widetilde{V}_1(t) y_0 \frac{\partial F_0}{\partial \xi} \right) d\xi.$$

The boundary conditions (3.20) for $\tau = 0$ lead to the equation

$$\widetilde{V}_0(t) y_{01}(t) - \widetilde{V}_1 y_0(t) = e^{\Phi_0} V_0 \int_\infty^0 e^{-F_0(\xi,t)} \xi \frac{dF_0}{d\xi}\, d\xi + (\widetilde{V}_0 y_{01} - \widetilde{V}_1 y_0)(1 - e^{\Phi_0}). \tag{3.23}$$

Similarly, equation (3.16) yields (3.11) and

$$V_0(t) y_{01}(t) - V_1(y_3(t) - y_0(t)) = e^{-\Phi_0} V_0 \int_\infty^0 e^{F_0(\tau,t)} \tau \frac{dF_0}{d\tau}\, d\tau + (V_0 y_{01} - V_1 y_0)(1 - e^{-\Phi_0}). \tag{3.24}$$

By construction, the boundary conditions for $\tau \to \infty$ hold identically.

The problem (3.12) is a consequence of equation (3.18). Let us consider equation (3.7) describing the motion of the interface between the phases. This gives two equations (we set the sums of coefficients at ε^0 and ε^1 equal to zero). By (3.11), we have

$$\frac{dy_0}{dt} = c^* \left[-\frac{\partial W_0}{\partial \tau} + W \frac{\partial F_0}{\partial \tau} + \widetilde{V}_0(\tau - y_{01}) \frac{dF_0}{d\tau} + \widetilde{V}_1 y_0 \frac{dF_0}{d\tau} + V_0(t) - \varepsilon \widetilde{V}_1(t) + \dots \right] \Bigg|_{\tau=0}.$$

By (3.14), we get

$$\frac{dy_0}{dt} = c^* V_0(t). \tag{3.25}$$

The sum of coefficients at ε^1 gives

$$\frac{dy_{01}}{dt} = -c^* \widetilde{V}_1(t). \tag{3.26}$$

Conditions (3.20), (3.21) imply

$$\widetilde{V}_0 = V_0 = \frac{1}{y_0(t)}. \tag{3.27}$$

Then, by (3.23), (3.24), we have the expressions for $\widetilde{V}_1$ and V_1:

$$\begin{aligned} V_1(t) &= -\frac{1}{y_0^2(t)} \int_\infty^0 e^{F_0(\tau,t)} \tau \frac{dF_0}{d\tau}\, d\tau + \frac{y_1}{y_0^2}, \\ \widetilde{V}_1(t) &= \frac{1}{y_0^2(t)} \int_\infty^0 e^{-F_0(\tau,t)} \tau \frac{dF_0}{d\tau}\, d\tau - \frac{y_1}{y_0^2}. \end{aligned} \tag{3.28}$$

The functions W_1, F_1, N_1 are determined by equations (3.15), (3.17), (3.19) with corresponding boundary conditions.

CHAPTER VII

THE FLOW AROUND A FLAT PLATE

7.1. Introduction

In this chapter we shall consider the steady laminar flow of incompressible "low-viscous" liquid around a semi-infinite flat plate placed parallel to the flow. In Cartesian coordinates the plate coincides with the positive half-axis Ox (see Figure 7.1).

FIGURE 7.1

As is known [93–95], in this problem the use of the parabolic coordinates (optimal according to Kaplun [94]) is most advantageous. In parabolic coordinates the boundary layer solution in the first approximation gives a uniformly suitable solution up to the second order. This solution describes not only the correct form of nonviscous flow of the first order in the external region (precisely as in Cartesian coordinates), but also the correct form of nonviscous flow.

First, we present our method for constructing an asymptotic solution of the problem about the flow around a smooth plate. In the last two sections of this chapter, an asymptotic solution describing the flow around a semi-infinite plate with small periodic irregularities is given. For the most part we consider steady flows. However, as it turns out, in the problem about the flow around a plate with small periodic irregularities, the stationary steady flow exists only for amplitudes of irregularities not greater than a certain critical magnitude. If the amplitude of irregularities is greater, there is no stationary stable steady flow. The flow becomes almost periodic with respect to time, i.e., vortices appear and disappear between the humps (at present, we cannot prove this fact, but we observed this effect in numerical experiments). As already noted, we shall use the dimensionless parabolic coordinates $x = \xi^2 - \eta^2$,

$y = 2\xi\eta$. In these coordinates the Navier–Stokes equation for the stream function has the form

$$2\left(\xi\frac{\partial\psi}{\partial\eta} - \eta\frac{\partial\psi}{\partial\xi}\right)\Delta\psi + (\xi^2+\eta^2)\left(\frac{\partial\psi}{\partial\xi}\frac{\partial}{\partial\eta} - \frac{\partial\psi}{\partial\eta}\frac{\partial}{\partial\xi}\right)\Delta\psi$$
$$+\varepsilon^2\left\{(\xi^2+\eta^2)\Delta\Delta\psi - 4\left(\xi\frac{\partial}{\partial\xi}+\eta\frac{\partial}{\partial\eta}-1\right)\Delta\psi\right\} = 0, \quad (1.1)$$

where $\Delta = \partial^2/\partial\xi^2 + \partial^2/\partial\eta^2$, $\xi \in R$, $\eta \in [0,\infty[$, $\psi = \psi(\xi,\eta,\varepsilon)$ is an unknown stream function and ε is a small parameter.

In this problem the boundary conditions have the form:

the conditions of adhesion on the plate

$$\psi = \frac{\partial\psi}{\partial\eta} = 0 \qquad \text{for} \quad \eta = 0, \tag{1.2a}$$

the flow symmetry conditions

$$\psi = \frac{\partial^2\psi}{\partial\xi^2} = 0, \qquad \text{for} \quad \xi = 0, \tag{1.2b}$$

the incident flow homogeneity conditions

$$\frac{1}{\eta}\frac{\partial\psi}{\partial\xi} \to 2 \qquad \text{as} \quad \eta\to\infty. \tag{1.2c}$$

Most investigations of the flow around a semi-infinite plate concentrate on finding asymptotic solutions of equation (1.1), describing the flow either near the leading edge, or downstream far from the leading edge. Usually, in this case, the small parameter in the Navier–Stokes equation is eliminated by the change of variables $\psi = \varepsilon^2\overline{\psi}$, $\xi = \varepsilon\tau_1$, $\eta = \varepsilon\tau_2$:

$$\left[2\left(\tau_1\frac{\partial\overline{\psi}}{\partial\tau_2} - \tau_2\frac{\partial\psi}{\partial\tau_3}\right) + (\tau_1^2+\tau_2^2)\left(\frac{\partial\overline{\psi}}{\partial\tau_1}\frac{\partial}{\partial\tau_2} - \frac{\partial\overline{\psi}}{\partial\tau_2}\frac{\partial}{\partial\tau_1}\right)\right.$$
$$\left.+(\tau_1^2+\tau_2^2)\Delta_\tau - 4\left(\tau_1\frac{\partial}{\partial\tau_1}+\tau_2\frac{\partial}{\partial\tau_2}-1\right)\right]\Delta_\tau\overline{\psi} = 0, \quad (1.3)$$

$$\Delta_\tau = \frac{\partial^2}{\partial\tau_1^2} + \frac{\partial^2}{\partial\tau_2^2}.$$

The solution of equation (1.3), describing the flow near the leading edge and satisfying the conditions of flow adhesion and flow symmetry, was obtained by G. I. Carrier and C. C. Lin [83] as an expansion of the stream function in terms of the distance from the leading edge (see also [108]). In parabolic coordinates this solution has the form

$$\overline{\psi} \sim \sqrt{A}\tau_1\tau_2^2 - \frac{B\sqrt{2}}{24}[\tau_1\tau_2^4 - 3\tau_1^2\tau_2^2]$$
$$+\frac{A^2}{64}\left[\frac{2}{5}\tau_1\tau_2^5 + \frac{1}{2}(\tau_1^2\tau_2^2 - \tau_1^4\tau_2^2)\arctan\frac{\tau_1}{\tau_2}\left\{\frac{1}{4}C\frac{2}{5}+\frac{1}{2}\log\frac{1}{2}(\tau_1^2+\tau_2^2)\right\}\tau_1^3\tau_2^3\right]$$
$$+\sum_{\alpha+\beta=7} O(\tau_1^\alpha\tau_2^\beta).$$

We note that the constant A is unknown and cannot be obtained, since there are no conditions at infinity.

For large τ_1, the asymptotic behavior of the solution of (1.3) was obtained by S. Goldstein [93], A. Amai [80], and K. Stewardson [106]. In the first approximation the solution coincides with the Blasius solution for a flat plate. To satisfy the condition that the vorticity rapidly decreases with the distance, they introduced logarithmic terms in the subsequent approximations. Actually, S. Goldstein proposed the following representation of the stream function in the potential flow:

$$\overline{\psi} = \operatorname{Im} W = \operatorname{Im}\Big\{ -\zeta^2 + \beta\zeta + \sum_{m=1} \frac{1}{\zeta^m}\Big[b_{m,m}(\ln\zeta)^m + b_{m,m-1}(\ln\zeta)^{m-1} + \cdots + b_{m,1}\ln\zeta - b_{m,0}\Big]\Big\}, \quad b_{i,j} \in R, \qquad \zeta = \tau_2 - i\tau_1. \tag{1.4}$$

For large τ_1 ($|\tau_1| > \tau_2$), he obtained the asymptotic solution in the boundary layer:

$$\overline{\psi} = \tau_1 f(\tau_2) + \tau_1^{-1}[f_1(\tau_2) + \ln\tau_1 g_1(\tau_2)] + \tau_1^{-2}[f_2(\tau_2) + \ln\tau_1 g_2(\tau_2) + (\ln\tau_1)^2 h_2(\tau_2)] + \dots, \tag{1.5}$$

where f is the Blasius function. The functions f_n, g_n and h_n and their first derivatives are solutions of ordinary differential equations, vanish at the origin and satisfy certain conditions obtained from expansion (1.4).

I. D. Murray [102] extended the expansion (1.5) up to the terms with τ_1^{-1}, so that all the above conditions were satisfied. However, the obtained functions contained two unknown constants, which cannot be determined, since the expansion is not global and has a singularity at $\tau_1 = 0$. Note that the representation of $\overline{\psi}$ in the form of a double series in $1/\xi$ and $\ln\xi/\xi$ has not been proved to be sufficiently general.

W. R. Dean [90], R. T. Davis [85], A. J. Van de Vooren and D. Dijkstra [109] and others, by different methods under various assumptions, tried to construct the solution of equation (1.3) so that it satisfy the equation at any distance from the leading edge. For example, A. J. Van de Vooren and D. Dijkstra obtained a numerical solution to equation (1.3) written for deviations of the stream function and the vorticity from their quantities that were obtained by means of the first-order theory of the boundary layer. In this case it was assumed that, as $\tau_1 \to \infty$, the stream function is determined by (1.5), and the solution $\overline{\psi}_1$ describing the deviations of the stream function from the Blasius solution has the following asymptotic expansion

$$\overline{\psi}_1 = \overline{\psi} - \tau_1 f(\tau_2) \sim \frac{1}{\tau_1}[f_1(\tau_2) + g_1(\tau_2)\ln\tau_1] \qquad \text{as} \quad \tau_1 \to \infty.$$

In Section 7.2 we pose the problem of constructing a global asymptotic solution to the Navier–Stokes equation (1.1) with a small parameter satisfying the conditions (1.2). Our method has something in common with the method for finding soliton-like solutions of nonstationary equations proposed in [55, 59]. Namely, the solution, depending on the "fast" (τ_1, τ_2) and "slow" (ξ, η) variables, is represented as the sum

$$\psi(\xi, \eta) = \psi(\tau_1, \tau_2, \xi, \eta, \varepsilon) = \Psi(\tau_1, \tau_2, \xi, \eta, \varepsilon) + \varphi(\tau_1, \tau_2, \varepsilon),$$

where Ψ is the "soliton" part of the solution decreasing as $\tau_2 \to \infty$. A function, concentrated near the plate, whose support is bounded by the parabola $\eta = \text{const} \cdot \varepsilon$, is called the *"soliton" part of the solution*, see Figure 7.2.

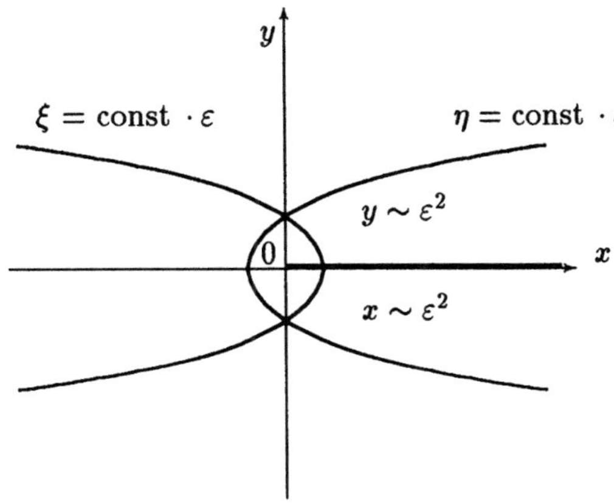

FIGURE 7.2

In the variables τ_1, τ_2, the support of the "soliton" part of the solution is bounded by the parabola $\tau = \text{const}$. As $\tau_2 \to \infty$ (i.e., $\eta \geqslant c \cdot \varepsilon^{1-\delta}$, $\delta > 0$), we come to the region of unperturbed flow outside the boundary layer. Thus, the boundary-layer region near the plate is the support of the "soliton" part of the solution. In contrast to the standard solution, the "soliton" part is not localized with respect to the variable τ_1.

On the other hand, as $\tau_1 \to \infty$, the initial equation yields the equation for the boundary layer, whose solution contains the Blasius solution as the "principal term". Subtracting this solution from the solution of the initial problem, we obtain a localized "soliton", i.e., a function decreasing with respect to τ_1 and τ_2. We assume that this function decreases like a power (the assumption that a localized "soliton" decreases like an exponential as $\tau_1 \to \infty$ yields corrections (in ε) to the Blasius solution, and these corrections are singular at $\xi = 0$.

In this case the technique of dealing with the operator of multiplication by a slow variable must be modified. In the standard situation, when the function $g(x, \varepsilon)$ decreases like an exponential function as $\tau = S(x)/\varepsilon \to \infty$ (i.e., faster than any power of $(S(x)/\varepsilon)^{-1}$), the operator of multiplication by any positive power of the phase $S(x)$ is "small", for example, $x^N \exp(-x/\varepsilon) = \varepsilon^N \tau^N \exp(-\tau^2)|_{\tau=x/\varepsilon} = O(\varepsilon^N)$ uniformly in x, since the function $\tau^N \exp(-\tau^2)$ is bounded. Conversely, in the case when the function decreases at any polynomial rate, the "smallness" of the operator of multiplication by the Nth power of the "slow variable" depends not only on N, but also on the rate of decrease, for example:

$$x \frac{1}{1 + (x/\varepsilon)^2} = \varepsilon \frac{\tau}{1 + \tau^2}\bigg|_{\tau = x/\varepsilon} = O(\varepsilon),$$

$$x^N \frac{1}{1+(x/\varepsilon)^2} = x^{N-2}\frac{\tau^2\varepsilon^2}{1+\tau^2}\varepsilon^2 = x^{N-2}O(\varepsilon^2), \quad N \geqslant 3,$$

since the functions $\tau/(1+\tau^2)$ and $\tau^2/(1+\tau^2)$ are bounded.

As already noted, one can change the scale of variables in the initial problem in order to eliminate the small parameter from the equations. Thus, the problem is reduced to solving the complete Navier–Stokes equation. On the other hand, it is well-known that the Blasius solution (of the boundary-layer equations) gives a good description of the flow far from the leading edge of the plate.

The above procedure for constructing the solution implies (see the formula in Theorem 2.1) that, under certain conditions, preserving the small parameter in the equations, one can indicate the structure of the solution that is in good agreement with the experimental data: with precision up to quantities of the order of $O(\varepsilon^2)$ the flow is described by the Blasius solution outside a $\varepsilon^{1-\delta}$ neighborhood ($\delta > 0$) of the leading edge of the plate. In the ε neighborhood all the summands in the formula given in the theorem are of the same order of smallness and the solution of the problem is reduced to the solution of the complete Navier–Stokes equation.

7.2. Uniformly suitable asymptotic solution to the problem about the flow of low-viscous liquid around a semi-infinite thin plate

To be more precise, in this section, following [29], we describe an algorithm for constructing an asymptotic solution that satisfies the equation up to any given precision in ε within any (independent of ε) region in R^2_{xy}, containing the point $(0, 0)$. As pointed out above, since the small parameter can be eliminated by a changing the scale of variables, this algorithm gives a natural result, namely, a certain representation of the exact solution.

However, in the subsequent section we use the obtained results for constructing an asymptotic solution describing the flow around a plate with small periodic irregularities. In the latter case the answer is not so obvious.

We assume that in parabolic coordinates the asymptotic solution of problem (1.1), (1.2) has the following form

$$\psi = \psi(\tau_1, \tau_2, \xi, \eta, \varepsilon)\Big|_{\substack{\tau_1=\xi/\varepsilon,\\ \tau_2=\eta/\varepsilon}} = \sum_{i=2}^{N} \varepsilon^i \chi_i(\tau_1, \tau_2, \xi, \eta, \varepsilon)\Big|_{\substack{\tau_1=\xi/\varepsilon,\\ \tau_2=\eta/\varepsilon}}$$
$$+ \sum_{i=1}^{N} \varepsilon^i \Phi_i(\tau_2, \xi, \eta)\Big|_{\tau_2=\eta/\varepsilon} + \sum_{i=0}^{N} \varepsilon^i \varphi_i(\xi, \eta) + O(\varepsilon^{N+1}), \quad (2.1)$$

where all the functions are infinitely differentiable with respect to their variables and possess the following properties:

- as $\tau_2 \to \infty$ the functions Φ_i, $i \geqslant 1$, together with all their derivatives decrease faster than any negative power of τ_2;
- the functions χ_i, $i \geqslant 2$ mainly decrease powerlike, namely, $\chi_i\big|_{\tau_1+\tau_2\to\infty} \to 0$.

The following estimates hold:

$$\chi_i = O(\tau_1^{-1}), \qquad \tau_1 \to \infty, \qquad \tau_2 = O(1),$$
$$\chi_i = O(\tau_2^{-1}), \qquad \tau_2 \to \infty, \qquad \tau_1 = O(1).$$

Thus the sum

$$\sum_{i=2}^{N} \varepsilon^i \chi_i + \sum_{i=1}^{N} \varepsilon^i \Phi_i$$

represents a "soliton" of general type, concentrated in a neighborhood of the semi-infinite plate, while the term

$$\sum_{i=2}^{N} \varepsilon^i \chi_i$$

is a "soliton" localized near the leading edge.

Before formulating the principal theorem of this section, we consider the following ordinary differential equation:

$$LF(\tau_2) = \frac{d}{d\tau_2}(\tau_2 f'(\tau_2) - f)^2, \tag{2.2}$$

where

$$L = \frac{d^4}{d\tau_2^4} + f\frac{d^3}{d\tau_2^3} + 3f'd^2 d^2\tau_2 + f''',$$

f is the Blasius function.

This equation appears in the theory of boundary layer after the separation of variables in the equation for the correction to the Blasius solution (for details, see [95, 102]), while in our case it determines the asymptotic (as $\tau_1 \to 0$) behavior of the localized "soliton" part of the solution of the Navier–Stokes equation.

As is known, the corresponding homogeneous equation has only two solutions, $y^{(1)}$ and $y^{(2)}$, satisfying homogeneous conditions of the following form

$$y^{(i)}\Big|_{\tau_2=0} = \frac{d}{d\tau_2}y^{(i)}\Big|_{\tau_2} = 0, \qquad i = 1, 2,$$

and the first of these solutions is expressed in terms of the Blasius function by the formula

$$y^{(1)} = \tau_2 f' - f.$$

Then the general solution of equation (2.2), satisfying the same boundary conditions, can be written as follows

$$F = C_1 y^{(1)} + C_2 y^{(2)} + y^{(3)}, \tag{2.3}$$

where $y^{(3)}$ is a partial solution of this problem; C_1 and C_2 are arbitrary constants.

Let us consider another principal equation:

$$\begin{aligned}
&2\tau_1^2 f'(D_F+\Delta_\tau\chi_2^1)+2(\tau_1^2FR-\tau_2 f)(F''R\tau_1+D_F+\Delta_\tau\chi_2^1)\\
&\quad+2\left[\tau_1\frac{\partial\chi_2^1}{\partial\tau_2}-3\tau_1^2\tau_2^2FR^2-\tau_2\frac{\partial\chi_2^1}{\partial\tau_1}+\tau_2FR-2\tau_2(\tau_2^3)+1FR^2\right]\\
&\qquad\qquad\times(\tau_1 f''+\tau_1F''R+D_F+\Delta_\tau\chi_2^1)\\
&\quad+(\tau_1^2+\tau_2^2)\Big\{\tau_1R^2(-3\tau_2^2fF''-2f'(\tau_2^3+1)F'')+\left(f\frac{\partial}{\partial\tau_2}-\tau_1\frac{\partial}{\partial\tau_1}\right)(D_F+\Delta_\tau\chi_2^1)\\
&\quad+R\left(-F\frac{\partial}{\partial\tau_2}-F'\tau_1\frac{\partial}{\partial\tau_1}\right)(\tau_1f''R+D_F+\Delta_\tau\chi_2^1)\\
&\quad+\left[\left(2(\tau_2^3+1)FR^2+\frac{\partial\chi_2^1}{\partial\tau_2}\right)\frac{\partial}{\partial\tau_2}+\left(3\tau_1\tau_2^2-\frac{\partial\chi_2^1}{\partial\tau_2}\right)\frac{\partial}{\partial\tau_1}\right](\tau_1f''+\tau_2F''R+D_F)\Big\}\\
&\quad+\tau_2^2\tau_1R(fF'''+f'F''-Ff'''-F'f'')\\
&=(\tau_1^2+\tau_2^2)(D_{F''}+\Delta_\tau D_F+\Delta_\tau\Delta_\tau\chi_2^1)\\
&\quad-4\left(\tau_1\frac{\partial}{\partial\tau_1}+\tau_2\frac{\partial}{\partial\tau_2}-1\right)(\tau_1F''R+D_F+\Delta_\tau\chi_2^1),
\end{aligned}\tag{2.4}$$

where $\chi_2^1=\chi_2^1(\tau_1,\tau_2)$ is an unknown function, $R=(\tau_1^2+\tau_2^3+1)^{-1}$, and

$$D_F=2\tau_1F\big((1-3\tau_2)R^2-(4\tau_2^3+4-9\tau_2^4)\big)-G\tau_1\tau_2^2F'R.$$

This partial differential equation in τ_1 and τ_2 is derived from the complete Navier–Stokes equation while an asymptotic solution is constructed and, under the assumption that the localized "soliton" part decreases powerlike, plays the part of the standard solution, i.e., it determines the dependence of the solution on the "fast" variables τ_1 and τ_2. The coefficients and the right-hand side of equation (2.4) contain the function F defined above, and thus there are two arbitrary constants C_1 and C_2. By varying these constants, we can try to construct a solution of required form. Under the assumption that this procedure can be performed, we obtain the following theorem.

Theorem 2.1. *Let there exist constants C_1 and C_2 such that equation (2.4) has the solution $\chi_2^1=\chi_2^1(\tau_1,\tau_2)$ satisfying the homogeneous boundary conditions of the form:*

$$\begin{aligned}
&\chi_2^1\Big|_{\tau_2=0}=\frac{\partial\chi_2^1}{\partial\tau_2}\Big|_{\tau_2=0}=0 &&\textit{adhesion condition},\\
&\chi_2^1\Big|_{\tau_1=0}=\frac{\partial\chi_2^1}{\partial\tau_1}\Big|_{\tau_1=0}=0 &&\textit{symmetry condition},\\
&\chi_2^1\Big|_{\tau_1\to\infty}\to0,\qquad \chi_2^1\Big|_{\tau_2\to\infty}\to0,
\end{aligned}\tag{2.5}$$

and possessing the following asymptotics:

$$\begin{aligned}
&\chi_2^1=O(\tau_1^{-2}), &&\tau_1\to\infty\quad \tau_2=O(1),\\
&\chi_2^1=O(\tau_2^{-2}), &&\tau_2\to\infty\quad \tau_1=O(1).
\end{aligned}$$

Then the asymptotic solution to problem (2.4), (2.5) *has the form* (2.1), *where*

$$\varphi_0 = 2\xi\eta, \qquad \Phi_1 + \varphi_1 = \xi f(\tau_2) - 2\xi\tau_2,$$

and the remaining terms φ_i *and* Φ_i *vanish identically for* $i \geqslant 2$,

$$\chi_2 = \frac{\tau_1 F(\tau_2}{\tau_1^2 + \tau_2^3 + 1} + \chi_2^1,$$

where the function F *is determined by* (2.3),

$$\chi_i \equiv 0, \qquad i \geqslant 3.$$

Remark. Under the assumptions of the theorem, the constructed solution

$$\psi = [\varepsilon^2\chi_2(\tau_1, \tau_2) + \varepsilon(\xi f(\tau_2) - 2\xi\tau_2) + 2\xi\eta]\Big|_{\substack{\tau_1=\xi/\varepsilon \\ \tau_2=\xi/\varepsilon}}$$

is the exact solution of the problem. In this expression the terms $2\xi'\eta$ and $-2\varepsilon\xi\tau_2$ can be cancelled out, but we retain them to preserve the structure of the asymptotic solution described above (as $\tau_2 \to \infty$ the dependence on the first variable τ_2 must vanish).

Proof. By substituting the asymptotic expansion (2.1) of the stream function into (1.1) and multiplying it by ε, we obtain the following equation in variables τ_1, τ_2, ξ and η:

$$\begin{aligned}
2\Bigg[&\xi\bigg(\frac{\partial(\widehat{\Phi}_N + \varepsilon\widehat{\chi}_N)}{\partial\tau_2} + \frac{\partial(\widehat{\varphi}_N + \varepsilon\widehat{\Phi}_N + \varepsilon^2\widehat{\chi}_N)}{\partial\eta}\bigg) \\
&- \eta\bigg(\frac{\partial(\widehat{\chi}_N)}{\partial\tau_1} + \frac{\partial(\widehat{\varphi}_N + \varepsilon\widehat{\Phi}_N + \varepsilon^2\widehat{\chi}_N)}{\partial\eta}\bigg)\Bigg]\widetilde{\Delta}(\widehat{\varphi}_N + \varepsilon\widehat{\Phi}_N + \varepsilon^2\widehat{\chi}_N) \\
&+ (\xi^2 + \eta^2)\Bigg[\bigg(\varepsilon\frac{\partial\widehat{\chi}_N}{\partial\tau_1} + \frac{\partial(\widehat{\varphi}_N + \varepsilon\widehat{\Phi}_N + \varepsilon^2\widehat{\chi}_N)}{\partial\xi}\bigg)\bigg(\frac{\partial}{\partial\tau_2} + \varepsilon\frac{\partial}{\partial\eta}\bigg) \\
&- \bigg(\frac{\partial(\widehat{\Phi}_N + \varepsilon\widehat{\chi}_N}{\partial\tau_2} + \frac{\partial(\widehat{\varphi}_N + \varepsilon\widehat{\Phi}_N + \varepsilon^2\widehat{\chi}_N)}{\partial\eta}\bigg)\bigg(\frac{\partial}{\partial\tau_1} + \varepsilon\frac{\partial}{\partial\xi}\bigg) + \varepsilon^2\widetilde{\Delta}\Bigg] \\
&\qquad \times \widetilde{\Delta}(\widehat{\varphi}_N + \varepsilon\widehat{\Phi}_N + \varepsilon^2\widehat{\chi}_N) \\
&- 4\varepsilon^2\bigg[\xi\bigg(\frac{\partial}{\partial\tau_1} + \varepsilon\frac{\partial}{\partial\xi}\bigg) + \eta\bigg(\frac{\partial}{\partial\tau_2} + \varepsilon\frac{\partial}{\partial\eta}\bigg)\bigg]\widetilde{\Delta}(\widehat{\varphi}_N + \varepsilon\widehat{\Phi}_N + \varepsilon^2\widehat{\chi}_N) = 0,
\end{aligned}$$

where

$$\widehat{\varphi}_N = \varepsilon^i\varphi_i, \qquad \widehat{\Phi}_N = \varepsilon^{i-1}\Phi_i, \qquad \widehat{\chi}_N = \varepsilon^{i-2}\chi_i,$$
$$\widetilde{\Delta} = \frac{1}{\varepsilon^2}\Delta_\tau + \frac{2}{\varepsilon}\bigg(\frac{\partial^2}{\partial\tau_1\partial\xi} + \frac{\partial^2}{\partial\tau_2\partial\eta}\bigg) + \Delta.$$

Let τ_1, τ_2 tend to infinity, i.e., let us pass to the external region, where ξ and η are variables of the order of infinity. Taking into account that $\lim_{\tau_2\to\infty}\Phi_i = 0$, $i \geqslant 1$, $\lim_{\substack{\tau_1\to\infty\\ \tau_2\to\infty}} \chi_i = 0$, $i \geqslant 2$, and that the function Φ_i decreases as described above, we set the coefficient at ε^1 equal to zero and obtain the equation for the function φ_0 in the external region:

$$2\left(\xi\frac{\partial\varphi_0}{\partial\eta} - \eta\frac{\partial\varphi_0}{\partial\xi}\right)\Delta\varphi_0 + (\xi^2+\eta^2)\left[\frac{\partial\varphi_0}{\partial\xi}\frac{\partial}{\partial\eta} - \frac{\partial\varphi_0}{\alpha\eta}\frac{\partial}{\partial\xi}\right]\Delta\varphi_0 = 0. \tag{2.6}$$

This equation coincides with the Euler equation for the stream function in parabolic coordinates. The boundary conditions for the function φ_0 follow from (1.5):

$$\varphi_0\Big|_{\eta=0} = 0, \qquad \text{the adhesion condition,} \tag{2.7a}$$

$$\varphi_0\Big|_{\xi=0} = 0, \quad \frac{\partial^2\varphi_0}{\partial\xi^2}\Big|_{\xi=0} = \frac{\partial^2\chi_2}{\partial\tau_1^2}\Big|_{\substack{\tau_1=0\\ \tau_2\to\infty}} = 0, \quad \text{the flow symmetry condition,} \tag{2.7b}$$

$$\frac{1}{\eta}\frac{\partial\varphi_0}{\partial\xi}\Big|_{\eta\to\infty} \to 2, \qquad \text{the incident flow homogeneity condition.} \tag{2.7c}$$

Note that the second adhesion condition in not posed for the function φ_0 since this condition contains an unknown function Φ_1, namely,

$$\frac{\partial\varphi_0}{\partial\eta}\Big|_{\eta=0} + \frac{\partial\Phi_1}{\partial\tau_2}\Big|_{\tau_2=0} = 0.$$

This relation serves as the boundary condition for the function φ_0. Obviously, the solution to problem (2.6), (2.7) is the homogeneous flow, i.e., in the principal term of the incident flow is not influenced by the flat plate.

Before proceeding further with construction of the solution, we point out some properties of this problem. As $\tau_2 \to \infty$, the functions Φ_i together with all their derivatives are assumed to decrease faster than any negative power of τ_2. As noted above, the operator of multiplication by a positive power of the phase η is "small" if it is applied to such a rapidly decreasing function, i.e.,

$$\eta^N\frac{\partial^l\Phi_i}{\partial\tau_2^l} = \varepsilon^N\tau_2^N\frac{\partial^l\Phi_i}{\partial\tau_2^l} = O(\varepsilon^N), \quad \forall N, \quad l \geqslant 0. \tag{2.8}$$

On the other hand, we assumed that the functions χ_i decrease powerlike as $\tau_1 \to \infty$ and $\tau_2 \to \infty$, namely,

$$\tau_1\chi_i \xrightarrow[\substack{\tau_1\to\infty\\ \tau_2=O(1)}]{} \chi_{i\,1}(\tau_2,\xi,\eta), \qquad \tau_2\chi_i \xrightarrow[\substack{\tau_2\to\infty\\ \tau_1=O(1)}]{} \chi_{i\,2}(\tau_1,\xi,\eta). \tag{2.9}$$

Then the following relations hold:

$$\begin{aligned} &\eta^N\chi_i = \eta^{N-1}\varepsilon\tau_2\chi_i = \eta^{N-1}O(\varepsilon),\\ &\xi^N\chi_i = \xi^{N-1}\varepsilon\tau_1\chi_i = \xi^{N-1}O(\varepsilon), \qquad N \geqslant 1,\\ &\xi^N\frac{\partial\chi_i}{\partial\tau_1} = \xi^{N-2}\varepsilon^2\tau_1^2\frac{\partial\chi_i}{\partial\tau_1} = \xi^{N-2}O(\varepsilon^2), \qquad N \geqslant 2, \quad \text{etc.} \end{aligned} \tag{2.10}$$

Taking into account the above considerations, return to equation (2.5). By substituting the function φ_0 in a certain specific form into this equation and expanding the coefficients by formulas (2.8) and (2.10) (note that, due to (2.8), any smooth function in the variable η, multiplied by the function Φ_i, $i \geqslant 1$, or its derivatives, can be expanded into a Taylor series at the point $\eta = 0$), we can write the principal term in (2.5)

$$\varepsilon^0: \quad 2\left[\xi\frac{\partial\Phi_1}{\partial\tau_2} + 2\xi\right]\frac{\partial^2\Phi_1}{\partial\tau_2^2} + \xi^2\left[\left(\frac{\partial\Phi_1}{\partial\xi} + \left.\frac{\partial\varphi_1}{\partial\xi}\right|_{\eta=0} + 2\tau_2\right)\right.$$
$$\left. + \frac{\partial^3\Phi_1}{\partial\tau_2^3} - \left(\frac{\partial\Phi_1}{\partial\tau_2} + 2\xi\right)\frac{\partial^3\Phi_1}{\partial\xi\partial\tau_2^2} + \frac{\partial^4\Phi_1}{\partial\tau_2^4}\right] = 0. \quad (2.11)$$

This is the Prandtl equation in parabolic coordinates written for the function $\Phi_1^* = \Phi_1 + \varphi_1\big|_{\eta=0} + 2\xi\tau_2$. Let us formulate the boundary conditions for Φ_1^*. The adhesion condition of the form

$$\left.\Phi_1\right|_{\tau_2=0} + \left.\varphi_1\right|_{\eta=0} = 0, \qquad \left.\frac{\partial\Phi_1}{\partial\tau_2}\right|_{\tau_2=0} + \left.\frac{\partial\varphi_0}{\partial\eta}\right|_{\eta=0} = 0$$

implies

$$\left.\Phi_1^*\right|_{\tau_2=0} = 0, \qquad \left.\frac{\partial\Phi_1^*}{\partial\tau_2}\right|_{\tau_2=0} = 0. \quad (2.12a)$$

The first of the flow symmetry conditions

$$(\Phi_1 + \varphi_1)\big|_{\xi=0} = 0$$

yields

$$\Phi_1^*\big|_{\xi=0} = 0 \quad (2.12b)$$

(we do not pose the second condition for the function Φ_1^*, since it contains additional unknown functions).

Finally, by definition, $\Phi_1 \to 0$ as $\tau_2 \to \infty$, therefore $\Phi_1^* \sim 2\xi\tau_2 + \varphi_1\big|_{\eta=0}$ and it is more convenient to pose the following condition for the derivative of the function Φ_1^*:

$$\frac{\partial\Phi_1^*}{\partial\tau_2} - 2\xi = O(\tau_2^{-N}), \quad \tau_2 \to \infty, \quad \forall N \gg 1. \quad (2.12c)$$

By simple separation of variables, we obtain the solution to problem (2.11), (2.12) in the form

$$\Phi_1^* = \xi f(\tau_2),$$

where $f(\tau_2)$ is the Blasius function satisfying the problem

$$ff'' + f''' = 0,$$
$$f(0) = f'(0) = 0, \qquad f'\big|_{\tau_2\to\infty} \to 2.$$

Now, the first-order boundary-layer solution is obtained, and we can pose the boundary condition at $\eta = 0$ for the function φ_1 in the external domain. By definition of the functions Φ_1^* and Φ_1, we get

$$\varphi_1\Big|_{\eta=0} = \lim_{\tau_2\to\infty}(\Phi_1^* - 2\xi\tau_2).$$

As $\tau_2 \to \infty$, we have the following asymptotic expansion of the Blasius function:

$$f = 2\tau_2 - \beta + \sigma(\tau_2^{-N}), \quad \forall N \gg 1, \quad \beta = \text{const} > 0.$$

Hence

$$\varphi_1\Big|_{\eta=0} = 0. \tag{2.13}$$

Let us write the equation for the function φ_1 in the domain of the external flow. This equation is obtained from the initial equation (2.5) as $\tau_1 \to \infty$, $\tau_2 \to \infty$:

$$2(\xi^2 - \eta^2)\Delta\varphi_1 + (\xi^2 + \eta^2)\left(\eta\frac{\partial}{\partial\eta} - \xi\frac{\partial}{\partial\xi}\right)\Delta\varphi_1 = 0.$$

The following boundary conditions must be satisfied, as well as conditions (2.13):

$$\varphi_1\Big|_{\xi=0} = -\Phi_1\Big|_{\xi=0} = 0, \qquad \frac{\partial^2\varphi_1}{\partial\xi^2}\Big|_{\xi=0} = -\frac{\partial^2\chi_3}{\partial\tau_1^2}\Big|_{\substack{\tau_1=0\\ \tau_2\to\infty}} = 0, \qquad \frac{1}{\eta}\frac{\partial\varphi_1}{\partial\xi}\Big|_{\eta\to\infty} \to 0.$$

The solution is given by the function

$$\varphi_1 = -\beta\xi,$$

which is the second term of the asymptotic expansion (as $\tau_2 \to \infty$) of the boundary-layer solution Φ_1^*. Precisely this fact reflects the remarkable advantage of optimal parabolic coordinates.

Further, in (2.5) transformed according to (2.8), (2.10), we set the summands of the order $O(\varepsilon)$ equal to zero and obtain the linear differential equation for the function $\Phi_2^* = \Phi_2 + \varphi_2\big|_{\eta=0}$:

$$\Phi_2^* = \Phi_2 + \varphi_2\Big|_{\eta=0}:$$

$$2f'\frac{\partial^2\Phi_2^*}{\partial\tau_2^2} + \frac{\partial\Phi_2^*}{\partial\tau_2}f'' + \frac{\partial\Phi_2^*}{\partial\xi}\xi f''' + f\frac{\partial^3\Phi_2^*}{\partial\tau_2^3} - \xi f'\frac{\partial^3\Phi_2^*}{\partial\xi\partial\tau_2^2} + \frac{\partial^4\Phi_2^*}{\partial\tau_2^4} = 0. \tag{2.14}$$

The boundary conditions are determined as above and have the form

$$\Phi_2^*\Big|_{\tau_2=0} = -\chi_2\Big|_{\substack{\tau_2=0\\ \tau_1\to\infty}} = 0,$$

$$\frac{\partial\Phi_2^*}{\partial\tau_2}\Big|_{\tau_2=0} = -\frac{\partial\chi_2}{\partial\tau_2}\Big|_{\substack{\tau_2=0\\ \tau_1\to\infty}} = 0,$$

$$\frac{\partial\Phi_2^*}{\partial\tau_2}\Big|_{\tau_2\to\infty} \to 0, \qquad \frac{\partial\Phi_2}{\partial\tau_2} = O(\tau_2^{-N}), \qquad \forall N \gg 1, \qquad \tau_2 \to \infty.$$

Note that in this case we do not impose both conditions on the function Φ_2^*, since for this order the solution already has a localized "soliton". If, by setting $\Phi_2^* = g(\tau_2)$, the variables in (2.14) can be separated, then the resulting ordinary differential equation has no solutions decreasing as exponentials. Since the boundary-value problem for Φ_2^* is homogeneous, we can take a trivial solution:

$$\Phi_2^* \equiv 0.$$

This yields the boundary condition for φ_2:

$$\varphi_2\big|_{\eta=0} = \lim_{\tau_2\to\infty} \Phi_2^* = 0,$$

and therefore

$$\Phi_2 \equiv 0.$$

Now let us consider the next approximation. We write the summands of the order $O(\varepsilon^2)$ in the transformed equation (2.5):

$$\begin{aligned}
&2\xi^2 f' \frac{\partial^2\Phi_3}{\partial\tau_2^3} + \xi^2\frac{\partial\Phi_3}{\partial\tau_2} f'' + \xi^2 \frac{\partial^3\Phi_3}{\partial\tau_2^3} f + \xi^3 f''' \frac{\partial(\Phi_3 + \varphi_3|_{\eta=0})}{\partial\xi} - \xi^3 f' \frac{\partial^3\Phi_3}{\partial\xi\partial\tau_2^2} \\
&+ \xi^2\frac{\partial^4\Phi_3}{\partial\tau_2^4} + \xi\left\{\tau_1\frac{\partial\chi_2}{\partial\tau_2} f'' + 2\tau_1 f' \frac{\partial^2\chi_2}{\partial\tau_2^2} - \xi\tau_1\frac{\partial^3\chi_2}{\partial\xi\partial\tau_2^2} f' + \xi\tau_1\frac{\partial\chi_2}{\partial\xi} f'''\right. \\
&\left. + \tau_1 f\frac{\partial^3\chi_2}{\partial\tau_2^3} + \tau_1^2\frac{\partial\chi_2}{\partial\tau_1} f''' - \tau_1^2 f' \frac{\partial^3\chi_2}{\partial\tau_1\partial\tau_2^2} + \tau_1\frac{\partial^4\chi_2}{\partial\tau_2^4}\right\} \\
&- 4\eta\tau_2\frac{\partial\chi_2}{\partial\tau_1^2} + \eta\tau_2^2 f\frac{\partial^3\chi_2}{\partial\tau_1^2\partial\tau_2} + 2\eta^2\tau_2\frac{\partial^3\chi_2}{\partial\eta\partial\tau_1^2} - 2\eta\tau_2\tau_1\frac{\partial^3\chi_2}{\partial\tau_1^3} + \tau_2\eta\frac{\partial^4\chi_2}{\partial\tau_1^4} \\
&= \xi\frac{d}{d\xi}(\tau_2 f' - f)^2.
\end{aligned} \tag{2.15}$$

In this equation, for finite values of τ_1, we pass to the limit as $\tau_2 \to \infty$ and obtain the equation for the function $\chi_{22} = \lim_{\tau_2\to\infty} \tau_2\chi_2$:

$$-6\frac{\partial^2\chi_{22}}{\partial\tau_1^2} + 2\eta\frac{\partial^3\chi_{22}}{\partial\eta\partial\tau_1^2} - 2\tau_1\frac{\partial^3\chi_{22}}{\partial\tau_1^3} + \frac{\partial^4\chi_{22}}{\partial\tau_1^4} = 0. \tag{2.16}$$

The flow symmetry conditions imply

$$\left.\frac{\partial^2\chi_{22}}{\partial\tau_1^2}\right|_{\tau_1=0} = 0.$$

Therefore we can set

$$\frac{\partial^2\chi_{22}}{\partial\tau_1^2} \equiv 0.$$

Hence $\chi_{22} = A + B\tau_1$, the other symmetry condition for the flow, $\chi_{22}|_{\tau_1=0} = 0$, implies

$$A = 0, \qquad \chi_{22} = B\tau_1,$$

where B is a certain (possibly zero) constant. In other words, the function χ_2 has the following asymptotic expansion

$$\chi_2 \sim B\tau_1/\tau_2, \quad \tau_2 \to \infty, \quad \tau_1 = O(1). \tag{2.17}$$

Now we consider another limit of equation (2.15). Let $\tau_1 \to \infty$ and τ_2 be finite. Since the functions f and Φ_3 are independent of τ_1 and

$$\lim_{\tau_1\to\infty} \tau_1\chi_2 = \chi_{21},$$

we have

$$\xi\left[2f'\frac{\partial^2\Phi_3^*}{\partial\tau_2^2} + \frac{\partial\Phi_3^*}{\partial\tau_2}f'' + \xi\frac{\partial\Phi_3^*}{\partial\xi}f''' + f\frac{\partial^3\Phi_3^*}{\partial\tau_2^3} - \xi f'\frac{\partial^3\Phi_3^*}{\partial\xi\partial\tau_2^2} + \frac{\partial^4\Phi_3^*}{\partial\tau_2^4}\right]$$
$$= -\left[3f'\frac{\partial^2\chi_{21}}{\partial\tau_2^2} + \frac{\partial\chi_{21}}{\partial\tau_2}f'' + \xi\frac{\partial\chi_{21}}{\partial\xi}f''' + f\frac{\partial^3\chi_{21}}{\partial\tau_2^3}\right.$$
$$\left. - \xi f'\frac{\partial^3\chi_{21}}{\partial\xi\partial\tau_2^2} + \frac{\partial^4\chi_{21}}{\partial\tau_2^4} - \chi_{21}f'''\right] + \frac{d}{d\tau_2}(\tau_2 f' - f)^2. \tag{2.18}$$

In this equation there are two unknown functions: χ_{21} and $\Phi_3^* = \Phi_3 + \varphi_3\big|_{\eta=0}$. If we want to solve this equation completely, then, in order the function Φ_3^* have no singularities at $\xi = 0$, the functions χ_{21} must contain a summand of the form $\xi^l h(\tau_2)$, $l \geqslant 1$. Suppose the function χ_{21} can be represented in the form $\chi_{21} = g(\tau_2)+\xi h(\tau_2)$, then $\chi_2 \sim \frac{g(\tau_2)+\xi h(\tau_2)}{\tau_1} = \frac{g(\tau_2)}{\tau_1} + \varepsilon h(\tau_2)$ as $\tau_1 \to \infty$. Here the last summand contains a small parameter and is independent of τ_1. Consequently, this summand is contained in the function Φ_3. Therefore we separately set the left- and right-hand sides of (2.18) equal to zero. For the function Φ_3^*, we obtain a linear homogeneous differential equation with homogeneous boundary conditions, thus $\Phi_3^* \equiv 0$ and

$$\Phi_3 \equiv 0, \qquad \varphi_3 \equiv 0.$$

For the function χ_{21}, we obtain the following equation with right-hand side:

$$3f'\frac{\partial^2\chi_{21}}{\partial\tau_2^2} + \frac{\partial\chi_{21}}{\partial\tau_2}f'' + \xi\frac{\partial\chi_{20}}{\partial\xi}f''' - \chi_{21}f''' + f\frac{\partial^3\chi_{21}}{\partial\tau_2^3} - \xi f'\frac{\partial^3\chi_{21}}{\partial\xi\partial\tau_2^2} + \frac{\partial^4\chi_{21}}{\partial\tau_2^4}$$
$$= \frac{d}{d\tau_2}(\tau_2 f' - f)^2.$$

By setting $\chi_{21} = F(\tau_2)$, we reduce this equation to an ordinary differential equation for the function F:

$$F^{(4)} + fF''' + 3f'F'' + f''F' - F'''F = \frac{d}{d\tau_2}(\tau_2 f' - f)^2.$$

The adhesion condition implies

$$F(0) = F'(0) = 0.$$

As already pointed out, the general solution to this boundary-value problem has the form

$$F = C_1 y^{(1)} + C_2 y^{(2)} + y^{(3)},$$

where $y^{(3)}$ is a partial solution, while $y^{(1)}$ and $y^{(2)}$ are the eigenfunctions of the corresponding homogeneous boundary-value problem. The functions $y^{(i)}$, $i = 1, 2, 3$, have been studied in detail (see [95, 102]). By using the above formulas, one can write the following asymptotic expansion of the function F as $\tau_2 \to \infty$:

$$F = (C_2/4 + \beta^2/4)\tau_2 + (C_2 a^{(2)} + C_1\beta + a^{(3)}) + \\ + (C_2 b^{(2)} + b^{(3)})E_1 + \sigma(\tau_2^{-N}), \qquad \forall N \gg 1, \quad (2.19)$$

where

$$E_1 = \frac{1}{\tau_2 - \beta/2} + O((\tau_2 - \beta/2)^{-3}), \qquad \tau_2 \to \infty,$$

$a^{(2)}$, $a^{(3)}$, $b^{(2)}$ and $b^{(3)}$ are nonzero constants determined numerically, C_1, C_2 are arbitrary constants, β is the constant from the asymptotic expansion of the Blasius function. Note that at this stage none of these constants can be fixed, since no conditions are imposed on the function F as $\tau_2 \to \infty$ (the function χ_{21} itself is the limit of the χ_2 as $\tau_1 \to \infty$ for finite values of τ_2).

To continue our construction, it is necessary to eliminate the slowly decreasing part form the localized "soliton" part of the solution, i.e., to subtract from χ_2 any function with asymptotic expansion at infinity similar to that of χ_2. By (2.17), (2.19), as such a function, we take

$$N(\tau_1, \tau_2) = \frac{\tau_1 F(\tau_2)}{\tau_1^2 + \tau_2^2 + M(\tau_1, \tau_2)},$$

where $M = \sigma(\tau_1^2)$, $\tau_1 \to \infty$, $M = \sigma(\tau_2^2)$, $\tau_2 \to \infty$, since this function satisfies the estimates

$$N \sim \frac{F(\tau_2)}{\tau_1}, \quad \tau_1 \to \infty, \quad \tau_2 = O(1),$$
$$N \sim C\frac{\tau_1}{\tau_2}, \quad \tau_2 \to \infty, \quad \tau_1 = O(1), \quad C = \text{const}.$$

Let us calculate the derivative:

$$\frac{\partial N}{\partial \tau_2} = \frac{F'\tau_1}{\tau_1^2 + \tau_2^2 + M} - \frac{\tau_1 F(2\tau_2 + \partial M/\partial \tau_2)}{(\tau_1^2 + \tau_2^2 + M)^2}.$$

Obviously, for the relation

$$\frac{\partial N}{\partial \tau_2} \sim \frac{F'}{\tau_1} + O(\tau_1^{-2}), \qquad \tau_1 \to \infty$$

to hold (in this case χ_2 has no additional summands decreasing slower than τ_1^{-2} as $\tau_1 \to \infty$, except $N(\tau_1, \tau_2)$), it is necessary that

$$\frac{\partial M}{\partial \tau_2} = O(\tau_1), \qquad M = O(\tau_1), \qquad \tau_1 \to \infty.$$

On the other hand,

$$\frac{\partial N}{\partial \tau_1} = \frac{F}{\tau_1^2 + \tau_2^2 + M} - \frac{\tau_1 F(2\tau_1 + \partial M/\partial \tau_1)}{(\tau_1^2 + \tau_2^2 + M)^2}$$

and since, in general, $F \sim C_2\tau_2/4$, $\tau_2 \to \infty$, the condition $\partial N/\partial \tau_1 \sim O(\tau_2^{-2})$, $\tau_2 \to \infty$, implies that $\partial M/\partial \tau_1 = O(\tau_2)$, i.e., $M = O(\tau_2)$, $\tau_2 \to \infty$. By summing the above estimates, we obtain $M = O(\tau_1\tau_2)$ as $\tau_1 \to \infty$, $\tau_2 \to \infty$. But in this case, $N \to$ const as $\tau_1 = \tau_2 \to \infty$. To avoid this effect, we reinforce the assumption on how the function χ_2 decreases as $\tau_2 \to \infty$, by setting $\chi_2 = O(\tau_2^{-2})$, $\tau_2 \to \infty$, $\tau_1 = O(1)$, and represent the function χ_2 in the form

$$\chi_2 = \frac{\tau_1 F(\tau_2)}{\tau_1^2 + \tau_2^3 + 1} + \chi_2^1$$

(unity has been added to the denominator in the first summand to eliminate any singularity at the origin). In this representation the second summand must satisfy the condition:

$$\chi_2^1\big|_{\tau_1^2+\tau_2^2\to\infty} \to 0,$$

and

$$\begin{aligned} &\chi_2^1 = O(\tau_1^{-2}), \qquad \tau_1 \to \infty, \qquad \tau_2 = O(1),\\ &\chi_2^1 = O(\tau_2^{-2}), \qquad \tau_2 \to \infty, \qquad \tau_1 = O(1). \end{aligned} \tag{2.20}$$

Then, assuming

$$\varphi_i \equiv 0, \quad \Phi_i \equiv 0, \quad i \geq 4, \qquad \chi_i \equiv 0, \quad i \geq 3,$$

we obtain that, according to the rule for applying the operator of multiplication by a phase ("slow variable"), the initial equation with a small parameter can be rewritten for the function $\chi_2^1 = \chi_2^1(\tau_1, \tau_2)$ in the = variables τ_1 and τ_2:

$$\begin{aligned} &2\tau_1^2 f'(D_F + \Delta_\tau \chi_2^1) + 2(\tau_1^2 FR - \tau_2 f)(F'' R\tau_1 + D_F + \Delta_\tau \chi_2^1)\\ &\quad + 2\left[\tau_1 \frac{\partial \chi_2^1}{\partial \tau_2} - 3\tau_1^2\tau_2^2 FR^2 - \tau_2 \frac{\partial \chi_2^1}{\partial \tau_1} + \tau_2 FR - 2\tau_2(\tau_2^3 + 1)FR^2\right]\\ &\qquad\qquad \times (\tau_1 f'' + \tau_1 F'' R + D_F + \Delta_\tau \chi_2^1)\\ &\quad + (\tau_1^2 + \tau_2^2)\Big\{\tau_1 R^2(-3\tau_2^2 f F'' - 2f'(\tau_2^3 + 1)F'')\\ &\quad + \left(f\frac{\partial}{\partial \tau_2} - \tau_1 f' \frac{\partial}{\partial \tau_1}\right)(D_F + \Delta_\tau \chi_2^1)\\ &\quad + R\left(-F\frac{\partial}{\partial \tau_2} - F'\tau_1 \frac{\partial}{\partial \tau_1}\right)(\tau_1 f'' R + D_F + \Delta_\tau \chi_2^1)\\ &\quad + \left[\left(2(\tau_2^3 + 1)FR^2 + \frac{\partial \chi_2^1}{\partial \tau_2}\right)\frac{\partial}{\partial \tau_2} + \left(3\tau_1\tau_2^2 FR^2 - \frac{\partial \chi_2^1}{\partial \tau_2}\right)\frac{\partial}{\partial \tau_1}\right] \end{aligned}$$

$$
\begin{aligned}
&\times (\tau_1 f'' + \tau_1 F'' R + D_F)\Big\} \\
&+ \tau_2^2 \tau_1 R(fF''' + fF'' - Ff''' - F'f'') \\
&= (\tau_1^2 + \tau_2^2)(D_{F''} + \Delta_\tau D_F + \Delta_\tau \Delta_\tau \chi_2^1) \\
&- 4\left(\tau_1 \frac{\partial}{\partial \tau_1} + \tau_2 \frac{\partial}{\partial \tau_2} - 1\right)(\tau_1 F'' R + D_F + \Delta_\tau \chi_2^1),
\end{aligned} \tag{2.21}
$$

where

$$
\begin{aligned}
R &= 1/(\tau_1^2 + \tau_2^3 + 1), \\
D_F &= 2\tau_1 F[(1 - 3\tau_2)R^2 - (4\tau_2^3 + 4 - 9\tau_2^4)] - 6\tau_1 \tau_2^2 F' R^2.
\end{aligned}
$$

This is the *standard equation* for the problem under consideration (it determines the dependence of the solution on "fast" variables). Since the function $F\tau_1/(\tau_1^2 + \tau_2^3 + 1)$ satisfies the adhesion and flow symmetry conditions and decreases at infinity, the boundary conditions for χ_2^1 are homogeneous

$$
\chi_2^1\Big|_{\tau_2=0} = \frac{\partial \chi_2^1}{\partial \tau_2}\Big|_{\tau_2=0} = 0, \quad \chi_2^1\Big|_{\tau_1=0} = \frac{\partial^2 \chi_2^1}{\partial \tau_1^2}\Big|_{\tau_1=0} = 0, \quad \chi_2^1\Big|_{\tau_1^2+\tau_2^2\to\infty} \to 0. \tag{2.22}
$$

In turn, since equation (2.21) is inhomogeneous, there exists a nontrivial solution to problem (2.21), (2.22). We also recall that equation (2.21) itself was written under the assumption that the asymptotic estimates (2.19) hold and the coefficients of this equation contain the function F linearly depending on two unknown constants. Thus if these constants can be chosen so that the standard equation (2.21) possess a solution satisfying the boundary conditions (2.22) and the asymptotic estimates (2.20), then the formula given in the theorem is the exact solution to the problem about the flow around a flat plate. Since equation (2.21) coincides with equation (2.4), the theorem is proved.

We have reduced our problem to a standard equation and shown that the unknown constants, determining the behavior of the localized "soliton" part of the solution as $\tau_1 \to \infty$, cannot be obtained before this equation (an analog of the complete Navier–Stokes equation) is solved. Only two constants were included into the solution and the "soliton" part was assumed to decrease powerlike ($\sim 1/\tau_1$).

7.3. Asymptotic behavior of the laminar flow around a plate with small periodic irregularities

In this section, following [88], we shall consider the system of Navier–Stokes equations with a small parameter

$$
\begin{aligned}
\langle U, \nabla \rangle U &= -\nabla P + \varepsilon^2 \Delta U, \\
\langle \nabla, U \rangle &= 0
\end{aligned} \tag{3.1}
$$

in $\mathbf{R}^2$. The boundary conditions have the form

$$U\Big|_{\substack{y=\varepsilon^{4/3}\mu(x/\varepsilon,x)\\ x>0}} = \begin{pmatrix}0\\0\end{pmatrix}, \qquad U_y\Big|_{\substack{y=0\\x<0}} = 0, \qquad \frac{\partial U_x}{\partial y}\Big|_{\substack{y=0\\x<0}} = 0,$$
$$U\Big|_{|y|\to\infty} \to \begin{pmatrix}1\\0\end{pmatrix}, \quad U\Big|_{x\to-\infty} \to \begin{pmatrix}1\\0\end{pmatrix}, \tag{3.2}$$

where $U = U(x,y,\varepsilon)$ is the velocity vector, $P = P(x,y,\varepsilon)$ is the pressure, $\mu = \mu(x/\varepsilon, x)$ is a given smooth function, 2π-periodic in $\zeta = x/\varepsilon$, with fixed mean value over the period, $\int_0^{2\pi} \mu(\zeta,x)\,d\zeta = 0$. This function vanishes at the point $x = 0$ together with all its derivatives up to a certain order $K > 1$: $\frac{\partial^k \mu}{\partial x^k}\big|_{x=0} = 0$, $k = 0,\dots,K$. This problem models the flow of incompressible low-viscous fluid around a semi-infinite plate with irregularities. The plate surface is given by the equation $y = \varepsilon^{4/3}\mu(x/\varepsilon, x)$, $x \geqslant 0$ (see Figure 7.3). A formal asymptotic solution to problem (3.1), (3.2) is constructed in a finite (independent of ε) domain $\Omega \subset \mathbf{R}^2$ containing the origin of coordinates. This solution describes the global structure of the fluid flow in a domain containing the front edge of the plate and a sufficiently large amount of irregularities uniformly situated on the surface of the plate. This is an essential distinction from the papers by F. T. Smith [105], O. S. Pyzhov and E. D. Terent'ev [62], and others, in which a local picture of the flow in a neighborhood of a single hump is investigated by matching the asymptotic expansions.

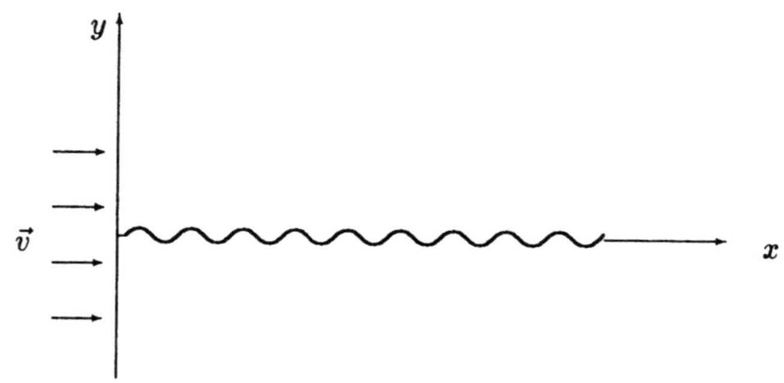

FIGURE 7.3

We introduce the stream function $\psi = \psi(x,y,\varepsilon)$ by the formulas $U_x = \partial\psi/\partial y$, $U_y = -\partial\psi/\partial x$ and rewrite equation (3.1) in the form

$$\left(\frac{\partial\psi}{\partial y}\frac{\partial\psi}{\partial x} - \frac{\partial\psi}{\partial x}\frac{\partial\psi}{\partial y}\right)\Delta\Psi = \varepsilon^2\Delta\Delta\psi. \tag{3.3}$$

For the sake of convenience, we complete the definition of the function μ by introducing a function identically equal to 0 for $x < 0$, and make the change of variables $z = y - \varepsilon^{4/3}\mu$, $x' = x$, $z = x/\varepsilon$ (in these variables the plate coincides with the positive half-axis x'). Then, by analogy to the case of a flat plate without irregularities, we pass to parabolic coordinates (optimal according to Kaplun [93, 95])

ξ, η: $x' = \xi^2 - \eta^2$, $z = 2\xi\eta$. In new coordinates, in terms of the stream function $\psi = \psi(\zeta, \xi, \eta, \varepsilon)$, the Navier–Stokes equation (3.3) has the form

$$\left[\left(\frac{\partial\psi}{\partial\eta} - \frac{2\eta}{\varepsilon}\frac{\partial\psi}{\partial\zeta}\right)\frac{\partial}{\partial\xi} - \left(\frac{\partial\psi}{\partial\xi} + \frac{2\xi}{\varepsilon}\frac{\partial\psi}{\partial\zeta}\right)\frac{\partial}{\partial\eta} + \frac{2}{\varepsilon}\left(\eta\frac{\partial\psi}{\partial\xi} + \xi\frac{\partial\psi}{\partial\eta}\right)\frac{\partial}{\partial\zeta}\right]\Delta\psi = \varepsilon^2(\xi^2 + \eta^2)\Delta\Delta\psi, \quad (3.4)$$

where

$$\begin{aligned}
\Delta\psi = {} & \frac{1}{4(\xi^2+\eta^2)}\left(\frac{\partial^2\psi}{\partial\xi^2} + \frac{\partial^2\psi}{\partial\eta^2}\right) + \frac{1}{\varepsilon(\xi^2+\eta^2)}\left(\xi\frac{\partial^2\xi}{\partial\zeta\partial\xi} - \eta\frac{\partial^2\xi}{\partial\zeta\partial\eta}\right) + \frac{1}{\varepsilon^2}\frac{\partial^2\psi}{\partial\zeta^2} \\
& - \frac{1}{(\xi^2+\eta^2)}\left[\frac{\varepsilon^{4/3}}{4(\xi^2+\eta^2)^3}\left\{\xi\frac{\partial\mu}{\partial\xi}(3\eta^2-\xi^2) + \eta\frac{\partial\mu}{\partial\eta}(3\xi^2-\eta^2)\right\}\right. \\
& + \frac{\varepsilon^{4/3}}{4(\xi^2+\eta^2)^2}\left\{\xi^2\frac{\partial^2\mu}{\partial\xi^2} - 2\xi\eta\frac{\partial^2\mu}{\partial\xi\partial\eta} + \eta^2\frac{\partial^2\mu}{\partial\eta^2}\right\} \\
& \left. + \frac{\varepsilon^{1/3}}{(\xi^2+\eta^2)}\left\{\xi\frac{\partial^2\mu}{\partial\zeta\partial\xi} - \eta\frac{\partial^2\mu}{\partial\zeta\partial\eta}\right\} + \frac{1}{\varepsilon^{2/3}\frac{\partial^2\mu}{\partial\zeta^2}}\right]\left(\eta\frac{\partial\psi}{\partial\xi} + \xi\frac{\partial\psi}{\partial\eta}\right) \\
& - \left[\frac{\varepsilon^{4/3}}{2(\xi^2+\eta^2)}\left(\xi\frac{\partial\mu}{\partial\xi} - \eta\frac{\partial\mu}{\partial\eta}\right) + \varepsilon^{1/3}\frac{\partial\mu}{\partial\zeta}\right] \\
& \quad \times\left[\frac{1}{2(\xi^2+\eta^2)^3}\left\{\eta\frac{\partial\psi}{\partial\xi}(\eta^2-3\xi^2) + \xi\frac{\partial\psi}{\partial\eta}(3\eta^2-\xi^2)\right\}\right. \\
& + \frac{1}{2(\xi^2+\eta^2)^2}\left\{\xi\eta\left(\frac{\partial^2\psi}{\partial\xi^2} - \frac{\partial^2\psi}{\partial\eta^2}\right) + (\xi^2-\eta^2)\frac{\partial^2\psi}{\partial\xi\partial\eta}\right\} \\
& \left. + \frac{1}{\varepsilon(\xi^2+\eta^2)}\left\{\eta\frac{\partial^2\psi}{\partial\zeta\partial\xi} + \xi\frac{\partial^2\psi}{\partial\zeta\partial\eta}\right\}\right] \\
& + \left[\frac{\varepsilon^{4/3}}{2(\xi^2+\eta^2)}\left(\xi\frac{\partial\mu}{\partial\xi} - \eta\frac{\partial\mu}{\partial\eta}\right) + \varepsilon^{1/3}\frac{\partial\mu}{\partial\zeta}\right]^2 \\
& \quad \times\left[\frac{1}{4(\xi^2+\eta^2)^3}\left\{\xi\frac{\partial\psi}{\partial\xi}(\xi^2-3\eta^2) + \eta\frac{\partial\psi}{\partial\eta}(\eta^2-3\xi^2)\right\}\right. \\
& \left. + \frac{1}{4(\xi^2+\eta^2)^2}\left\{\eta^2\frac{\partial^2\psi}{\partial\xi^2} + 2\xi\eta\frac{\partial^2\psi}{\partial\xi\partial\eta} + \xi^2\frac{\partial^2\psi}{\partial\eta^2}\right\}\right].
\end{aligned}$$

The boundary conditions (3.2) can be reformulated as follows. The fluid adhesion conditions on the plate:

$$\left.\frac{\partial\psi}{\partial\eta}\right|_{\eta=0} = 0, \qquad \left.\frac{2\xi}{\varepsilon}\frac{\partial\psi}{\partial\zeta}\right|_{\eta=0} + \left.\frac{\partial\psi}{\partial\xi}\right|_{\eta=0} = 0. \quad (3.5.a)$$

The symmetry condition for the incident flow:

$$-\left.\frac{2\eta}{\varepsilon}\frac{\partial\psi}{\partial\zeta}\right|_{\xi=0} + \left.\frac{\partial\psi}{\partial\eta}\right|_{\xi=0} = 0, \quad \left.\frac{\partial\psi}{\partial\eta}\right|_{\xi=0} + \left.\eta\frac{\partial^2\psi}{\partial\xi^2}\right|_{\xi=0} = 0. \quad (3.5.b)$$

The homogeneity condition for the incident flow:

$$\frac{1}{2\eta}\left(\frac{\partial\psi}{\partial\xi}+\frac{\xi}{\varepsilon}\frac{\partial\psi}{\partial\zeta}\right)\bigg|_{\eta\to\infty}\to 1. \tag{3.5.c}$$

According to the character of our problem, we must add the periodicity condition (with respect to the variable ζ): $\psi|_{\zeta=0}=\psi|_{\zeta=2\pi}$.

In the previous section an algorithm was proposed for constructing a formal asymptotic solution uniformly satisfying the problem (3.4), (3.5) in the case $\mu\equiv 0$, $\partial\psi/\partial\zeta\equiv 0$. The solution constructed there (this solution is exact, since the series with respect to a small parameter ε has a finite number of terms) has the following structure in parabolic coordinates: ε^0 (the potential flow) $+\varepsilon^1$ (the boundary layer + the potential flow) $+\varepsilon^2$ (the flow near the front edge). The boundary layer is of width $\sim\varepsilon$ and is bounded by a parabola. Near the front edge there is a transition region of width $\sim\varepsilon^1$. Inside this region all the terms of the expansion are of the same order of smallness ε^2, while outside this region all effects related to the existence of the front edge are small in contrast to the Blasius flow.

Let us consider a plate with rough surface. We note that the characteristic change of the function $\varepsilon^{4/3}\mu$ describing this surface takes place at the distance $\sim\varepsilon\gg\varepsilon^2$. If we extend the function μ to the domain $x<0$ and assume that this function is sufficiently smooth at the point $x=0$, then the character of the flow in a ε^2-neighborhood of the front edge does not differ essentially from the situation of a flat plate.

Outside this neighborhood the order of perturbations caused by the front edge does not increase. Thus, when constructing an asymptotic solution to our problem with precision up to terms of the second order of smallness, we can neglect the influence of the front edge. The irregularities on the plate are responsible for the summands of fractional orders in the asymptotic expansion, these summands begin with the order $\varepsilon^{4/3}$ (the principal term is the solution to the flat problem). Then, inside the boundary layer, a thin sublayer of width $\sim\varepsilon^{4/3}$ appears and its contribution to the solution is the term $O(\varepsilon^{5/3})$, determined by equations of Prandtl type with nonzero pressure gradient.

Below we write out an asymptotic solution with properties described above (see (3.25)). The solution is constructed in terms of the stream function in parabolic coordinates. This allows not to consider the singularities at the origin, which is inevitable when we operate in Cartesian coordinates in terms of velocities.

For convenience, the following theorem presents the solution describing the flow along a plate with irregularities in terms of velocities expressed as functions of the derivatives of the stream function.

Theorem 3.1. *The asymptotic solution to problem* (3.1), (3.2) *has the form*

$$U_x(x,y,\varepsilon)=\left[\frac{1}{2}f'(\tau_2)+\varepsilon^{1/3}\frac{\xi}{2(\xi^2+\eta^2)}\left\{\frac{\partial F_3^*}{\partial\theta}(\zeta,\xi,\theta)+\frac{\partial\Phi_2^*}{\partial\tau_2}(\zeta,\xi,\tau_2)\right.\right.$$

$$- \frac{\partial \Phi_2^*}{\partial \tau_2}(\zeta, \xi, 0) - \theta \xi f''(0) \Big\} + O(\varepsilon^{2/3}) \Big] \Bigg| _{\substack{\zeta = x/\varepsilon \\ \xi = \pm \frac{1}{\sqrt{2}} \sqrt{x + \sqrt{x^2 + (y - \varepsilon^{4/3}\mu)^2}} \\ \eta = \frac{1}{\sqrt{2}} \sqrt{-x + \sqrt{x^2 + (y - \varepsilon^{4/3}\mu)^2}} \\ \tau_2 = \eta/\varepsilon, \quad \theta = \eta/\varepsilon^{4/3}}}$$

$$U_y(x, y, \varepsilon) = \Big[-\varepsilon^{2/3} \frac{\partial}{\partial \zeta} \{ F_3^*(\zeta, \xi\theta) + \widetilde{\Phi}_3(\zeta, \xi, \tau_2) - \widetilde{\Phi}_3(\zeta, \xi, 0) \}$$

$$+ \frac{\xi}{2(\xi^2 + \eta^2)} \frac{\partial \mu(\zeta, \xi^2)}{\partial \zeta} \Big\{ \frac{\partial F_3^*}{\partial \theta}(\zeta, \xi, \theta) + \frac{\partial \Phi_2^*}{\partial \tau_2}(\zeta, \xi, \tau_2) - \frac{\partial \Phi_2^*}{\partial \tau_2}(\zeta, \xi, 0) \Big\}$$

$$+ O(\varepsilon) \Big] \Bigg| _{\substack{\zeta = x/\varepsilon \\ \xi = \pm \frac{1}{\sqrt{2}} \sqrt{x + \sqrt{x^2 + (y - \varepsilon^{4/3}\mu)^2}} \\ \eta = \frac{1}{\sqrt{2}} \sqrt{-x + \sqrt{x^2 + (y - \varepsilon^{4/3}\mu)^2}} \\ \tau_2 = \eta/\varepsilon, \quad \theta = \eta/\varepsilon^{4/3}}}$$

Here $f(\tau_2)$ is the Blasius function (see (3.20)), the function $\Phi_2^*(\zeta, \xi, \tau_2)$ has the form $\Phi_2^* = \overline{\Phi}_2^*(\xi, \tau_2) + \frac{1}{2}\mu(\zeta, \xi^2) f'(\tau_2)$, where the nonoscillating function $\overline{\Phi}_2^*$ satisfies the boundary-value problem (3.30), (3.31) for the linearized Prandtl equation, the function $\widetilde{\Phi}_3(\zeta, \xi, \tau_2)$ is the solution of the nonlinear inhomogeneous equation (3.32) with boundary conditions (3.33), the function $F_3^*(\zeta, \xi, \theta)$ satisfies the following boundary-value problem

$$2\xi \left(\frac{\partial F_3^*}{\partial \theta} \frac{\partial^2 F_3^*}{\partial \zeta \partial \theta} - \frac{\partial F_3^*}{\partial \zeta} \frac{\partial^2 F_3^*}{\partial \theta^2} \right) = \frac{\partial^3 F_3^*}{\partial \theta^3} - \frac{\partial P_3}{\partial \zeta},$$

$$F_3^* \Big|_{\theta=0} = \frac{\partial F_3^*}{\partial \theta} \Big|_{\theta=0} = 0, \qquad F_3^* \Big|_{\xi=0} = 0,$$

$$\frac{\partial^2 F_3^*}{\partial \theta^2} \Big|_{\theta \to \infty} \to \xi f''(0), \qquad \frac{\partial^2 F_3^*}{\partial \theta \partial \zeta} \Big|_{\theta \to \infty} \to \frac{1}{2} \frac{\partial \mu(\zeta, \xi^2)}{\partial \zeta} f''(0),$$

$$F_3^* \Big|_{\zeta=0} = F_3^* \Big|_{\zeta=2\pi},$$

where the function $\partial P_3 / \partial \zeta$ has the form

$$\frac{\partial P_3}{\partial \zeta}(\zeta, \xi) = 2\xi^2 f''(0) \lim_{\theta \to \infty} \left(\frac{\partial F_3^*}{\partial \zeta} - \frac{1}{2} \theta \frac{\partial \mu(\zeta, \xi^2)}{\partial \zeta} f''(0) \right)$$

$$- \xi \frac{\partial \mu(\zeta, \xi^2)}{\partial \zeta} f''(0) \lim_{\theta \to \infty} \left(\frac{\partial F_3^*}{\partial \theta} - \xi f''(0) \theta \right).$$

The function F_3^* describes the fluid flow in the thin sublayer. We do not prove the solvability of the boundary-value problem in general, but perform numerical calculations of the flow along certain specific surfaces. Some results of these calculations are presented below.

The above boundary-value problem for the function F_3^* with the notation $u = 2\xi\partial F_3^*/\partial\theta$, $v = -2\xi\partial F_3^*/\partial\zeta$, $p = 2\xi P_3$ can be rewritten in the form of the system of Prandtl equations with nonzero pressure:

$$\begin{cases} u\dfrac{\partial u}{\partial\zeta} + v\dfrac{\partial u}{\partial\theta} = \dfrac{\partial^2 u}{\partial\theta^2} - \dfrac{\partial p}{\partial\zeta}, \\ \dfrac{\partial u}{\partial\zeta} + \dfrac{\partial v}{\partial\theta} = 0, \end{cases} \tag{3.6}$$

$$\begin{gathered} u\big|_{\theta=0} = v\big|_{\theta=0} = 0, \qquad u\big|_{\zeta=0} = u\big|_{\zeta=2\pi}, \\ u \sim 2\xi^2 f''(0)\theta + \xi\mu f''(0) + A(\xi), \qquad \theta \to \infty, \\ v \sim -\left(\xi\frac{\partial\mu}{\partial\zeta}f''(0)\theta + \frac{\partial B}{\partial\zeta}(\zeta,\xi)\right), \qquad \theta \to \infty, \\ \frac{\partial p}{\partial\zeta} = \left[2\xi^2\frac{\partial B}{\partial\zeta} - (\xi\mu f''(0) + A(\xi))\xi\frac{\partial\mu}{\partial\zeta}\right] f''(0). \end{gathered} \tag{3.7}$$

Problem (3.6), (3.7) differs from the standard problem of the theory of boundary layer by the periodicity condition and some unusual conditions as $\theta \to \infty$ (the "velocity" increases with the distance from the plate). We can draw a more precise analogy to problems with self-induced pressure (in particular, see [62, 105], where the flows in a neighborhood of localized perturbations are considered). In this case, there is no periodicity condition, equation (3.6) holds in the internal sublayer of other width and, as $\theta \to \infty$, the boundary conditions and the formula for pressure have another form. These distinctions appear because the size of irregularities is different and because, in fact, we consider the "superposition" of effects in the flow along a large number of single humps.

Note that the height of irregularites is taken to be equal to $O(\varepsilon^{4/3})$ and, for a given longitudinal dimension of a single hump $O(\varepsilon)$, this height is critical in the following sense: if the height of irregularities is less than critical, there is a thin sublayer of width $O(\varepsilon^{4/3})$ on the plate, so that the solution can be obtained by the perturbation theory presented in this section, and if the height is greater than critical, then the solution has another form.

A formal asymptotic solution uniformly satisfying the boundary-value problem (3.4), (3.5) will be constructed by the same methods as in the previous section for $\mu \equiv 0$. By this method, the desired solution must simultaneously depend on slow variables ξ, η and various fast variables corresponding to the problem (for a flat plate, $\tau_1 = \xi/\varepsilon$ and $\tau_2 = \eta/\varepsilon$ are fast variables).

Assuming that, for $\mu \not\equiv 0$, the general structure of boundary layers is preserved, we put the dependence on fast variables τ_1, τ_2 in the solution similar to that in the flat case.

In addition, due to the character of our problem, we introduce a 2π-periodic dependence on a fast variable $\zeta = (\xi^2 - \eta^2)/\varepsilon$ and expand the solution into a series with respect to a small parameter ε (as is easy to see, beginning from ε^2, the expansion is in powers of $\varepsilon^{1/3}$). Thus the asymptotic solution to the problem about the flow around a plate with irregularities (in terms of the stream function) has the form

$$\Psi(\zeta,\xi,\eta,\varepsilon) = \Big(\varphi_0(\xi,\eta) + \sum_{i=1}^{3N} [\Phi_i(\zeta,\xi,\tau_2) + \varphi_i(\zeta,\xi,\eta)]\varepsilon^{(i+2)/3} + \sum_{i=4}^{3N} \chi_i(\zeta,\tau_1,\tau_2)\varepsilon^{(i+2)/3}\Big)\Big|_{\substack{\tau_1=\xi/\varepsilon \\ \tau_2=\eta/\varepsilon}} + O(\varepsilon^{N+1}). \tag{3.8}$$

Here the functions Φ_i, $i \geqslant 1$, decrease as $\tau_2 \to \infty$ faster than any negative power of τ_2; χ_i, $i \geqslant 4$, decrease as $\tau_1^2 + \tau_2^2 \to \infty$ (in other words, the functions Φ_i do not vanish only within the boundary layer, and the functions χ_i do not vanish only in a neighborhood of the front edge of the plate).

The functions φ_i, Φ_i, $i \geqslant 1$, and χ_i, $i \geqslant 4$, are periodic in the variable ζ (with period 2π) and can be represented as the sum

$$V(\zeta,\cdot) = \widetilde{V}(\zeta,\cdot) + \overline{V}(\cdot), \tag{3.9}$$

where V is an arbitrary function, 2π-periodic in ζ, $\bar{V}(\cdot) = \frac{1}{2\pi}\int_0^{2\pi} V(\zeta,\cdot)\,d\zeta$ is its mean value over the period, and the oscillating part $\widetilde{V}$ satisfies the condition $\int_0^{2\pi} \widetilde{V}(\zeta,\cdot)\,d\zeta = 0$, i.e., $\widetilde{V} = V - \bar{V}$.

Proof of Theorem 3.1. Let us substitute the expansion (3.9) into the initial Navier–Stokes equation (3.4) (here and below we multiply the equation by the common denominator). In the obtained complete equation we let τ_1 and τ_2 tend to infinity for finite ξ and η and, beginning from ε^{-2}, set the coefficients at equal powers of ε equal to zero. We obtain

$$\begin{aligned}
\varepsilon^{-2}: &\quad \partial\varphi_1/\partial\zeta^3 = 0,\\
\varepsilon^{-5/3}: &\quad \partial^3\varphi_2/\partial\zeta^3 = \left(\eta\frac{\partial\varphi_0}{\partial\xi} + \xi\frac{\partial\varphi_0}{\partial\eta}\right)\frac{\partial^3\mu}{\partial\zeta^3}\Big/\left(2(\xi^2+\eta^2)\right),\\
\varepsilon^{-4/3}: &\quad \partial^3\varphi_3/\partial\zeta^3 = 0,\\
\varepsilon^{-1}: &\quad \partial^3\varphi_4/\partial\zeta^3 = 0.
\end{aligned}$$

Since the mean values of the functions $\widetilde{\varphi}_i$ and μ are equal to zero, the latter relations imply

$$\begin{gathered}
\widetilde{\varphi}_1 \equiv \widetilde{\varphi}_3 \equiv \widetilde{\varphi}_4 \equiv 0,\\
\widetilde{\varphi}_2 = \left(\eta\frac{\partial\varphi_0}{\partial\xi} + \xi\frac{\partial\varphi_0}{\partial\eta}\right)\mu\Big/\left(2(\eta^2+\xi^2)\right).
\end{gathered} \tag{3.10}$$

Note that the expressions for oscillating functions in the region of the external flow can be obtained before any solution in the internal (boundary layer) region is known. This fact does not coincide with the standard procedure in the theory of boundary layer, when the equations are solved first in the external and then in the internal regions and, for $\eta = 0$, one of the boundary conditions for the function in slow variables ξ, η is generated by a boundary layer function of the same order. Moreover, to derive formula (3.10), no boundary conditions were used except the periodicity condition.

We shall show that, in this case, the conditions that are consequences of (3.5) and contain only the functions $\tilde{\varphi}_i$ are satisfied. Expression (3.9) implies that the flow symmetry condition (3.5b) splits into two following conditions:

$$\overline{\Psi}\Big|_{\xi=0} = 0, \qquad -\frac{2\eta}{\varepsilon}\frac{\partial \Psi}{\partial \zeta}\Big|_{\xi=0} + \frac{\partial \widetilde{\Psi}}{\partial \eta}\Big|_{\xi=0} = 0.$$

By substituting the stream function of the form (3.8) into the latter relation, we obtain $\tilde{\varphi}_i\big|_{\xi=0} = 0$, $i = 1, 2, \ldots$. Similarly, by the homogeneity condition for the incident flow, we get

$$\frac{1}{2\eta}\frac{\partial \varphi_0}{\partial \xi}\Big|_{\eta\to\infty} \to 1, \qquad \frac{1}{\eta}\varphi_i\Big|_{\eta\to\infty} \to 0.$$

In the nontrivial case ($i = 2$) these conditions hold by the very properties of the extended function μ. Actually, the value $\xi = 0$ corresponds to the half-plane $x \leqslant 0$, where $\mu = 0$, so that

$$\widetilde{\varphi}_2\Big|_{\xi=0} = \frac{1}{2\eta}\frac{\partial \varphi_0}{\partial \xi}\mu\Big|_{\xi=0} = 0,$$

and the condition for $\eta \to \infty$ is a consequence of the boundedness of the function μ and the condition imposed on φ_0:

$$\frac{1}{\eta}\widetilde{\varphi}_2\Big|_{\eta\to\infty} = \frac{1}{2\eta^2}\frac{\partial \varphi_0}{\partial \xi}\mu\Big|_{\eta\to\infty} \to 0.$$

One can show that these conditions also hold for $i \geqslant 5$. The other boundary conditions, obtained from (3.5), contain, together with the functions $\tilde{\varphi}_i$, certain functions in fast variables τ_1, τ_2, and can be found in the boundary-value problems posed in the internal domain. As we shall see, this leads to overdetermined problems for oscillating boundary layer functions, and hence an internal boundary layer of lesser width must exist.

Precisely as above, considering the equations for the external flow, one obtains expressions for the functions $\tilde{\varphi}_5, \tilde{\varphi}_6, \ldots$, which we do not write here. Besides of this, for the order φ^0, we first obtain a nontrivial averaged equation of the form

$$\left[2\left(\xi\frac{\partial \varphi_0}{\partial \eta} - \eta\frac{\partial \varphi_0}{\partial \xi}\right) + (\eta^2+\xi^2)\left(\frac{\partial \varphi_0}{\partial \xi}\frac{\partial}{\partial \eta} - \frac{\partial \varphi_0}{\partial \eta}\frac{\partial}{\partial \xi}\right)\right]\left(\frac{\partial^2 \varphi_0}{\partial \xi^2} + \frac{\partial^2 \varphi_0}{\partial \eta^2}\right) = 0.$$

This equation for φ_0, coinciding with the Euler equation in parabolic coordinates, also takes place in the case of a flat plate (see the preceding section). Since for φ_0 the boundary conditions remain the same, i.e.,

$$\varphi_0\Big|_{\eta=0} = 0, \quad \varphi_0\Big|_{\xi=0} = 0, \quad \frac{1}{2\eta}\frac{\partial \varphi_0}{\partial \xi}\Big|_{\eta\to\infty} \to 1,$$

the solution is again a homogeneous flow: $\varphi_0 = 2\xi\eta$. The equation for $\overline{\varphi}_1$ (obtained by taking the coefficient at ε^1 and averaging it) is the linearization of the above Euler equation and coincides with the case $\mu \equiv 0$. However, the solution cannot yet be found, because the first adhesion condition $\overline{\varphi}_1\big|_{\eta=0} + \overline{\Phi}_1\big|_{\tau_2=0} = 0$ contains an unknown function $\overline{\Phi}_1$ (i.e., for mean values we have the standard successive Prandtl procedure described above).

By passing to finite values of the variable τ_2 (describing the boundary layer), we note that the above equations have been derived on the basis of the following rule formulated above:

at the point $\eta = 0$ smooth functions in the variable η can be expanded into the Taylor series with respect to powers of $\varepsilon\tau_2$ if they are multiplied by functions rapidly decreasing as $\tau_2 \to \infty$, for example,

$$\eta^N \frac{\partial \Phi_1}{\partial \tau_2} = \varepsilon^N \tau_2^N \frac{\partial \Phi_1}{\partial \tau_2} = O(\varepsilon^N), \qquad \forall N > 0.$$

By setting the coefficient at ε^{-2} in the complete Navier–Stokes equation equal to zero, we obtain the equation for the function $\Phi_1^*(\zeta, \xi, \tau_2) = \Phi_1 + \overline{\varphi}_1\big|_{\eta=0} + 2\tau_2\xi$ that holds in the boundary layer

$$\left(- \frac{\partial \Phi_1^*}{\partial \zeta} \frac{\partial}{\partial \tau_2} + \frac{\partial \Phi_1^*}{\tau_2} \frac{\partial}{\partial \zeta} \right) \left(\frac{\partial^2 \Phi_1^*}{\tau_2^2} + 4\xi^2 \frac{\partial^2 \Phi_1^*}{\partial \zeta^2} \right) = 0. \tag{3.11}$$

With precision up to the factor $4\xi^2$, equation (3.11) coincides in form with the Euler equation in Cartesian coordinates ζ and τ_2. The solution of this equation must satisfy the following additional boundary condition:

$$\begin{aligned} \Phi_1^*\Big|_{\tau_2=0} &= 0, & \frac{\partial \Phi_1^*}{\tau_2}\Big|_{\tau_2=0} &= 0, \\ \frac{\partial \Phi_1^*}{\tau_2}\Big|_{\tau\to\infty} &\to 2\xi, & \frac{\partial \Phi_1^*}{\partial \zeta}\Big|_{\tau_2\to\infty} &\to 0. \end{aligned} \tag{3.12}$$

The boundary-value problem (3.11), (3.12) is overdetermined, however, one can see that equation (3.11) is satisfied either for $\partial \Phi_1^*/\partial \tau_2 = 0$ or for $\partial \Phi_1^*/\partial \zeta = 0$. Since the first relation contradicts the boundary condition as $\tau_2 \to \infty$, we set $\widetilde{\Phi}_1 \equiv 0$ (the function $\overline{\Phi}_1^*$, satisfying conditions (3.12), is defined later). Thus we have shown that the asymptotic solution to problem (3.4), (3.5) about the flow around a plate with irregularities does not oscillate in the leading term.

Continuing the construction, we write the equations in the boundary layer up to the order ε^{-1} inclusively:

$$\varepsilon^{-5/3}: \; - \frac{\partial \Phi_2^*}{\partial \zeta} \frac{\partial^3 \Phi_1}{\partial \tau_2^3} + \frac{\partial \Phi_1^*}{\partial \tau_2} \frac{\partial \mathcal{L}_0}{\partial \zeta} = 0, \tag{3.13}$$

$$\varepsilon^{-4/3}: \; - \frac{\partial \Phi_3}{\partial \zeta} \frac{\partial^3 \Phi_1}{\partial \tau_2^3} + \left(- \frac{\partial \Phi_2^*}{\partial \zeta} \frac{\partial}{\partial \tau_2} + \frac{\partial \Phi_2^*}{\partial \tau_2} \frac{\partial}{\partial \zeta} \right) \mathcal{L}_0 + \frac{\partial \Phi_1^*}{\partial \tau_2} \frac{\partial \mathcal{L}_1}{\partial \zeta} = 0, \tag{3.14}$$

$$\varepsilon^{-1}: \quad \frac{\partial\Phi_1^*}{\partial\tau_2}\left(\xi\frac{\partial^3\Phi_1^*}{\partial\xi\partial\tau_2^2} - 2\frac{\partial^2\Phi_1^*}{\partial\tau_2^2}\right) - \xi\frac{\partial\Phi_1^*}{\partial\xi}\frac{\partial^3\Phi_1^*}{\partial\tau_2^3} - \xi\frac{\partial^4\Phi_1^*}{\partial\tau_2^4}$$
$$+ 2\xi^2\left[-\frac{\partial\Phi_4}{\partial\zeta}\frac{\partial^3\Phi_1}{\partial\tau_2^3} + \left(-\frac{\partial\Phi_3}{\partial\zeta}\frac{\partial}{\partial\tau_2} + \frac{\partial\Phi_3}{\partial\tau_2}\frac{\partial}{\partial\zeta}\right)\mathcal{L}_0\right.$$
$$\left. + \left(-\frac{\partial\Phi_2^*}{\partial\zeta}\frac{\partial}{\partial\tau_2} + \frac{\partial\Phi_2^*}{\partial\tau_2}\frac{\partial}{\partial\zeta}\right)\mathcal{L}_1 + \frac{\partial\Phi_1^*}{\partial\tau_2}\frac{\partial\mathcal{L}_2}{\partial\zeta}\right] = 0, \tag{3.15}$$

where $\Phi_2^* = \Phi_2 + \varphi_2\big|_{\eta=0}$ and the function Φ_1^* is defined above. Introduce the notation

$$\mathcal{L}_0(\zeta,\xi,\tau_2) = \frac{\partial^2\Phi_2^*}{\partial\tau_2^2} + 4\xi^2\frac{\partial^2\Phi_2^*}{\partial\zeta^2} - 2\xi\frac{\partial^2\mu}{\partial\zeta^2}\bigg|_{\eta=0}\frac{\partial\Phi_1^*}{\partial\tau_2},$$
$$\mathcal{L}_k(\zeta,\xi,\tau_2) = \frac{\partial^2\Phi_{k+2}}{\partial\tau_2^2} + 4\xi^2\frac{\partial^2\Phi_{k+2}}{\partial\zeta^2} - 2\xi\frac{\partial^2\mu}{\partial\zeta^2}\bigg|_{\eta=0}\frac{\partial\Phi_{k+1}}{\partial\tau_2}$$
$$+ \left(\frac{\partial\mu}{\partial\zeta}\right)^2\bigg|_{\eta=0}\frac{\partial^2\Phi_k}{\partial\tau_2^2} - 4\xi\frac{\partial\mu}{\partial\zeta}\bigg|_{\eta=0}\frac{\partial^2\Phi_{k+1}}{\partial\zeta\partial\tau_2}, \qquad k = 1, 2. \tag{3.16}$$

First, consider each of the equations (3.11)–(3.15) in the mean. Since the function Φ_1^* is independent of ζ, the averaged equation (3.13) is satisfied. By averaging the equation (3.14), we get

$$\int_0^{2\pi}\left(-\frac{\partial\Phi_2^*}{\partial\zeta}\frac{\partial}{\partial\tau_2} + \frac{\partial\widetilde{\Phi}_2^*}{\partial\tau_2}\frac{\partial}{\partial\zeta}\right)\widetilde{\mathcal{L}}_0\,d\zeta = 0. \tag{3.17}$$

By expressing the function $\widetilde{\mathcal{L}}_0$ from (3.13),

$$\widetilde{\mathcal{L}}_0 = \widetilde{\Phi}_2^*\frac{\partial^3\Phi_1}{\partial\tau_2^3}\bigg/\frac{\partial\Phi_1^*}{\partial\tau_2},$$

and substituting this function into (3.17), we arrive at the identity

$$-\frac{\partial}{\partial\tau_2}\left(\frac{\partial^3\Phi_1}{\partial\tau_2^3}\bigg/\frac{\partial\Phi_1^*}{\partial\tau_2}\right)\int_0^{2\pi}\widetilde{\Phi}_2^*\frac{\partial\Phi_2^*}{\partial\zeta}\,d\zeta = 0.$$

The averaged equation (3.15) yields

$$\frac{\partial\Phi_1^*}{\partial\tau_2}\left(\xi\frac{\partial^3\Phi_1}{\partial\tau_2^3} - 2\frac{\partial^2\Phi_1}{\partial\tau_2^2}\right) - \xi\frac{\partial\Phi_1^*}{\partial\xi}\frac{\partial^3\Phi_1}{\partial\tau_2^3} - \xi\frac{\partial^4\Phi_1}{\partial\tau_2^4}$$
$$+ \frac{\xi^2}{\pi}\int_0^{2\pi}\left[\left(-\frac{\partial\Phi_2^*}{\partial\zeta}\frac{\partial}{\partial\tau_2} + \frac{\partial\widetilde{\Phi}_2^*}{\partial\tau_2}\frac{\partial}{\partial\zeta}\right)\widetilde{\mathcal{L}}_1\right.$$
$$\left. + \left(-\frac{\partial\Phi_3}{\partial\zeta}\frac{\partial}{\partial\tau_2} + \frac{\partial\widetilde{\Phi}_3}{\partial\tau_2}\frac{\partial}{\partial\zeta}\right)\left(\widetilde{\Phi}_2^*\frac{\partial^3\Phi_1}{\partial\tau_2^3}\bigg/\frac{\partial\Phi_1^*}{\partial\tau_2}\right)\right]d\zeta = 0. \tag{3.18}$$

From (3.14) we get the following representation of the factor $\widetilde{\mathcal{L}}_1$:

$$\widetilde{\mathcal{L}}_1 = \left\{ \widetilde{\Phi}_3 \frac{\partial^3 \Phi_1}{\partial \tau_2^3} + \widetilde{\Phi}_2^* \frac{\partial^3 \overline{\Phi}_2}{\partial \tau_2^3} - \widetilde{\Phi}_2^* \frac{\partial \overline{\Phi}_2}{\partial \tau_2} \frac{\partial^3 \Phi_1}{\partial \tau_2} \Big/ \frac{\partial \Phi_1^*}{\partial \tau_2} + \frac{1}{2} (\widetilde{\Phi}_2^*)^2 \frac{\partial}{\partial \tau_2} \left(\frac{\partial^3 \Phi_1}{\partial \tau_2} \Big/ \frac{\partial \Phi_1^*}{\partial \tau_2} \right) \right\} \Big/ \frac{\partial \Phi_1^*}{\partial \tau_2}.$$

By substituting this expression into (3.18) and performing the necessary cancellations, we get the integral summand in the form

$$\frac{\xi^2}{\pi} \left[- \frac{\partial}{\partial \tau_2} \left(\frac{\partial^3 \Phi_1}{\partial \tau_2^3} \Big/ \frac{\partial \Phi_1^*}{\partial \tau_2} \right) \int_0^{2\pi} \left(\frac{\partial \Phi_2^*}{\partial \zeta} \widetilde{\Phi}_3 + \widetilde{\Phi}_2^* \frac{\partial \Phi_3}{\partial \zeta} \right) d\zeta - \frac{\partial}{\partial \tau_2} \left(\frac{\partial^3 \overline{\Phi}_2}{\partial \tau_2^3} \Big/ \frac{\partial \Phi_1^*}{\partial \tau_2} \right) \int_0^{2\pi} \widetilde{\Phi}_2^* \frac{\partial \Phi_2^*}{\partial \zeta} \, d\zeta - \frac{1}{2} \frac{\partial}{\partial \tau_2} \left(\frac{\partial}{\partial \tau_2} \left(\frac{\partial^3 \Phi_1}{\partial \tau_2^3} \Big/ \frac{\partial \Phi_1^*}{\partial \tau_2} \right) \Big/ \frac{\partial \Phi_1^*}{\partial \tau_2} \right) \int_0^{2\pi} (\Phi_2^*)^2 \frac{\partial \Phi_2^*}{\partial \zeta} \, d\zeta \right].$$

Obviously, since the functions $\widetilde{\Phi}_2^*$ and $\widetilde{\Phi}_3$ are periodic, all the integrals are equal to zero, and thus equation (3.18) takes the final form

$$\frac{\partial \Phi_1^*}{\partial \tau_2} \left(2 \frac{\partial^2 \Phi_1^*}{\partial \tau_2^2} - \xi \frac{\partial^3 \Phi_1^*}{\partial \xi \partial \tau_2^2} \right) + \xi \frac{\partial \Phi_1^*}{\partial \xi} \frac{\partial^3 \Phi_1^*}{\partial \tau_2^3} + \xi \frac{\partial^4 \Phi_1^*}{\partial \tau_2^4} = 0. \tag{3.19}$$

This equation coincides with the Prandtl equation in parabolic coordinates, obtained earlier for the problem about the flow around a flat plate. Together with the boundary conditions

$$\Phi_1^* \Big|_{\tau_2=0} = \frac{\partial \Phi_1^*}{\partial \tau_2} \Big|_{\tau_2=0} = 0, \qquad \Phi_1^* \Big|_{\xi=0} = 0, \quad \frac{\partial \Phi_1^*}{\partial \tau_2} \Big|_{\tau \to \infty} \to 2\xi,$$

this equation determines the solution $\Phi_1^* = \xi f(\tau_2)$, where f is the Blasius function ~~satisfying the~~ problem

$$f f'' + f''' = 0, \qquad f(0) = f'(0) = 0, \qquad f' \big|_{\tau_2 \to \infty} \to 2. \tag{3.20}$$

Now one can say that the function $\varphi_1 \equiv \overline{\varphi}_1$ is determined similarly to the flat case, i.e., $\varphi_1 = -\beta \xi$, where $\beta = \lim_{\tau_2 \to \infty} (f(\tau_2) - 2\tau_2)$. Thus, the irregularities on the plate do not influence the asymptotic solution to this problem up to the terms of order ε inclusively.

Let us return to equation (3.13) for the function Φ_2^*. By integrating it with respect to ζ with zero constant of integration and substituting into it the obtained expression for the function Φ_1^*, we arrive at the following equation for the oscillating function : $\widetilde{\Phi}_2^* = \widetilde{\Phi}_2 + \mu\big|_{\eta=0}$,

$$f'''\widetilde{\Phi}_2^* - f'\left(\frac{\partial^2\widetilde{\Phi}_2^*}{\partial\tau_2^2} + 4\xi^2\frac{\partial^2\Phi_2^*}{\partial\zeta^2} - 2\xi\frac{\partial^2\mu}{\partial\zeta^2}\bigg|_{\eta=0} f'\right) = 0. \tag{3.21}$$

For the sake of convenience, introducing the function

$$\widetilde{\Phi}_{21}(\zeta,\xi,\tau_2) = \widetilde{\Phi}_2^* - \frac{1}{2}f'\mu\bigg|_{\eta=0},$$

we transform (3.21) into the homogeneous equation

$$f'''\widetilde{\Phi}_{21} - f'\left(\frac{\partial^2\widetilde{\Phi}_{21}}{\partial\tau_2^2} + 4\xi^2\frac{\partial\widetilde{\Phi}_{21}}{\partial\zeta^2}\right) = 0. \tag{3.22}$$

By taking the properties of the functions μ, $\widetilde{\Phi}_2^*$, f' into account, we can write the boundary conditions in the form

$$\widetilde{\Phi}_{21}\bigg|_{\tau_2=0} = 0, \tag{3.23a}$$

$$\frac{\partial\widetilde{\Phi}_{21}}{\partial\tau_2}\bigg|_{\tau_2=0} = -\frac{1}{2}f''(0)\mu\bigg|_{\eta=0} \neq 0, \tag{3.23b}$$

$$\widetilde{\Phi}_{21}\bigg|_{\tau_2\to\infty} \to 0, \tag{3.23c}$$

$$\widetilde{\Phi}_{21}\bigg|_{\xi=0} = 0, \qquad \widetilde{\Phi}_{21}\bigg|_{\zeta=0} = \widetilde{\Phi}_{21}\bigg|_{\zeta=2\pi}. \tag{3.23d}$$

Obviously, the boundary-value problem (3.22), (3.23) is overdetermined and has no solution. Actually, the zero conditions for the function $\widetilde{\Phi}_{21}$ (periodic with respect to ζ) for $\tau_2 = 0$ and $\tau_2 \to \infty$ uniquely imply the trivial solution $\widetilde{\Phi}_{21} \equiv 0$ that does not satisfy (3.23b). To improve the situation, it is necessary to eliminate the superfluous condition for the function $\widetilde{\Phi}_{21}$. For this purpose we assume that, inside the common boundary layer on the plate with irregularities (where equation (3.22) holds), there exists a thin sublayer of width $\sim \varepsilon^{1+\gamma}$ (the constant $\gamma > 0$ will be calculated below). Denoting by θ the "superfast" variable that takes finite values in this sublayer ($\theta = \eta/\varepsilon^{1+\gamma} = \tau_2/\varepsilon^{\gamma}$), we add to the asymptotic expansion (3.8) a new summand of the form $\sum \varepsilon^{\gamma_k} F_k(\zeta,\xi,\theta)$, where the functions F_k, $k = 3, 4, \dots$, are periodic in ζ with period 2π and decrease faster than any negative power of θ as $\theta \to \infty$, $\gamma_k = \text{const} \geqslant 1 + \gamma$. If the asymptotics of the highest added term has the order $\gamma_3 = \frac{4}{3} + \gamma$, then, instead of (3.23b), the first adhesion condition (3.5a) implies

$$\frac{\partial\widetilde{F}_3}{\partial\theta}\bigg|_{\theta=0} + \frac{\partial\widetilde{\Phi}_{21}}{\partial\tau_2}\bigg|_{\tau_2=0} + \frac{1}{2}\mu\bigg|_{\eta=0} f''(0) = 0, \tag{3.23b$'$}$$

which is one of the boundary conditions for $\widetilde{F}_3$. In this case the above results remain to hold, and the function $\widetilde{\Phi}_{21} \equiv 0$ is uniquely determined by problem (3.22), (3.23a), (3.23b), i.e., $\widetilde{\Phi}_2^* = \frac{1}{2}\mu\big|_{\eta=0} f'(\tau_2)$.

The form of the equation for the function F_3 depends on γ. We shall choose it so that the boundary-value problem for F_3 be solvable. For this purpose consider the highest terms on both sides of the Navier–Stokes equation (3.4) inside the thin sublayer, i.e., for $\theta = O(1)$. Note that here we can apply a rule similar to that formulated before equation (3.11) and expand the functions, depending smoothly on η and τ_2, into the Taylor series with respect to powers of $\varepsilon^{1+\gamma}\theta$ and $\varepsilon^\gamma\theta$ respectively. On the left-hand side of the equation, the magnitude of γ determines whether

$$\varepsilon^{-\gamma-4/3} 2\xi \left\{ \left(\frac{\partial F_3}{\partial \theta} + \frac{\partial \Phi_2}{\partial \tau_2}\bigg|_{\tau_2=0} \right) \frac{\partial^3 F_3}{\partial \theta^2 \partial \zeta} - \left(\frac{\partial}{\partial \zeta}\left(F_3 + \Phi_3\bigg|_{\tau_2=0} \right) + \theta \frac{\partial^2 \Phi_2}{\partial \tau_2 \partial \zeta}\bigg|_{\tau_2=0} \right) \frac{\partial^3 F_3}{\partial \theta^3} \right\}$$

or

$$\varepsilon^{-5/3} 2\xi \frac{\partial^2 \Phi_1}{\partial \tau_2^2}\bigg|_{\tau_2=0} \theta \frac{\partial^3 F_3}{\partial \theta^2 \partial \zeta}$$

is the leading term with respect to ε. On the right-hand side, the leading term has the form

$$\varepsilon^{-3\gamma-2/3} \frac{\partial^4 F_3}{\partial \theta^4}.$$

For $\gamma = 1/3$, all the above terms are of the same order. Therefore, for the function F_3, we obtain the nonlinear differential equation of the fourth order

$$2\xi \left\{ \left(\frac{\partial F_3}{\partial \theta} + \frac{\partial \Phi_2}{\partial \tau_2}\bigg|_{\tau_2=0} + \theta \frac{\partial^2 \Phi_1}{\partial \tau_2^2}\bigg|_{\tau_2=0} \right) \frac{\partial^3 F_3}{\partial \theta^2 \partial \zeta} - \left(\frac{\partial}{\partial \zeta}\left(F_3 + \Phi_3\bigg|_{\tau_2=0} \right) + \theta \frac{\partial^2 \Phi_2}{\partial \tau_2 \partial \zeta}\bigg|_{\tau_2=0} \right) \frac{\partial^3 F_3}{\partial \theta^3} \right\} = \frac{\partial^4 F_3}{\partial \theta^4}. \quad (3.24)$$

For other values of γ, we have the equation for F_3:

$$2\xi \frac{\partial^2 \Phi_1}{\partial \tau_2^2}\bigg|_{\tau_2=0} \theta \frac{\partial^3 F_3}{\partial \theta^2 \partial \zeta} = 0 \qquad \left(\gamma < \frac{1}{3}\right)$$

or

$$\frac{\partial^4 F_3}{\partial \theta^4} = 0 \qquad \left(\gamma > \frac{1}{3}\right).$$

Since the function F_3 decreases, $F_3 \equiv 0$ as $\theta \to \infty$. The condition (3.23b) is not satisfied. Thus the width of the internal boundary layer formed on the plate is of the same order as the height of irregularities $O(\varepsilon^{4/3})$.

Taking the above result into account, we can write the asymptotic solution uniformly satisfying the problem (3.4), (3.5) in the form

$$\Psi(\zeta,\xi,\eta,\varepsilon) = \Big[2\xi\eta + \varepsilon\big(\xi f(\tau_2) - 2\xi\tau_2\big) + \varepsilon^{4/3}\big(\overline{\Phi}_2(\xi,\tau_2) + \overline{\varphi}_2(\xi,\eta) + \frac{1}{2}\mu\Big|_{\eta=0}(f'(\tau_2) - 2) + \mu\big) + \varepsilon^{5/3}\big(F_3(\zeta,\xi,\theta) + \Phi_3(\zeta,\xi,\tau_2) + \varphi_3(\xi,\eta)\big)\Big]\Big|_{\substack{\tau_2=\eta/\varepsilon \\ \theta=\eta/\varepsilon^{4/3}}} + O(\varepsilon^2). \tag{3.25}$$

By introducing the function $F_3^* = F_3^*(\zeta,\xi,\theta)$ according to the formula

$$F_3^* = F_3 + \Phi_3\Big|_{\tau_2=0} + \overline{\varphi}_3\Big|_{\eta=0} + \left(\frac{1}{2}f''(0)\mu\Big|_{\eta=0} + \frac{\partial\overline{\Phi}_2}{\partial\tau_2}\Big|_{\tau_2=0}\right)\theta + \xi f''(0)\frac{\theta^2}{2},$$

we rewrite equation (3.24):

$$2\xi\left(\frac{\partial F_3^*}{\partial\theta}\frac{\partial^3 F_3^*}{\partial\zeta\partial\theta^2} - \frac{\partial F_3^*}{\partial\zeta}\frac{\partial^3 F_3^*}{\partial\theta^3}\right) = \frac{\partial^4 F_3}{\partial\theta^4}. \tag{3.26}$$

The boundary conditions for the function F_3^* follow from the adhesion conditions

$$F_3^*\Big|_{\theta=0} = \frac{\partial F_3^*}{\partial\theta}\Big|_{\theta=0} = 0, \tag{3.27a}$$

the first of the flow symmetry conditions

$$F_3^*\big|_{\xi=0} = 0, \tag{3.27b}$$

the definition of the function F_3^* (the decrease of the function F_3 as $\theta\to\infty$ is taken into account):

$$\frac{\partial^2 F_3^*}{\partial\theta^2}\Big|_{\theta\to\infty} \to \xi f''(0), \qquad \frac{\partial^2 F_3^*}{\partial\zeta\partial\theta}\Big|_{\theta\to\infty} \to \frac{1}{2}f''(0)\frac{\partial\mu}{\partial\zeta}\Big|_{\eta=0}, \tag{3.27c}$$

and the periodicity condition

$$F_3^*\big|_{\zeta=0} = F_3^*\big|_{\zeta=2\pi}. \tag{3.27d}$$

By integrating (3.26) with respect to θ, we obtain

$$2\xi\left(\frac{\partial F_3^*}{\partial\theta}\frac{\partial^2 F_3^*}{\partial\zeta\partial\theta} - \frac{\partial F_3^*}{\partial\zeta}\frac{\partial^2 F_3^*}{\partial\theta^2}\right) = \frac{\partial^3 F_3^*}{\partial\theta^3} - \frac{\partial P_3}{\partial\zeta}, \tag{3.28}$$

where the "constant" of integration $\partial P_3(\zeta,\xi)/\partial\zeta$ (an analog of the self-induced pressure gradient) is related to the asymptotic expansion of the function F_3^* as $\theta\to\infty$:

$$\frac{\partial P_3}{\partial\zeta} = 2\xi^2 f''(0)\frac{\partial\Phi_3}{\partial\zeta}\Big|_{\tau_2=0} - \xi f''(0)\frac{\partial\mu}{\partial\zeta}\Big|_{\eta=0}\left(\frac{\partial\overline{\Phi}_2}{\partial\tau_2}\Big|_{\tau_2=0} + \frac{1}{2}f''(0)\mu\Big|_{\eta=0}\right).$$

The form of equation (3.28) differs only by the factor 2ξ from that of the equation in Cartesian coordinates ζ and θ describing the boundary layer with zero pressure gradient. But, in the whole, the setting of problem (3.27), (3.28) is far from being standard not only because of the periodicity condition, but also because of nonstandard conditions as $\theta \to \infty$. Some problems with similar conditions at infinity were posed in [62, 105] when solutions were constructed in a neighborhood of a single hump. In the problem about the flow around a plate with irregularities, equation (3.28), which holds in a thin sublayer outside a neighborhood of the front edge, plays the role of another standard equation (as the first standard equation, we take the Prandtl equation (3.19) in the wider boundary layer, this equation holds for a flat plate). Equation (3.28) can be solved numerically, some results of calculations are given below.

If the solution to problem (3.27), (3.28) is known, one can pose the boundary-value problems for the functions $\overline{\Phi}_2$ and $\widetilde{\Phi}_3$. In the Navier–Stokes equation, let $\theta \to \infty$, and in the coefficient at $\varepsilon^{-2/3}$ let $\tau_2 = O(1)$, $\xi = O(1)$. Then, after averaging, we obtain the following equation for the function $\overline{\Phi}_2^* = \overline{\Phi}_2 + \overline{\varphi}_2\big|_{\eta=0}$:

$$\begin{aligned}
&\frac{\partial \Phi_1^*}{\partial \tau_2}\left(-2\frac{\partial^2 \Phi_2^*}{\partial \tau_2^2} + \xi\frac{\partial^3 \Phi_2}{\partial \xi \partial \tau_2^2}\right) + \frac{\partial \Phi_2^*}{\partial \tau_2}\left(-2\frac{\partial^2 \Phi_1}{\partial \tau_2^2} + \xi\frac{\partial^3 \Phi_1}{\partial \xi \partial \tau_2^2}\right) \\
&\quad - \xi\frac{\partial \Phi_1^*}{\partial \xi}\frac{\partial^3 \Phi_2^*}{\partial \tau_2^3} - \xi\frac{\partial \Phi_2^*}{\partial \xi}\frac{\partial^3 \Phi_1}{\partial \tau_2^3} - \xi\frac{\partial^4 \Phi_2^*}{\partial \tau_2^4} \\
&\quad - \frac{1}{\pi}\int_0^{2\pi}\left\{\left(\frac{\partial \Phi_2^*}{\partial \zeta}\frac{\partial}{\partial \tau_2} - \frac{\partial \widetilde{\Phi}_2^*}{\partial \tau_2}\frac{\partial}{\partial \zeta}\right)\widetilde{\mathcal{L}}_2 + \left(\frac{\partial \Phi_3}{\partial \zeta}\frac{\partial}{\partial \tau_2} - \frac{\partial \widetilde{\Phi}_3}{\partial \tau_2}\frac{\partial}{\partial \zeta}\right)\widetilde{\mathcal{L}}_1\right. \\
&\quad \left. + \left(\frac{\partial \Phi_4}{\partial \zeta}\frac{\partial}{\partial \tau_2} - \frac{\partial \widetilde{\Phi}_4}{\partial \tau_2}\frac{\partial}{\partial \zeta}\right)\left(\widetilde{\Phi}_2^*\frac{\partial^3 \Phi_1}{\partial \tau_2^3}\Big/\frac{\partial \Phi_1^*}{\partial \tau_2}\right)\right\} d\zeta = 0,
\end{aligned} \tag{3.29}$$

where the functions $\mathcal{L}$, $k = 1, 2$, are determined above (see (3.16)). By using equations (3.14), (3.15) for $\widetilde{\Phi}_3$ and $\widetilde{\Phi}_4$, we can transform the integrand so that the integral (3.29) be equal to zero. Thus the function $\overline{\Phi}_2^*$ satisfies the linearized Prandtl equation. Taking into account the form of the function Φ_1^*, we can write this equation as follows

$$f'\left(2\frac{\partial^2 \overline{\Phi}_2^*}{\partial \tau_2^2} - \xi\frac{\partial^3 \overline{\Phi}_2^*}{\partial \xi \partial \tau_2^2}\right) + f\frac{\partial^3 \overline{\Phi}_2^*}{\partial \tau_2^3} + \xi f'''\frac{\partial \overline{\Phi}_2^*}{\partial \xi} + f''\frac{\partial \overline{\Phi}_2^*}{\partial \tau_2} + \frac{\partial^4 \overline{\Phi}_2^*}{\partial \tau_2^4} = 0 \tag{3.30}$$

and pose the boundary conditions

$$\begin{aligned}
&\Phi_2^*\Big|_{\tau_2=0} = 0, \qquad \frac{\partial \Phi_2^*}{\partial \tau_2}\Big|_{\tau_2=0} = \lim_{\theta\to\infty}\left(\frac{\partial \overline{F}_3^*}{\partial \theta} - \theta\xi f''(0)\right), \\
&\overline{\Phi}_2^*\Big|_{\xi=0} = 0, \qquad \frac{\partial \overline{\Phi}_2^*}{\partial \tau_2}\Big|_{\tau_2\to\infty} \to 0.
\end{aligned} \tag{3.31}$$

The equation for $\widetilde{\Phi}_3$ has been written above (see (3.14)). By substituting expressions for Φ_1^* and $\widetilde{\Phi}_2^*$ into it and integrating with respect to ζ, we get the following linear inhomogeneous equation

$$\xi f'''\widetilde{\Phi}_3 - \xi f'\left(\frac{\partial^2\widetilde{\Phi}_3}{\partial\tau_2^2} + 4\xi^2\frac{\partial^2\widetilde{\Phi}_3}{\partial\zeta^2}\right) = \frac{1}{2}\mu\Big|_{\eta=0}\left(f'''\frac{\partial\overline{\Phi}_2}{\partial\tau_2} - f'\frac{\partial^3\overline{\Phi}_2}{\partial\tau_2^3}\right)$$
$$- \frac{1}{8}\widetilde{\mu}^2\Big|_{\eta=0}(f^{(4)}f' - f'''f'') - \xi f'\left(\frac{\partial^2}{\partial\zeta^2}2\xi\mu\Big|_{\eta=0}\frac{\partial\overline{\Phi}_2}{\partial\tau_2} + \frac{1}{2}\xi\mu^2\Big|_{\eta=0}f''\right). \quad (3.32)$$

Together with the boundary conditions

$$\widetilde{\Phi}_3\Big|_{\tau_2=0} = \lim_{\theta\to\infty}\left(\widetilde{F}_3^* - \frac{1}{2}\theta\mu\Big|_{\eta=0}f''(0)\right), \qquad \widetilde{\Phi}_3\Big|_{\tau_2\to\infty} \to 0, \quad (3.33)$$

= and the periodicity condition in ζ, this equation uniquely determines the function $\widetilde{\Phi}_3$. The obtained results prove Theorem 3.1. An asymptotic solution to problem (3.4), (3.5) in terms of the stream function can be extended by using the above scheme. Then, in the next step, the linearized equation for the correction in the thin layer must be solved first of all.

The numerical solving of problem (3.27), (3.28) is complicated by the unboundedness of the unknown function F_3^* and the domain of integration. To overcome the first difficulty, we represent the function F_3^* in the form:

$$F_3^* = F(\zeta,\xi,\theta) + \widetilde{a}_1(\zeta,\xi)\theta + \overline{a}_2(\xi)\frac{\theta^2}{2},$$
$$\widetilde{a}_1 = \frac{1}{2}f''(0)\mu\Big|_{\eta=0}, \qquad \overline{a}_2 = \xi f''(0),$$

and take the function $\widetilde{F}(\zeta,\xi,\theta)$ as a new dependent variable. Obviously, this function is bounded (by definition, $\widetilde{F} = \widetilde{F}_3 + \widetilde{\Phi}_3|_{\tau_2=0}$ and $\widetilde{\Phi}_3|_{\theta\to\infty} \to 0$). The boundary-value problem for $\widetilde{F}$ can be formulated as follows

$$2\xi\left[\left(\frac{\partial(\widetilde{F}+\overline{F})}{\partial\theta} = +\widetilde{a}_1 + \bar{a}_2\theta\right)\frac{\partial^2\widetilde{F}}{\partial\zeta\partial\theta} + \left(\frac{\partial(\widetilde{F}+\overline{F})}{\partial\theta} - \lim_{\theta\to\infty}\frac{\partial\overline{F}}{\partial\theta}\right)\frac{\widetilde{a}_1}{\partial\zeta}\right.$$
$$\left.- \left(\frac{\partial\widetilde{F}}{\partial\zeta} - \lim_{\theta\to\infty}\frac{\partial\widetilde{F}}{\partial\zeta}\right)\bar{a}_2 - \left(\frac{\partial\widetilde{F}}{\partial\zeta} + \theta\frac{\partial\widetilde{a}_1}{\partial\zeta}\right)\frac{\partial^2(\widetilde{F}+\overline{F})}{\partial\theta^2}\right] = \frac{\partial^3(\widetilde{F}+\overline{F})}{\partial\theta^3}, \quad (3.34)$$

$$\widetilde{F}\big|_{\theta=0} = 0, \qquad \widetilde{F}\big|_{\theta=0} = 0, \qquad \widetilde{F}\big|_{\theta=0} = 0, \qquad \widetilde{F}\big|_{\theta=0} = 0,$$
$$\frac{\partial\widetilde{F}}{\partial\theta}\Big|_{\theta=0} = -\widetilde{a}_1, \qquad \frac{\partial\widetilde{F}}{\partial\theta}\Big|_{\theta\to\infty} \to 0.$$

The function $\overline{F}$ of the form

$$\overline{F} = \overline{F}_3(\xi,\theta) + \overline{\Phi}_3\Big|_{\tau_2=0} + \overline{\varphi}_3\Big|_{\eta=0} + \theta\frac{\partial\overline{\Phi}_2}{\partial\tau_2}\Big|_{\tau_2=0}$$

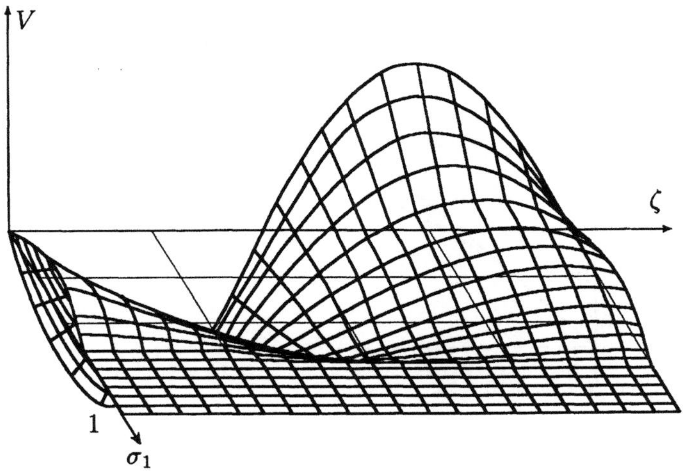

FIGURE 7.4A

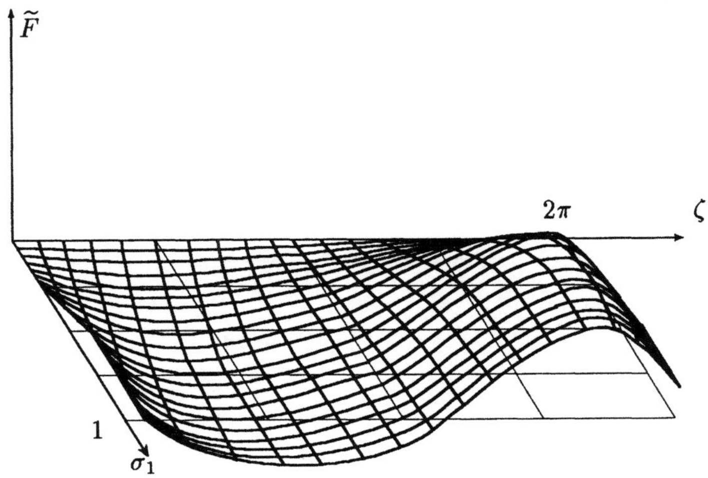

FIGURE 7.4B

is determined by the equation

$$\frac{\partial^3 \overline{F}}{\partial \theta^3} = -\frac{\xi}{\pi} \int_0^{2\pi} \left(\frac{\partial \widetilde{F}}{\partial \zeta} + \theta \frac{\partial \widetilde{a}_1}{\partial \zeta} \right) \frac{\partial^2 \widetilde{F}}{\partial \theta^2} \, d\zeta \tag{3.35}$$

and the boundary conditions

$$\overline{F}\Big|_{\theta=0} = 0, \qquad \frac{\partial \overline{F}}{\partial \theta}\Big|_{\theta=0} = 0, \qquad \frac{\partial^2 \overline{F}}{\partial \theta^2}\Big|_{\theta \to \infty} \to 0.$$

By using the transformations $\zeta_1 = \zeta/2\pi$, $\sigma_r = \theta/(\theta + 1/r)$ (it is convenient to vary the parameter r with respect to the value of ζ), we transform the unbounded domain

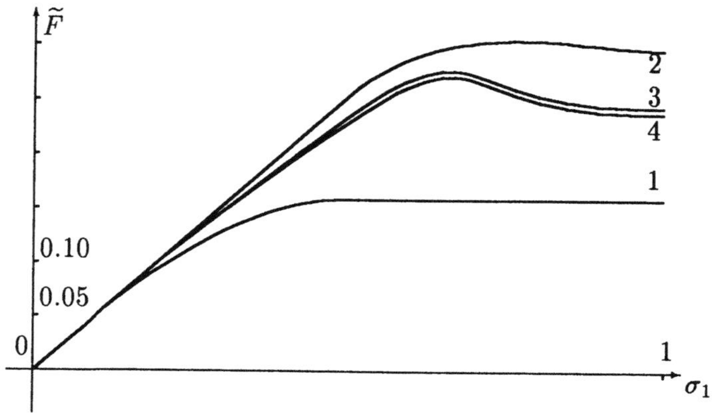

FIGURE 7.5A

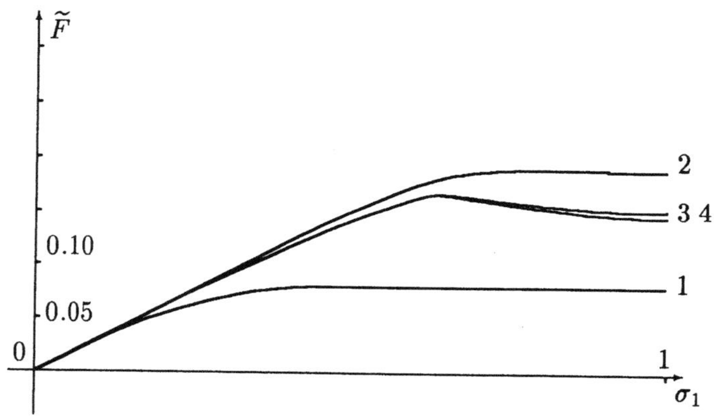

FIGURE 7.5B

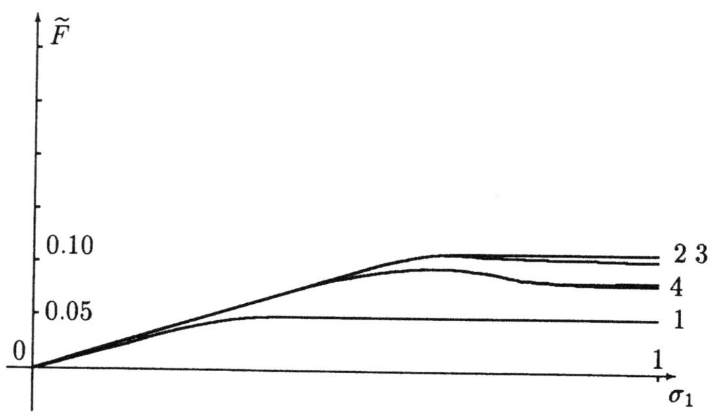

FIGURE 7.5C

of integration $\Pi = [0, 2\pi] \times [0, \infty) \ni (\zeta, \theta)$ into the unit square $K = [0, 1]^2 \ni (\zeta_1, \sigma_r)$. Similarly to the considerations used in the minimax principle (see [63]), we can construct a difference approximation of equation (3.34) (rewritten in the variables ζ_1, σ_r) on the lattice $\xi_1^i = i/N$, $i = 0, \dots, N$; $\sigma_r^j = j/L$, $j = 0, \dots, L$; $N, L = \text{const}$. For any fixed ζ, the solutions $\widetilde{F}_{ij} = \widetilde{F}(\zeta_1^i, \xi, \sigma_r^j)$ and $\overline{F}_j = \overline{F}(\xi, \sigma_r^j)$ to the difference analogs of equation (3.3) and the related equation (3.35) can be obtained by an iterated method (the procedure begins from a certain distribution $\widetilde{F}_{ij}^{(1)}$, for which $\overline{F}_j^{(1)}$ is calculated, then $\widetilde{F}_{ij}^{(2)}$ is calculated, etc.). Figures 7.4 and 7.5 illustrate the results obtained in the model case $\widetilde{a}_1 = A(\xi^2)\sin\zeta$ (i.e., $\mu(\zeta, x) = 2A\sin\zeta/f''(0)$), $A(x)\big|_{x\geqslant 1} = 1$ for $N = L = 20$. Figures 7.4A and 7.4B present the distributions of functions $\widetilde{F}_{ij}$, $V_{ij} = (\partial\widetilde{F}/\partial\theta)_{ij}$ for $\xi = 1$ (for convenience, the plane $((\zeta, \sigma_1$ is drawn). This figure clearly shows that all boundary conditions are satisfied and the functions are being stabilized near the boundary $\sigma_1 = 1$. In Figures 7.5A, 7.5B, and 7.5C the functions $\widetilde{F}$ are plotted in the cross-section $\zeta = 2\pi/5$ for three values of $\xi = 1, 2, 3$ respectively (the calculations were performed for $r = \xi$). This figure shows that the iterated process converges (the curves are drawn at regular amounts of iterations and numbered sequentially, the first curve corresponds to the first iteration).

7.4. Critical amplitude and vortices in the flow around a plate with small periodic irregularities

In the preceding section the flow around a plate with small periodic irregularities was studied. This process is described by the system of equations

$$\begin{aligned} \langle U, \nabla\rangle U &= -\nabla P + \varepsilon^2 \Delta U, \\ \langle \nabla, U\rangle &= 0, \end{aligned} \tag{4.1}$$

where $U = (U_x, U_y)$ is the velocity vector, P is the pressure, and ε^2 is a small parameter (namely, $\varepsilon = \mathrm{Re}^{-1/2}$, where Re is the classical Reynolds number). We assume that the plate surface is described by the equation (see Figure 7.3)

$$y = \varepsilon^{4/3}\mu(x/\varepsilon, x),$$

where

$$\mu(\zeta + 2\pi, x) = \mu(\zeta, x), \qquad \int_0^{2\pi} \mu(\zeta, x)\, d\zeta = 0, \qquad \frac{\partial^k \mu}{\partial x^k}(\zeta, 0) = 0, \quad k \leqslant 2.$$

The boundary conditions have the form

$$\begin{aligned} &U\Big|_{\substack{y=x^{4/3}\mu(x/\varepsilon,x)\\ x>0}} = \begin{pmatrix}0\\0\end{pmatrix}, \quad U\Big|_{|y|\to\infty} \to \begin{pmatrix}1\\0\end{pmatrix}, \\ &\frac{\partial U_x}{\partial y}\Big|_{\substack{y=0,\\ x<0}} = 0, \quad U_y\Big|_{\substack{y=0\\ x<0}} = 0, \quad U\Big|_{x\to-\infty} \to \begin{pmatrix}1\\0\end{pmatrix}. \end{aligned} \tag{4.2}$$

From the viewpoint of asymptotic methods, this is to be solved by constructing rapidly oscillating solutions encountered in the theory of nonlinear waves and by constructing boundary layer type solutions. Therefore, this problem is a rare example in which oscillations and boundary layers are naturally combined in the solution.

In the preceding chapter the solution to problem (4.1)–(4.2) was given in parabolic coordinates. This permits one to construct an asymptotic solution taking into account the influence of the front edge of the plate. However, the use of the parabolic coordinates complicates the formulas and is not essential far from the front edge. There the solution can be rewritten in Cartesian coordinates. The analysis of the solution in this domain [89] shows that there exists a critical amplitude μ_* of the irregularities such that for $|\mu| > \mu_*$ the flow is laminar, whereas for $|\mu| < \mu_*$ vortices appear and disappear between the humps.

The main result of this section is the following theorem [89].

Theorem 4.1. *Let $x \geqslant \delta > 0$. Then the asymptotic solution to problem* (4.1), (4.2) *has the form*

$$\begin{aligned}
U_x &= \Big[f'\Big(\frac{\tau}{\sqrt{x}}\Big) + \varepsilon^{1/3}\Big(u(\theta,\zeta,x) + \frac{\partial\Phi(\tau,x)}{\partial\tau}\Big) + O(\varepsilon^{2/3})\Big]\Big|_{\substack{\theta=-\mu+y/\varepsilon^{4/3}\\ \tau=y/\varepsilon\\ \zeta=x/\varepsilon}},\\
U_y &= \big[\varepsilon^{2/3}\big(v(\theta,\zeta,x) + V(\tau,\zeta,x)\big) + O(\varepsilon)\big]\big|_{\substack{\theta=-\mu+y/\varepsilon^{4/3}\\ \tau=y/\varepsilon\\ \zeta=x/\varepsilon}},\\
P &= \big[p_0 + \varepsilon^{2/3}P_1(\tau,\zeta,x) + O(\varepsilon)\big]\big|_{\substack{\tau=y/\varepsilon\\ \zeta=x/\varepsilon}},
\end{aligned} \tag{4.3}$$

where f is the Blasius function,

$$u = \frac{\partial F}{\partial\theta} - \frac{\partial\Phi}{\partial\tau}\Big|_{\tau=0} - \frac{f''(0)}{\sqrt{x}}(\theta+\mu), \quad v = -\frac{\partial F}{\partial\zeta} + \frac{\partial\mu}{\partial\zeta}\frac{\partial F}{\partial\theta} - V(0,\zeta,x),$$

and the stream function $F = F(\theta,\zeta,x)$ near the surface is the solution of the problem

$$\frac{\partial F}{\partial\theta}\frac{\partial^2 F}{\partial\theta\partial\zeta} - \frac{\partial F}{\partial\zeta}\frac{\partial^2 F}{\partial\theta^2} = \frac{\partial^3 F}{\partial\theta^3} - \frac{\partial P_1}{\partial\zeta}\Big|_{\tau=0}, \tag{4.4}$$

$$\begin{aligned}
&F\Big|_{\theta=0} = \frac{\partial F}{\partial\theta}\Big|_{\theta=0} = 0, \qquad \frac{\partial^2 F}{\partial\theta^2}\Big|_{\theta\to\infty} \to \frac{f''(0)}{\sqrt{x}},\\
&\frac{\partial^2 F}{\partial\theta\partial\zeta}\Big|_{\theta\to\infty} \to \frac{f''(0)}{\sqrt{x}}\frac{\partial\mu}{\partial\zeta}, \qquad F\big|_{\zeta=0} = F\big|_{\zeta=2\pi}.
\end{aligned} \tag{4.5}$$

The function V is the solution of the following problem posed on a cylinder:

$$\begin{aligned}
&f'\Big(\frac{\tau}{\sqrt{x}}\Big)\Big(\frac{\partial^2 V}{\partial\tau^2} + \frac{\partial^2 V}{\partial\zeta^2}\Big) - \frac{1}{x}f'''\Big(\frac{\tau}{\sqrt{x}}\Big)V = 0,\\
&V\big|_{\tau=0} = \lim_{\theta\to\infty}\Big(-\frac{\partial F}{\partial\zeta} + \frac{\partial\mu}{\partial\zeta}\frac{\partial F}{\partial\theta}\Big),\\
&V\Big|_{\tau\to\infty} \to 0, \quad V\Big|_{\zeta=0} = V\Big|_{\zeta=2\pi}, \quad \int_0^{2\pi} V\,d\zeta = 0.
\end{aligned} \tag{4.6}$$

The function Φ satisfies the Prandtl equation linearized on the stream function of the Blasius flow,

$$-f'\frac{\partial^2\Phi}{\partial x\partial\tau}+\frac{\tau}{2(\sqrt{x})^3}f''\frac{\partial\Phi}{\partial x}+\frac{1}{\sqrt{x}}f''\frac{\partial\Phi}{\partial\tau}+\frac{1}{2}\Big(\frac{1}{\sqrt{x}}f-\frac{\tau}{x}f'\Big)\frac{\partial^2\Phi}{\partial\tau^2}+\frac{\partial^3\Phi}{\partial\tau^3}=0, \quad (4.7)$$

and the boundary conditions

$$\Phi\Big|_{\tau=0}=0,\qquad \frac{\partial\Phi}{\partial\tau}\Big|_{\tau=0}=\lim_{\theta\to\infty}\Big(\frac{\partial F}{\partial\theta}-\frac{f''(0)}{\sqrt{x}}(\theta+\mu)\Big),\qquad \Phi\Big|_{\tau\to\infty}\to 0.$$

The pressure is determined by the relations

$$\frac{\partial P_1}{\partial\zeta}=f'\Big(\frac{\tau}{\sqrt{x}}\Big)\frac{\partial V}{\partial\tau}-\frac{1}{\sqrt{x}}f''\Big(\frac{\tau}{\sqrt{x}}\Big)V,\qquad \int_0^{2\pi}P_1\,d\zeta=0,\qquad p_0=\mathrm{const}$$

The proof of this theorem is similar to the proof of the corresponding theorem in Section 7.3.

Note that by solving the problem in terms of the stream function, we get the answer in the form of the Blasius function $+\,\varepsilon^{4/3}(\dots)+\dots$, i.e., the leading term of the asymptotics is known.

However, if we describe the solution of this problem in terms of velocities, then, according to (4.3), the longitudinal component of the velocity vector does not coincide with the longitudinal component of the Blasius flow velocity already in the leading term. Thus, if we solve the problem in terms of velocities, already the leading term of the asymptotics in the boundary layer is not a well-known expression.

Let us describe the characteristic properties of the obtained solution.

The irregularities on the plate generate the second, thin, boundary layer described by the variable $\theta=y\varepsilon^{-4/3}-\mu(\zeta,x)$. Note that the exponent 4/3 is critical in the sense that if we consider a surface given by the equation $y=\varepsilon^{\alpha}\mu(x/\varepsilon,x)$ for $\alpha>4/3$, then the stream function that defines the flow in the thin boundary layer satisfies a linear (linearized) equation, and whenever $\alpha<4/3$, the principal term of the asymptotic expansion of the stream function is not the Blasius function.

The existence of the critical exponent and of the thin boundary layer is characteristic of problems modeling the flow over a single hump [9, 62, 105]. In contrast to the papers mentioned above, we consider the flow over numerous humps. Problem (4.4), (4.5), which plays the role of a standard equation, is also similar to the corresponding problems in [62, 105]. In our paper we consider different boundary conditions (4.5). Just as in [62, 105], the existence of the solution to problem (4.4), (4.5) is not proved as yet, but this problem can be studied numerically. In particular, the numerical analysis shows that there is a critical amplitude of roughness and for amplitudes greater than the critical one, an inverse flow (or vortices) appears behind the humps.

Note that the solution to problem (4.4), (4.5) in fact describes a near-surface flow (in the thin boundary layer). Indeed, according to (4.3), the velocity components in the thin boundary layer ($\theta\sim$ const, $y\sim\varepsilon^{4/3}$, $\tau=\varepsilon^{1/3}(\theta+\mu)$) have the form

$$U_x\sim\varepsilon^{1/3}\Big[u(\theta,\zeta,x)+\frac{\partial\Phi}{\partial\tau}\Big|_{\tau=0}+\frac{f''(0)}{\sqrt{x}}(\theta+\mu)\Big]=\varepsilon^{1/3}\frac{\partial F}{\partial\theta},$$

$$U_y\sim\varepsilon^{2/3}\big[v(\theta,\zeta,x)+V(0,\zeta,x)\big]=\varepsilon^{2/3}\Big[-\frac{\partial F}{\partial\zeta}+\frac{\partial\mu}{\partial\zeta}\frac{\partial F}{\partial\theta}\Big].$$

These formulas allow for the contribution of the velocity components of the classical boundary layer of width $\sim \varepsilon$ to the thin boundary layer. This contribution is represented by the sum

$$\left.\frac{\partial \Phi}{\partial \tau}\right|_{\tau=0} + \frac{f''(0)}{\sqrt{x}}(\theta+\mu)$$

and by the function $V(\tau,\zeta,x)\big|_{\tau=0}$.

Problem (4.4), (4.5) can be rewritten as a system of Prandtl equations with pressure

$$u^*\frac{\partial u^*}{\partial \zeta} + v^*\frac{\partial u^*}{\partial \theta} = \frac{\partial^2 u^*}{\partial \theta^2} - \frac{\partial P^*}{\partial \zeta}, \qquad \frac{\partial u^*}{\partial \zeta} + \frac{\partial v^*}{\partial \theta} = 0,$$

where

$$u^*(\theta,\zeta,x) = \frac{\partial F}{\partial \theta}, \qquad v^*(\theta,\zeta,x) = -\frac{\partial F}{\partial \zeta}, \qquad P^*(\zeta,x) = P_1\big|_{\tau=0},$$

and the boundary conditions have the form

$$u^*\big|_{\theta=0} = v^*\big|_{\theta=0} = 0, \qquad u^*\big|_{\zeta=0} = u^*\big|_{\zeta=2\pi}, \qquad v^*\big|_{\zeta=0} = v^*\big|_{\zeta=2\pi},$$

$$\left.\frac{\partial u^*}{\partial \theta}\right|_{\theta\to\infty} \to \frac{f''(0)}{\sqrt{x}}, \qquad \left.\frac{\partial u^*}{\partial \zeta}\right|_{\theta\to\infty} \to \frac{f''(0)}{\sqrt{x}}\frac{\partial \mu}{\partial \zeta}.$$

Obviously, the existence of the inverse flow does not contradict Oleinik's conditions [60, 61].

Prior to describing the numerical solution of the standard problem (4.4), (4.5), we point out that the other problems cited in Theorem 4.1 are obviously solvable. In particular, problem (4.6) can be reduced to a system of one-dimensional problems in the variable τ, and problem (4.7) has actually been studied in [60, 61].

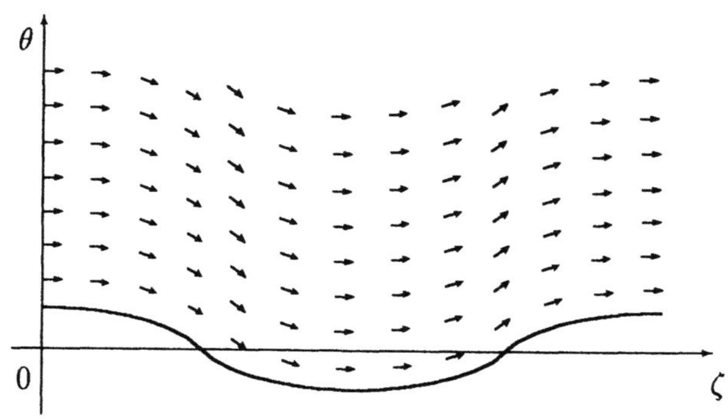

FIGURE 7.6

The difficulties in solving problem (4.4), (4.5) numerically are caused, first, by the unboundedness of the domain ($0 \leqslant \theta < \infty$) and, second, by the unboundedness of F. To overcome the latter difficulty, we take the bounded function

$$W = \frac{\partial F}{\partial \theta} - \frac{f''(0)}{\sqrt{x}}(\theta+\mu).$$

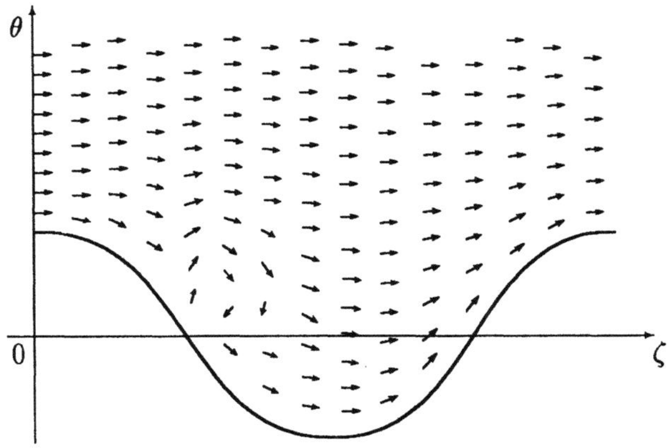

FIGURE 7.7A. Vortex formation behind a hump, $\mu = 3\cos\zeta$, $t = t_0$

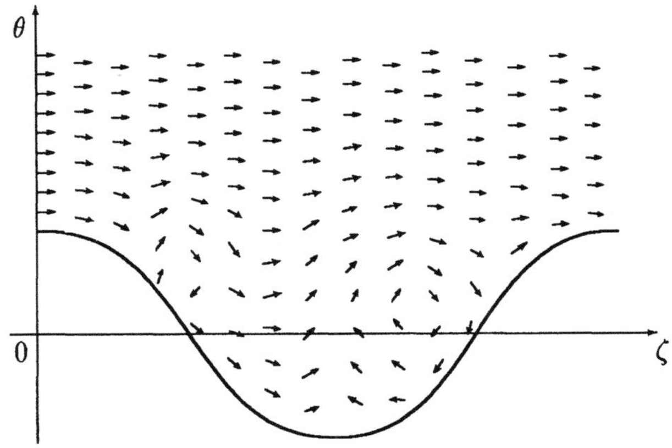

FIGURE 7.7B. Formation of the second vortex, evolution of the first vortex, $\mu = 3\cos\zeta$, $t = t_0 + 200\Delta t$

as the unknown function. The equation for W follows from the equation for F. It reads

$$\frac{\partial W}{\partial \zeta}\left(W + \frac{f''(0)}{\sqrt{x}}(\theta+\mu)\right) + \frac{f''(0)}{\sqrt{x}}\frac{\partial \mu}{\partial \zeta}(W - \lim_{\theta\to\infty} W) - \frac{f''(0)}{\sqrt{x}}(D - \lim_{\theta\to\infty} D)$$
$$- \frac{\partial W}{\partial \theta}\left(D + \frac{f''(0)}{\sqrt{x}}\theta\frac{\partial \mu}{\partial \zeta}\right) = \frac{\partial^2 W}{\partial \theta^2}, \qquad D(\theta,\zeta,x) = \int_0^\theta \frac{\partial W}{\partial \zeta}\, d\theta. \quad (4.8)$$

In what follows it is convenient to denote the mean value of W by $\overline{W}$ and the oscillator part by $\widetilde{W}$,

$$\overline{W} = \frac{1}{2\pi}\int_0^{2\pi} W\, d\zeta, \qquad \widetilde{W} = W - \overline{W}.$$

Since the boundary condition for $W|_{\theta\to\infty}$ is unknown, we take a function $\widetilde{W}$ for

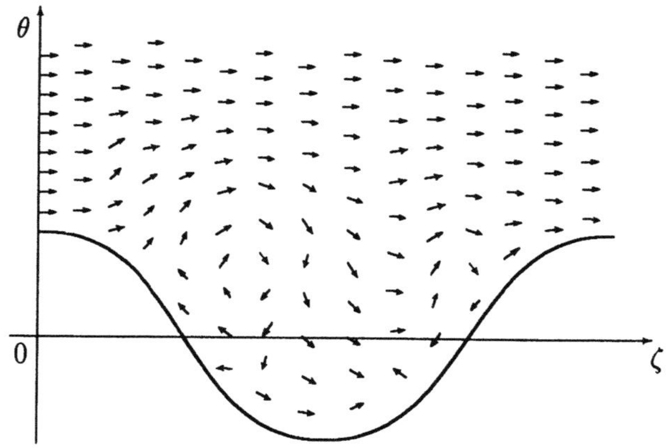

FIGURE 7.7C. Evolution of the second vortex, destruction of the first vortex, $\mu = 3\cos\zeta$, $t = t_0 + 300\Delta t$

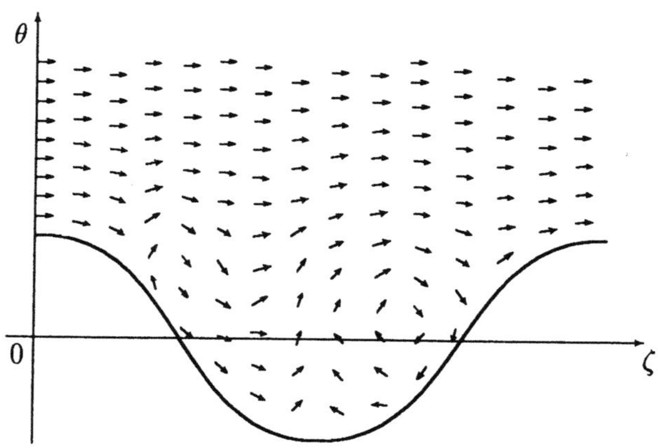

FIGURE 7.7D. The first vortex destroyed, $\mu = 3\cos\zeta$, $t = t_0 + 375\Delta t$

which $\lim_{\theta\to\infty} \widetilde{W} = 0$. Then $\overline{W}$ is determined by the formula

$$\frac{\partial^2 \overline{W}}{\partial\theta^2} = -\frac{1}{2\pi}\int_0^{2\pi}\left(\int_0^{\theta}\frac{\partial\widetilde{W}}{\partial\zeta}\,d\theta' + \frac{f''(0)}{\sqrt{x}}\theta\frac{\partial\mu}{\partial\zeta}\right)\frac{\partial\widetilde{W}}{\partial\theta}\,d\zeta$$

with the boundary conditions

$$\overline{W}\Big|_{\theta=0} = 0, \qquad \frac{\partial\overline{W}}{\partial\theta}\Big|_{\theta\to\infty} \to 0.$$

The expression for $\overline{W}$ suggests the following iterative algorithm. The functions $\overline{W}^k$, D^k, and $\widetilde{W}^{k+1}$ are determined by the given function $\widetilde{W}^k$. The equation for $\widetilde{W}^{k+1}$

has the form

$$\frac{\widetilde{W}^{k+1}-\widetilde{W}^{k}}{\Delta t}-\frac{\partial^2\widetilde{W}^{k+1}}{\partial\theta^2}-\left(D^k+\frac{f''(0)}{\sqrt{x}}\theta\frac{\partial\mu}{\partial\zeta}\right)\frac{\partial\widetilde{W}^{k+1}}{\partial\theta}$$
$$=-\frac{\partial\widetilde{W}^{k}}{\partial\zeta}\left(W^k+\frac{f''(0)}{\sqrt{x}}(\mu+\theta)\right)-\frac{f''(0)}{\sqrt{x}}\frac{\partial\mu}{\partial\zeta}\left(W^k-\lim_{\theta\to\infty}\overline{W}^k\right)$$
$$+\frac{f''(0)}{\sqrt{x}}(D^k-\lim_{\theta\to\infty}D^k)+\frac{\partial^2\overline{W}^k}{\partial\theta^2}+\left(D^k+\frac{f''(0)}{\sqrt{x}}\theta\frac{\partial\mu}{\partial\zeta}\right)\frac{\partial\overline{W}^k}{\partial\theta}. \tag{4.9}$$

To make the integration domain compact, we change the variables $\sigma=\theta/(\theta+1)$, thus taking the ray $[0,\infty)$ in θ into the interval $[0,1]$. Finally, the difference scheme has the form

$$\frac{1}{\Delta t}\widetilde{W}_{i,j}^{k+1}-(1-\sigma_j)^4\frac{1}{(\Delta\sigma)^2}(\widetilde{W}_{i,j-1}^{k+1}-2\widetilde{W}_{i,j}^{k+1}+\widetilde{W}_{i,j+1}^{k+1})$$
$$-|b|\frac{1}{\Delta\sigma}(\widetilde{W}_{i,j+l}^{k+1}-\widetilde{W}_{i,j}^{k+1})=\mathcal{F}^k,$$
$$\widetilde{W}_{i,0}=-\frac{f''(0)}{\sqrt{x}}\mu_i,\qquad \widetilde{W}_{i,N}=0,$$

where

$$b=\left(D_{i,j}^k+\frac{\sigma_j}{1-\sigma_j}\frac{1}{\sqrt{x}}\frac{\mu_{i+1}-\mu_{i-1}}{2\Delta\zeta}f''(0)\right)(1-\sigma_j)^2-2(1-\sigma_j)^3,$$
$$\zeta_i=i\cdot\Delta\zeta,\qquad \sigma_j=j\cdot\Delta\sigma,\qquad N=\frac{1}{\Delta\sigma},\qquad l=\begin{cases}1, & b>0\\ -1, & b<0\end{cases}$$
$$\mathcal{F}^k=-\frac{\widetilde{W}_{i+1,j}^k-\widetilde{W}_{i-1,j}^k}{2\Delta\zeta}\left(\widetilde{W}_{i,j}^k+\overline{W}_j^k+\left(\mu_i+\frac{\sigma_j}{1-\sigma_j}\right)\frac{1}{\sqrt{x}}f''(0)\right)$$
$$-\frac{f''(0)}{\sqrt{x}}\frac{\mu_{i+1}-\mu_{i-1}}{2\Delta\zeta}(\widetilde{W}_{i,j}^k+\overline{W}_{i,j}^k-\overline{W}_{i,N}^k)+b\frac{\overline{W}_{i,j+1}^k-\overline{W}_{i,j}^k}{\Delta\sigma}$$
$$+\frac{f''(0)}{\sqrt{x}}(D_{i,j}^k-D_{i,N}^k)+(1-\sigma_j)^4\frac{\overline{W}_{i,j+1}^k-2\overline{W}_{i,j}^k+\overline{W}_{i,j-1}^k}{(\Delta\sigma)^2}+\frac{\widetilde{W}_{i,j}^k}{\Delta t}.$$

The numerical results are shown in Figures 7.6 and 7.7A–D. One can see that the flow in the thin boundary layer (the velocity field) depending on the irregularity magnitude is essentially characterized. For the initial data shown in Figure 7.6 in the case $\mu=\cos x/\varepsilon$, near $x\sim 1$, there exists a stationary laminar flow similar to that shown in Figure 7.6. But in the case $\mu=3\cos x/\varepsilon$, near the same point $x\sim 1$ there is no stable stationary flow, and vortices appear between the humps (see Figures 7.7A–D). Thus in this example we have $1<\mu_*<3$. Note that Figure 7.7A and Figure 7.7D are alike, namely, a periodic space structure generates a process "almost" periodic in time, in which vortices appear and disappear.

Thus, for $\mu>\mu_*$ we may consider the nonstationary problem corresponding to the difference equation (4.9) as the original problem. The fact that the derivative

$\partial W/\partial t$ enters equation (4.8) for velocities is equivalent to the presence of $\varepsilon^{-2/3}\partial U/\partial t$ in the original system of Navier–Stokes equations. In its turn, this implies that the term $\varepsilon^{1/3}\partial^2/\partial t\partial\zeta$ appears in the Rayleigh-type equation (4.6) for oscillations in the classical boundary layer and the term $\varepsilon^{-2/3}\partial^2\Phi/\partial t\partial\tau$ appears in the Ohr–Sommerfeld-type equation (4.7) for nonoscillating velocities in the classical boundary layer.

Obviously, the solution of the equation for oscillations must be bounded for $t \sim \varepsilon^{-1/3}$, i.e., we consider in the stability problem for solutions of equation (4.6). However, in our case, this problem is trivial. In the domain where we solve equation (4.6) (in variables τ, ζ) the Blasius flow is planar, since the stream function of the main flow is independent of ζ. Thus we can use the well-known fact [3] that a planar parallel flow of ideal liquid without points of inflection is stable.

From the viewpoint of nonstationary problems, a surface with irregularities first generates velocity oscillations in the classical boundary layer, and then vortices appear in the thin boundary layer near this surface.

More precise calculations performed together with Professor A. Giovannini (P. Sabatier University, Toulouse, France) showed that, for supercritical amplitudes, the solution is not periodic with respect to t.

We shall try to explain these results by analogy to the behavior of the solution of a finite-dimensional dynamical system near the stationary point.

If the amplitude of irregularities is less than critical, we observe the fast (without oscillations) stabilization of the solution. By analogy to finite-dimensional dynamic systems, we say that the laminar stationary flow is a singular point of the stable node type. If the amplitude of irregularities is greater than critical, oscillations arise. However, if t increases, the solution becomes a stationary steady flow. The more the difference between the real and critical values of amplitude, the greater the oscillations become and the more time is needed for the solution to stabilize. This means that, if the real amplitude takes the critical value, the type of singularity changes. The singular point becomes a stable focus, and the imaginary parts of eigenvalues grow with the increase of the difference between the real and critical values of the amplitude of irregularities.

REFERENCES

1. E. I. Andriankin, *Propagation of a non-selfsimilar heat wave*, Zh. Eksper. Teoret. Fiz. **29:11** (1959), 1368–1372. (Russian)
2. ______, *Heat wave with energy radiating front*, Zh. Eksper. Teoret. Fiz. **35:2** (1978), 428–432 (Russian); English transl. in Soviet Phys. JETP.
3. V. I. Arnold, *Mathematical Methods of Classical Mechanics*, 3rd ed., Nauka, Moscow, 1989; English transl. of 2nd ed. Springer-Verlag, Berlin and New York, 1989.
4. Pai Shih-i, *Viscous Flow Theory. II — Turbulent Flow*, Nostrand, Princeton, 1957.
5. G. I. Barenblatt, *Unsteady motion of gas and fluid in porous media*, Prikl. Mat. Mekh. **16:1** (1952), 67–78. (Russian)
6. V. S. Berman, *Nonstationary propagation of combustion waves in a medium with slowly varying properties*, Prikl. Mat. Mekh. **42:3** (1978), 450–457 (Russian); English transl. in J. Appl. Math. Mekh..
7. ______, *Solution of a nonstationary problem about the propagation of a chemical reaction front*, Dokl. Akad. Nauk SSSR **242:2** (1978), 265–267 (Russian); English transl. in Soviet Math. Dokl..
8. ______, *Instability of front propagation for one-stage chemical reactions*, Fizika Goreniya i Vzryva (Physics of Combustion and Explosion) **22:2** (1986), 76–83. (Russian)
9. V. V. Bogolepov and V. Ya. Neiland, *Investigation of local perturbations in viscous supersonic flows*, Aeromechanics, Collection of Papers Dedicated to V. V. Struminsky's 60th Anniversary, Nauka, Moscow, 1976, pp. 104–118. (Russian)
10. P. P. Volosevich and E. I. Levanov, *Different regimes of heat transfer in two- and three-dimensional plasma*, Mathematical Modeling. Transport Processes in Nonlinear Media, Nauka, Moscow, 1986. (Russian)
11. K. A. Volosov, *Transformation of approximate solutions of linear parabolic equations into asymptotic solutions of quasilinear parabolic equations*, Mat. Zametki **56:6** (1994), 122–126 (Russian); English transl. in Math. Notes.
12. ______, *Temperature waves in moving media*, Inzh.-Fiz. Zh. **26:5** (1981), 929–930 (Russian); English transl. in J. Engrg. Phys..
13. K. A. Volosov and A. I. Fedotov, *Asymptotic representations of the solution of quasilinear parabolic equation close to the front*, Zh. Vychisl. Mat. i Mat. Fiz. **23:5** (1983), 93–101 (Russian); English transl. in U.S.S.R. Comput. Math. and Math. Phys..
14. K. A. Volosov, V. G. Danilov, and V. P. Maslov, *Combustion wave asymptotics in nonlinear inhomogeneous media with slowly varying properties*, Dokl. Akad. Nauk SSSR **290:5** (1986), 1089–1094 (Russian); English transl. in Soviet Math. Dokl..
15. ______, *Mathematical Simulation of Technological Processes in LSI Production*, VINITI, Moscow, 1986. (Russian)
16. ______, *Weak discontinuity structure of solutions of quasilinear degenerate parabolic equations*, Mat. Zametki **43:6** (1988), 829–838 (Russian); English transl. in Math. Notes.
17. K. A. Volosov and V. G. Danilov, *Model of thermal silicon oxidation*, Mat. Model **1:11** (1989), 58–67. (Russian)
18. Y. H. Kuo, *J. Aeronaut. Sci* **23:2** (1956), 125–136. (Russian)
19. M. F. Gryukanov and N. K. Koktabaev, *Nonlinear waves arising in compression of a θ-pinch between faces*, Fizika Plasmy (Physics of Plasma) **7:6** (1981). (Russian)

20. V. G. Danilov, *Global formulas for solutions of quasilinear parabolic equations and the property of being ill-posed*, Mat. Zametki **46:1** (1989), 115–117 (Russian); English transl. in Math. Notes.
21. ———, *Asymptotic solutions of traveling wave type satisfying semilinear parabolic equations with a small parameter*, Mat. Zametki **48:2** (1990), 148–151 (Russian); English transl. in Math. Notes.
22. ———, *Well-posedness of h-ppseudodifferential equations in the class of asymptotics of exponential type*, Mat. Zametki **44:3** (1988), 406–409 (Russian); English transl. in Math. Notes.
23. ———, *Asymptotic finite solutions of degenerating quasilinear parabolic equations with small diffusion*, Mat. Zametki **50:2** (1991), 77–87 (Russian); English transl. in Math. Notes.
24. V. G. Danilov and P. Yu. Subochev, *Kink interaction in the KPP-Fisher equation*, Mat. Zametki **50:3** (1991), 152–154 (Russian); English transl. in Math. Notes.
25. ———, *Exact one- and two-phase solutions of semilinear parabolic equations*, Preprint N1, Math. Inst. Acad. Sci. SSSR, Moscow (1988), 22p. (Russian)
26. V. G. Danilov, V. P. Maslov, and P. Yu. Subochev, *Asymptotic (with respect to smoothness) solutions to quasilinear hyperbolic and parabolic equations*, Deposited in VINITI, 02.01.1989, no.3 (1989), 1–32. (Russian)
27. V. G. Danilov and P. Yu. Subochev, *Wave solutions of semilinear parabolic equations*, Teoret. Mat. Fiz. **89:1** (1991), 25–47 (Russian); English transl. in Theoret. and Math. Phys..
28. V. G. Danilov and S. M. Frolovichev, *Reflection method in construction of the Green function asymptotics for parabolic equations with variable coefficients*, Mat. Zametki **57:3** (1995) (Russian); English transl. in Math. Notes.
29. V. G. Danilov, M. V. Makarova, and V. A. Tsupin, *A problem about the flow of a viscous incompressible liquid around a semi-infinite plate*, Trudy Inst Mat. Akad. Nauk Ukr.SSR (1989), 53–64. (Russian)
30. S. Yu. Dobrokhotov and V. P. Maslov, *Multi-dimensional Dirichlet series in problems about spectral series of nonlinear elliptic operators*, Itogi Nauki i Tekhniki: Sovremennye Problemy Mat., vol. 23, VINITI, Moscow, 1983, pp. 137–222. (Russian)
31. ———, *Finite-gap almost periodic solutions in WKB-approximations*, Itogi Nauki i Tekhniki: Sovremennye Problemy Mat., vol. 15, VINITI, Moscow, 1980, pp. 3–94. (Russian)
32. L. A. Zhukova, *Theory of Static and Dynamic Precipitation and Coprecipitation of Ions*, Energoizdat, Moscow, 1981. (Russian)
33. Ya. B. Zeldovich, *About the theory of flame propagation*, Zh. Fizich. Khimii (Journal of Physical Chemistry) **22:1** (1948), 27–48. (Russian)
34. ———, *Approximate theory of chain reactions*, Kinetika i Kataliz (Kinetics and Catalysis) **2** (1961), 305–314. (Russian)
35. Ya. B. Zeldovich and A. S. Komponeets, *About the theory of heat propagation when thermal conductivity depends on temperature*, To A.F.Ioffe's 70th Anniversary, Izdat. AN SSSR, Moscow, 1950, pp. 61–71. (Russian)
36. Ya. B. Zeldovich, G. I. Barenblatt, V. B. Librovich, and G. M. Makhviladze, *Mathematical Theory of Combustion and Explosion*, Nauka, Moscow, 1980. (Russian)
37. M. V. Karasev, *Calculus of almost-generating ordered operators*, Asymptotic Methods in the Theory of Partial Differential Equations. Trudy MIEM, vol. 38, Izdat. MIEM, Moscow, 1974, pp. 73–115. (Russian)
38. L. A. Kalyakin, *Perturbation of the Korteweg-de Vries soliton*, Teoret. Mat. Fiz. **92:1** (1992), 62–76 (Russian); English transl. in Theoret. and Math. Phys..
39. Ya. I. Kanel', *Stabilization of solutions of the Cauchy problem for equations used in the theory of combustion*, Mat. Sb. **59** (1962), 245–288. (Russian)
40. V. V. Kucherenko, *Necessary conditions for stability of difference schemes*, Mat. Zametki **16:4** (1976), 345-552 (Russian); English transl. in Math. Notes.
41. A. N. Kolmogorov, I. G. Petrovskii, and I. S. Piskunov, *A study of the diffusion equation with increase in the quantity of matter, and its application to a biological problem*, Bull. Moscow State Univ., section A **1:6** (1937), 1–25. (Russian)
42. V. P. Korobeinikov, *Problems of the Theory of Point Explosion*, Nauka, Moscow, 1985. (Russian)

43. S. P. Kurdyumov, *Eigenfunctions of combustion of a nonlinear medium*, Preprint N29, Inst. Appl. Math. Acad. Sci. SSSR, Moscow (1979), 64p. (Russian)
44. L. D. Landau and E. M. Lifshits, *Hydrodynamics*, Nauka, Moscow, 1986. (Russian)
45. A. V. Lykov and B. M. Berkovskii, *Convection and Heat Waves*, Energiya, Moscow, 1974. (Russian)
46. L. K. Martinson, *Investigation of a mathematical model for the nonlinear heat conduction in media with bulk absorption*, Mathematical Modeling. Transport Processes in Nonlinear Media, Nauka, Moscow, 1986. (Russian)
47. ———, *Localized heat structures in media with bulk heat absorption*, Prikl. Mekh. i Tekhn. Fiz. **21:2** (1981), 70–75. (Russian)
48. ———, *Nonlinear effects in evolution of heat structures*, Zh. Vychisl. Mat. i Mat. Fiz. **24:3** (1984), 464–468 (Russian); English transl. in U.S.S.R. Comput. Math. and Math. Phys..
49. V. P. Maslov, *Global exponential asymptotics of solutions of tunnel type equations and the problem about large deviations*, Trudy Mat. Inst. Steklov **163** (1984), 150–180 (Russian); English transl. in Proc. Steklov Inst. Math..
50. ———, *Three algebras corresponding to nonsmooth solutions of quasilinear hyperbolic equations*, Uspekhi Mat. Nauk, **35** (1980) (Russian); English transl. in Russian Math. Surveys.
51. V. P. Maslov, V. G. Danilov, and K. A. Volosov, *Mathematical Modelling of Process of Heat-Mass Transfer*, Nauka, Moscow, 1987. (Russian)
52. V. P. Maslov, *The Complex WKB Method for Nonlinear Equations*, Nauka, Moscow, 1977 (Russian); English transl.: *The Complex WKB Method for Nonlinear Equations* I. Linear Theory, Birkhäuser, Basel–Boston–Berlin, 1994.
53. ———, *Nonstandard characteristics in asymptotic problems*, Uspekhi Mat. Nauk **38:6** (1983), 3–36; English transl. in Russian Math. Surveys.
54. ———, *Operational Methods*, Nauka, Moscow, 1973; English transl. MIR, Moscow, 1976.
55. V. P. Maslov and G. A. Omel'yanov, *Asymptotic soliton-like solutions of equations with small dispersion*, Uspekhi Mat. Nauk **36:3** (1981), 63–126; English transl. in Russian Math. Surveys.
56. ———, *On Hugoniot-type conditions for infinitely narrow solutions of equations for simple waves*, Sibirsk Mat. Zh. **24:5** (1983), 50–64; English transl. in Siberian Math. J..
57. V. P. Maslov and M. V. Fedoryuk, *Quasi-classical Approximation for the Equations of Quantum Mechanics*, Nauka, Moscow, 1976; English transl., *Semiclassical Approximation in Quantum Mechanics*, Reidel, Dordrecht, 1981.
58. V. P. Maslov, G. A. Omel'yanov, and V. A. Tsupin, *Asymptotics of some differential and pseudodifferential equations and dynamic systems with small dispersion*, Mat. Sb. **122:2** (1983), 197–219; English transl. in Math. USSR–Sb..
59. V. P. Maslov and V. A. Tsupin, *Generalized by Sobolev δ-like solutions of quasilinear equations*, Uspekhi Mat. Nauk **34:1** (1979), 235–236; English transl. in Russian Math. Surveys.
60. O. A. Oleynik, *A system of boundary layer equations*, Zh. Vychisl. Mat. i Mat. Fiz. **3:3** (1963), 489–507. (Russian)
61. ———, *Mathematical problems in the boundary layer theory*, Uspekhi Mat. Nauk **23:3** (1968), 3–65.
62. O. A. Ryzhov and E. D. Terent'ev, *Nonstationary boundary layer with self-induced pressure*, Prikl. Mat. Mekh. **41:5** (1977), 1007–1023 (Russian); English transl. in J. Appl. Math. Mekh..
63. A. A. Samarskii, *Theory of Difference Schemes*, Nauka, Moscow, 1977.
64. A. A. Samarskii, G. V. Zmitrienko, S. P. Kurdyumov, and A. P. Mikhailov, *Heat structures and fundamental length in a medium with nonlinear thermal conductivity and bulk heat sources*, Dokl. Akad. Nauk SSSR **227:2** (1976), 321–324 (Russian); English transl. in Soviet Math. Dokl..
65. A. A. Samarskii and I. M. Sobol', *Examples of numerical calculation of temperature waves*, Zh. Vychisl. Mat. i Mat. Fiz. **3:4** (1963), 702–719. (Russian)
66. A. A. Samarskii, S. P. Kurdyumov, and P. P. Volosevich, *Traveling waves in a medium with nonlinear thermal conductivity*, Zh. Vychisl. Mat. i Mat. Fiz. **5:6** (1965), 199–217. (Russian)
67. A. A. Samarskii, S. P. Kurdyumov, A. P. Mikhailov, and G. V. Zmitrienko, *Metastable heat localization in a medium with nonlinear thermal conductivity and conditions under which this*

effect can be observed in experiments Preprint Inst. Appl. Math. Acad. Sci. SSSR, Moscow (1979), 67p. (Russian)

68. A. A. Samarskii, *About new methods for investigation of asymptotic properties of parabolic equations*, Trudy Mat. Inst. Steklov **158** (1981), 153–162 (Russian); English transl. in Proc. Steklov Inst. Math..
69. A. A. Samarskii, G. V. Zmitrienko, S. P. Kurdyumov, and A. P. Mikhailov, *Effect of metastable heat localization in a medium with nonlinear thermal conductivity*, Dokl. Akad. Nauk SSSR **223:6** (1975), 1344–1347 (Russian); English transl. in Soviet Math. Dokl..
70. V. N. Smagin, E. A. Lukashov, V. O. Sheftel', L. A. Kvitka, and V. M. Khiryaev, *Technological investigation of water purification from pesticides by domestic sets ED-500 and KPV-1000*, Khimiya and Tekhnologiya Vody (Chemistry and Water Technology) **14:10** (1992), 769–776. (Russian)
71. V. N. Smagin, E. A. Lukashov, and L. A. Kvitka, *Investigation of equilibrium adsorption of chlororganic compounds by granulated porous carbons in water purification*, Khimiya and Tekhnologiya Vody (Chemistry and Water Technology) **15:11–12** (1993), 726–731. (Russian)
72. S. N. Taranenko and L. D. Pokrovskii, *Conditions of spatial localization for solutions of nonlinear heat equation*, Zh. Vychisl. Mat. i Mat. Fiz. **22:3** (1982), 747–751 (Russian); English transl. in U.S.S.R. Comput. Math. and Math. Phys..
73. E. B. Tolubinskii, *Theory of Transport Processes*, Naukova Dumka, Kiev, 1969.
74. A. A. Frank-Kamenetskii, *Diffusion and Heat Transfer in Chemical Kinetics*, Nauka, Moscow, 1967.
75. G. Ekker, *Theory of Fully Ionized Plasma*, MIR, Moscow, 1977.
76. M. J. Ablowitz and A. Zeppetella, *Explicit solutions of Fisher's equation for a special wave speed*, Bull. Math. Biol. **41** (1979), 835–840.
77. D. G. Aranson, *Regularity properties of flows through porous media*, SIAM J. Appl. Math. **17** (1969), 461–467.
78. ———, *Regularity properties of flows through porous media: a counterexample*, SIAM J. Appl. Math. **19** (1970), 299–307.
79. ———, *Regularity properties of flows through porous media: the interface*, SIAM J. Appl. Math. **37** (1970), 1–10.
80. A. Amai, *Second aproximation to the laminar boundary layer flow over a flat plate*, J. Aeronat. Sci. **24:2** (1957), 155–156.
81. J. Butler, *Ionic Equilibrium States, a Mathematical Approach*, Addison–Wesley, Reading (Mass.), 1964.
82. M. R. Booty, R. Haberman, and A. A. Minzoni, *The accomodation of traveling waves of Fisher's type to the dynamics of the leading tail*, SIAM J. Appl. Math. **53:3** (1993).
83. G. I. Carrier and C. C. Lin, *On the nature of the boundary layer near the leading edge of a flat plate*, Quart. Appl. Math. **6:1** (1948), 63–68.
84. F. Cariello and M. Tabor, *Painleve expansions for nonintegrable evolution equations*, Phys. D **32** (1989), 77–89.
85. R. T. Davis, *Laminar incompressible flow past a semi-infinite flat plate*, J. Eng. Math. **4:1** (1970), 9–27.
86. V. G. Danilov and K. A. Volosov, *Asymptotic solutions in mathematical models of nonlinear heat transfer*, Proc. Int. Conf. on Plasma Phys., Kiev, Apr. 6–12, 1987, vol. 4, 1987, pp. 332–335.
87. V. G. Danilov, *Asymptotic solutions describing localized heat structures in plasma*, Proc. of the III Int. Workshop, Kiev, Apr. 13–26, 1987, vol. 2, 1987, pp. 124–127.
88. V. G. Danilov and M. V. Makarova, *Asymptotics of a laminar flow along a plate with small periodic roughness*, Siberian J. Diff Eq. **3** (1993), . Nova Science, Novosibirsk–New York, 17–39.
89. ———, *Asymptotic and numerical analysis of the flow around a plate with small periodic roughness*, Russ. J. Math. Phys. **2:1** (1994), 49–56.
90. W. R. Dean, *On the steady motion of viscous liquid past a flate plate*, Mathematica **1:2** (1954), 43–156.

91. P. De Mottoni and M. Schatzman, *Evolution Geometrique d'Interfaces*, C. R. Acad. Sc. Paris **309** (1989), 453–458.
92. H. Engler, *Relations between traveling wave solutins of quasi-linear parabolic equations*, Proc. Amer. Math. Soc. **93:2** (1985), 297–302.
93. S. Goldstein, *Lectures on Fluid Mechanics*, Wiley (Interscience), New York, 1960.
94. R. Grimshaw, *Slowly varying solitory waves. I, Korteweg-de Vries equation*, Proc. Roy. Soc. London Ser. A **368** (1979), 359–375.
95. S. Kaplun, *The role of coordinate systems in boundary layer theory*, J. Appl. Math. Phys. **75:2** (1954), 111–135.
96. T. Kawahara and M. Tanaka, *Interaction of traveling fronts on exact solution of nonlinear diffusion equation*, Phys Lett A **97** (1983), 311–319.
97. P. O. Kershner, *On certain properties of generalized solutions of quasilinear degenerate parabolic equations*, Acta Math. Acad. Sci. Hungaricae **32:3–4** (1978), 301–330.
98. J. C. Luke, *A perturbation method for nonlinear dispersive wave problems*, Proc. Roy. Soc. London Ser. A **292** (1966), 403–412.
99. V. P. Maslov and G. A. Omel'yanov, *Soliton-type asymptotic behavior of interior waves in stratified liquid with weak dispersion*, Differential Equations **21:10** (1985), 1197–1204.
100. V. P. Maslov, *Asymptotic solutions of equations with slowly varying coefficients*, Proceedings of Ecole Politechnique, Paris, 1988–89.
101. J. D. Murray, *Lectures on Nonlinear Differential Equations. Models in Biology*, Claredon Press, Oxford, 1977.
102. J. D. Murray, *Incompressible viscous flow past a semi-infinite flat plate*, J. Fluid Mech. **21** (1965), 337–344.
103. A. C. Newell and J. A. Whithead, *Finite bandwidth, finite amplitude convection*, J. Fluid Mech. **38** (1969), 279–286.
104. A. Scott, *Active and Nonlinear Wave Propagation Media in Electronics*, Wiley Interscience, New York, 1970.
105. F. T. Smith, *Laminar flow over a small hump on a flat plate*, J. Fluid Mech. **51** (1973), 4.
106. K. Stewardson, *On asymptotic expansions in the theory of boundary layers*, J. Math. Phys. **36** (1957), 173–191.
107. K. Ushiyama, *The behavior of solutions of some nonlinear diffusion equations for large time*, J. Math. Kioto Univ. **18:3** (1978), 453–511.
108. M. Van Dyke, *Perturbation Methods in Fluid Mechanics*, Acad. Press, New York–London, 1964.
109. A. J. Van de Vooren and D. Dijkstra, *The Navier–Stokes solutions for laminar flow past a semi-infinite plate*, J. Eng. Math. **4:1** (1970), 9–27.
110. J. Weiss, M. Tabor, and G. Carnevale, *The Painlevé property for partial differential equations*, J. Math. Phys. **24** (1983), 522–528.
111. B. Witham, *Nonlinear dispersion of water waves*, J. Fluid Mech. **27** (1967), 399–415.
112. ———, *Two-timing, variational principles and waves*, J. Fluid Mech. **44** (1970), 373–395.
113. M. Freidlin, *Functional Integration and Partial Differential Equations*, Ann. Math. Studies, Princeton University Press, 1985.

APPENDIX

JUSTIFICATION OF ASYMPTOTIC SOLUTIONS

S. A. VAKULENKO

In this section we shall consider the problem of justifying some asymptotic solutions constructed in this book. The progress in this field is more conservative, nevertheless the obtained results allow conclusions important both in pure mathematics and in applications.

The greatest progress was made in solving one-dimensional problems for parabolic equations and systems. It turns out that, under certain conditions of stability, asymptotic solutions are close (in corresponding norms) to some actual solutions.

These stability conditions arise when we study the standard problems. Moreover, the asymptotic solutions are related to local attractors, i.e., to families of solutions that attract all solutions lying in (small) neighborhoods of these families.

It is much more difficult to prove such properties in multidimensional situations. For the most part, it has been possible to consider only the case of constant coefficients [10–12]. Here we shall describe some results for equations and systems of a certain class. Curiously, there is an outstanding class of systems (the so-called monotone or cooperative systems) that was thoroughly investigated in the theory of attractors [13–16].

Below (f, g) denotes the integral $\int_{-\infty}^{\infty} fg\,dx$, $\|f\|$ is the standard L_2-norm $(f, f)^{1/2}$, $|f|_p$ is the C^p-norm for integer p and the Hölder norm for noninteger p.

1. One-dimensional scalar case

Consider the Cauchy problem for the Ginzburg–Landau type equation describing the propagation of nonlinear waves in homogeneous and nonhomogeneous media:

$$\varepsilon u_t = \varepsilon^2 u_{xx} + \alpha^2(x) f(u), \qquad u(x, 0, \varepsilon) = \varphi(x/\varepsilon, \varepsilon), \quad x \in \mathbb{R}. \tag{1.1}$$

By a simple change of variables $t' = t/\varepsilon$, $x' = x/\varepsilon$, this equation can be reduced to the equation with slowly varying coefficients:

$$u_{t'} = u_{x'x'} + \alpha^2(\varepsilon x') f(u), \qquad u(x', 0, \varepsilon) = \varphi(x', \varepsilon), \quad x' \in \mathbb{R}. \tag{1.2}$$

Note that more general nonlinearities of the form $f = f(u, x)$ can be also considered, but for simplicity, we shall examine only the case of equation (1.1).

Assume that the following conditions hold:

(A1). The even potential $F \in C^{m+1}$, $m \geq 1$,

$$F = \int_0^u f(s)\,ds, \tag{1.3}$$

is bistable, i.e., F has precisely two nondegenerate maxima $u^{\pm} = \pm v$, such that

$$F''(v) < 0, \qquad F'(u) = f(u) < 0, \qquad u > v. \tag{1.4}$$

(A2). The initial data φ lie in the interval

$$-v < \varphi(x/\varepsilon, \varepsilon) < v \quad \text{for all} \quad x \in \mathbb{R}, \tag{1.5}$$

and the derivatives of φ are bounded

$$\varphi(z, \varepsilon) \in C^{2+\nu}, \qquad |\varphi|_{2+\nu} < C, \tag{1.6}$$

where $C^{2+\nu}$ is the Hölder space with exponent $2+\nu$, $0 < \nu < 1$, [7, 8, 17].

In applications (especially in the theory of phase transitions) we often met the case

$$f = 2(u - u^3), \qquad F = u^2 - u^4/2. \tag{1.7}$$

Here $v = 1$.

The property of being even means that the two states $v = \pm 1$ are completely equivalent (the effect of bistability). Thus, in this case, equation (1.1) describes the arising of two phases with equal properties.

Condition (A1) implies that $f(u)$ is an odd function and the solution of the standard equation, corresponding to the following equation

$$\varepsilon u_t = \varepsilon^2 u_{xx} + \alpha^2 f(u), \tag{1.8}$$

is a fixed kink (i.e., a traveling wave with velocity $b = 0$, see Chapter I)

$$u = Y(\alpha x/\varepsilon), \tag{1.9}$$

where the function $Y(z)$ satisfies the problem ($z \in \mathbb{R}$):

$$Y'' + \alpha^2 f(Y) = 0, \qquad Y(\pm\infty) = \pm v. \tag{1.10}$$

As is easy to see, such a solution exists, is unique up to translations, uneven, and

$$|Y(z) - v|, |Y'(z)| < C\exp(-k|z|), \quad k, C > 0. \tag{1.11}$$

Further, we shall assume that the initial data are close to the "displaced" kink.

(A3). There exists s such that

$$J(u(x,0), s) = \int_{-\infty}^{\infty} [u(x,0) - Y(\alpha(s)(x-s)/\varepsilon)]^2\,dx < \delta\varepsilon^{1/2}.$$

In other words, the initial data are close to the one-dimensional manifold M_ε consisting of the functions

$$Y\big(\alpha(s)(x-s)\varepsilon^{-1}\big).$$

(Below, for simplicity, we shall follow the usual agreement: positive constants independent of ε as $\varepsilon \to 0$ are denoted by C, k, δ, etc.)

Finally, we assume

(A4). The function $\alpha(x)$ satisfies the conditions

$$\alpha \in C^{m+1}, \qquad \sup_{x_1,x_2} |\alpha(x_1)-\alpha(x_2)| < \delta_1, \qquad \alpha(x) > \delta_2, \tag{1.12}$$

$$|\alpha|_2 < C. \tag{1.13}$$

By (A1), (A2), the solutions $u(x,t)$ exist on the whole interval $t \in (0,\infty)$. Actually, to obtain an estimate a priori, we can use the method of comparing solutions [7] (the minimax principle). If we find a function $u^+(x,t)$ such that

$$u^+(x,0) > u(x,0) \tag{1.14}$$

$$\varepsilon u_t^+ > \varepsilon^2 u_{xx}^+ + \alpha^2(x) f(u^+), \tag{1.15}$$

then

$$u^+(x,t) > u(x,t) \qquad (x \in \mathbb{R},\ t > 0). \tag{1.16}$$

In this case, $u^+ = v + \delta$ is an upper solution (for any small $\delta > 0$). Hence,

$$u(x,t) \le v. \tag{1.17}$$

Similarly, one can obtain the inequality $-v \le u(x,t)$. Thus

$$\sup_x |u(x,t)| \le v, \qquad t > 0. \tag{1.18}$$

Hence (see [7]), for $t > 0$, the derivatives of the solutions are also bounded

$$\sup_x \left|\frac{\partial u}{\partial x}(x,t)\right| \le C\varepsilon^{-1}, \qquad t > 0. \tag{1.19}$$

Actually, by changing the variables $t' = t/\varepsilon$, $x' = x/\varepsilon$, we get equation (1.2). For (1.2), our estimate (1.19) takes the form

$$\sup \left|\frac{\partial u}{\partial x'}(x',t')\right| \le C, \qquad t' > 0. \tag{1.20}$$

This estimate is standard for equations of type (1.2) with sufficiently smooth coefficients, see [7].

For example, to (1.2) we can apply the estimate ([17], Theorem 4.2, Chapter IV; [7], Chapter IV, §7):

$$|u|_{2+\nu} \le C\big(|f|_\nu + |u(x,0)|_{2+\nu} + |u|_0\big), \tag{1.21}$$

where the norms u, f and $u(x,0)$ are taken from the subspaces of functions defined on $\mathbb{R} \times [0,T]$, and the constant C is independent of ν and T. By (1.21), we see that if $|u|_\nu$ is bounded, the norm $|u|_{2+\nu}$ is also bounded. The boundedness of solutions, as well as their existence, can be proved as follows (see [18]). It is sufficient to prove this statement only on an arbitrary interval $(t_0, t_0+\delta)$, where δ is sufficiently small. Write (1.2) as the integral equation

$$u(x',t') = \int_{t_0}^{t'} ds([\Gamma * \alpha^2 f(\,\cdot\,,s)](x',s)) + \Gamma * u(\,\cdot\,,t_0), \tag{1.22}$$

where $\Gamma * g(\,\cdot\,,s)$ is the convolution between $g(x',s)$ and the heat potential

$$(\Gamma * g)(x',s) = \frac{1}{\sqrt{4\pi|t'-s|}} \int_{-\infty}^{\infty} \exp\Big(-\frac{(x'-y)^2}{4(t'-s)}\Big) g(y,s)\,dy. \tag{1.23}$$

For the operator Γ, we have (see [17])

$$|\Gamma * u|_\nu < c|u|_\nu, \tag{1.24}$$

where c is independent of t' and s, $t' > s$. Therefore, by using the principle of contracting mappings, one can prove the boundedness of $|u|_\nu$ and the existence of solutions [18]. There are some other methods for estimating, see [22, 19].

Below it is also convenient to use the inequality of Galyard-Nierenberg type:

$$\sup|w| < (\sup|w_x|)^{1/3}\|w\|^{2/3}. \tag{1.25}$$

Hence, we obtain the estimates for the functions w such that $|w_x| < C\varepsilon^{-1}$:

$$\sup|w(x,t)| < C\varepsilon^{-1/3}\|w\|^{2/3}. \tag{1.26}$$

The fundamental property of being stable can be formulated as follows.

Let us consider the operator $A(q,\varepsilon)$ that arises after $f(u)$ is linearized near the kink Y. This operator has the form

$$A(q,\varepsilon) = \varepsilon^2\frac{\partial}{\partial x^2} + \alpha^2(q) f'\Big(Y\Big(\alpha(q)\frac{x-q}{\varepsilon}\Big)\Big). \tag{1.27}$$

For any function $w \in C^2$, we have (see [19]):

$$(A(q,\varepsilon)w, w) \le -\beta\|w\|^2, \qquad \beta > 0, \tag{1.28}$$

if

$$(w, Y') = 0, \qquad Y' = Y'(z), \qquad z = \alpha(q)(x-q)\varepsilon^{-1}, \tag{1.29}$$

where the constant β is independent of ε.

The property of being stable allows to show (in problems with constant coefficients and $\varepsilon = 1$) that the kink is stable. As $t \to \infty$, a small perturbation of the kink results in another kink that differs from the first only by a shift q.

Our aim is to justify the formal asymptotic formulas for the solution of problem (1.1).

A formal asymptotic solution can be written in the form

$$u_{as}(x,t,\varepsilon) = Y\Big(\alpha(q)\frac{x-q(t,\varepsilon)}{\varepsilon}\Big) + \varepsilon Y_1 + \dots. \tag{1.30}$$

By using the methods of Chapter IV, we get a simple equation for the kink coordinate $q(t)$:

$$\frac{dq}{dt} = -\varepsilon\alpha'(q)\alpha^{-1}(q). \tag{1.31}$$

To justify these formulas, let us consider a special construction known under different names (for example, the method of "nonlinear initialization", the "slow manifold" method, SLEP-method, etc. [3, 5]).

The idea of this construction originates from the classical Lyapunov–Schmidt construction in the theory of bifurcations [9].

First we make the change $u \to (q,w)$ and instead of $u(x,t,\varepsilon)$ we look for a pair of functions $q(t,\varepsilon)$ and $w(x,t,\varepsilon)$. Actually, the new unknown function $q = q(t,\varepsilon)$ determines the coordinate (shift) of the kink. For each t, we define this function by the relation

$$\int_{-\infty}^{\infty} w(x,t,\varepsilon)Y'\Big(\alpha(q)\frac{x-q}{\varepsilon}\Big)\,dx = 0, \tag{1.32}$$

where

$$w(x,t,\varepsilon) = u(x,t,\varepsilon) - Y\Big(\alpha(q)\frac{x-q}{\varepsilon}\Big). \tag{1.33}$$

By standard methods (e.g., by using the implicit function theorem [9]), one can prove that the transformation $u \to (q,w)$ is uniquely determined in a certain (sufficiently small) $\delta\varepsilon^{1/2}$-neighborhood U_ε of the one-dimensional manifold

$$M = \Big\{u\colon u = Y\Big(\alpha(q)\frac{x-q}{\varepsilon}\Big),\quad q\in\mathbb{R}\Big\}. \tag{1.34}$$

(The distance in defined in the L_2-norm). Actually, on the manifold M we can find an element $Y\big(\alpha(q)\frac{x-q}{\varepsilon}\big)$ that is the closest to a certain function u (in the L_2 sense). In other words, we can find $Y\big(\alpha(q)\frac{x-q}{\varepsilon}\big)$ (or q) so that the distance

$$J(q,\varepsilon) = \Big\|u - Y\Big(\alpha(q)\frac{x-q}{\varepsilon}\Big)\Big\|^2 = \int_{-\infty}^{\infty}\Big[u(x,t,\varepsilon) - Y\Big(\alpha(q)\frac{x-q}{\varepsilon}\Big)\Big]^2\,dx \tag{1.35}$$

be minimal. Obviously, such an element exists if u lies in the $\delta\varepsilon^{1/2}$-neighborhood of M, since $\lim_{q\to\infty} J(q,\varepsilon) = \infty$.

On the other hand, after $\alpha(q)/\varepsilon$ is cancelled out in the condition that the distance is minimal (i.e., the approximation of $u(x,t,\varepsilon)$ by a kink is the best in the L_2 sense), we get

$$\int_{-\infty}^{\infty}\Big[(u-Y)\Big\{Y'(z)\Big(1+\varepsilon\alpha'(q)\alpha^{-1}\frac{x-q}{\varepsilon}\Big)\Big\}\Big]\,dx = 0. \tag{1.36}$$

The properties of the kink (1.9) imply that the function $Y'\big(\alpha(q)\frac{x-q}{\varepsilon}\big)(x-q)\varepsilon^{-1}$ is bounded:

$$\sup_x |Y'(x-q)\varepsilon^{-1}| = \alpha(q)^{-1}\sup_z |Y'(z)z| < C. \tag{1.37}$$

Obviously, the solution of (1.35) exists. On the other hand, as we have already seen, equation (1.32) differs from (1.36) by terms of the order ε. Hence, as can be verified directly, by the implicit function theorem, equation (1.32) also has a solution. Besides of this, in a certain $\delta\varepsilon^{1/2}$-neighborhood, this solution can be taken uniquely by defining the one-to-one mapping $u \to (q, w)$. Actually, let q, $\tilde{q}$ be two different minima of J. The analysis of (1.36), based on the Taylor formula, shows that $|q-\tilde{q}| > c\varepsilon$, where c is constant. Then

$$\begin{aligned} 2\delta\varepsilon^{1/2} &\geq \left\| u - Y\Big(\alpha(q)\frac{x-q}{\varepsilon}\Big)\right\| + \left\| u - Y\Big(\alpha(q)\frac{x-\tilde{q}}{\varepsilon}\Big)\right\| \\ &\geq \left\| Y\Big(\alpha(q)\frac{x-q}{\varepsilon}\Big) - Y\Big(\alpha(q)\frac{x-\tilde{q}}{\varepsilon}\Big)\right\| \geq c|q-\tilde{q}|\varepsilon^{-1}\cdot c_1\varepsilon^{1/2} \geq c_2\varepsilon^{1/2}. \end{aligned} \tag{1.38}$$

If $2\delta < c_2$, these inequalities lead to a contradiction. Hence, our procedure actually determines the required one-to-one mapping, and if $\partial u/\partial t$ exists, then the derivative dq/dt also exists.

Now let $P(q)$ be the projection operator

$$P(q)v = \varepsilon\alpha^{-1}(q)r(v, Y')Y'\Big(\alpha(q)\frac{x-q}{\varepsilon}\Big), \qquad r = \|Y'\|^{-2} = \Big(\int_{-\infty}^{\infty} Y'^2(z)\,dz\Big)^{-1}. \tag{1.39}$$

Let us substitute

$$u = Y\Big(\alpha(q)\frac{x-q}{\varepsilon}\Big) + w \tag{1.40}$$

into (1.1) and expand $\alpha^2(x)f(u)$ into a series in w and $(x-q)$ in a neighborhood of $u = Y$, $x = q$. (Here and below, as in (1.37), we again use the boundedness of the operator of multiplication by $((x-q)/\varepsilon)^k$). Then (denoting $z = \alpha(q)\frac{x-q}{\varepsilon}$) we obtain

$$\begin{aligned} \{Y'(z)(-\alpha + \alpha'(x-q))\}\frac{dq}{dt} + \varepsilon\frac{\partial w}{\partial t} \\ = A(q,\varepsilon)w + g(w,q,\varepsilon) + 2\alpha'(q)\alpha(q)(x-q)f(Y), \end{aligned} \tag{1.41}$$

where

$$g = g_1 + g_2 + g_3, \tag{1.42}$$

and

$$g_1 = \big(\alpha^2(x) - 2\alpha'(q)\alpha(q)(x-q) - \alpha^2(q)\big)f(Y), \tag{1.43}$$

$$g_2 = \alpha^2(q)\big(f(Y+w) - f(Y) - f'(Y)w\big), \tag{1.44}$$

$$g_3 = \big(\alpha^2(x) - \alpha^2(q)\big)\big(f(Y+w) - f(Y)\big). \tag{1.45}$$

For g_2, we have the simple estimate

$$|g_2| < Cw^2. \tag{1.46}$$

Further, we apply the projectors P and $Q = I - P$ to both sides of (1.41). The action of the projector $P(q)$ is equivalent to the multiplication by Y' and integration over all x.

Finally, after cumbersome calculations, we get the system for two unknown variables (q, w):

$$r_1(w, q, \varepsilon)\frac{dq}{dt} = R(q, w, \varepsilon), \tag{1.47}$$

$$\frac{\partial w}{\partial t} = A(q, \varepsilon)w + Qg + Q\big[(x - q)\{2\alpha'\alpha(q)f(Y) - Y'\alpha'(q)\}\big], \tag{1.48}$$

where

$$R = \varepsilon^{-1}(g, Y') + \big(2\alpha\alpha'(q)\varepsilon^{-1}(x - q)f(Y), Y'\big). \tag{1.49}$$

For r_1 we have:

$$|r_1 - r| \le C(\varepsilon + \varepsilon^{-1/2}\|w\|). \tag{1.50}$$

For $\|w(\cdot, t)\| < \delta\varepsilon^{1/2}$, the system (1.47), (1.48) is equivalent to the initial equation. By using Assumptions (AI)–(A4) and the simplest a priopi estimates, we can show that if this inequality holds for $t = 0$, it holds for all $t > 0$ provided that δ is sufficiently small and $0 < \varepsilon < \varepsilon_0$. Multiplying both sides of (1.48) by w, integrating, and using the fact that the kink (1.28) is stable, we get

$$\varepsilon\frac{1}{2}\frac{d}{dt}\|w\|^2 < -\beta\|w\|^2 + |(Qg, w)| + C\varepsilon^{3/2}\|w\|. \tag{1.51}$$

Now let us estimate the second term on the right-hand side. By (1.46) and (1.26), we immediately obtain

$$\big|(Qg_2, w)\big| \le C\varepsilon^{-1/2}\|w\|^{8/3}. \tag{1.52}$$

Further, since $f(Y)$ decreases exponentially as $x \to \infty$, we have

$$\big|(Qg_1, w)\big| \le \varepsilon^{5/2}\|w\|. \tag{1.53}$$

To derive (1.53), we have used the Schwartz inequality and

$$\|g_1\| \le c\left[\varepsilon^5 \int_{-\infty}^{\infty} f^2\Big(Y\Big(\alpha(q)\frac{x-q}{\varepsilon}\Big)\Big)\Big(\frac{x-q}{\varepsilon}\Big)^4 d\Big(\frac{x-q}{\varepsilon}\Big)\right]^{1/2} \le C\varepsilon^{5/2}. \tag{1.54}$$

Finally, by the mean-value theorem and Assumption (A4), we obtain

$$\big|(Qg_3, w)\big| \le \delta_1\|w\|^2. \tag{1.55}$$

If δ_1 is sufficiently small (for example, $\delta_1 < \beta/2$), then by (1.52), (1.53) and (1.55), we get the differential inequality

$$\frac{d}{d\tau}\|w\| \le -C_1\|w\| + C_2(\varepsilon^{-1/3}\|w\|^{5/3} + \varepsilon^{3/2}), \qquad \|w(\cdot, 0)\| \le \delta\varepsilon^{1/2}, \tag{1.56}$$

where $\tau = \varepsilon t$.

Clearly,

$$\|w(\cdot,\tau)\| \le \varepsilon^{1/2} W(\tau), \tag{1.57}$$

where the majorizing function W satisfies the following differential equation and the initial conditions:

$$\frac{dW}{d\tau} = -C_1 W + C_2(W^{5/3} + \varepsilon), \qquad W(0) = \delta. \tag{1.58}$$

If δ is small (however, δ is independent of ε as $\varepsilon \to 0$), we have

$$W(\tau) \le \delta. \tag{1.59}$$

(One can take any arbitrary δ such that $-C_1\delta + C_2\delta^{5/3} < 0$).

By (1.59), we can write $\delta^{2/3}W$ instead of W on the right-hand side. Then, by solving the new linear equation, we get

$$\|w(\cdot,t)\| < C\varepsilon^{1/2}\big(\delta \exp(-ct/\varepsilon) + \varepsilon\big). \tag{1.60}$$

Moreover, (1.59) implies

$$\|w(\cdot,t)\| < \delta\varepsilon^{1/2}. \tag{1.61}$$

The obtained estimate shows that system (1.47), (1.48) is equivalent to the initial equation for all $t > 0$. Moreover, for $t > k\varepsilon^{1-s}$, where $s \in (0,1)$, (1.26) gives

$$\|w(\cdot,t)\| < C\varepsilon^{3/2}, \qquad |w| < C\varepsilon^{2/3}. \tag{1.62}$$

So, we see that the solution $u(x,t,\varepsilon)$ of the Cauchy problem converges at an exponential rate to the leading term Y of the asymptotics. However, it remains to be proved that the coordinate $q(t)$ is determined by an equation close to equation (1.31). Note that in our system the first equation (1.47) is exact and, moreover, if we put $w = 0$ in the exact equation, we get our equation (1.31) obtained from asymptotic formulas (this can be verified by integration by parts).

In order to estimate the precision of the asymptotic equation (1.31), it is sufficient to estimate $\varepsilon^{-1}(g, Y')$. This estimate follows from formulas (1.43)–(1.45). Actually, after the integration, we have

$$\varepsilon^{-1}|(g_1, Y')| < C\varepsilon^2. \tag{1.63}$$

Further,

$$\varepsilon^{-1}|(g_2, Y')| < \varepsilon^{-1} C \sup|w|^2 \int_{-\infty}^{\infty} Y'\,dx \le C\varepsilon^{4/3}, \tag{1.64}$$

and, similarly,

$$\varepsilon^{-1}|(g_3, Y')| < C\varepsilon \sup|w| \le C\varepsilon^{5/3}. \tag{1.65}$$

Finally,

$$\frac{dq}{dt} = -\varepsilon\alpha'\alpha^{-1}(q) + \gamma(\varepsilon), \qquad |\gamma| < C\varepsilon^{4/3}, \tag{1.66}$$

and thus the leading term of our asymptotics is justified.

Remark 1. The case when the velocity of an (unperturbed) kink is everywhere greater than zero was considered by P. C. Fife[1]. Both these cases deal with Semenov-type waves. Apparently, it is much more difficult to analyse the Kolmogorov–Piskunov–Petrovskii case and the situation when the unperturbed velocity of the kink vanishes at certain isolated points.

Remark 2. There are situations when such simple estimates are of no use. Then we must apply the methods developed in Henry's book [8]. Our (q, w)-system can be exactly divided into fast (related to the w-component) and slow (related to the q-component) motions. The equation for w can be written in the integral form, and the existence of its solutions (for any $q(t)$, $|q|_1 < C$) can be proved by the method of contracting mappings. When $u(\cdot, t, \{q(\cdot)\})$ is obtained, we substitute it into the q-equation and prove that this equation has solutions.

This method can be also used to prove the existence of invariant manifolds

$$u = \sigma(q, \varepsilon).$$

Actually, our asymptotic expansion (1.30) is the expansion of σ with respect to powers of ε.

2. Complete description of the solution behavior in a neighborhood of the manifold M

In the preceding section it was shown that a tube $\delta\varepsilon^{1/2}$-neighborhood U_ε of the manifold M, composed of smooth terms of asymptotic solutions, is an attracting domain (for appropriate δ and small ε). Therefore, on U_ε we can consider the global half-flow $\{S^t\}$ generated by our Cauchy problem (1.1). Thus we have a half-flow $\{S^t_\varepsilon\}$. The obtained results completely describe the motion in U_ε if we assume that all zeros q_i of the function $\alpha'(q)$ are simple. In other words we assume that all extremum points of $\alpha(q)$ are nondegenerate. (For the functions α in the "general position", this assumption is satisfied).

Let us consider equation (1.31). For any initial value $q(0)$ the solution $q(t)$ tends to one of the zeros of $\alpha'(q)$, which are the extreme points of $\alpha(q)$. If q_i corresponds to a local minimum of α, then the corresponding trivial solution $q \equiv q_i$ is stable (this solution describes the stationary kink). But if the zero point corresponds to a local maximum of α, then the solution is unstable.

This result has a simple physical meaning. Namely, our equation (1.1) has the Lyapunov functional

$$L[u] = \int_{-\infty}^{\infty} \left[\tfrac{1}{2}\varepsilon^2 u_x^2 + \alpha^2(x)\Phi(u)\right] dx, \tag{2.1}$$

where $\Phi' = f$ and $\Phi(v) = 0$. This functional can be interpreted as the free energy of the system. We have

$$L[u_{\mathrm{as}}(x, t, \varepsilon)] = C\varepsilon\alpha^2(q) + O(\varepsilon^2). \tag{2.2}$$

Thus, the "energy" L takes its local minima on stable stationary asymptotic solutions.

Proposition 2.1. *Let $\alpha \in C^2$ be bounded above in the norm C^2 and possess only nondegenerate extreme points. To each root q_i of the equation*

$$\alpha'(q) = 0, \tag{2.3}$$

corresponds a certain stationary solution

$$u_i = Y\left(\alpha(\tilde{q}_i)\frac{x - \tilde{q}_i}{\varepsilon}\right) + O(\varepsilon),$$

where $|\tilde{q}_i - q_i| = o(\varepsilon)$, $(\varepsilon \to 0)$.

The solution u_i is stable for minimal q_i and unstable for maximal q_i. All solutions $u(x,t)$ of the Cauchy problem that start from a certain neighborhood U_ε (with L_2-radius $\delta\varepsilon^{1/2}$) of the manifold M tend to one of the stationary solutions u_i as $t \to \infty$.

We shall not prove this statement in detail. The proof consists of some stages. The first step is to prove that system (1.48), (1.49) has an invariant manifold

$$u = Y\left(\alpha(q)\frac{x-q}{\varepsilon}, q, \varepsilon\right) + W(q,\varepsilon), \tag{2.4}$$

where $\|W\| < C\varepsilon$. This can be proved by using Theorem 9.1.1 from [8]. Then the flow $\{S^t\}$ is reduced to the flow determined by the ordinary differential equation

$$r(q,\varepsilon)\frac{dq}{dt} = R(q,\varepsilon). \tag{2.5}$$

All the attractors $\{S^t\}$ lie on a stable invariant one-dimensional manifold $\widetilde{M}$ determined by (2.4). This manifold attracts all solutions of (1.1), starting from U_ε. The infinite-dimensional dynamical system determined by (1.1) is equivalent to the simple equation (2.5). The estimate (1.66) implies that, in turn, this equation possesses stable stationary points $\tilde{q}_i$ close to the stationary points q_i of equation (1.31).

A similar method related to the theory of invariant manifolds allows to obtain some results about the kink interaction [3–6], as well as to prove the existence of chaotic and periodically oscillating kinks for systems of equations.

3. Zeldovich waves

Let us consider the Cauchy problem for the following equation with a small parameter:

$$\varepsilon u_t = \varepsilon^2 \Delta u + \alpha^2(x) f(u), \qquad 0 < \varepsilon \ll 1, \quad x \in \mathbb{R}^n. \tag{3.1}$$

As above, we assume that $\alpha \in C^2$, $|\alpha|, |\Delta\alpha| < C$, $\alpha > \delta > 0$. But now, in contract to the preceding section, the nonlinearity f satisfies the conditions:

$$f'(0) = 0, \qquad f(0) = 0, \qquad f'(1) < 0. \tag{3.2}$$

For example, it may take the form

$$f = u^\nu(1-u), \qquad \nu \geq 2. \tag{3.3}$$

This is a Zeldovich-type nonlinearity, and the standard equation

$$bu_\tau = u_{\tau\tau} + f(u) \tag{3.4}$$

has the unique solution in the kink form

$$u = Y(b\tau) \tag{3.5}$$

with a certain uniquely determined velocity V satisfying the estimates

$$\begin{aligned} |Y(\tau)| &= O\big(\exp(c\tau)\big), \qquad \tau \to -\infty, \\ |Y(\tau) - 1| &= O\big(\exp(-c\tau)\big), \qquad \tau \to +\infty, \end{aligned} \tag{3.6}$$

(for precise estimates, see Chapter I).

Assume that we have constructed an asymptotic solution of the form

$$u_{\mathrm{as}} = Y\Big(\frac{S(x,t,\varepsilon)}{\varepsilon}, x, \varepsilon\Big), \tag{3.7}$$

where the phase S has the form

$$S = S_0(x,t) + \varepsilon S_1 + \dots. \tag{3.8}$$

An asymptotic method for constructing solutions of this type is given in Chapter IV.

Assume that the following general conditions (B.1–B.3) hold.

(B1). The function $f \in C^\infty$, the standard equation

$$bu_\tau = u_{\tau\tau} + f(u)$$

has a monotone ($Y'_\tau > 0$) solution $Y(\tau)$ with asymptotic properties (3.6), and relations (3.2) hold.

(B2). For $0 < t < T_\varepsilon$, there is a smooth wave front P, determined by the equation

$$S(x,t,\varepsilon) = 0,$$

such that the derivative along the normal to P does not vanish:

$$\nabla S \cdot \mathbf{e} = S_n(x,t,\varepsilon) > \delta > 0,$$

for all x satisfying $S(x,t,\varepsilon) = 0$, where $\mathbf{e}$ is the unit vector normal to the front. This front is a hypersurface dividing $\mathbb{R}^n$ into two domains Ω^+ and Ω^-.

(B3). In the neighborhood $\Omega_{\varepsilon,p} = \{x : |S| < c\varepsilon^p\}$ of the front, where $0 < p < 1$, the asymptotic solution $u_{\mathrm{as}}\big(S(x,t,\varepsilon)/\varepsilon, x, \varepsilon\big) \in C^\infty$, satisfying equation (3.1) on the interval $[0, T_\varepsilon]$, is determined with the correction $h(x,t,\varepsilon) \le C\varepsilon^{m+1}$.

Note that, by condition (B2), we have

$$\left|(u_{\rm as})_\tau S_n + \varepsilon\langle\nabla_x u, \mathbf{e}\rangle\right| > \delta_2 > 0$$

in the domain $\Omega_{\varepsilon,p}$. Outside this domain we can extend the solution into the domain $\Omega^+ \subset \mathbb{R}^n$ with constant $u = 1_+$, and into the domain Ω^- with constant $u = 0_-$. Then, by using the unity partition, we match this extended solution together with $u_{\rm as}$ in $\Omega_{\varepsilon,p}$ and thus obtain the correction $h \le C\varepsilon^{m+1}$ uniformly in the whole $\mathbb{R}^n$.

We shall look for the upper solution u^+ in the form

$$u^+ = Y\left(S\Big(\frac{S(\mathbf{x}+\mathbf{e}b(t),t,\varepsilon)}{\varepsilon}\Big), \mathbf{x}+\mathbf{e}b, \varepsilon\right) + v, \tag{3.11}$$

where $\mathbf{e}(x,t)$ is normal to the front at the point (x,t), $S(x,t,\varepsilon)=0$, and the functions $v(t)$, $b(t)>0$ can be found from the inequality:

$$\varepsilon\frac{\partial}{\partial t}u^+ > \varepsilon^2\Delta u^+ + \alpha^2(x)f(u^+). \tag{3.12}$$

The function Y with displaced argument satisfies equation (3.1) with displaced argument and correction h. This fact implies that inequality (3.12) is a consequence of a stronger inequality

$$\begin{aligned}\varepsilon v_t + \big(Y'_\tau\langle\nabla S,\mathbf{e}\rangle + \varepsilon\langle\nabla_x Y,\mathbf{e}\rangle\big)b + \varepsilon\langle\nabla_x Y,\mathbf{e}\rangle b_t&\\ > f(Y)\big\{\alpha^2(\mathbf{x}) - \alpha^2(\mathbf{x}+\mathbf{e}b)\big\} + \alpha^2(x)\max_{0\le\xi\le 1} f'(Y+\xi v)v + h.&\end{aligned} \tag{3.13}$$

Below we shall use the following properties of the standard solution

$$|f'(Y)| < c|Y'| \quad \text{if} \quad |Y|, |Y'| < \delta, \tag{3.14}$$

$$|f'(Y)| < c_1|Y| \quad \text{otherwise}. \tag{3.15}$$

We set

$$b(t) = B\varepsilon^\rho\exp(\sigma t), \qquad v(t) = A\varepsilon^\rho\exp(\sigma t), \tag{3.16}$$

where the constants ρ, σ will be defined below. Now it is necessary to divide the domain $\Omega_{\varepsilon,p}$ into three regions Ω_1, Ω_2, Ω_3.

I. Ω_1: $0 < Y'_\tau \le \delta_1$, but $f'(Y) \le -c_f$, $c_f > 0$. Here Y is close to 1.

II. Ω_2: $Y'_\tau > \delta_1 > 0$.

III. Ω_3: $0 < Y'_\tau \le \delta_1$, but Y is close to zero since $|f'(Y)| < C|Y'_\tau|$.

Substituting (3.16) into (3.13), we obtain three inequalities. These inequalities are sufficient for (3.13) to hold. In Ω_2 we have

$$\sigma A\varepsilon^{\rho+1} + \delta_1 B\sigma\varepsilon^\rho - Bc_1\sigma\varepsilon^{\rho+1} > c_2\,\mathrm{Lip}\,\alpha B\varepsilon^\rho + Ac_3\varepsilon^\rho + C\varepsilon^{m+1}\exp(-\sigma t). \tag{3.17}$$

In order these conditions hold on $[0,T_0]$, it is sufficient to take $o < \rho < m+1$, $0 < \varepsilon < \varepsilon_0$, where ε_0 is small enough, and then to put

$$\sigma\delta_1 > c_2\tilde{\delta} = c_2\,\mathrm{Lip}\,\alpha. \tag{3.18}$$

(Here $\tilde{\delta}$ estimates $\operatorname{Lip}\alpha$).

In the domain Ω_1, we get

$$A\varepsilon^{\rho+1}\sigma - Bc_1\sigma\varepsilon^{\rho+1} > c_2 \operatorname{Lip}\alpha B\varepsilon^{\rho} - \tfrac{1}{2}c_f A\varepsilon^{\rho} + C\varepsilon^{m+1}\exp(-\sigma t). \tag{3.19}$$

This implies that (under the above conditions imposed on ε, ρ, and t) we must have

$$4B\tilde{\delta}c_2 < Ac_f. \tag{3.20}$$

Finally, let us consider the domain Ω_3. Here, using the estimate

$$|f'(Y+\xi v) - f'(Y)| < c|v|,$$

we expand $f'(Y+\xi v)$ into a Teylor series. We get

$$\begin{aligned}&A\sigma\varepsilon^{\rho+1} - cB\varepsilon^{\rho+1}(\sigma+1)\\ &\quad > c\exp(-\sigma t)\varepsilon^{m+1} + c_7 c_f A^2\varepsilon^{2\rho} + f(Y)\tilde{\delta}B\varepsilon^{\rho} - Y'_S c_5 B\varepsilon^{\rho} + c_6 f'(Y)A\varepsilon^{\rho}.\end{aligned} \tag{3.21}$$

If $\tilde{\delta} = \sup|\operatorname{Lip}\alpha|$ is sufficiently small (but independent of ε as $\varepsilon \to 0$), then this inequality holds for $0 < \rho < m$ and

$$\sigma A > c_5 B(\sigma+1), \qquad B\sigma > cA + c_6 B\delta_1 \sup|\operatorname{Lip}\alpha| \tag{3.22}$$

We see that all inequalities (3.18), (3.20), and (3.22) can hold simultaneously for sufficiently small $\tilde{\delta} = \operatorname{Lip}\alpha$.

Since the point $u = 0$ is unstable, some difficulties may arise when we look for the lower solution in the domain Ω_3. We show that the lower solution can be constructed similarly in the form:

$$u^- = Y\left(\frac{S(\mathbf{x}+\mathbf{e}b(t),t,\varepsilon)}{\varepsilon}, \mathbf{x}+\mathbf{e}b(t), t, \varepsilon\right) + v(t).$$

As above, we put $\tau = S/\varepsilon$. After the substitution, we see that it is sufficient to verify the inequality

$$\begin{aligned}\varepsilon v_t + b_t\big[Y'_\tau\langle\nabla_x S, \mathbf{e}\rangle + \varepsilon\langle\nabla_x Y, \mathbf{e}\rangle\big] + b\varepsilon\langle\nabla_x Y, \mathbf{e}_t\rangle &< f(Y)\big[\alpha^2(x) - \alpha^2(\mathbf{x}+\mathbf{e}b)\big]\\ &\quad + |\alpha^2(x)| \sup_{0<\xi<1} f'(Y+\xi v)v + h, \quad h < C\varepsilon^{m+1}.\end{aligned} \tag{3.23}$$

Since Y decreases exponentially in τ, due to properties of the germ f, we have

$$|f'(Y)| < c|Y'_\tau|, \qquad |f(Y)| < c|Y|^2, \qquad |Y| < c|Y'|$$

in the domain Ω_3 close to 0. We shall look for b, v in the form (we simply change the sign)

$$b = -B\varepsilon^{\rho}\exp(\sigma t), \qquad v = -A\varepsilon^{\rho}\exp(\sigma t), \qquad A, B, \sigma > 0.$$

If the following inequality holds:

$$-\sigma A\varepsilon^{\rho+1} + c_1 B\varepsilon^{\rho+1} < -c_2 A\varepsilon^{\rho} Y'_\tau(1+o(\varepsilon)) - c_3|Y'_\tau|^2 B\varepsilon^{\rho} - c_4\varepsilon^{m+1} + c_5 y'_\tau B\sigma,$$

where $c_3 = \operatorname{Lip}\alpha$, we get (3.23). So, since the obtained inequalities are linear in b, $v(t)$ with precision up to the terms vanishing as $\varepsilon \to 0$, we see that they are the same as above, but with the opposite sign:

$$a_1 > b_1 \Longleftrightarrow -a_1 < -b_1.$$

Now let us formulate

Proposition 3.1. *Let the Lipschitz constant* Lip α *be sufficiently small:*

$$0 < \operatorname{Lip}\alpha < \tilde{\delta}. \tag{3.23}$$

Let the asymptotic solution u_{as} *of equation* (3.1) *exist* $\operatorname{mod}\varepsilon^{m+1}$, $m \geq 1$, *on the interval* $(0, T_\varepsilon)$.

Then the exact solution of our problem also exists for all ε *such that* $0 < \varepsilon < \varepsilon_0$, *where* ε_0 *is sufficiently small, and*

$$|u - u_{\mathrm{as}}| < C\varepsilon^m \tag{3.24}$$

on the interval of length at least equal to $\min(T_\varepsilon, T_0)$, *where* T_0 *is independent of* ε. *The constants* T_0 *and* $\tilde{\delta}$ *depend on* f *and* m.

References

1. P. C. Fife, *Diffusive waves in inhomogeneous media*, Proc. of the Edinburgh Math. Soc. (ser. 2, part 2), vol. 32, 1989, pp. 291–316.
2. S. A. Vakulenko, *Justification of asymptotic solutions for one-dimensional parabolic equations*, Matem. Zametki **52:3** (1992), 10–17. (Russian)
3. G. Fusco, *A geometric approach to the dynamics of* $u_t = \varepsilon^2 u_{xx} + f(u)$ *for small* ε, Lecture Notes in Physics, Problems Involving Change of Type, vol. 359, Springer, Berlin–Heidelberg, 1990, pp. 53–73.
4. H. Ikeda, M. Mimura, and Y. Nishiura, *Global bifurcation phenomena of traveling wave solutions for bistable and reaction–diffusion systems*, Nonlinear Anal., TMA **13** (1989), 507–526.
5. J. Carr and R. Pego, *Very slow phase separation in one dimension*, Lecture Notes in Physics, (M. Rasele et al., eds.), vol. 344, Springer, 1989, pp. 216–226.
6. J. Carr and R. Pego, *Metastable patterns in solution of* $u_t = \varepsilon^2 u_{xx} - f(u)$, Comm. Pure Appl. Math. **42** (1989), 523–576.
7. A. Friedman, *Partial Differential Equations of Parabolic Type*, Holt, Rinehart and Winston, New York, 1969.
8. D. Henry, *Geometric Theory of Semilinear Parabolic Equations*, Springer, Berlin, 1991.
9. L. Nirenberg, *Topics in Nonlinear Functional Analysis*, Courant Institute, New York, 1974.
10. P. de Mottoni and P. Schatzman, *Evolution geometrique d'interface*, Comptes Rendus Acad. Sci. Paris, ser I Math. **309** (1989), 453–458.
11. J. X. Xin, *Multidimensional stability of traveling waves in bistable reaction-diffusion equations* I; Comm. Partial Differential Equations **17** (1992), 1889–1899.
12. D. Levermore and J. X. Xin, *Multidimensional stability of traveling waves in bistable reaction-diffusion equations* II; Comm. Partial Differential Equations **17** (1992), 1901–1924.
13. M. W. Hirsch, *Equations and convergence almost everywhere in strongly monotone flows*, Contemp. Math., AMS **17** (1983), 267–285.
14. P. Polacik, *Complicated dynamics in scalar semilinear parabolic equations in higher space dimensions*, J. Differential Equations **89** (1991), 244–271.
15. J. Mierezynski, *On a genetic behavior in strongly cooperative differential equations*, Colloqua Mathematica Societatis Janos Boyai, vol. 53, North Holland, 1990, pp. 402–406.
16. H. L. Smith and H. R. Thieme, *Quasiconvergence and stability for order-preserving semiflows*, SIAM J. Math. Anal. **21** (1990), 673–692.
17. O. A. Ladyzhenskaya, V. A. Solonnikov, and N. N. Ural'tseva, *Linear and Quasilinear Equations of Parabolic Type*, Nauka, Moscow, 1967. (Russian)
18. J. A. Smoller, *Shock Waves and Reaction–Diffusion Equations*, Springer, New York, 1983.
19. P. C. Fife and J. B. Macleod, *The appproach of solutions of nonlinear differential equations to traveling front solutions*, Arch. Rational Mech. Anal. **65:4** (1977).

Other *Mathematics and Its Applications* titles of interest:

J. Chaillou: *Hyperbolic Differential Polynomials and their Singular Perturbations.* 1979, 184 pp. ISBN 90-277-1032-5

S. Fucik: *Solvability of Nonlinear Equations and Boundary Value Problems.* 1981, 404 pp. ISBN 90-277-1077-5

V.I. Istratescu: *Fixed Point Theory. An Introduction.* 1981, 488 pp. *out of print,* ISBN 90-277-1224-7

F. Langouche, D. Roekaerts and E. Tirapegui: *Functional Integration and Semiclassical Expansions.* 1982, 328 pp. ISBN 90-277-1472-X

N.E. Hurt: *Geometric Quantization in Action. Applications of Harmonic Analysis in Quantum Statistical Mechanics and Quantum Field Theory.* 1982, 352 pp. ISBN 90-277-1426-6

F.H. Vasilescu: *Analytic Functional Calculus and Spectral Decompositions.* 1983, 392 pp. ISBN 90-277-1376-6

W. Kecs: *The Convolution Product and Some Applications.* 1983, 352 pp. ISBN 90-277-1409-6

C.P. Bruter, A. Aragnol and A. Lichnerowicz (eds.): *Bifurcation Theory, Mechanics and Physics. Mathematical Developments and Applications.* 1983, 400 pp. *out of print,* ISBN 90-277-1631-5

J. Aczel (ed.) : *Functional Equations: History, Applications and Theory.* 1984, 256 pp. *out of print,* ISBN 90-277-1706-0

P.E.T. Jorgensen and R.T. Moore: *Operator Commutation Relations.* 1984, 512 pp. ISBN 90-277-1710-9

D.S. Mitrinovic and J.D. Keckic: *The Cauchy Method of Residues. Theory and Applications.* 1984, 376 pp. ISBN 90-277-1623-4

R.A. Askey, T.H. Koornwinder and W. Schempp (eds.): *Special Functions: Group Theoretical Aspects and Applications.* 1984, 352 pp. ISBN 90-277-1822-9

R. Bellman and G. Adomian: *Partial Differential Equations. New Methods for their Treatment and Solution.* 1984, 308 pp. ISBN 90-277-1681-1

S. Rolewicz: *Metric Linear Spaces.* 1985, 472 pp. ISBN 90-277-1480-0

Y. Cherruault: *Mathematical Modelling in Biomedicine.* 1986, 276 pp. ISBN 90-277-2149-1

R.E. Bellman and R.S. Roth: *Methods in Approximation. Techniques for Mathematical Modelling.* 1986, 240 pp. ISBN 90-277-2188-2

R. Bellman and R. Vasudevan: *Wave Propagation. An Invariant Imbedding Approach.* 1986, 384 pp. ISBN 90-277-1766-4

A.G. Ramm: *Scattering by Obstacles.* 1986, 440 pp. ISBN 90-277-2103-3

Other *Mathematics and Its Applications* titles of interest:

C.W. Kilmister (ed.): *Disequilibrium and Self-Organisation.* 1986, 320 pp.
ISBN 90-277-2300-1

A.M. Krall: *Applied Analysis.* 1986, 576 pp.
ISBN 90-277-2328-1 (hb), ISBN 90-277-2342-7 (pb)

J.A. Dubinskij: *Sobolev Spaces of Infinite Order and Differential Equations.* 1986, 164 pp. ISBN 90-277-2147-5

H. Triebel: *Analysis and Mathematical Physics.* 1987, 484 pp.
ISBN 90-277-2077-0

B.A. Kupershmidt: *Elements of Superintegrable Systems. Basic Techniques and Results.* 1987, 206 pp. ISBN 90-277-2434-2

M. Gregus: *Third Order Linear Differential Equations.* 1987, 288 pp.
ISBN 90-277-2193-9

M.S. Birman and M.Z. Solomjak: *Spectral Theory of Self-Adjoint Operators in Hilbert Space.* 1987, 320 pp. ISBN 90-277-2179-3

V.I. Istratescu: *Inner Product Structures. Theory and Applications.* 1987, 912 pp.
ISBN 90-277-2182-3

R. Vich: *Z Transform Theory and Applications.* 1987, 260 pp.
ISBN 90-277-1917-9

N.V. Krylov: *Nonlinear Elliptic and Parabolic Equations of the Second -Order.* 1987, 480 pp. ISBN 90-277-2289-7

W.I. Fushchich and A.G. Nikitin: *Symmetries of Maxwell's Equations.* 1987, 228 pp. ISBN 90-277-2320-6

P.S. Bullen, D.S. Mitrinovic and P.M. Vasic (eds.): *Means and their Inequalities.* 1987, 480 pp. ISBN 90-277-2629-9

V.A. Marchenko: *Nonlinear Equations and Operator Algebras.* 1987, 176 pp.
ISBN 90-277-2654-X

Yu.L. Rodin: *The Riemann Boundary Problem on Riemann Surfaces.* 1988, 216 pp.
ISBN 90-277-2653-1

A. Cuyt (ed.): *Nonlinear Numerical Methods and Rational Approximation.* 1988, 480 pp. ISBN 90-277-2669-8

D. Przeworska-Rolewicz: *Algebraic Analysis.* 1988, 640 pp. ISBN 90-277-2443-1

V.S. Vladimirov, YU.N. Drozzinov and B.I. Zavialov: *Tauberian Theorems for Generalized Functions.* 1988, 312 pp. ISBN 90-277-2383-4

G. Morosanu: *Nonlinear Evolution Equations and Applications.* 1988, 352 pp.
ISBN 90-277-2486-5

Other *Mathematics and Its Applications* titles of interest:

A.F. Filippov: *Differential Equations with Discontinuous Righthand Sides*. 1988, 320 pp. ISBN 90-277-2699-X

A.T. Fomenko: *Integrability and Nonintegrability in Geometry and Mechanics*. 1988, 360 pp. ISBN 90-277-2818-6

G. Adomian: *Nonlinear Stochastic Systems Theory and Applications to Physics*. 1988, 244 pp. ISBN 90-277-2525-X

A. Tesar and Ludovt Fillo: *Transfer Matrix Method*. 1988, 260 pp. ISBN 90-277-2590-X

A. Kaneko: *Introduction to the Theory of Hyperfunctions*. 1989, 472 pp. ISBN 90-277-2837-2

D.S. Mitrinovic, J.E. Pecaric and V. Volenec: *Recent Advances in Geometric Inequalities*. 1989, 734 pp. ISBN 90-277-2565-9

A.W. Leung: *Systems of Nonlinear PDEs: Applications to Biology and Engineering*. 1989, 424 pp. ISBN 0-7923-0138-2

N.E. Hurt: *Phase Retrieval and Zero Crossings: Mathematical Methods in Image Reconstruction*. 1989, 320 pp. ISBN 0-7923-0210-9

V.I. Fabrikant: *Applications of Potential Theory in Mechanics. A Selection of New Results*. 1989, 484 pp. ISBN 0-7923-0173-0

R. Feistel and W. Ebeling: *Evolution of Complex Systems. Selforganization, Entropy and Development*. 1989, 248 pp. ISBN 90-277-2666-3

S.M. Ermakov, V.V. Nekrutkin and A.S. Sipin: *Random Processes for Classical Equations of Mathematical Physics*. 1989, 304 pp. ISBN 0-7923-0036-X

B.A. Plamenevskii: *Algebras of Pseudodifferential Operators*. 1989, 304 pp. ISBN 0-7923-0231-1

N. Bakhvalov and G. Panasenko: *Homogenisation: Averaging Processes in Periodic Media. Mathematical Problems in the Mechanics of Composite Materials*. 1989, 404 pp. ISBN 0-7923-0049-1

A.Ya. Helemskii: *The Homology of Banach and Topological Algebras*. 1989, 356 pp. ISBN 0-7923-0217-6

M. Toda: *Nonlinear Waves and Solitons*. 1989, 386 pp. ISBN 0-7923-0442-X

M.I. Rabinovich and D.I. Trubetskov: *Oscillations and Waves in Linear and Nonlinear Systems*. 1989, 600 pp. ISBN 0-7923-0445-4

A. Crumeyrolle: *Orthogonal and Symplectic Clifford Algebras. Spinor Structures*. 1990, 364 pp. ISBN 0-7923-0541-8

V. Goldshtein and Yu. Reshetnyak: *Quasiconformal Mappings and Sobolev Spaces*. 1990, 392 pp. ISBN 0-7923-0543-4

Other *Mathematics and Its Applications* titles of interest:

I.H. Dimovski: *Convolutional Calculus.* 1990, 208 pp. ISBN 0-7923-0623-6

Y.M. Svirezhev and V.P. Pasekov: *Fundamentals of Mathematical Evolutionary Genetics.* 1990, 384 pp. ISBN 90-277-2772-4

S. Levendorskii: *Asymptotic Distribution of Eigenvalues of Differential Operators.* 1991, 297 pp. ISBN 0-7923-0539-6

V.G. Makhankov: *Soliton Phenomenology.* 1990, 461 pp. ISBN 90-277-2830-5

I. Cioranescu: *Geometry of Banach Spaces, Duality Mappings and Nonlinear Problems.* 1990, 274 pp. ISBN 0-7923-0910-3

B.I. Sendov: *Hausdorff Approximation.* 1990, 384 pp. ISBN 0-7923-0901-4

A.B. Venkov: *Spectral Theory of Automorphic Functions and Its Applications.* 1991, 280 pp. ISBN 0-7923-0487-X

V.I. Arnold: *Singularities of Caustics and Wave Fronts.* 1990, 274 pp. ISBN 0-7923-1038-1

A.A. Pankov: *Bounded and Almost Periodic Solutions of Nonlinear Operator Differential Equations.* 1990, 232 pp. ISBN 0-7923-0585-X

A.S. Davydov: *Solitons in Molecular Systems. Second Edition.* 1991, 428 pp. ISBN 0-7923-1029-2

B.M. Levitan and I.S. Sargsjan: *Sturm-Liouville and Dirac Operators.* 1991, 362 pp. ISBN 0-7923-0992-8

V.I. Gorbachuk and M.L. Gorbachuk: *Boundary Value Problems for Operator Differential Equations.* 1991, 376 pp. ISBN 0-7923-0381-4

Y.S. Samoilenko: *Spectral Theory of Families of Self-Adjoint Operators.* 1991, 309 pp. ISBN 0-7923-0703-8

B.I. Golubov A.V. Efimov and V.A. Scvortsov: *Walsh Series and Transforms.* 1991, 382 pp. ISBN 0-7923-1100-0

V. Laksmikantham, V.M. Matrosov and S. Sivasundaram: *Vector Lyapunov Functions and Stability Analysis of Nonlinear Systems.* 1991, 250 pp. ISBN 0-7923-1152-3

F.A. Berezin and M.A. Shubin: *The Schrödinger Equation.* 1991, 556 pp. ISBN 0-7923-1218-X

D.S. Mitrinovic, J.E. Pecaric and A.M. Fink: *Inequalities Involving Functions and their Integrals and Derivatives.* 1991, 588 pp. ISBN 0-7923-1330-5

Julii A. Dubinskii: *Analytic Pseudo-Differential Operators and their Applications.* 1991, 252 pp. ISBN 0-7923-1296-1

V.I. Fabrikant: *Mixed Boundary Value Problems in Potential Theory and their Applications.* 1991, 452 pp. ISBN 0-7923-1157-4

Other *Mathematics and Its Applications* titles of interest:

A.M. Samoilenko: *Elements of the Mathematical Theory of Multi-Frequency Oscillations.* 1991, 314 pp. ISBN 0-7923-1438-7

Yu.L. Dalecky and S.V. Fomin: *Measures and Differential Equations in Infinite-Dimensional Space.* 1991, 338 pp. ISBN 0-7923-1517-0

W. Mlak: *Hilbert Space and Operator Theory.* 1991, 296 pp. ISBN 0-7923-1042-X

N.Ja. Vilenkin and A.U. Klimyk: *Representation of Lie Groups and Special Functions. Volume 1: Simplest Lie Groups, Special Functions, and Integral Transforms.* 1991, 608 pp. ISBN 0-7923-1466-2

N.Ja. Vilenkin and A.U. Klimyk: *Representation of Lie Groups and Special Functions. Volume 2: Class I Representations, Special Functions, and Integral Transforms.* 1992, 630 pp. ISBN 0-7923-1492-1

N.Ja. Vilenkin and A.U. Klimyk: *Representation of Lie Groups and Special Functions. Volume 3: Classical and Quantum Groups and Special Functions.* 1992, 650 pp. ISBN 0-7923-1493-X
(Set ISBN for Vols. 1, 2 and 3: 0-7923-1494-8)

K. Gopalsamy: *Stability and Oscillations in Delay Differential Equations of Population Dynamics.* 1992, 502 pp. ISBN 0-7923-1594-4

N.M. Korobov: *Exponential Sums and their Applications.* 1992, 210 pp. ISBN 0-7923-1647-9

Chuang-Gan Hu and Chung-Chun Yang: *Vector-Valued Functions and their Applications.* 1991, 172 pp. ISBN 0-7923-1605-3

Z. Szmydt and B. Ziemian: *The Mellin Transformation and Fuchsian Type Partial Differential Equations.* 1992, 224 pp. ISBN 0-7923-1683-5

L.I. Ronkin: *Functions of Completely Regular Growth.* 1992, 394 pp. ISBN 0-7923-1677-0

R. Delanghe, F. Sommen and V. Soucek: *Clifford Algebra and Spinor-valued Functions. A Function Theory of the Dirac Operator.* 1992, 486 pp. ISBN 0-7923-0229-X

A. Tempelman: *Ergodic Theorems for Group Actions.* 1992, 400 pp. ISBN 0-7923-1717-3

D. Bainov and P. Simenov: *Integral Inequalities and Applications.* 1992, 426 pp. ISBN 0-7923-1714-9

I. Imai: *Applied Hyperfunction Theory.* 1992, 460 pp. ISBN 0-7923-1507-3

Yu.I. Neimark and P.S. Landa: *Stochastic and Chaotic Oscillations.* 1992, 502 pp. ISBN 0-7923-1530-8

H.M. Srivastava and R.G. Buschman: *Theory and Applications of Convolution Integral Equations.* 1992, 240 pp. ISBN 0-7923-1891-9

Other *Mathematics and Its Applications* titles of interest:

A. van der Burgh and J. Simonis (eds.): *Topics in Engineering Mathematics*. 1992, 266 pp. ISBN 0-7923-2005-3

F. Neuman: *Global Properties of Linear Ordinary Differential Equations*. 1992, 320 pp. ISBN 0-7923-1269-4

A. Dvurecenskij: *Gleason's Theorem and its Applications*. 1992, 334 pp. ISBN 0-7923-1990-7

D.S. Mitrinovic, J.E. Pecaric and A.M. Fink: *Classical and New Inequalities in Analysis*. 1992, 740 pp. ISBN 0-7923-2064-6

H.M. Hapaev: *Averaging in Stability Theory*. 1992, 280 pp. ISBN 0-7923-1581-2

S. Gindinkin and L.R. Volevich: *The Method of Newton's Polyhedron in the Theory of PDE's*. 1992, 276 pp. ISBN 0-7923-2037-9

Yu.A. Mitropolsky, A.M. Samoilenko and D.I. Martinyuk: *Systems of Evolution Equations with Periodic and Quasiperiodic Coefficients*. 1992, 280 pp. ISBN 0-7923-2054-9

I.T. Kiguradze and T.A. Chanturia: *Asymptotic Properties of Solutions of Nonautonomous Ordinary Differential Equations*. 1992, 332 pp. ISBN 0-7923-2059-X

V.L. Kocic and G. Ladas: *Global Behavior of Nonlinear Difference Equations of Higher Order with Applications*. 1993, 228 pp. ISBN 0-7923-2286-X

S. Levendorskii: *Degenerate Elliptic Equations*. 1993, 445 pp. ISBN 0-7923-2305-X

D. Mitrinovic and J.D. Kečkić: *The Cauchy Method of Residues, Volume 2*. Theory and Applications. 1993, 202 pp. ISBN 0-7923-2311-8

R.P. Agarwal and P.J.Y Wong: *Error Inequalities in Polynomial Interpolation and Their Applications*. 1993, 376 pp. ISBN 0-7923-2337-8

A.G. Butkovskiy and L.M. Pustyl'nikov (eds.): *Characteristics of Distributed-Parameter Systems*. 1993, 386 pp. ISBN 0-7923-2499-4

B. Sternin and V. Shatalov: *Differential Equations on Complex Manifolds*. 1994, 504 pp. ISBN 0-7923-2710-1

S.B. Yakubovich and Y.F. Luchko: *The Hypergeometric Approach to Integral Transforms and Convolutions*. 1994, 324 pp. ISBN 0-7923-2856-6

C. Gu, X. Ding and C.-C. Yang: *Partial Differential Equations in China*. 1994, 181 pp. ISBN 0-7923-2857-4

V.G. Kravchenko and G.S. Litvinchuk: *Introduction to the Theory of Singular Integral Operators with Shift*. 1994, 288 pp. ISBN 0-7923-2864-7

A. Cuyt (ed.): *Nonlinear Numerical Methods and Rational Approximation II*. 1994, 446 pp. ISBN 0-7923-2967-8

Other *Mathematics and Its Applications* titles of interest:

G. Gaeta: *Nonlinear Symmetries and Nonlinear Equations*. 1994, 258 pp.
ISBN 0-7923-3048-X

V.A. Vassiliev: *Ramified Integrals, Singularities and Lacunas*. 1995, 289 pp.
ISBN 0-7923-3193-1

N.Ja. Vilenkin and A.U. Klimyk: *Representation of Lie Groups and Special Functions*. Recent Advances. 1995, 497 pp. ISBN 0-7923-3210-5

Yu. A. Mitropolsky and A.K. Lopatin: *Nonlinear Mechanics, Groups and Symmetry*. 1995, 388 pp. ISBN 0-7923-3339-X

R.P. Agarwal and P.Y.H. Pang: *Opial Inequalities with Applications in Differential and Difference Equations*. 1995, 393 pp. ISBN 0-7923-3365-9

A.G. Kusraev and S.S. Kutateladze: *Subdifferentials: Theory and Applications*. 1995, 408 pp. ISBN 0-7923-3389-6

M. Cheng, D.-G. Deng, S. Gong and C.-C. Yang (eds.): *Harmonic Analysis in China*. 1995, 318 pp. ISBN 0-7923-3566-X

M.S. Livšic, N. Kravitsky, A.S. Markus and V. Vinnikov: *Theory of Commuting Nonselfadjoint Operators*. 1995, 314 pp. ISBN 0-7923-3588-0

A.I. Stepanets: *Classification and Approximation of Periodic Functions*. 1995, 360 pp. ISBN 0-7923-3603-8

C.-G. Ambrozie and F.-H. Vasilescu: *Banach Space Complexes*. 1995, 205 pp.
ISBN 0-7923-3630-5

E. Pap: *Null-Additive Set Functions*. 1995, 312 pp. ISBN 0-7923-3658-5

C.J. Colbourn and E.S. Mahmoodian (eds.): *Combinatorics Advances*. 1995, 338 pp. ISBN 0-7923-3574-0

V.G. Danilov, V.P. Maslov and K.A. Volosov: *Mathematical Modelling of Heat and Mass Transfer Processes*. 1995, 330 pp. ISBN 0-7923-3789-1

GPSR Compliance

The European Union's (EU) General Product Safety Regulation (GPSR) is a set of rules that requires consumer products to be safe and our obligations to ensure this.

If you have any concerns about our products, you can contact us on ProductSafety@springernature.com

In case Publisher is established outside the EU, the EU authorized representative is:

Springer Nature Customer Service Center GmbH
Europaplatz 3
69115 Heidelberg, Germany

Batch number: 10607320

Printed by Printforce, the Netherlands